# Nutritional Elements and Clinical Biochemistry

# Nutritional Elements and Clinical Biochemistry

Edited by

**Marge A. Brewster**

*Departments of Pathology, Biochemistry, and Pediatrics,*
*University of Arkansas for Medical Sciences and*
*Metabolic Laboratory, Arkansas Children's Hospital*
*Little Rock, Arkansas*

and

**Herbert K. Naito**

*Department of Biochemistry, Division of Laboratory Medicine,*
*Department of Atherosclerosis and Thrombosis Research, Division of Research,*
*Cleveland Clinic Foundation and Department of Chemistry,*
*Cleveland State University, Cleveland, Ohio*

PLENUM PRESS · NEW YORK AND LONDON

Library of Congress Cataloging in Publication Data

Main entry under title:

Nutritional elements and clinical biochemistry.

"Proceedings of the third annual meeting of the National Academy of Clinical Biochemistry, held in New Orleans, Louisiana, July 13–14, 1979."
Includes bibliographies and index.
1. Nutrition disorders–Congresses. 2. Biological chemistry–Congresses. 3. Nutrition–Congresses. I. Brewster, Marge A. II. Naito, Herbert K. III. National Academy of Clinical Biochemistry (U.S.)
RC620.N86 616.3'9 80-21136
ISBN-13:978-1-4613-3170-4 e-ISBN-13:978-1-4613-3168-1
DOI: 10.1007/978-1-4613-3168-1

Proceedings of the Third Annual Meeting of the National Academy of Clinical Biochemistry, held in New Orleans, Louisiana, July 13–14, 1979


Softcover reprint of the hardcover 1st edition 1980

A Division of Plenum Publishing Corporation
227 West 17th Street New York, N.Y. 10011

Preparation of this volume has been an effort of the National Academy of Clinical Biochemistry, derived in part from the program of the third annual meeting, held in New Orleans, Louisiana, July 13-14, 1979.

# FOREWORD

The important role that the nutritional status exerts in determining the course of life from birth to death in the human being and especially its impact in disease states is only partially appreciated at this time. Nutritional deficiencies are usually considered to be major problems only in under-privileged or developing populations, except for those occurring in specific diseases. This attitude is incorrect as indicated by reports of Bestrian et al (1974, 1976) and Merritt and Suskin (1979) and others who found evidence of nutritional depletion in as much as 50% of the patients in varied groups of hospitalized patients in the United States. Other studies, some of which are included in this book, emphasized the existence of deficiencies of certain specific nutrients.

Despite evidence of nutritional deficiencies occurring more frequently than previously appreciated, there is no well established protocol of laboratory studies that the clinical chemist or scientist should provide to help the physician detect lack of essential nutrients before extensive and possibly irreparable damage has occurred to the individual patient.

Considerable research data are needed to determine the best biologic material (i.e., erythrocytes, leucocytes, plasma, serum, urine, cerebrospinal fluid, lymph) for analysis to determine accurately at an early stage metabolic deficiencies due to a specific nutritional element. Improved techniques for analysis of key metabolites and nutrients are available in the research laboratory and many of them can be adapted to the clinical laboratory.

Need for accurate nutritional evaluation of the individual patient has been greatly escalated in recent years. For example, patients can be maintained for many years on hemodialysis programs, for weeks or months by parenteral nutrition and on other programs where adequacy of nutrition either due to inadequate absorption or excessive metabolic waste may occur. Wisely selected laboratory studies could be an aid to the physician in determining the type and adequacy of the nutritional program of these difficult cases.

The scientific program of the third national meeting of the National Academy of Clinical Biochemistry was devoted to a symposium on Nutritional Elements in Clinical Biochemistry. Seven of these speakers (D. Hill, J. Jones, H. Naito, P. Tocci, A. Dubin, R. Shamberger and L. Lewis) expanded their symposium presentations, and other authors kindly agreed to contribute greater breadth of nutritional information.

It is hoped that the monograph, the first produced under NACB auspices, will serve as a basic reference for current nutritional knowledge in clinical biochemistry, and will help catalyze the much needed research in nutritional assessment methodologies and especially in the evaluation of alterations in nutritional status secondary (or primary) to disease states and/or to their therapies.

Marge A. Brewster

Herbert K. Naito,

Co-Editors

Bistrian, B. R., Blackburn, G. L., Hallowell, E., Heddle, R., 1974, Protein status of general surgical patients, J. Am. Med. Assoc., 230:858.

Bistrian, B. R., Blackburn, G. L., Vitalle, J., Cochran, D., Naylor, J., 1976, Prevalence of malnutrition in general medical patients, J. Am. Med. Assoc. 235:1567.

Merritt, R. J., and Suskind, R. M., 1979, Nutritional survey of hospitalized pediatric patients, Am. J. Clin. Nutr., 32:1320.

# CONTENTS

# CLINICAL CORRELATES IN NUTRITIONAL DISEASE

Donald E. Hill, M.D.

Professor of Pediatrics and Physiology, University of Arkansas for Medical Sciences, Little Rock, Arkansas

## INTRODUCTION

Many of the nutritional disorders of infants and children such as rickets, scurvey, marasmus and kwashiorkor have been recognized by their clinical and laboratory findings. Less well appreciated are the abnormalities in growth due to major disturbances in macronutrient intake, absorption and/or utilization that result in failure-to-thrive at one end of the spectrum or obesity at the other end. The origins of these growth abnormalities related to nutrition are frequently manifest in-utero or in infancy and may be corrected by appropriate diagnosis and intervention. The purpose of this article is to highlight some clinical examples of failure-to-thrive and to contrast these with excessive growth and obesity.

## FETAL GROWTH

Factors that regulate fetal growth can be divided initially into genetic and environmental influences. The genetic influence can be examined by estimates of the effect on birth weight variance. Karn and Penrose (1951) have reported as much as 40% of the variance due to the genetic effect with approximately one-half due to the maternal genotype and one-half due to the fetal genotype. The paternal influences are mediated through the fetal autosomal genes and sex of the infant. Another approach that provides an index of genetic influence is to examine populations of infants with chromosome anomalies. Both autosomal anomalies and some sex chromosome anomalies result in a lower average birth weight than in controls (Table I). Polani (1974) emphasizes the strong influence of the X chromosome and the mild influence of the Y

Table I. Mean Birth Weights for Chromosomal Disorders

| | |
|---|---|
| Triploidy | 2.1 kg (36 weeks gestation) |
| Trisomy 18 syndrome | 2.26 kg |
| Trisomy 13 syndrome | 2.6 kg |
| 5p - syndrome | 2.66 kg |
| 18q - syndrome | 2.86 kg |
| 19p - syndrome | 2.87 kg |
| XO Turner syndrome | 2.9 kg |
| Down's syndrome | 3.06 kg |
| XXY syndrome | 3.15 kg |
| XYY syndrome | 3.36 kg |

(Data from Polani, 1974)

chromosome on fetal growth. The absence of an X chromosome may reduce the birth weight by as much as 500 grams or, if an extra X chromosome is present, the birth weight may be increased by up to 300 grams.

The remaining 60% of the variance in birth weight can be accounted for by the environment in which the fetus grows. Further subdivided; maternal size, prepregnant weight, weight gain during pregnancy, age, parity, and general health may account for 20-30% of this variance (Elliot and Knight, 1974). We are left with approximately 30% of the variance in birth weight due to unknown factors in any one pregnancy. Some of the external influences, such as smoking, alcohol, other drugs, infection, and severe undernutrition, may be recognized in a specific instance as the major etiologic agent producing significant fetal growth retardation.

## Vascular Factors

If one attempts to classify the causes of fetal growth retardation, it is evident that the majority of known causes are related to vascular disorders in the mother (Table II) (Elliot and Knight, 1974). The emphasis on etiology has not promoted extensive investigation of the mechanism involved in reducing fetal growth. One obvious need is to examine carefully the vascular supply to the fetus since most of the nutrient transfer is flow related. The utero-placental blood flow can be influenced by a number of factors, and reductions in maternal systemic perfusion produce a proportional disease in uterine blood flow. During pregnancy, the uterine vascular bed seems maximally dilated and is relatively unresponsive to further stimuli such as reduced perfusion pressure, hypoxia, or hyperoxia. In addition, estrogens have little effect

Table II. Suggested Percentage Distribution of Causes of Small-for-Dates Babies in the UK

| CAUSE | |
|---|---|
| Normal variation | 10 |
| Chromosomal and other congenital anomalies | 10 |
| Infections (maternal and fetal) | 5 (perhaps < 5?) |
| Poor uterus | 1 |
| Placenta and cord | 2 |
| Vascular disease in the mother | 35 |
| Drugs, medicaments and smoking | 5 |
| Other | 32 |

(From Elliot and Knight, 1974)

late in pregnancy on uterine vasculature. There is little response to beta-adrenergic stimulation, but stimulation of the sympathetic nervous system results in vasoconstriction and reduced uterine flow (Burd, 1976).

The presumed cause of reduced fetal growth in maternal vascular disease is compromised uteroplacental perfusion and a decrease in trophoblastic surface area. Since the placenta acts as an organ of nutrient exchange, and flow and permeability influence the placental clearance of molecules across the placenta, more attention needs to be directed toward the assessment of nutrient transfer to the fetus. Except under unusual circumstances, substrate availability in the maternal circulation could not be viewed as the rate limiting process in fetal growth. More likely, a reduction in uterine blood flow and therefore placental transfer is operant in the majority of instances. Placental clearance can be measured in experimental animals by calculating (see Table III) umbilical blood flow and determining the arterial-venous concentration difference of the particular substrate (Burd, 1976).

However, good estimates of placental clearance or capacity are difficult to obtain in humans. In some laboratories, the clearance of dehydroepiandrosterone sulphate (DHEAS) is used as an assessment of placental function (Worly, 1975) and has been helpful in identifying those patients with significant reduction in placental function prior to overt clinical signs of pregnancy hypertension. There are a number of limitations to this test and clearly there is a need for development of a good, safe, reliable, non-invasive technique for assessing placental function.

Table III. Placental Clearance for a Compound that is Transferred from Fetus to Mother

$$C_{\text{Placental Clearance}} = \frac{F_{\text{Umbilical Blood Flow}} \times (A_{1\,\text{Umbilical Arterial Conc.}} - V_{1\,\text{Umbilical Venous Conc.}})}{A_{1\,\text{Umbilical Arterial Conc.}} - A_{2\,\text{Uterine Arterial Conc.}}}$$

(From Burd, L.I., 1976)

## Hormonal Factors

In addition to an adequate blood supply, the fetus must produce a number of polypeptide hormones to grow adequately. Most of the attention has been directed toward fetal thyroid hormone and growth hormone or somatomedin. These hormones have little influence on overall somatic growth in utero, while they are important for maturation and possibly altering substrate supply. More recently insulin has gained prominence as a major regulator of fetal growth, particularly during the latter part of the third trimester (Hill, 1976, 1978). Both clinical and experimental data support a role for insulin during fetal life as essential for fetal growth (Table IV). The classic example of the insulin effect on fetal growth is the infant-of-the-diabetic-mother (IDM).

Table IV. Insulin Effects on Fetal Growth

| Hyperinsulinemia or Increased Sensitivity | Hypoinsulinemia or Decreased Sensitivity |
|---|---|
| Infant of the Diabetic Mother | Transient Diabetes Mellitus |
| Insulinoma and Nesidioblastosis | Pancreatic Agenesis |
| Beckwith-Wiedemann Syndrome | End Organ Defect (Receptor) |
| Infant of the Gestational Diabetic | Intrauterine Growth Retardation |

In this instance, poor control of the diabetes in the mother results in large oscillations in blood glucose that are directly transferred to the fetus. The fetal beta cell responds to the glucose stimulus by hypertrophy and hyperplasia and increased insulin production. As the insulin output increases, there is overgrowth of almost all fetal tissues with the most striking changes occurring in the adipose tissue, liver, and muscle (Fig. 1). The infant becomes dependent on the relatively high blood gluocse and when delivered, has a precipitious fall in glucose and frequently has symptomatic hypoglycemia. It is important to recognize that the fetal hypersomatism and the induced metabolic changes can be abolished by meticulous control of the maternal diabetes. Prognosis for normal growth in these infants is good and a slowing in growth velocity occurs during the first two years of postnatal life. These infants are good examples of how alterations in substrate supply (glucose) can induce hormonal changes and altered growth in the fetus, as well as indicating how excessive fetal growth can be transient and return to normal. In severe maternal diabetes with vascular disease (White Class F), the reduction in uterine blood flow and placental transfer of substrate is such that the majority of infants from these patients are growth retarded.

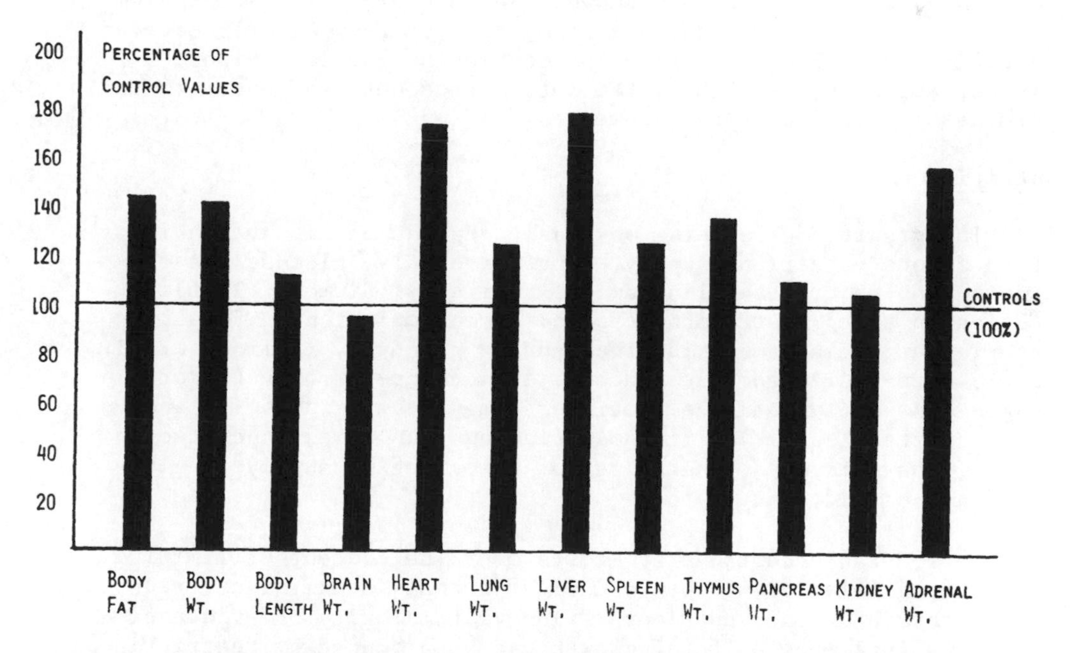


Fig. 1. The body weight and organ weight are shown for some infants of diabetic mothers compared with controls. (Re-drawn from Osler and Petersen, 1960 and Naeye, (1965).

Other examples of fetal hypersomatism are listed in Table IV and are relatively rare, but frequently have a very similar clinical appearance to the IDM and have excess de novo insulin production. In these instances, neonatal hypoglycemia usually persists until the pancreatic overproduction of insulin is ablated through drugs or surgery. The prognosis for normal growth is good.

Contrasted with the IDM and other forms of insulin overproduction is the group of patients who have insufficient or total absence of insulin and intrauterine growth retardation. The classic cases of total pancreatic agenesis (Sherwood et al., 1974) offer the opportunity to examine how fetal growth is arrested in the absence of fetal insulin production. These infants obtain a size of approximately 30 weeks gestation and do not have adequate development of muscle, adipose tissue, and liver. They behave as diabetics once they are born and can no longer depend on maternal glucose homeostasis for control. In addition, a large number of growth retarded infants with maturational lag in pancreatic islet function have been described. These infants develop diabetes mellitus and may die if not treated appropriately. With good care and insulin therapy, catch-up growth is dramatic and most infants recover their own pancreatic function.

The above cases are not known to have any particular problem in vascular supply or substrate transfer, but they clearly have an inability to utilize substrate for developing muscle, adipose tissue, and glycogen. Thus, the interaction between substrate and hormones for optimal fetal growth.

<u>Nutrition</u>

The growth of the fetus obviously depends on the availability of an adequate nutrient supply. Until recently, glucose was considered the major metabolic fuel of the fetus (Simmons, 1976). Studies in sheep where catheters have been implanted in fetal arteries and veins and maintained during pregnancy (Simmons et al., 1974), have indicated that glucose may account for only 45% of oxygen consumption and the remainder comes from amino acids and from lactate (Table V). If these findings can be extrapolated to humans, there is need, therefore, of a continuing supply of glucose, amino acid, and lactate.

Considerable controversy exists over the concept of limiting or reducing fetal growth by reducing maternal intake of calories and protein and other nutrients. The vast majority of evidence that some influence on fetal growth can be produced by restricting maternal intake comes from data on rats (Winick, 1972). In humans and non-human primates, it is difficult to demonstrate a direct relationship between maternal nutrient intake and fetal growth (Cheek, 1975). In North America, where undernutrition is usually

Table V. Contributions of Various Substrates to Fetal Oxidative Catabolism

| | |
|---|---|
| Glucose | 0.45 |
| Amino acids | 0.25 |
| Glycerol | 0.01 |
| Fructose | 0.01 |
| Free fatty acids (C > 5) | 0.01 |
| Lactate | 0.25 |
| | 1.00 |

(From Simmons, 1976)

coincident with poor socioeconomic conditions, it is nearly impossible to separate nutrient intake as a single variable and demonstrate a rate-limiting effect on fetal growth. Similarly, if one examines the number of growth retarded infants associated with significant undernutrition in the mother, the total number is fewer than 1%. There appears to be a margin of safety for the fetus as exemplified in some cases of severe undernutrition with normal or slightly reduced birth weights (Stein and Susser, 1975). In studies on the effects of dietary supplementation, there is a modest (40-60 gram) increase in the average birth weight and only in those women who were thin and undernourished at the beginning of the study (Stein et al., 1978). The Dutch famine study supports the notion that a sparse but well balanced diet had little or no relation to fetal growth until there was severe food deprivation.

These findings do not negate the possibility that a fetus may be marginally nourished or malnourished in-utero in the presence of a normally or optimally nourished mother. As emphasized in the section on vascular or hormonal factors, the basic nutrient supply to the fetus may be compromised by poor uterine and umbilical flow and by alterations in assimilation of the substrate.

## Classification of Altered Fetal Growth

The first good documentation of the undernourished full-term infant appeared in an article by McBurney (1947). Since that time there have been numerous growth curves established for the fetus related to gestational age. The most familiar of these is that produced from the Denver population (Fig. 2) (Battalgia and Lubchenco, 1967). While this graph is frequently used, others from Baltimore (Gruenwald, 1964) and Montreal (Usher and McLean, 1969) as well as local standards should be used since there are definite regional and ethnic differences in birth weights. In any of these

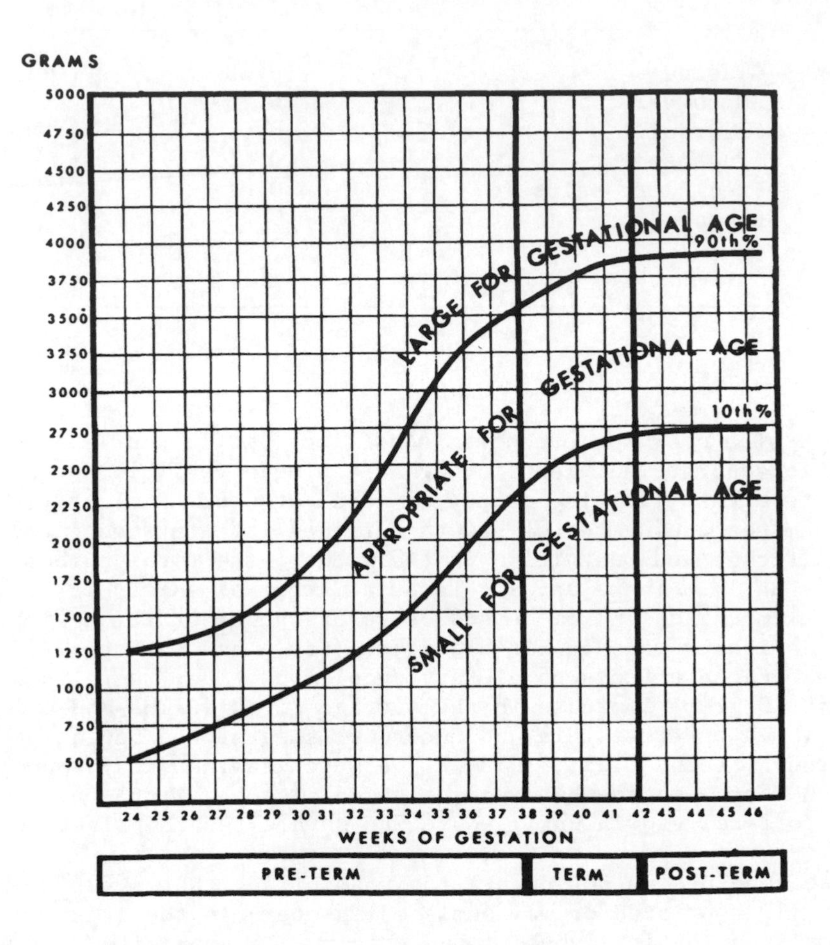


Fig. 2. The classification of newborns by birth weight and gestational age commonly used in newborn nurseries. (Battaglia and Lubchenco, 1967).

growth grids, the important point is to properly determine the gestational age of the newborn and to determine if the infant is preterm, term, or post-term and then to determine if the infant is large-for-gestational-age (LGA), appropriate-for-gestational-age (AGA), or small-for-gestational-age (SGA). This initial classification should be done for weight, length, and head circumference. Once it has been determined that the infant is SGA, then a search for the etiology can proceed. In approximately 30% of patients, the pattern of growth established prenatally will persist postnatally and the infant and child continue to exhibit low profile growth at or below the 10th percentile. Usually, no specific cause can be found for their poor growth. The remaining infants have variable degrees of catch-up growth often accompanied by an excessive nutrient intake for body size or age. These infants are no longer under constraint and consume nutrients to satisfy the increased demand for growth. Many of these infants cross percentile lines and return to their individual genetic pattern or channel for growth.

It is evident, therefore, that the SGA infants have heterogenous etiologies that produce growth failure in-utero and the eventual prognosis for postnatal growth is very much determined by the specific etiologic agent. In those instances where cell replication is inhibited or arrested for a significantly long period, such as in rubella or cytomegalic inclusion disease, the growth retardation is usually permanent. Conversely, if the constraint to growth is due to a reduced blood supply late in gestation, there is usually rapid and complete postnatal catch-up growth. If more effort is expended on properly classifying newborn infants according to body proportions and birth weight for gestational age, then prognostication on ultimate growth and performance of the child will become easier.

## POSTNATAL GROWTH

One of the most frequent reasons for a child to be referred for assessment is because of an abnormality in his or her postnatal growth. As mentioned in the first few paragraphs, these growth abnormalities may be at either end of the spectrum, i.e., failure-to-thrive or obesity. Usually, a detailed history including the obstetric and neonatal history will provide some evidence of growth abnormality prior to birth. In those infants who fail-to-thrive after a normal prenatal and birth history, there is frequently a history of disturbed nutritional intake or assimilation (Table VI). When these infants are fed in a controlled environment, they will often experience dramatic catch-up growth. Seldom is there an organic cause for their failure-to-thrive, but if, in fact, they have a major system disease, the prognosis for normal growth is closely related to the severity of the disease. In addition, disorders such as renal failure or severe asthma are often treated with growth retarding drugs such as corticosteroids. When the corticosteroids are reduced or discontinued, there is frequently some catch-up growth. There is little evidence that nutrition, per se, is the rate-limiting

Table VI. Causes of Failure to Thrive in a Hospital Population

| | No. | Percent |
|---|---|---|
| Suboptimal nutrition | 174 | 87 |
| Other cases: | | |
| Cystic fibrosis | 6 | 3 |
| Hypothyroid | 5 | 2.5 |
| Subdural hematoma | 5 | 2.5 |
| Glycogen storage disease | 1 | 0.5 |
| Celiac syndrome | 1 | 0.5 |
| Methylmalonic acidemia | 1 | 0.5 |
| Maple-syrup urine disease | 1 | 0.5 |
| Unknown | 6 | 3 |
| TOTAL | 200 | 100 |

factor although some improved growth can be witnessed following caloric supplementation in chronic renal disease. True endocrine abnormalities such as hypothyroidism or hypopituitarism are rare causes of growth failure in children. Usually, a good history plus the appearance of the child and his or her bone age will suggest one of these endocrinopathies. The nutrient intake in these children is commensurate with the rate of growth of the child. Abnormalities of insulin such as juvenile diabetes mellitus can produce growth retardation if poorly controlled.

Excessive growth during the first few years postnatally is usually manifest by an abnormal pattern of weight gain due to adipose tissue. This has become a major concern in pediatrics since there is evidence that a relatively high percent of obese infants become obese children and obese adults (Heald and Hollander, 1965). The factors causing this increase in the incidence of obesity in infants and children are not clear. Inherited factors play an important role, but clear-cut inheritance patterns have not been established. Some studies claim that bottle feeding plus the early introduction of solid foods have a major role in the production of obesity in infants. Work completed in our clinic (Dubois et al., 1979) does not support this contention and suggests that individual differences in energy storage and utilization must be important since obese infants had a lower caloric intake per kilogram than the non-obese infants. The same finding has been reported in older obese children and adults.

It would appear that individual disturbances of hypothalamic regulation may result in morbid obesity or conversely, in anorexia and cachexia. There are both human and experimental data illustrating how damage to the ventromedial hypothalamus can produce hyperphagia and obesity (Martin et al., 1974). These conditions are usually associated with excessive insulin secretion that precedes the hyperphagia and obesity (Martin et al., 1974). A peptide isolated from the ventrolateral hypothalamus of rats and monkey has a very specific stimulatory effect on pancreatic insulin production (Hill et al., 1977). This peptide has not yet been characterized, but in crude form is present in the plasma of obese animals. It may mediate the lipogenic effect through insulin stimulation.

While there is evidence of central hypothalamic disturbance in some children, these are the rarer causes of obesity. It must be concluded that numerous factors enter into the equation resulting in obesity, but each child who becomes obese has exceeded the energy requirement for normal basal metabolism and growth during a significant period of development.

SUMMARY

Abnormalities in both prenatal and postnatal growth have been highlighted. Disturbances in the intake, assimilation, or metabolism of macronutrients appear to be the common denominator in a number of these disorders. The challenge to the biochemist and clinical investigator is in at least four areas:

1) Development of non-invasive, safe techniques for the assessment of placental transfer and function.

2) Further elucidation of the role of insulin in both prenatal and postnatal growth.

3) Assessment of the rate of intrinsic cell growth and alterations with nutritional manipulations.

4) Clearer definition of the individual metabolic differences in the obese versus the non-obese child.

REFERENCES

Battalgia, F. C. and Lubchenco, L. O., 1967, A practical classification of newborn infants by birth weight and gestational age, J. Pediatr., 71:159.

Burd, I., 1976, "Placental insufficiency syndromes" in The Neonate (D. S. Young and J. M. Hicks, eds.), pp. 47-58, John Wiley and Sons, New York.

Cheek, D. B. (ed.), 1975, Fetal and Postnatal Cellular Growth, John Wiley and Sons, New York.

DuBois, S., Hill, D. E., and Beaton, H., 1979, An examination of factors believed to be associated with infantile obesity, Amer. J. Clin. Nutr., 32:1979.

Elliot, K. and Knight, J., 1974, General Discussion, in Size at Birth, Ciba Foundation Symposium 27 (new series, p. 393, Associated Scientific Publishers, Amsterdam.

Gruenwald, P., 1964, Infants of low birth weight among 5,000 deliveries, Pediatrics, 34:157.

Hill, E., 1976, "Insulin and fetal growth" in Diabetes and Other Endocrine Disorders During Pregnancy and in the Newborn, (M. I. New and R. H. Fiser, eds.), Alan R.Liss, Inc., New York.

Hill, E., 1978, Effect of insulin on fetal growth, Sem. in Perinatol., 2(4):319.

Hill, D. E., Mayes, S., DiBattista, D., Lockhart-Ewart, R., and Martin, J. M., 1977, Hypothalamic regulation of insulin release in Rhesus monkey, Diabetes, 26:726.

Karn, M. N. and Penrose, L. S., 1951, Birth weight and gestation time in relation to maternal age, parity and infant survival, Ann. Engen., 16:147.

Lockhart-Ewart, R. B., Mok, C., and Martin, J. M., 1976, Neuro-endocrine control of insulin secretion, Diabetes, 25:96.

Martin, J. M., Konijnendijk, W., and Bouman, P. R., 1974, Insulin and growth hormone secretion in rats with ventromedial hypothalamic lesions maintained on restricted food intake, Diabetes, 23:203.

Martin, J. M., Mok, C. C., Penfold, J., Howard, N. J., and Crowne, D., 1974, Hypothalamic stimulation of insulin release, J. Endocrinol., 58:681.

McBurney, R. D., 1947, The undernourished full-term infant, West. J. Surg. Obstet. Gynecol., 5:363.

Naeye, R. L., 1965, Infants of diabetic mothers. A quantitative morphologic study, Pediatrics, 35:980.

Osler, M. and Pederson, J., 1960, The body composition of newborn infants of diabetic mothers, Pediatrics, 26:985.

Polani, P. E., 1974, "Chromosomal and other genetic influences on birth weight variation" in Size at Birth, Ciba Founcation Symposium 27 (new series), pp. 127 - 164, Associated Scientific Publishers, Amsterdam.

Sherwood, W. G., Chance, G. W., Hill, D. E., 1974, A new syndrome of pancreatic agenesis. The role of insulin and glucagon in somatic and cell growth. Pediatr. Res., 8:360 (Abstract)

Simmons, A., 1976, "Normal physiological development of the fetus" in The Neonate (D. S. Young and J. M. Hicks, eds.), pp. 27 - 38, John Wiley and Sons, New York.

Simmons, M. A., Meschia, G., Makowski, E. L., and Battaglia, F. C., 1974, Fetal metabolic response to maternal starvation, Pediatr. Res., 8:830.

Stein, Z. A., Susser, M. W., 1975, The Dutch famine, 1944/45, and the reproductive process: I. Effects on six indices at birth, Pediatr. Res., 9:70.

Stein, Z., Susser, M. and Rush, D., 1978, Prenatal nutrition and birth weight: experiments and quasi-experiments in the past decade, J. Repro. Med., 21(5):287.

Usher, R., and McLean, F., 1969, Intrauterine growth of live-born Caucasian infants at sea level: standards obtained from measurements in 7 dimensions of infants born between 25 and 44 weeks of gestation, J. Pediatr., 72:901.

Winick, M., 1972, Nutrition and Development, John Wiley and Sons, New York.

Worley, R. J., Everett, R. B., MacDonald, P. C. and Grant, N. F., 1975, Placental clearance of dehydroisoandrosterone sulfate and pregnancy outcome in three categories of hospitalized patients with pregnancy induced hypertension, Gynceol. Invest., 6:28.

# CLINICAL SIGNIFICANCE OF RIBOFLAVIN DEFICIENCY

Surat Komindr, M.D.

Department of Medicine, Ramathibodi Hospital, Mahidol University, Bangkok, Thailand

George E. Nichoalds, Ph.D.

Department of Obstetrics and Gynecology, University of Tennessee Center for the Health Sciences, Memphis Tennessee 38163

## INTRODUCTION

The clinical consequences of riboflavin deficiency in humans were first described by Sebrell and Butler (1938). On a diet which furnished only 0.5 mg of riboflavin daily, 10 of 18 adult women developed oral lesions that disappeared following riboflavin supplementation. Now, over 40 years later, riboflavin deficiency is recognized as a global health problem that can occur in persons of any race, sex or age. Much new knowledge about riboflavin has been obtained during the last 10 years due to major improvements in methods for measuring riboflavin status and a concomitant renewed interest in nutrition. This paper will review the clinical significance of riboflavin deficiency with a special emphasis being given to recently acquired information.

## STRUCTURES AND CHEMICAL-PHYSICAL PROPERTIES

The history of riboflavin discovery, isolation and characterization has been reviewed in detail (Gyorgy, 1954; Wagner-Jauregg, 1954; Gyorgy, 1967). Karren et al. (1935) and Kuhn et al. (1935) independently synthesized riboflavin and proved its structure to be 7,8-dimethyl-10-(1'-D-ribityl) isoalloxazine (Figure 1). The name, riboflavin, thus reflects the presence of D-ribitol and flavin. Riboflavin is commonly referred to as vitamin $B_2$ and was earlier called vitamin G in the United States. Other previous names for riboflavin include lactoflavin, ovoflavin, uroflavin and

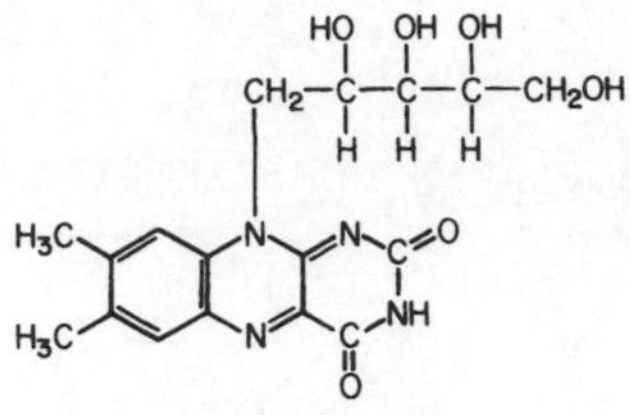

RIBOFLAVIN

Figure 1.

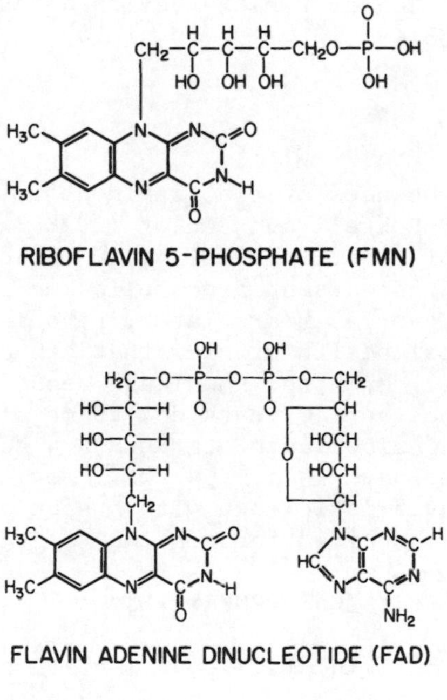

RIBOFLAVIN 5-PHOSPHATE (FMN)

FLAVIN ADENINE DINUCLEOTIDE (FAD)

Figure 2.

hepatoflavin which reflect the source of the vitamin. The principal forms of riboflavin found in nature are riboflavin 5'-phosphate (FMN) and flavin adenine dinucleotide (FAD) (Figure 2).

Riboflavin, FMN and FAD have molecular weights of 376, 456 and 786, respectively. Flavins have a characteristic yellow color resulting from strong absorption in the violet and blue regions. In crystalline form flavins are stable to heating up to 100-120$^{o}$. Riboflavin is odorless but has a bitter taste. It is poorly soluble in water (0.03 - 0.15 g/100 ml), perhaps accounting for its low toxicity. A saturated solution has a pH of approximately 6. FMN and its sodium salt are highly soluble and FAD is very hygroscopic and is freely soluble. All three compounds have multiple absorption peaks and exhibit intense greenish-yellow fluoresence (riboflavin and FMN) and orange fluoresence (FAD).

In aqueous solution flavins are relatively heat stable in the absence of light. Flavins present in food are unaffected by baking in an oven below 500$^{o}$F (Williams and Cheldelin, 1942; Mayfield and Hedrick, 1949) or short-term sterilization by autoclaving (Wagner-jauregg, 1972). In contrast, the riboflavin content of milk is reduced by more than 50% within two hours of exposure to sunlight (Ziegler, 1944). Flavins are relatively stable against common oxidizing agents. Generally, acids have no destructive influence on the isoalloxazine nucleus but the ester bond of the nucleotides is readily hydrolyzed. FAD can be fully hydrolyzed to FMN by allowing it to stand overnight at 38$^{o}$C in 10% trichloroacetic acid although no destruction occurs at 0$^{o}$C for 30 minutes. All flavins undergo decomposition in the presence of alkalies; if irradiated by light, lumiflavin and lumichrome appear as photolytic products. Several excellent reviews of the physical and chemical properties of flavins are available (Yagi, 1962; Wagner-Jauregg, 1972; Merrill and McCormick, 1980).

## METABOLISM AND FUNCTIONS

Riboflavin is consumed primarily as FMN- and FAD-protein complexes (McCormick, 1975). Prior to absorption the flavins are released by gastric acidification and proteolysis. Flavins are absorbed from the proximal region of the small intestine by a specialized site-specific and saturable transport process (Campbell and Morrison, 1963; Levy and Jusko, 1966; Jusko and Levy, 1967). Before flavins can be taken up by the intestine, it is necessary that they be dephosphorylated in the intestinal lumen. Prior to their release in the blood, the flavins are rephosphorylated in the mucosal cells.

Man is fairly efficient in absorbing riboflavin but concomitant ingestion with meals provides maximal absorption. The effect of food upon enhancing absorption may be due to the release of bile salts, agents that appear to facilitate the absorption of flavins (Mayersohn et al., 1969). When low levels of flavins are fed to normal persons, the amounts recovered in the urine are proportional to the amounts ingested. There is, however, an upper limit in intestinal absorption of riboflavin of about 25 mg (Jusko and Levy, 1967). When 30 mg of FMN or more are given orally, only about 50% is absorbed. Neither the mechanism underlying this latter observation nor its possible regulatory role are known.

Flavins are present in all biological fluids and tissues. After entering the circulation a large fraction of riboflavin and FMN is bound to albumin in a rather weak association (Jusko and Levy, 1969). Numerous other plasma protein fractions also bind riboflavin. Very recent work has described the purification of riboflavin-binding proteins from bovine plasma that migrate electrophoretically with the globulin fraction (Merrill et al., 1979). Preliminary work indicates the presence of comparable binding proteins in human plasma (Merrill and McCormick, 1980).

The flavins are quickly transported to the various tissues with the majority being taken up by the liver and kidneys leaving very little in the blood (Baker et al., 1966; Yagi et al., 1966; Fazekas and Sandor, 1973). Most tissues contain 1 μg flavin/g tissue or less (Koziol, 1971). In the tissues riboflavin is converted by flavokinase to FMN (McCormick, 1961; Merrill et al., 1978) which is subsequently converted to FAD by a pyrophosphorylase (McCormick, 1964). Some studies have indicated that thyroid hormones are involved in maintaining flavin coenzyme levels (Rivlin et al., 1968). FAD is the predominant flavin in tissues (Cerletti and Ipata, 1960; Yagi, 1962). The interconversions of riboflavin, FMN and FAD have been extensively reviewed by McCormick (1975).

Renal excretion of riboflavin involves glomerular filtration, tubular secretion and reabsorption (Morrison and Campbell, 1960a). The amount excreted depends not only on the amount ingested but also on the levels present in the body. When radioactive riboflavin is injected, about 81% of the radioactivity is retained after 24 hours (Yang and McCormick, 1967). Free riboflavin is the only flavin excreted from the body in significant amounts. The main excretory route is the urine with very minor amounts being excreted in the bile, feces, sweat and breast milk. There are trace levels of flavin catabolites present in the urine (West and Owen, 1969) which probably arise from intestinal flora (Yang and McCormick, 1967).

FMN and FAD form the prosthetic groups of several different enzyme systems which are collectively known as flavoproteins. They participate in numerous reactions (Horwitt and Witting, 1972; Merrill and McCormick, 1980). The oxidation of D- and L-amino acids by the corresponding amino acid oxidases requires FAD and FMN, respectively. In carbohydrate metabolism FMN acts in concert with pyridine nucleotides to oxidize glucose-6-phosphate and produce energy. Xanthine oxidase and aldehyde dehydrogenase are additional examples of enzymes involved in aerobic oxidation that require flavoproteins. FMN and FAD are also cofactors in several mitochondrial anaerobic dehydrogenation reactions. Examples of the latter include the transfer of hydrogen from NADH to the cytochromes, succinic dehydrogenase in the tri-carboxylic acid cycle and the β-oxidation of fatty acids via acyl-CoA dehydrogenases.

Most of the flavins found in blood are present in the erythrocytes with FAD being the predominant form (Burch et al., 1948). Since erythrocytes lack mitochondria, there are only a limited number of flavin enzymes present in these cells. One important role of FAD is as a coenzyme of glutathione reductase (EGR) which ensures proper maintenance of reduced glutathione (GSH) levels. The other FAD flavoprotein in red cells is methemoglobin reductase.

In pregnancy the FAD present in the maternal circulation does not cross the placenta but is degraded to free riboflavin in the placenta for secretion into the fetal blood (Lust et al., 1954). Riboflavin is also present in amniotic fluid (Clarke, 1973a). Umbilical cord blood levels at delivery are related to maternal blood levels but are always higher in the neonates - even in face of maternal hypovitaminosis (Kaminetzky et al., 1973, 1974). Several workers have confirmed the finding that neonates invariably have better riboflavin status than their mothers (Bamji, 1976; Baker et al., 1977; Clarke, 1977b). This disproportionate distribution of flavins between the mother and her fetus may be due to a carrier-mediated transport mechanism (Nutr. Rev., 1979). Although not yet demonstrated in the human, pregnancy-specific riboflavin-binding proteins have been shown to occur in chicks (Murthy and Adiga, 1977a,b,c, 1978), rats (Adiga and Muniyappa, 1978) and cows (Merrill et al., 1979). Furthermore, estrogen appears to induce their synthesis in liver and modulate their levels in plasma.

Accompanying the weight loss that normally occurs in a neonate during the first 3 days of life after birth is a high urinary excretion of riboflavin (Hamil et al., 1947). These large amounts excreted early in life may partly reflect the intrauterine storage as well as the negative nitrogen balance that usually occurs in the first few days of life. Urinary excretion of riboflavin decreases as the neonate undergoes transition to the extrauterine environment and milk intake is increased.

## REQUIREMENTS

Man must have a constant supply of riboflavin for the optimal maintenance of the many metabolic processes involving the flavin coenzymes (Tillotson and Baker, 1972). This need is met in adults when dietary riboflavin intake is sufficient to keep the body tissues near-saturated by replacing riboflavin lost from the body. During physiological stress periods of growth, pregnancy and lactation, riboflavin requirements are increased. Recommendations for the dietary riboflavin intake of healthy normal people are set at levels which will not only prevent the deficiency syndrome of ariboflavinosis but will ensure maximal functional capacity and meet any needs due to physiological stress.

The bulk of the data used to establish riboflavin requirements has been derived from balance studies. An important consideration in this regard is that meaningful measurements of urinary riboflavin excretion cannot be obtained in individuals with negative nitrogen balance (Oldham et al., 1947; Pollack and Bookman, 1951; Bro-Rasmussen, 1958; Smith et al., 1959). Inadequate protein intake and disease are invariably accompanied by negative nitrogen balance. This is a major reason why requirements are known only in normal, healthy individuals. Other factors which can result in altered nitrogen metabolism are discussed in the section on Assessment (Biochemical, Assay Methods) and some of the diseases known to decrease the levels of riboflavin are summarized in the section on Occurrence of Riboflavin Deficiency. There is little evidence, however, that disease per se causes an increased need for water-soluble vitamins (Coon, 1965).

There have been several suggestions for the best way to express riboflavin requirements. Arguments for estimating riboflavin needs in terms of caloric consumption (Bro-Rasmussen, 1958) and protein requirements (Horwitt, 1966) have been critically reviewed. The World Health Organization (1965, 1967 and 1974) has chosen to relate riboflavin needs to caloric intake at a level of 0.55 mg/1000 kcal and makes no additional adjustments for riboflavin requirements during physiological stresses. This value is similar to the riboflavin intake which maximizes the level of EGR (Beutler, 1969). The Food and Nutrition Board of the National Acacemy of Sciences (RDA, 1980) expresses riboflavin requirement in terms of metabolic body sizes, e.g., the adult requirement is 0.07 mg per day per $Kg^{0.75}$ If related to calorie intake, the RDA are higher than the WHO recommendations.

The early studies of Horwitt et al., (1949, 1950) suggested that the riboflavin requirement of the resting adult is 1.1 to 1.6 mg per day. Intakes below 0.5 mg per day are insufficient to support normal tissue repair and will result in ariboflavinosis (Bro-Rasmussen, 1958). The needs of infants and young children were initially evaluated to be approximately 0.4 mg per day (Oldham et al., 1944; Snyderman et al., 1949). These values for human riboflavin requirements have stood the test of time well (See Table I).

Some studies suggest no increased requirement for riboflavin with the progress of pregnancy (Oldham et al., 1950) but other reports of decreased urinary excretion (Dubrausky and Blazso, 1943; Brzezinski et al., 1952), reduced whole blood levels (Clarke, 1971),

Table I. Daily Riboflavin Requirement

| Group | Age (Year) | Weight (Kg) | Energy (Kcal) | Recommended Dietary Allowances (1980)* mg/daily |
|---|---|---|---|---|
| Infants | 0-0.5 | 6 | Kg. X 115 | 0.4 |
| | 0.5-1.0 | 9 | Kg. X 105 | 0.6 |
| Children | 1-3 | 13 | 1300 | 0.8 |
| | 4-6 | 20 | 1700 | 1.0 |
| | 7-10 | 28 | 2400 | 1.4 |
| Males | 11-14 | 45 | 2700 | 1.6 |
| | 15-18 | 66 | 2800 | 1.7 |
| | 19-22 | 70 | 2900 | 1.7 |
| | 23-50 | 70 | 2700 | 1.6 |
| | 51-75 | 70 | 2400 | 1.4 |
| | 76+ | 70 | 2050 | 1.4 |
| Females | 11-14 | 46 | 2200 | 1.3 |
| | 15-18 | 55 | 2100 | 1.3 |
| | 19-22 | 55 | 2100 | 1.3 |
| | 23-50 | 55 | 2000 | 1.2 |
| | 51 | 55 | 1800 | 1.2 |
| | 76+ | 55 | 1600 | 1.2 |
| Pregnancy | | | + 300 | + 0.3 |
| Lactation | | | + 500 | + 0.5 |

*These allowances are intended to provide for individual variations among most normal persons as they live in the United States under usual environmental stresses.

enzyme tests indicating riboflavin deficiency (Heller et al., 1974) and an increased incidence of deficiency syndromes (Brzezinski et al., 1952) clearly indicate an increased need. Lactation also increases riboflavin requirements (Roderuck et al., 1946). The increase in requirement for B-vitamins during pregnancy, particularly in the final trimester, may be due partly to sequestration of vitamins by the fetus and placenta, and partly to the metabolic effects of high circulating levels of female sex hormones (Bamji, 1976).

The best food sources of riboflavin are milk, liver, eggs, meat and some of the green leafy vegetables (Horwitt, 1972; Foy and Mbaya, 1977). One quart of milk contains about 2.0 mg of riboflavin which is equivalent to approximately two pounds of lean meat. Yeast is the richest natural source but does not contribute significantly to dietary intake. Unenriched grains and legumes are poor sources but supply important amounts to many regimens due to large consumption of these foods. Grains have become valuable riboflavin sources in the United States because of bread enrichment programs. The riboflavin intake of many diets is related to the amount of ingested animal protein. Riboflavin consumption in the United States increased from 1.9 in 1935 to 2.3 mg/person/day in 1959, mostly due to greater consumption of beef (Foy and Mbaya, 1977). Today's current market prices for meat have undoubtedly reduced riboflavin intake from this source. Non-food sources of riboflavin include vitamin synthesis by intestinal flora (Najjar et al., 1944) but the amounts produced are small and may not be taken up by the body.

ASSESSMENT OF RIBOFLAVIN STATUS

The methods used for assessing nutritional riboflavin status in man are comparable to those used for other vitamins. There are several very good summaries concerning nutritional assessment techniques (Goldsmith, 1959, 1964, 1975; Jelliffee, 1966; Ten State Nutrition Survey, 1968-1970; Christakis, 1973). A diagnosis of riboflavin deficiency is usually made on the basis of a prolonged history of poor riboflavin intake which may or may not be accompanied by clinical signs and symptoms of ariboflavinosis. Final confirmation is accomplished by biochemical evaluation. In those instances where laboratory tests are not available, therapeutic diagnosis might be considered. It is essential to know nutritional requirements, be familiar with nutrient functions, be cognizant of the potential influence of disease on nutrient requirements and utilization and to evaluate possible biochemical disturbances in the body.

In population groups with very uniform food patterns, which is often the case in developing countries, relatively simple methods yield good results (Brubacher, 1974). In this situation the aim is to decide if there is a public health problem. Methods can thus be used which are applicable to screening large population groups. In highly industrialized countries, however, more sophisticated techniques are needed which can be applied to individuals in a clinical setting. It is unfortunate but there is generally a poor correlation between dietary, clinical and biochemical evaluation of riboflavin nutriture (Axelrod et al., 1941; Brocklehurst et al., 1968; Thurnham et al., 1970; Buzina et al., 1971, 1973; Sharada and Bamji, 1972; Schorah and Messenger, 1975; Kaufmann and Guggenheim, 1977). Only when the riboflavin deficiency is prolonged and severe is there a very good correlation between these variables (Bamji, 1969; Beutler, 19 9; Tillotson and Baker, 1972; Vir and Love, 1977).

## Dietary

An estimate of riboflavin status can be made by determining dietary intake of the vitamin and comparing this with an individual's needs as discussed in the previous section. An indication of adequate riboflavin intake does not insure that there are adequate body stores but does imply minimal risk for developing ariboflavinosis. Dietary methods are laborious and expensive. They are particularly useful in nutrition surveys for screening large numbers of people and for monitoring hospitalized patients.

There are several methods by which to evaluate dietary intake. The most commonly utilized approach is the diet record. The subject records types and amounts of food consumed over some specified period of time. Alternately, this same information can be obtained by interviews. More exacting information results when home visits are included in the evaluation. Each of these methods requires the use of food composition tables which often contain inadequate or questionable values. The most precise approach involves the actual determination of the amount of riboflavin present in 24-hour food composites by the same basic assay methods in the Biochemical Section. Availability of this latter method is, however, extremely limited.

## Clinical

Few clinical signs have pathognomic significance. Unlike some of the other nutrient deficiencies, ariboflavinosis is not a life threatening situation. Only the general aspects of the clinical evaluation of riboflavin status will now be discussed. Specific clinical manifestations of ariboflavinosis are described in the section on Effects of Riboflavin Deficiency.

The clinical assessment of riboflavin nutriture in man is inexpensive, can be done quickly and requires minimal training. There are, however, several problems in the clinical evaluation of nutritional status such as examiner variation that must be borne in mind at all times (Plough and Bridgeforth, 1960). It is rare to find pure, uncomplicated riboflavin deficiency in clinical practice. Most patients with ariboflavinosis will also be deficient in other nutrients which may alter riboflavin status. For example, it has been stated that deficiencies of pyridoxine and niacin cause the same lesions of the mouth typically associated with riboflavin deficiency (Goldsmith, 1956; Jelliffe, 1966; Krishnaswamy, 1971; Iyenger, 1973). The rationale for this statement is discussed in the section on Pathophysiology.

It is also important to realize that clinical manifestations are late reflections of metabolic disturbances that occur only after nutrient body stores have become depleted. Brin et al. (1965) have described the progressive stages of vitamin deficiency in man. Most nutrition surveys done in the United States reveal a rather low incidence of clinical evidence of riboflavin deficiency. This should not be taken as an indication that subclinical ariboflavinosis is not present. Clinical observations should only be utilized as a quick screening method to identify those patients at greatest risk of being riboflavin-deficient. Confirmation by dietary evaluation and biochemical assessment is absolutely necessary to ensure proper treatment.

## Biochemical

There are several biochemical methods for measuring riboflavin and guidelines for interpretation of the results (Pearson, 1962, 1967a; O'Neal et al., 1970; Sauberlich et al., 1974). Biological and chemical assays can be used with samples of tissue, urine or blood. These tests measure changes reflecting dietary riboflavin intake or tissue storage and saturation of riboflavin, FMN and FAD (Kaufmann and Guggenheim, 1977). Other analytical techniques have been described but are not used often (Couch and Davies, 1963). Recently, an enzymatic functional test to assess riboflavin status has been developed which appears to be more valid than other measurements. All of these assays are sensitive to small changes in experimental conditions but are reproducible and accurate when done by capable and experienced laboratory personnel.

## Sample Types

Tissue. When we attempt to measure riboflavin status, we are, in effect, seeking a measure of body stores (Pearson, 1967a). These stores are not markedly affected by normal dietary riboflavin intakes because they are not mobilized until the circulating riboflavin in blood has been depleted (Tillotson and Baker, 1972). Thus, tissue flavin levels are an ideal measurement. Most of the riboflavin in tissues is present as FMN and FAD bound to protein. These flavins must be extracted, liberated and separated prior to analysis according to any of several published procedures (Yagi, 1951, 1962; Yagi and Okuda, 1958; Pearson, 1967b; Koziol, 1971; Fazekas and Kokai, 1971; Baker and Frank, 1975). The release of covalently-bound flavin, e.g., FAD present in succinic dehydrogenase, requires enzyme and acid digestion for liberation

(Singer et al., 1971). Tissue measurements are rarely made in man because they are very difficult to obtain and gram quantities of tissue are required.

Urine. Riboflavin status has been evaluated extensively by measuring its levels in urine (Horwitt, 1966). For all practical purposes the only flavin present in urine is riboflavin. Despite the lack of human data on the relationship between urinary excretion and total body stores of riboflavin (Pearson, 1967a), the clinical signs of ariboflavinosis and dietary intake of the vitamin have been shown to correlate with urinary riboflavin levels under carefully controlled studies (see Requirements Section).

For survey purposes non-fasting casual, 2-hour or 6-hour collections can be used successfully (Lowry, 1952; Hegsted et al., 1956; Plough and Consolazio, 1959). In clinical instances where riboflavin deficiency is suspected, a 24-hour urine collection should be obtained if at all possible (Sauberlich et al., 1974). Expressing urinary riboflavin excretions per gram of creatinine tends to correct for variations due to body size. When this correlation is applied, children excrete more riboflavin per gram of creatinine than do adults (Pearson, 1962). A riboflavin load test had been used occasionally in which the urinary excretion of riboflavin is measured in the 4-hour period following the oral administration of a 5-mg test dose of the vitamin (Lossy et al., 1951). Several considerations must be taken into account when using urine to assess riboflavin nutriture.

Urinary riboflavin levels give no indication of the body's stores of FMN and FAD, the functional flavin forms. There can be large variations in urinary excretion due to differences in dietary intake (Tillotson and Baker, 1972) and physiological circumstances. Riboflavin excretion is increased under conditions of negative nitrogen balance, fasting, heat, stress, prolonged hard physical work and enforced bed rest (Pollack and Bookman, 1951; Bro-Rasmussen, 1958; Smith et al., 1959; Tucker et al., 1960; Coon, 1965). Sleep and short periods of hard physical work decrease vitamin excretion (Tucker et al., 1960).

Blood. Riboflavin nutritional status has also been evaluated by measuring riboflavin, FMN and FAD levels in whole blood, plasma and erythrocytes (Pearson, 1962). As was the case with urine, plasma concentrations of flavins tend to reflect recent dietary intake and exhibit considerable daily fluctuation (Burch et al., 1948; Suvarnakich et al., 1952; Coon, 1965; Bamji, 1969). Erythrocyte flavin levels are more sensitive than plasma and have been used in several studies as an index of riboflavin status (Horwitt et al., 1949; Bessey et al., 1956; Bamji, 1969; Clarke, 1969b). Blood levels are, nevertheless, of limited value since concentrations do not change until more easily determined parameters have been altered (Sauberlich et al., 1974).

## Assay Methods

Fluorometric. Most fluorometric assays are based on the method of Slater and Morell (1946). Procedures are available for use with tissues (Bessey et al., 1949; Burch et al., 1956), blood (Burch et al., 1948) and urine (Albright and Degner, 1967; Mellor and Maass, 1972). Either the fluorescence of the flavins can be measured directly or they can be converted to lumiflavin and its fluorescence determined (Fujita and Matsuura, 1950; Clarke, 1969a, 1977a). Fluorometric methods of analysis are used more often than microbiological procedures as they tend to give higher and more consistent values (Bamji et al., 1973).

Biologic. Aside from the lactic acid bacteria, few free-living bacteria require exogenous riboflavin. Optimal assay conditions for using *Lactobacillus casei* to determine flavin levels are well documented (Snell and Strong, 1939; Strong and Carpenter, 1942; Clegg et al., 1952; Prager et al., 1958). Growth of the *Lactobacillus casei* can be measured turbidimetrically or by the level of acid production. Although not used as frequently, *Leuconostoc mesenteroides* is claimed to be 50 times more sensitive than *L. casei* for assaying flavins (Kornberg et al., 1948). There are, however, certain disadvantages in using microbiological assays. Antibiotic therapy in patients often precludes their use. The growth response of *L. casei* is different for riboflavin, FMN and FAD (Snell and Strong, 1939; Langer and Charoensiri, 1966). *L. casei* is also known to grow in response to fatty acids (Pearson, 1962) and flavin analogs that may or may not have biologic activity (Yang et al., 1964). Microbiological assays tend to be more satisfactory for urine determinations than they are for blood measurments (Pearson, 1967b). A method based on the protozoan, *Tetrahymena pyriformis*, has been successfully used by Baker et al., (1966) for several years.

Enzymatic. In recent years EGR has been used as an enzymatic index of the riboflavin nutriture of man. The enzyme catalyzes the reduction of oxidized glutathione (GSSG) in the following manner:

$$\text{NADPH} + \text{H}^+ + \text{GSSG} \xrightarrow{\text{FAD}} \text{NADP}^+ + 2\text{GSH}$$

EGR activity is altered *in vivo* by dietary riboflavin (Bamji, 1969; Beutler, 1969; Glatzle et al., 1970) and *in vitro* by FAD using purified enzyme (Buzard and Kopko, 1963; Scott et al., 1963; Icen, 1967; Staal et al., 1969) and red cell hemolysates (Glatzle et al., 1968; Beutler, 1969; Glatzle et al., 1970). The degree of *in vitro* stimulation of EGR activity depends on the degree of apoenzyme saturation by FAD, which in turn depends upon the availability of riboflavin. The enzyme is usually not saturated with FAD (Beutler, 1969), but a relatively constant percentage of FAD saturation of the EGR is found in persons consuming adequate amounts of riboflavin (Tillotson and Baker, 1972). A direct measurement of EGR activity can be used to determine riboflavin status, but to do so requires a carefully standardized assay procedure and the use of a reference such as hemoglobin to eliminate the effects of factors such as anemia. Much more frequently EGR activity is measured in the presence and absence of *in vitro* FAD. This stimulatory effect is then expressed as an activity coefficient - EGR AC.

$$\text{EGR AC} = \frac{\text{Enzyme Activity with added FAD}}{\text{Enzyme Activity without added FAD}}$$

The question of what specific assay conditions give maximal EGR activity has received considerable attention (Staal and Veeger, 1969; Staal et al., 1969; Hornbeck and Bradley, 1974; Schorah and Messenger, 1975; Bayoumi and Rosalki, 1976). Enzyme activity can be measured by following the oxidation of NADPH at 340 mμ (Beutler, 1969; Glatzle et al., 1970; Brewster et al., 1974; Nichoalds et al., 1974) or the production of GSH (Tatt et al., 1975; Garry and Owen, 1976). Whole blood (Weber et al., 1973; Glatzle et al., 1974) and white blood cells (Muller and Bates, 1977) can be used instead of red cells.

The EGR AC test has numerous advantages. It requires only small amounts of blood and is independent of age and sex. Fasting blood samples are not needed and the enzyme is relatively stable. The usefulness of the test for identifying individuals receiving marginal dietary riboflavin over prolonged periods of time as well as individuals suffering from severe riboflavin deficiency has been well documented (Thurnham et al., 1970, 1972; Buzina et al., 1971; Flatz, 1971a,b,c; Tillotson and Baker, 1972; Sauberlich et al., 1972, 1974; Cooperman et al., 1973; Weber et al., 1973).

While the EGR AC responds rapidly to riboflavin deficiency, it plateaus quickly and does not continue to change in response to progressive tissue depletion (Sterner and Price, 1973). The test cannot be used in persons with G-6-P-D deficiency due to a greater avidity of EGR for FAD (Flatz, 1970; Thurnham, 1972); the prevalence of G-6-P-D deficiency is about 10% among American Blacks (Schrier et al., 1958; Frischer et al., 1973). Other diseases may also alter EGR activity in a manner unrelated to riboflavin status (Beutler, 1969; Yawata and Tanaka, 1971) and more information in this area is needed.

At the moment the measurement of the EGR AC appears to be the method of choice for evaluating riboflavin nutriture in man. Some of the areas which certainly warrant closer investigation include luminescence and pyridoxamine 5'-phosphate (PMP) oxidase. The application of the luciferase reaction (Strehler and Cormier, 1954) in the clinical laboratory has recently been reviewed (Gorus and Schram, 1979; Whitehead et al., 1979) and both FMN and FAD can be measured using this technique (Mitchell and Hastings, 1969; Chappele and Picciolo, 1971; Stanley, 1971; Meighen and MacKenzie, 1973). PMP oxidase activity has been found to decrease significantly in liver, kidney and brain tissue of riboflavin-deficient rats (Rasmussen et al., 1979). Importantly, PMP oxidase activity is more sensitive than the ERG AC to the degree of riboflavin depletion.

## OCCURRENCE OF RIBOFLAVIN DEFICIENCY

Riboflavin status can be evaluated by assessing the adequacy of dietary intake, the presence or absence of clinical signs associated with ariboflavinosis, and by several different biochemical determinations of riboflavin levels in the body. Using one or more of these criteria, indications of inadequate riboflavin nutriture have commonly observed in nutrition surveys (Davis et al., 1969; Kelsay, 1969; May, 1961-1972; Ten-State Nutrition Survey, 1968-1970).

The largest series of international nutrition surveys were conducted by the Interdepartmental Committee on Nutrition for National Defense. The incidence of riboflavin deficiency is extremely high in some of the developing countries such as India, Thailand and Trinidad. In situations where a population subsists on insufficient quantities of food which are poor sources of riboflavin, over 25% of the people may suffer from riboflavin deficiency (Fernandez et al., 1965). The frequency of ariboflavinosis is seasonal in developing countries and can vary 10-fold (Thurnham et al., 1970; Buzina et al., 1973).

Except for particularly vulnerable segments of the American population such as patients with disease, riboflavin deficiency is an unexpected entity in a country such as the United States where an abundance of food is available. There is, however, a propensity of riboflavin deficiency in the socio-economically deprived groups of the population (Ten-State Nutrition Survey, 1968-1970) which has been attributed to the poor quality of dietary intake. Potential health problems are particularly manifested among impoverished Blacks and Spanish Americans of all ages. It has been pointed out earlier (Leverton, 1964) that the nutritive value of a nation's food supply cannot be used to assess the nutritional status of an individual.

Ariboflavinosis often develops as a consequence of pregnancy (Brzezinski et al., 1952). Oral lesions of the mouth associated with inadequate riboflavin nutriture are commonly observed in malnourished pregnant women (Clarke, 1971; Iyengar, 1973; Bamji, 1976; Clarke, 1976). Some investigators have not found riboflavin deficiency to be a problem in pregnancy (Baker et al., 1975), but the largest survey done to date reported an incidence of riboflavin depletion among apparently healthy women of up to 40% using the EGR AC test (Heller et al., 1974). Generally, the frequency of riboflavin deficiency increases with the progress of pregnancy.

Pregnant adolescents are the most vulnerable segment of the population to generalized malnutrition. Data on American teenagers are limited but indicate rates of riboflavin deficiency ranging from 10% to 30% (McGanity et al., 1969; Kaminetzky et al., 1973). In view of the recent tremendous increase in the occurrence of teenage pregnancy in this country (Stickle, 1975), more nutritional studies on this particular group are direly needed.

The Ten-State Nutrition Survey (1968-1970) identified riboflavin deficiency as a potential problem for infants and children. Preschool children in Mississippi (Owen et al., 1969) and children under 12 years of age from both lower and upper income groups in New York City (Newman and Martin, 1971) had unacceptable urinary riboflavin values ranging from 6 to 9%. In a more recent survey (Lopez et al., 1975) 11 out of 100 children ranging in age from a few days to 14 years who resided in economically depressed areas of New York City had biochemical evidence of inadequate riboflavin nutriture.

Undesirably low levels of riboflavin in the urine of teenagers have been reported to occur at a rate of 7% in Iowa (Hodges and Krehl, 1965), 5% in Vermont (Morse et al., 1965) and 21% in New York (Dibble et al., 1965). Sauberlich et al. (1972) evaluated the riboflavin nutriture of 431 rural Tennessee high school students by several criteria: 1) 47% of the adolescents had dietary riboflavin intakes less than 50% of the National Research Council's Recommended Dietary Allowance (RDA, 1980); 2) unacceptable urinary excretion of riboflavin was found in 22% and; 3) the incidence of inadequate riboflavin nutriture was 11% using the EGR AC assay. In all these studies riboflavin deficiency was more common in girls and probably reflects greater food consumption by the boys.

Information concerning riboflavin status in the aged is meager (Department of Health and Social Security, 1972) and has resulted in contradictory conclusions. Evaluations of riboflavin nutriture by measuring whole blood levels or urinary excretion have indicated a rather low incidence of deficiency (Brin et al., 1965; Brocklehurst et al., 1968; Thackray et al., 1972; Morgan et al., 1975) in elderly subjects despite a relatively high incidence of clinical signs associated with ariboflavinosis. This discrepancy is likely due to the insensitivity of whole blood measurements as is borne out by the observation that riboflavin supplementation does alleviate the clinical situation (Dymock and Brocklehurst, 1973). Failure to confirm the positive therapeutic effects of riboflavin (Berry and Darke, 1972) probably means that factors other than riboflavin deficiency caused the observed clinical signs and symptoms (see Clinical Signs and Symptoms Section). When riboflavin status has been assessed using the EGR AC technique, deficiency in geriatric subjects has been found to occur in 7% to 32% of this age group (Glatzle et al., 1970; Hoorn et al., 1975; Kaufmann and Guggenheim, 1977; Vir and Love,1977).

Diseases can precipitate and/or exacerbate riboflavin deficiency. Ariboflavinosis has been observed in prolonged febrile illnesses such as rheumatic fever, tuberculosis and sub-acute bacterial endocarditis, in malignancy, hyperthyroidism, cardiac failure, diabetes mellitus and diseases of the gastroentestinal tract (Goldsmith, 1975). Table II summarizes some of the diseases reported to aggravate riboflavin nutriture.

Perhaps the most common disease known to contribute to malnutrition in developed countries is alcoholism. Chronic consumption of relatively large quantities of alcohol can result in riboflavin deficiency. Low blood levels of riboflavin have been found to occur in 15% of alcoholics having no liver damage (Leevy et al., 1965a). Those with fatty livers have essentially the same incidence of deficiency (17%) but the prevalance increases to 25% in those with cirrhosis. A second complicating illness superimposed on the alcoholism can increase the rate of riboflavin deficiency to 50% (Rosenthal et al., 1973).

Recent reports (Butterworth, 1974; Butterworth and Blackburn, 1975) have presented rather alarming evidence that many hospitalized patients suffer from malnutrition, some of which is acquired after being hospitalized. Significant episodes of riboflavin deficiency have been found in patients in hospitals (Brocklehurst et al., 1968; Vir and Love,1977; Tanphaichitr et al., in press). In a random sample of 120 hospital patients in New Jersey, 12% had low levels of riboflavin in their blood (Leevy et al., 1965b).

## CAUSES OF DEFICIENCY

The causes of ariboflavinosis can be delineated as being due to: 1) inadequate intake; 2) decreased assimilation; 3) defective utilization; 4) increased requirement; 5) increased destruction and 6) increased excretion. The following examples of causes of riboflavin deficiency are applicable to nutrients in general.

Good riboflavin nutriture is initially dependent upon the availability of riboflavin-rich foods and then upon the capability to obtain those foods in sufficient quantities for consumption. In developing countries there is often a paucity of food for reasons

Table II. Riboflavin Deficiency in Disease

| Disease | References |
|---|---|
| Cancer | Rivlin, 1975a |
| Cardiac | Steier et al., 1976 |
| Diabetes mellitus | Pollack and Brookman, 1951<br>Cole et al., 1976 |
| Gastrointestinal and Biliary Obstruction | Jusko and Levy, 1967<br>Rivlin, 1970b<br>Jusko et al., 1971<br>Lai and Ransome, 1971<br>Tautt, 1976<br>Rosenberg et al., 1977<br>Sleisenger and Brandborg, 1977<br>Hines, 1978 |
| Infection | Beisel et al., 1972<br>Feigin, 1977 |
| Liver | Baker et al., 1964<br>Leevy et al., 1965a<br>Clarke, 1969b<br>Leevy et al., 1970<br>Rosenthal et al., 1973<br>Mezey, 1978 |
| Thyroid | Rivlin, 1968, 1970a, 1970b, 1975b, 1979 |

of nature such as drought, floods and insect infestation. However, in developed countries such as the United States, the difficulty of being able to purchase an adequate diet is much more of a problem than availability of food.

Inadequate riboflavin consumption may also be the result of food habits, traditions, customs and religious beliefs which lead to the avoidance of riboflavin-rich foods. Pica is one example of a relatively prevalent food habit (Kaminetzky et al., 1973) that can reduce riboflavin intake. One of the more popular and less radical food faddism movements in the United States has been vegetarianism. Most evaluations indicate that vegetarian diets can be adequate if sufficient thought goes into their planning (Register and Sonnenberg, 1973). In fact women on vegetarian diets have been reported to undergo pregnancy without any adverse effects on themselves or their offspring (Thomas and Ellis, 1977). Many traditional dietary restrictions apply most often to the nutritionally vulnerable groups - infants, toddlers, pregnant women, lactating women, and the sick. In addition, if anorexia is present, there can be a substantial decrease in food intake.

Commonly employed food processing techniques and storage practices can result in appreciable losses of riboflavin (Foy and Mbaya, 1977). Even cooking habits can decrease the availability of riboflavin present in food. Extensive riboflavin losses of up to 20% occur when foods are cooked for long periods of time in large volumes of water which are subsequently discarded (Krehl and Winters, 1950).

Ineffective digestion and/or absorption of food results in diminished assimilation. Defective digestive enzymes can also dramatically reduce nutrient assimilation. A lucid example found in most Asian people and Blacks is lactase deficiency which leads to an intolerance of milk, a particularly good source of riboflavin. Lactase is often the first digestive enzyme to be affected during illnesses such as diarrhea and protein-calorie malnutrition.

Conditions that cause increased motility and decreased gastrointestinal passage time can also result in malassimilation. Exemplary illustrations of some of these conditions include ingestion of laxative agents, antibiotic therapy, infectious enteritis, blockage of the biliary pathway and certain diseases such as irritable bowel syndrome and malabsorption syndrome (see Table II). Antibiotics may also cause riboflavin deficiency by altering the intestinal flora and inhibiting bacterial synthesis of riboflavin (Tuszynsky et al., 1958).

Malabsorption occurs whenever there is appreciable structural damage and/or a decrease in mucosal cells of the small intestine. This can occur as the result of ingestion of caustic agents, maladies such as tropical sprue and celiac disease, malignancy and resection of the small bowel (Hines, 1978). Finally, systemic diseases without primary involvement of the gastrointestinal tract, such as chronic liver disease and systemic bacterial infection, can also result in failure of riboflavin absorption (Jusko and Levy, 1967; Feigin, 1977).

Once riboflavin is present inside the body, the vitamin may not be maximally utilized. This can be caused by disturbances in hormonal production, e.g. thyroid, ACTH and aldosterone (Rivlin, 1979) as well as insulin. Certain drugs such as oral contraceptive pills (Briggs and Briggs, 1974; Sanpitak and Chayutimonkul, 1974; Ahmed et al., 1976b; Newman et al., 1978) and phenothiazine derivatives (Gabay and Harris, 1966, 1967; Horvath, 1976) also appear to impair riboflavin utilization.

The body's need for riboflavin becomes greater during the periods of physiological stress that occur during growth, pregnancy and lactation. Since physiological stress is usually construed as being normal, its importance is frequently neglected or underestimated in clincial practice. Pathological stress due to diseases can also be expected to increase nutrient requirements but, this subject has received little attention and definitive data are not available.

The treatment of neonatal jaundice with phototherapy has been observed to result in riboflavin deficiency by causing accelerated decomposition of the vitamin (Rubaltelli et al., 1974; Gromisch et al., 1977). Indirect evidence of riboflavin deficiency has similarly been noted in patients with β-thalassemia, another hematologic disease, during hemolytic episodes (Anderson et al., 1976).

Increased riboflavin excretion occurs in all conditions which produce increased nitrogen excretion, the cause of which may be attributed solely to the catabolic process of disease such as burns, uncontrolled diabetes mellitus, hyperthyroidism, systemic infections, starvation or even simple bed rest (Bro-Rasmussen, 1958; Tucker et al., 1960; Clarke, 1969b; Beisel et al., 1972; Feigin, 1977). Increased excretion of riboflavin can also be caused by certain antibiotics and phenothiazine drugs used in treatment processes (Goldsmith, 1975; Pinto et al., 1979).

The foregoing clearly points out that the etiology of riboflavin deficiency may range from simple inadequate dietary intake to complex and multiple causes, complicated by disease. For example, protein-calorie malnutrition diminishes both riboflavin absorption and utilization. The ariboflavinosis seen in alcoholics may be due to generally poor dietary intake or specific effects of alcohol upon the absorption, metabolic transformations or excretion of this vitamin. Systemic infections, even without gastrointestinal tract involvement, can cause inadequate intake, defective absorption, increased requirement, poor utilization and increased excretion.

## EFFECTS OF RIBOFLAVIN DEFICIENCY

### Clinical Signs and Symptoms

Since the first report of human riboflavin deficiency (Sebrell and Butler, 1938), there have been many experimental studies to elucidate the clinical symptoms and signs associated with ariboflavinosis. The following sections describe the known clinical manifestations of riboflavin deficiency in man. Other clinical descriptions include those by Jolliffe et al. (1939), Brocklehurst et al. (1968) and Goldsmith (1975).

Oral Mucosal and Tongue Alterations. Early symptoms of riboflavin deficiency include soreness of the lips, tongue and mouth (Lane et al., 1964) and are usually accompanied by difficulty in eating and swallowing. Pallor develops at the lip mucosa in the angles of the mouth followed by maceration. The lesions are bilateral and will be covered by a honey-colored crust. They extend downwardly from the angles of the mouth to the normal skin producing superficial transverse fissures known as angular stomatitis or perleche in the older literature. Any lesion resembling angular stomatitis that occurs only on one side is probably due to infection from herpes simplex rather than being attributable to riboflavin deficiency. Ill-fitting dentures can also cause similar lesions (Goldsmith, 1959), particularly in the elderly (Exton-Smith, 1975). At about the same time the lips may become swollen and denudation of the mucosa along the closure line of the mouth often occurs causing vertical fissures which are called cheilosis. These lesions may be the result of exposure to weather and have been found in persons with combined niacin and pyridoxine deficiency (Mueller and Vilter, 1950).

Glossitis is often seen in riboflavin-deficient persons. In the early stages the tongue which is normally pink in color becomes red accompanied by swelling and a burning sensation. It may be deeply fissured with prominent fungiform papillae at the tip. The patient may experience sore throat. As the riboflavin deficiency worsens, the filiform papillae begin shedding in small patches which gives the tongue a pebbled appearance (Figure 3). In the next progressive stage the filiform papillae continue shedding until the tongue becomes smooth (Figure 4). Taste acuity is diminished at the same time. If the patient does not receive treatment, the tongue will finally become atrophic. Glossitis is common in deficiency of many of the B-complex vitamins.

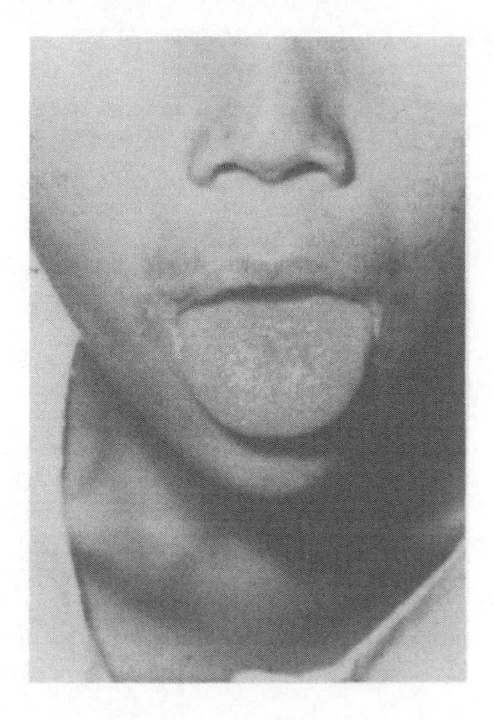

Fig. 3. Early stage of angular stomatitis. Note the patchy shedding of the filiform papillae and prominent fungiform papillae producing a pebbled appearance of the tongue.

Ocular Changes. Ocular changes due to riboflavin deficiency were first noted in man by Spies et al. (1939). Sydenstricker et al. (1940) beautifully described the progress and regress of the occular manifestation due to ariboflavinosis in patients. Photophobia and blepharospasm are always the first eye symptoms followed by conjunctival conjestion. The patients experience a burning sensation of the eyeballs and a dimness of vision uncorrectable by adjustment of the refractive errors. Circumcorneal injection can be grossly visualized and is usually accompanied by conjunctivitis. Examination by ophthalmoscopes and slit lamps is helpful in the early diagnosis of the disorder. In severe instances, capillary invasion from the circumcorneal plexus into the cornea will be observed. The capillary growth into the normally avascular cornea is probably an attempt to compensate for the breakdown of oxidation processes in the corneal cells (Bessey and Wolbach, 1939). Occasionally, frank iritis can be seen. It is noteworthy that comparable abnormalities can also be found in association with trauma, infection and allergic conditions. Some animal studies have implicated riboflavin deficiency as a causative

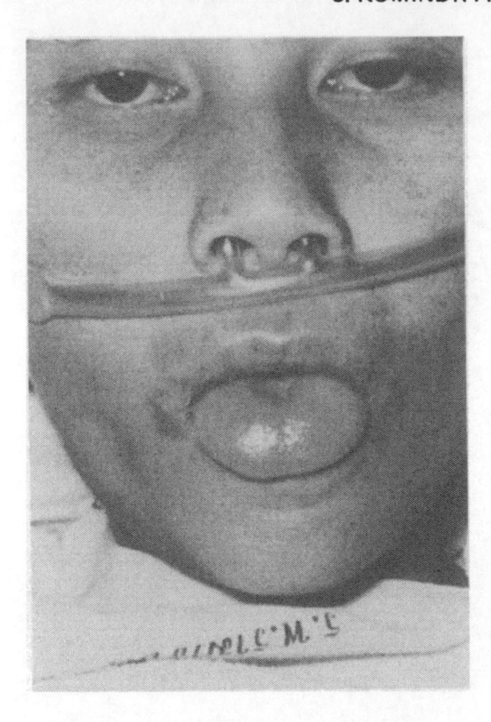

Fig. 4. Advanced stage of angular stomatitis and glossitis. Note the smooth tongue with prominent fugiform papillae.

factor in cataract formation (Day et al., 1931; Eckardt and Johnson, 1939; Srivastava and Beutler, 1970) but such a relationship has not been demonstrated in man (Nutr. Rev., 1976).

Dermatological Changes. Another common clinical manifestation of ariboflavinosis in man is seborrheic dermatitis. Frequently there is a greasy desquamation in the facial area that occurs in the naso-labial fold, alae nasi, nasal vestibule and on the ears. Sometimes the skin around the eye lids near the inner and outer canthi is also involved and is called angular palpebritis. Seborrheic dermatitis may occur at the scrotum in males and at the vulva in females (Horwitt et al., 1949). Similar changes often occur in the adjacent areas of the thighs characterized by redness, scaling, desquamation and oozing of serum from the superficial skin. Seborrheic dermatis may also be related to a lack of pyridoxine.

The interpretation of the above dermatological lesions is complicated by the appearance of similar lesions that have been described in patients with zinc deficiency. Particular attention has been drawn to patients who have received long-term total parenteral nutrition unsupplemented with zinc (Kay et al., 1976; Okada et al., 1976). In addition to dietary zinc deficiency, there is a genetically inherited disorder seen in infants and young children called acrodermatitis enteropathica that also results in zinc deficiency (Moynahan, 1974) and gives rise to dermatological lesions. However, there appears to be no metabolic interaction between riboflavin and zinc deficiencies (Pallauf and Kirchgessner,1973).

Neurological Alterations. Burning feet syndrome has been reported in one patient diagnosed as having malabsorption; the sensory abnormalities were alleviated within one week by intramuscular injections of riboflavin (Lai and Ransome, 1970). A decrement of hand grip strength has been noted in ariboflavinosis (Sterner and Price, 1973). Retarded intellectual development and electro-encephalographic changes have been observed in riboflavin-deficient children (Arakawa et al., 1968a). In adults, experimental hyporiboflavinosis may produce behavioral changes such as hypochondriasis, depression, hysteria, psychopathic-deviate and hypomania (Sterner and Price, 1973). The effects of dietary deprivation on human behavior have been reviewed by Brozek and Vaes (1961).

Hematological Dyscrasia. A few studies have indicated an indirect role for riboflavin in hematological dyscrasia. Normocytic-normochromic anemia and reticulocytopenia have been induced in man using a dietary regimen that was not only riboflavin-deficient but also contained a riboflavin antagonist (Lane and Alfrey, 1965). Similar hematological findings have also been found to occur in African children with marasmus and kwashiorkor (Foy et al., 1961; Kondi et al., 1967). All symptoms of the hematological dyscrasia that occur in riboflavin-deficient adults, and in children with protein-calorie malnutrition, regress after treatment with either riboflavin or corticosteroid, suggesting the possibility of an interrelationship between riboflavin deficiency, corticosteroids and the hematopoietic system. Details concerning the pathophysiology of these abnormalities are discussed in the Pathophysiology Section.

## Pathophysiology

Much recent data have clearly shown that riboflavin nutriture in man is influenced by numerous factors. At the same time it is also evident that riboflavin deficiency alters several metabolic pathways. Figure 5 summarizes the currently recognized interrelationships of riboflavin with other nutrients and the hormonal control of its conversion to active coenzymes. The following examples of the pathophysiology of riboflavin deficiency are not intended to constitute an exhaustive review of the literature but will serve to highlight major areas of concern where some data are available.

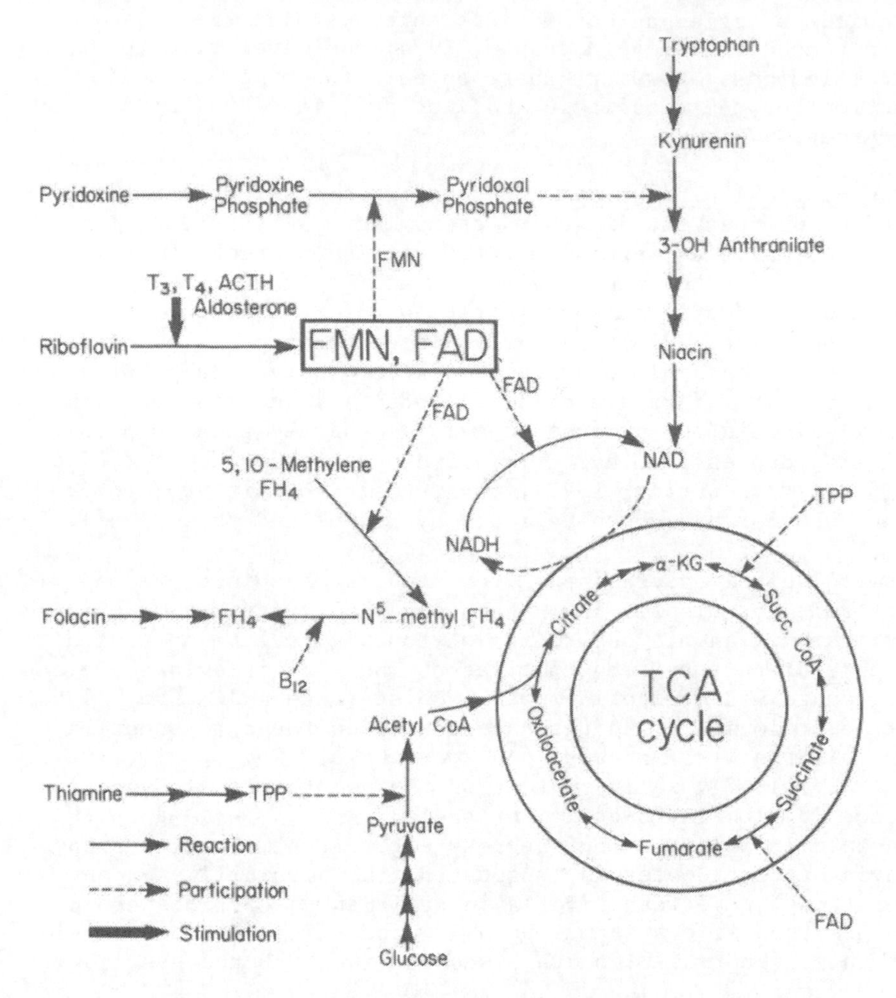


Fig. 5. Interrelationships of riboflavin with other nutrients and hormonal control.

Hormonal Control. The role of hormones in controlling riboflavin metabolism has just recently been reviewed (Rivlin, 1979). Several lines of evidence suggest that thyroid hormones augment flavokinase and FAD pyrophosphorylase activities (Rivlin et al., 1968; Rivlin, 1970a,b, 1975b), thereby enhancing the conversion of riboflavin to FMN and FAD. Changes in the activities of the above two enzymes and flavin coenzyme levels during hypothyroidism further support the concept that thyroid hormones play a role in controlling riboflavin metabolism. This control may extend to that fraction of flavins bound covalently to tissue proteins (Rivlin, 1979). This interaction is cyclic in that riboflavin deficiency impairs the synthesis of thyroid hormones.

ACTH and aldosterone have also been shown to enhance the formation of FMN and FAD in adrenal cortex, liver and kidney (Fazekas and Sandor, 1971, 1973; Tan and Trachewsky, 1975; Trachewsky, 1978). As is the case with thyroid hormones, ariboflavinosis diminishes steroid biosynthesis by the adrenal glands. Additional research is needed to clarify the consequences of hormonal abnormalities concerning riboflavin metabolism.

Vitamin Interrelationships. The clinical signs and symptoms of ariboflavinosis have often been associated with pyridoxine and niacin deficiencies (Spies et al., 1939; Mueller and Vilter, 1950; Vilter et al., 1953; Krishnaswamy, 1971; Iyengar, 1973; Lakshmi and Bamji, 1974). Patients diagnosed with clinical and biochemical riboflavin deficiency have been noted to respond positively to pyridoxine supplementation (Mueller and Vilter, 1950; Iyengar, 1973) even though such therapy does not improve the EGR AC (Lakshmi and Bamji, 1974). The interrelationships of riboflavin, pyridoxine and niacin are summarized in Figure 5.

Riboflavin is necessary for the aerobic conversion of pyridoxine to pyridoxal phosphate (PLP), the active coenzyme form, as shown in the following reaction (Wada and Snell, 1961; Sharada and Bamji, 1972; Anderson et al., 1976).

$$\left.\begin{array}{l}\text{Pyridoxamine } 5'PO_4\\ \text{Pyridoxine } 5'\ PO_4\end{array}\right] - \xrightarrow[\text{PMP Oxidase}]{\text{FMN}} \text{Pyridoxal } 5'PO_4$$

PMP oxidase has been postulated to be a key control point in pyridoxine metabolism (Merrill et al., 1978). Riboflavin deficiency causes a reduction in PMP oxidase activity which in turn results in diminished conversion of pyridoxine to PLP (Lakshmi and Bamji, 1974, 1976). In very pronounced riboflavin deficiency there is an accumulation of pyridoxine in the liver but liver and blood levels of PLP are unaffected. The latter may be due to decreased PLP phosphatase activity or there may be a FMN-dependent pathway for the formation of PLP (Lakshmi and Bamji, 1974, 1976).

By virtue of its role in the production of PLP, riboflavin is also important in tryptophan metabolism and the formation of NAD. Riboflavin deficiency causes an increased urinary excretion of tryptophan metabolites (Mason, 1953; Verjee, 1975; Rose and McGinty, 1968). Depleted body stores of niacin found in pellagrins show improvement with riboflavin therapy (Spies et al., 1939).

All of these findings are consistent with the suggestion that a cellular deficiency of PLP is the basis of the clinical lesions (cheilosis and angular stomatitis) seen in ariboflavinosis (Sharada and Bamji, 1972).

Numerous investigations point towards a close metabolic link between thiamin and riboflavin. In thiamin deficiency there is an increased urinary excretion of riboflavin (Tucker et al., 1960) and an increased accumulation of riboflavin in the liver (Singher et al., 1944; Bamji and Sharada, 1972; Sharada and Bamji, 1972). In similar fashion riboflavin deficiency results in enhanced liver thiamin levels (Singher et al., 1944). Thiamin supplementation of rats fed diets deficient in both thiamin and riboflavin increases EGR activity and decreases the EGR AC value (Vo-Khactu et al., 1976). Thus, it appears that a deficiency of either vitamin will alter the utilization of the other vitamin (see Figure 5).

Interactions between riboflavin and folacin also exist. Riboflavin deficiency causes derangements in folic acid metabolism in tissue, blood and urine (Miller et al., 1962; Foy et al., 1966; Honda, 1968; Narisawa et al., 1968; Bovina et al., 1969). A number of enzymes involving folic acid are also diminished in activity (Arakawa et al., 1968b; Narisawa et al., 1968). These derangements in folacin metabolism are probably attributable to reduced activity of flavoprotein enzymes required for normal utilization of folic acid (Honda, 1968; Narisawa et al., 1968). Folacin deficiency, on the other hand, has no effect on blood levels of riboflavin or on EGR AC values (Bamji and Sharada, 1972).

Serum levels of vitamin $B_{12}$ remain unchanged or are elevated during ariboflavinosis (Lane et al., 1964; Foy et al., 1966; Foy and Kondi, 1968). The increased blood levels of vitamin $B_{12}$ return to normal following riboflavin administration and likely reflect decreased utilization of vitamin $B_{12}$.

General Metabolic Derangements. Flavins are involved in a variety of oxidation-reduction processes in living cells and are particularly important in the utilization of carbohydrates, lipids and proteins (Foy and Mbaya, 1977). Many of the effects of riboflavin deficiency can be explained by abnormalities in hormonal control of riboflavin metabolism and interaction with other vitamins as just discussed. Forker and Morgan (1954) noted impaired gluconeogenesis in riboflavin deficiency which was normalized by riboflavin supplementation. Such changes in gluconeogenesis accompany a number of enzymatic and metabolic disturbances and, although important, cannot be regarded as specific and are probably only one of several different responses the body makes to a variety of stresses (Garthoff et al., 1973). There is at the same time an increased accumulation of liver glycogen (Chatterjee and Ghosh, 1967) during ariboflavinosis, some of which can be attributed to enhanced alanine transaminase activity and conversion of alanine to glycogen (Chatterjee et al., 1966). Increased glycogen stores may also, in part, be the result of diminished glycolysis due to reduced TCA cycle metabolism and decreased utilization of thiamin.

Lipid metabolism is particularly sensitive to riboflavin status. A deficiency of the vitamin prevents the beta-oxidation of fats and inhibits their utilization. This results in a greatly increased liver content of triglycerides (Burch et al., 1956; Sugioka et al., 1969). A reduction in the dehydrogenation of fatty acids occurs after prolonged riboflavin deficiency and causes diminished levels of linoleic, linolenic and arachidonic acids in serum and liver (Mookerjea and Hawkins, 1960; Koyanagi and Oikawa, 1965). Pronounced decreases in fatty acid oxidation occur prior to the morphological changes in hepatic mitochondrial cristae induced by riboflavin deficiency (Tandler et al., 1968; Hoppel and Tandler, 1976).

Protein metabolism is also affected by riboflavin deficiency. When dietary levels of protein are inadequate, the body cannot retain riboflavin and it is excreted in the urine (Sarette and Perlzweig, 1943; Bro-Rasmussen, 1958). The ensuing low body levels of riboflavin in turn result in deranged amino acid metabolism (see Vitamin Interrelationships Section). Hence, there is a vicious cycle between protein and riboflavin metabolism. Protein deficiency results in increased excretion of riboflavin and, in turn, riboflavin deficiency interferes with protein metabolism (Foy and Mbaya, 1977). In keeping with these findings, prolonged ariboflavinosis causes decreases in the levels of DNA and RNA in liver (Chatterjee et al., 1969).

While the level of protein in tissues is unaltered by riboflavin deficiency (Mookerjea and Hawkins, 1960), the concentration of the flavin coenzymes and the activities of enzymes that require FMN and FAD are depressed in most tissues and in the blood (Burch et al., 1956; Goldsmith, 1975). Some flavin-requiring enzymes are more sensitive to depletion than others and appear to parallel the reduction in levels of FMN and FAD. In erythrocytes riboflavin deficiency reduces both EGR and methemoglobin reductase (Beutler, 1969). In contrast, liver GR activity is only marginally affected by ariboflavinosis (Bamji and Sharada, 1972). Many non-flavin enzymes are unchanged or actually increased in activity as a consequence of riboflavin deficiency (Mookerjea and Jamdar, 1962; Chatterjee and Ghosh, 1967).

Riboflavin levels in liver diminish at a faster rate than that in red cells during progressive ariboflavinosis (Bamji and Sharada, 1972) while kidney flavin levels remain high (Glatzle et al., 1973). Concentrations of FMN are decreased proportionately more than concentrations of FAD. This difference in flavin levels is likely due to the decreased flavokinase activity and increased FAD pyrophosphorylase activity that occurs in riboflavin deficiency (Rivlin et al., 1968; Rivlin, 1970b). Since FAD is required by more enzymes than is FMN, Rivlin (1970b) has proposed that the body conserves FAD at the expense of FMN when riboflavin is limiting.

Hematologic Dyscrasia. Anemia is another consequence of riboflavin deficiency. Most data have shown that ariboflavinosis is associated with normocytic, normochromic anemia with decreased erythroid formation in the bone marrow (Foy et al., 1961; Lane et al., 1964). Since there is no alteration in red blood cell size or white blood cell morphology, it is unlikely that deranged folic acid or vitamin $B_{12}$ metabolism contribute to this hematologic defect (Lane et al., 1964, 1965).

Increased red cell fragility has been noted *in vitro*, but there is no evidence of a shortened red cell life span *in vivo* (Hassan and Thurnham, 1977). The anemia due to riboflavin deficiency is characterized by a selective hypoplasia of the erythroid series of blood cells (Foy et al., 1961; Lane et al., 1964; Lane and Alfrey, 1965; Foy et al., 1968, 1972; Lane et al., 1975). In this disorder there is a decreased incorporation of iron into erythrocytes which is normalized by riboflavin therapy accompanied by remission of the aplasia.

Interestingly, the same defects have also been found in African children with protein-calorie malnutrition (PCM). These hematologic abnormalities respond either to corticosteroid or riboflavin therapy (Kondi et al., 1967; Foy and Kondi, 1968). Examination of the adrenal glands in both riboflavin-deficient baboons and children with PCM reveal similar pathological findings regarding the presence of focal hemorrhages, fibrosis and cellular atrophy in various zones (Foy and Kondi, 1968). It appears that ineffective erythropoiesis in riboflavin deficiency and PCM are related to defective steroid metabolism. Further study has demonstrated that corticosteroid levels in blood and urine are lower in riboflavin deficiency and do not respond normally to ACTH stimulation. All of these findings point to the possibility that riboflavin deficiency causes erythroid hypoplasia by affecting the production of corticosteroids which in turn, are responsible for the stimulation of erythropoiesis of the bone marrow.

Neurological Abnormalities. There is some evidence that riboflavin status is important both in peripheral and central nervous system functions. Observations concerning burning feet syndrome suggest that riboflavin deficiency may be a factor in causing diminished nerve conduction as evidenced by reduced sensory perception of temperature, touch and vibration (Lai and Ransome, 1970; Lane et al., 1975) as well as decrement of the hand-grip strength (Sterner and Price, 1973). Microscopic and ultrastructural studies in experimental animals have shown a degeneration of the myelin sheath of peripheral nerves (Shaw and Phillips, 1941; Street et al., 1941; Norton et al., 1976) and the posterior columns of the spinal cord (Street et al., 1941); cellular organelles both of the myelinated and non-myelinated nerve fibers remained intact. These abnormalities are possibly due to the deranged β-oxidase of fatty acids that occurs during riboflavin deficiency; if such is the case, these abnormalities are not caused by reduced coenzyme $B_{12}$ status (Frenkel et al., 1979).

Inadequate riboflavin nutriture also appears to alter central nervous system functions by causing defects in the cellular metabolism of the brain (Nutr. Rev., 1975). These defects manifest themselves as behavioral and/or electrophysiological abnormalities (Brozek and Vaes, 1961; Arakawa et al., 1968a; Sterner and Price, 1973). If riboflavin deficiency occurs during the critical period of cellular proliferation in young growing animals, there is a decrease in the DNA content of brain (Fordyce and Driskell, 1975) which is not corrected by repletion during or after weaning. Other studies indicate that riboflavin-deficient rats score lower in tests of learning ability and that this impairmant is even more pronounced if the ariboflavinosis transpires *in utero* (Muenzinger et al., 1937).

*Congential Malformation*. A popular dictum is that malnutrition in general and ariboflavinosis in particular can complicate pregnancy and greatly compromise fetal outcome (Balfour, 1944; Chaudhuri, 1971; Heller et al., 1974; Smithells et al., 1976). Extensive studies (Warkany and Nelson, 1942; Warkany and Schraffenberger, 1943, 1944; Warkany, 1975) have shown that riboflavin deficiency in rats provokes widespread metabolic disturbances and teratogenesis. Severe riboflavin deficiency may prevent conception while moderate deficiency can cause prematurity, congenital abnormalities, fetal resorption and stillbirths. A slight depletion of riboflavin will not necessarily have any deleterious effect on the embryo. It is not known if these effects are caused by the reduced electron transport system activity that occurs in riboflavin-deficient rats during the critical period of organogenesis (Aksu et al., 1968).

A similar relationship between maternal riboflavin deficiency and fetal malformation has not so far been recognized in man (Brzezinski et al., 1952; McGanity et al., 1969; Heller et al., 1974; Rosenthal, 1978). Smithells et al. (1976) have reported the occurrence of neural tube defects (anenecephaly, meningocele, hydrocephalus or microcephalus) in six infants who were born to mothers that had had low blood riboflavin, folacin and vitamin C levels during the first trimester of pregnancy. The role of pure riboflavin deficiency in causing the neural tube defects remains to be clarified.

Carcinogenesis and Mutagenesis. Riboflavin deficiency has been suggested to be both a preventive and causative factor in carcinogenesis (Rivlin, 1975; Foy and Mbaya, 1977). Urinary excretion of riboflavin tends to be lower in cancer patients (Bamji and Ahmed, 1978) and may reflect poor dietary intake and/ or increased utilization by the tumor cells. Depriving rapidly growing tumors of flavin coenzymes vital for metabolism by intentionally producing riboflavin deficiency may prove useful in the treatment of certain malignancies such as polycythemia vera and lymphoma (Lane, 1971). In contrast, riboflavin deficiency in baboons can produce cellular metaplasia which is strikingly similar to the lesions of primary esophageal cancer in man (Foy and Mbaya, 1977).

Riboflavin may also play a role in mutagenesis. Recall that phototherapy of hyperbilirubinemia in infants can cause riboflavin deficiency (Causes of Deficiency Section) except in those neonates with G-6-P-D deficiency (Lopez et al., 1978). This finding, coupled with the observation that riboflavin potentiates the effect of blue light phototherapy by enhancing the rate of bilirubin decomposition (Kostenbauder and Sanvordeker, 1973; Sanvordeker and Kostenbauder, 1974; Pascale et al., 1976), provides rather substantial rationale for riboflavin supplementation. However, the addition of a physiological dose or riboflavin to either calf thymus DNA or cultured He La cells which are subsequently irradiated by blue light of the same wavelengths used in phototherapy results in alterations of the physiological properties of the DNA (Speck et al., 1975).

Immunology. Animal studies have shown that riboflavin deficiency is accompanied by a reduction in the levels of IgM and IgG and the number of antibody-forming cells of the spleen (Pruzansky and Axelrod, 1955; Kumar and Axelrod, 1978). The clinical significance of these observations during periods of infectious illnesses and convalescence with concomitant ariboflavinosis is not clear (Beisel et al., 1972; Feigin, 1977). It seems unlikely that a relatively brief febrile illness in a well-nourished adult will produce a long-term derangement in vitamin metabolism (Feigin, 1977) and require vitamin therapy. Chronic illness, however, is likely to deplete body stores of riboflavin. More knowledge about the role of riboflavin deficiency in altering cell mediated immunity is required to understand the relationship of ariboflavinosis to infection and even tumor immunology.

Drug Interactions. Certain drugs are known to interfere with riboflavin metabolism. Examples include phenothiazine (Pinto et al., 1979), oral contraceptive agents (Briggs and Briggs, 1974; Sanpitak and Chayutimonkul, 1974), diphenylhydantoin (Saad and

Shehata, 1978), and boric acid (Roe, 1971; Roe et al., 1972). Phenothiazine, a major tranquilizer, inhibits the conversion of riboflavin to FAD by virtue of its structural similarity to riboflavin. This drug causes increased urinary riboflavin excretion and alters overall riboflavin status resulting in an increased EGR AC. Since phenothiazine is widely applied in the treatment of psychotic patients, it is important that more information about its interaction with riboflavin be obtained.

The effect of oral contraceptive agents (OCA) on riboflavin status is controversial (Briggs and Briggs, 1974; Sanpitak and Chayutimonkul, 1974; Ahmed and Bamji, 1976b; Guggenheim and Segal, 1977; Briggs, 1979; Carrigan et al., 1979). This discrepancy may relate to the women's nutritional status prior to OCA usage. Decreased blood levels in women taking OCA is apparently not the result of diminished absorption (Carrigan et al., 1979) but may be due to a preferential redistribution of riboflavin to the liver where it is incorporated into flavoproteins (Ahmed and Bamji, 1976a).

Diphenylhydantoin can cause reduced hepatic riboflavin stores (Saad and Shehata, 1978) and ingestion of boric acid increases urinary riboflavin excretion. Finally, alcohol ingestion can cause riboflavin depletion. In none of these examples is the mechanism known with certainty. It is likely that alcohol interferes with riboflavin metabolism in several ways (see Causes of Deficiency Section).

Dietary riboflavin deficiency is known to diminish liver drug metabolizing capability (Zannoni and Sato, 1976). Thus persons subsisting on suboptimal amounts of dietary riboflavin, and at the same time are using one of the above drugs, can be expected to be at increased risk of developing ariboflavinosis.

## TREATMENT

When the presence of riboflavin deficiency has been diagnosed, treatment by oral administration of 5 to 10 times the daily RDA for riboflavin usually gives satisfactory results. Larger doses can be administered with impunity but may not greatly reduce recovery time. In those instances where absorption is impaired, riboflavin may be given intramuscularly or diluted in intravenous fluid.

The common practice among clinicians to recommend taking 5 mg of the vitamin two or three times a day is really not warranted. Regarding therapeutic regimens, multiple doses of riboflavin are no more effective than a single dose (Garber et al., 1949; Chapman and Campbell, 1955; Morrison and Campbell, 1960a). Divided doses produce sustained urinary riboflavin excretion but have no effect on the total amount excreted, i.e., dose size has no effect on the percent of riboflavin absorbed by the body (Melnick et al., 1945; Morrison et al., 1959).

There is considerable evidence that the pharmaceutical form of riboflavin has a marked influence on vitamin availability. Sugar-coated tablets which do not disintegrate in the body within 60 minutes are not completely available for absorption (Chapman et al., 1954; Morrison et al., 1959). Enteric-coated tablets behave similarly (Morrison and Campbell, 1960b). Even the light coatings used on chewable preparations can reduce riboflavin availability by 50% (Morrison and Campbell, 1962). Despite advertising claims, time-release vitamin formulations do not maintain blood levels of riboflavin because most of the vitamin has moved past its site of intestinal absorption within ninety minutes from the time of ingestion (Morrison et al., 1960).

Without treatment the clinical lesions of ariboflavinosis heal poorly (Horwitt et al., 1949). Improvement of the clinical lesions is generally seen within a few days of therapy followed by complete healing within 10 days to a few weeks. Biochemical abnormalities have been found to improve within 72 hours to a few days (Greene, 1972; Rosenthal et al., 1973). Actual recovery time is dependent upon the extent of depletion of total body stores and the absence or presence of precipitating causes. Maintenance dosage equal to the RDA requirement should be given to assure optimal intake for a while after the lesions have healed. As mentioned previously, since pure riboflavin deficiency is rarely observed in clinical practice, other nutrients should often be prescribed concomitantly.

The increased riboflavin requirements of the pregnant woman can generally be met simply by the consumption of additional quantities of food to satisfy her extra caloric needs (Committee on Maternal Nutrition, 1970). Opinions regarding the value of any further and specific riboflavin supplementation during pregnancy are divided. On the one hand, a significant percentage of pregnant women have subclinical riboflavin deficiency (Heller et al., 1974) and supplementation of deficient diets has improved the clinical outcome of pregnancy (Brzezinski et al., 1952). On the other hand, Kaminetzky et al. (1973) did not find that vitamin supplementation influenced the incidence of riboflavin deficiency and found that mothers with clinical symptoms of ariboflavinosis can have normal pregnancies and deliveries with no deleterious effects on their infants (Brzezinski et al., 1947). Another reason for not supplementing is that the crucial period of organogenesis is usually over by the time that most women seek medical attention; any benefit of remedial therapy is unlikely (Clarke, 1976). Riboflavin therapy may be advisable when anemia in pregnancy is being treated (Clarke, 1973b, Ramachandran and Iyer, 1974).

The usefulness of providing riboflavin supplements to women using OCA is equivocal. There is certainly no compelling evidence to indicate that well-nourished women taking OCA develop riboflavin deficiency. In malnourished women the administration of multivitamins containing 2 mg of riboflavin daily or twice this dose only during the 7 'non pill' days in the 28 day cycle, prevented the OCA effect on riboflavin status but did not correct the initial state of riboflavin deficiency that existed prior to the use of OCA (Bamji et al., 1979). Higher doses of riboflavin may be required to achieve the latter.

The wisdom of riboflavin supplementation in cases of neonatal hyperbilirubinemia being treated with phototherapy is questionable at this time (see Pathophysiology Section). A final decision must consider the potential risk of mutagenesis.

Treatment of disease is a clinican's primary goal but nutritional support must always be an integral component in the practice of good patient care. Perhaps the most significant attribute of good nutrition is the increased resistance to disease and the promotion of optimal growth, repair and maintenance of good health. Equally important are efforts to minimize and hopefully eliminate ariboflavinosis. Prevention is always a better alternative than treatment. Knowledge of food sources, food preparation, absorption, transport, storage and function of riboflavin is indispensable as a preventive measure in protecting against severe ariboflavinosis or reappearance of chronic riboflavin depletion.

REFERENCES

Adiga, P. R., and Muniyappa, K., 1978, Estrogen induction and functional importance of carrier proteins for riboflavin and thiamin in the rat during gestation, J. Steroid Biochem., 9:829.

Ahmed, F., and Bamji, M. S., 1976a, Biochemical basis for the "riboflavin defect" associated with the use of oral contraceptives. A study in female rats, Contraception, 14:297.

Ahmed, F., and Bamji, M. S., 1976b, Vitamin supplements to women using oral contraceptives (studies of Vitamins $B_1$, $B_2$, $B_6$ and A), Contraception, 14:309.

Aksu, O., Mackler, B., Shepard, T. H., and Lemire, R. J., 1968, Studies of the development of congenital anomalies in embryos of riboflavin deficient, galactoflavin-fed rats, II. Role of the terminal electron transport systems, Teratology, 1:93.

Albright, B. E., and Degner, E. F., 1967, Advances in automated vitamin analyses, Technicon Symposium on "Automation in Analytical Chemistry", October, pp. 461-465, New York.

Anderson, B. B., Saary, M., Stephens, A. D., Perry, G. M., Lersundi, I. C., and Horn, J. E., 1976, Effect of riboflavin on red-cell metabolism of vitamin $B_6$, Nature, 264:574.

Arakawa, T., Mizuno, T., Chiba, F., Sakai, K., Watanabe, S., Tamura, T., Tatsumi, S., and Coursin, D. B., 1968a, Frequency analysis of electroencephalograms and latency of photically induced average evoke response in children with ariboflavinosis, Tohoku J. Exp. Med., 94:327.

Arakawa, T., Tamura, T., Tada, K., and Hirono, H., 1968b, Methionine and glycine levels in the liver of riboflavin deficient rats, Tohoku J. Exp. Med., 95:203.

Axelrod, A. E., Spies, T. D., and Elvehjem, C. A., 1941, Riboflavin content of blood and muscle in normal and in malnourished humans, Proc. Soc. Exptl. Biol. Med., 46:146.

Baker, H., and Frank, O., 1975, Analysis of riboflavin and its derivatives in biological fluids and tissues, in "Riboflavin", R. S. Rivlin, ed., pp. 49-79, Plenum Press, New York.

Baker, H., Frank, O., Thompson, A. D., Langer, A., Munves, E. D., DeAngelis, B., and Kaminetzky, H. A., 1975, Vitamin profile of 174 mothers and newborns at parturition, Am. J. Clin. Nutr., 28:59.

Baker, H., Frank, O., Ziffer, H., Goldfarb, S., Leevy, C. M., and Sobotka, H., 1964, Effect of hepatic disease on liver B-complex vitamin titers, Am. J. Clin. Nutr., 14:1.

Baker, H., Frank, P., Feingold, S., Gellene, R. A., Leevy, C. M., and Hunter, S. H., 1966, A riboflavin assay suitable for clinical use and nutritional surveys, Am. J. Clin. Nutr., 19:17.

Baker, H., Thind, I. S., Frank, O., DeAngelis, B., Caterni, H., and Louria, D. B., 1977, Vitamin levels in low-birth-weight newborn infants and their mothers, Am. J. Obstet. Gynecol., 129:521.

Balfour, M. I., 1944, Supplementary feeding in pregnancy, Lancet, 1:208.

Bamji, M. S., 1969, Glutathione reductase activity in red blood cells and riboflavin nutritional status in humans, Clin. Chim. Acta, 26:263.

Bamji, M. S., 1976, Enzymic evaluation of thiamin, riboflavin and pyridoxine status of parturient women and their newborn infants, Br. J. Nutr., 35:259.

Bamji, M. S., and Ahmed, F., 1978, Vitamin deficiency in man - recent studies on fat-soluble vitamins, thiamin, riboflavin, pyridoxine and vitamin C, J. Sci. Ind. Res., 37:42.

Bamji, M. S., Prema, K., Lakshmi, B. A. R., Ahmed, F., and Jacob, C. M., 1979, Oral contraceptive used and vitamin nutrition status of malnourished women - effects of continuous and intermittent vitamin supplements, J. Steroid Biochem., 11:487.

Bamji, M. S., and Sharada, D., 1972, Hepatic glutathione reductase and riboflavin concentrations in experimental deficiency of thiamin and riboflavin in rats, J. Nutr., 102:443.

Brozek, J., and Vaes, G., 1961, Experimental investigations on the effects of dietary deficiencies on animal and human behavior, Vit. Horm., 19:43.

Brubacher, G., 1974, Biochemical studies for assessment of vitamin status in man, Nutr. Diet (Karger, Basel), 20:31.

Brzezinski, A., Bromberg, Y. M., and Braun, K., 1947, Riboflavin deficiency in pregnancy; Its relationship to the course of pregnancy and to the condition of the foetus, J. Obstet. Gyn. Br. Empire, 54:182.

Brzezinski, A., Bromberg, Y. M., and Braun, K., 1952, Riboflavin excretion during pregnancy and early lactation, J. Lab. Clin. Med., 39:84.

Burch, H. B., Bessey, O. A., and Lowry, O. H., 1948, Fluorometric measurements of riboflavin and its natural derivatives in small quantities of blood serum and cells, J. Biol. Chem., 175:457.

Burch, H. B., Lowry, O. H., Padilla, A. M., and Combs, A. M., 1956, Effects of riboflavin deficiency and realimentation of flavin enzymes of tissues, J. Biol. Chem., 223:29.

Butterworth, Charles E., Jr., 1974, The skeleton in the hospital closet, Nutrition Today, 9:4.

Butterworth, Charles E., and Blackburn, George L., 1975, Hospital malnutrition, Nutrition Today, 10:8.

Buzard, J. A., and Kopko, F., 1963, The flavin requirement and some inhibition characteristics of rat tissue glutathione reductase, J. Biol. Chem., 238:464.

Buzina, R., Brodarec, A., Jusic, M., Milanovic, N., Kolombo, V., and Brubacher, G., 1973, Epidemiology of angular stomatitis and bleeding gums, Internatl. J. Vit. Nutr. Res., 43:401.

Buzina, R., Jusic, A., Broderec, A., Milanovic, N., Brubacher, G., Vuilleumier, J. P., Wiss, O., and Christeller, S., 1971, The assessment of dietary vitamin intake of 24 Istrian farmers: II. Comparison between the dietary intake and biochemical status of ascorbic acid, vitamin A, thiamine, riboflavin and niacin, Internatl. J. Vit. Nutr. Res., 41:289.

Campbell, J. A., and Morrison, A. B., 1963, Some factors affecting the absorption of vitamins, Am. J. Clin. Nutr., 12:162.

Carrigan, P. J., Machinist, J., and Kershner, R. P., 1979, Riboflavin nutritional status and absorption in oral contraceptive users and nonusers, Am. J. Clin. Nutr., 32:2047.

Cerletti, P., and Ipata, P., 1960, Determination of riboflavin and its coenzymes in tissues, Biochem. J., 75:119.

Chapman, D. G., and Campbell, J. A., 1955, Reliability of the riboflavin excretion technique in determining availability of coated tablets, Canad. J. Biochem. Physiol., 33:753.

Chapman, D. G., Crisafio, R., and Campbell, J. A., 1954, The relation between in vitro disintegration time of sugar-coated tablets and physiological availability of riboflavin, J. Am. Pharm. Assoc., 43:297.

Chappelle, E. W., and Picciolo, G. L., 1971, Assay of flavin mononucleotide (FMN) and the flavin adenine dinucleotide (FAD) using the bacterial luciferase reaction, in "Methods in Enzymology, Volume XVIII, Vitamins and Coenzymes, Part B", D. B. McCormick and L. D. Wright, eds., pp. 381-385, Academic Press, New York.

Chatterjee, A. K., and Ghosh, B. B., 1967, Effect of riboflavin deficiency on in vivo incorporation of $C^{14}$ from labelled alanine into liver glycogen, Experientia, 23:633.

Chatterjee, A. K., Jamdar, S. C., and Ghosh, B. B., 1966, Changes in alanine transaminase activity in the liver of riboflavin-deficient rats, Experientia, 22:794.

Chatterjee, A. K., Roy, A. D., and Ghosh, B. B., 1969, Effect of riboflavin deficiency on nucleic acid metabolism of liver in the rat, Br. J. Nutr., 23:657.

Chaudhuri, S. K., 1971, Role of nutrition in the etiology of toxemia of pregnancy, Am. J. Obstet. Gyn., 110:46.

Christakis, G., 1973, Nutritional assessment in health programs, Am. J. Pub. Health, 63:November supplement.

Clarke, H. C., 1969a, A photodecomposition fluorimetric method for the determination of riboflavin in whole blood, Internatl. J. Vit. Nutr. Res., 39:182.

Clarke, H. C., 1969b, The relationship between whole blood riboflavin levels and results of riboflavin saturation tests in normal and pathological conditions in man, Internatl. J. Vit. Nutr. Res., 39:238.

Clarke, H. C., 1971, The riboflavin deficiency syndrome of pregnancy, Surg. Forum, 22:394.

Clarke, H. C., 1973a, B-vitamins in human amniotic fluid, Internatl. J. Vit. Nutr. Res., 43:324.

Clarke, H. C., 1973b, In pregnancy: Effect of iron and folic acid on riboflavin status, Intl. J. Vit. Nutr. Res., 43:438.

Clarke, H. C., 1976, In Trinidad: Angular stomatitis and pregnancy, Internatl. J. Vit. Nutr. Res., 46:366.

Clarke, H. C., 1977a, A photodecomposition fluorometric method for determination of riboflavin in the various constituents of blood, Internatl. J. Vit. Nutr. Res., 47:356.

Clarke, H. C., 1977b, Distribution of riboflavin in blood: in women and in prenates, Internatl. J. Vit. Nutr. Res., 47:361.

Clegg, K. M., Kodicek, E., and Mistry, S. P., 1952, A modified medium for Lactobacillus casei for the assay of B vitamins, Biochem. J., 50:326.

Cole, H. S., Lopez, R., and Cooperman, J. M., 1976, Riboflavin deficiency in children with diabetes mellitus, Acta Diabetol. Latin., 13:25.

Committee on Maternal Nutrition, 1970, "Maternal Nutrition and the Course of Pregnancy", Food and Nutrition Board, National Research Council, National Academy of Sciences, Washington, D.C.

Coon, W. W., 1965, Riboflavin metabolism in surgical patients, Surg. Gynecol. Obst., 120:1289.

Cooperman, J. M., Cole, H. S., Gordon, M., and Lopez, R., 1973, Erythrocyte glutathione reductase as a measure of riboflavin nutritional status of pregnant women and newborns, Proc. Soc. Exptl. Biol. Med., 143:326.

Couch, J. R., and Davies, R. E., 1963, Vitamins $B_1$, $B_2$, $B_6$, niacin, and ascorbic acid, in "Newer Methods of Nutritional Biochemistry, Vol. 1", A. A. Albanese, ed., pp. 199-234, Academic Press, New York.

Davis, T. R. A., Gershoff, S. N., and Gamble, D. F., 1969, Review of studies of vitamin and mineral nutrition in the United States (1950-1968), J. Nutr. Ed. (Suppl. I), 1:41.

Day, P. L., Langston, W. C., and O'Brien, C. S., 1931, Cataract and other ocular changes in vitamin G deficiency, Am. J. Ophth., 14:1005.

Department of Health and Social Security, 1972, Report by the Panel on Nutrition of the Elderly, Her Majesty's Stationery Office, London No. 3

Dibble, M. V., Brin, M., McMullen, E., Peel, A., and Chen, N., 1965, Some preliminary biochemical findings in junior high school children in Syracuse and Onondaga County, New York, Am. J. Clin. Nutr., 17:218.

Dubrauszky, V., and Blazso, S., 1943, Vitamin $B_2$ - stoffwechsel wahrend der schwangerschaft., Z. Vitaminforsch., 14:2.

Dymock, S. M., and Brocklehurst, J. C., 1973, Clinical effects of water soluble vitamin supplementation in geriatric patients, Age and Ageing, 2:172.

Eckardt, R. E., and Johnson, L. V., 1939, Nutritional cataract and relation of galactose to appearance of senile suture line in rats, Arch. Ophth., 19:315.

Exton-Smith, A. N., 1975, Problems of diet in old age, J. Roy. Coll. Phys. (Lond.), 9:148.

Fazekas, A. G., and Kokai, K., 1971, Extraction, Purification, and Separation of Tissue Flavins for Spectrophotometric Determination, in "Methods in Enzymology, Vol. XVIII, Vitamins and Coenzymes, Part B", D. B. McCormick and L. D. Wright, eds., pp. 385-398, Academic Press, New York.

Fazekas, A. G., and Sandor, T., 1971, The in vivo effect of adrenocorticotropin on the biosynthesis of flavin nucleotides in rat liver and kidney, Canad. J. Biochem., 49:987.

Fazekas, A. G., and Sandor, T., 1973, Studies on the biosynthesis of flavin nucleotides from 2-$^{14}$C-riboflavin by rat liver and kidney, Canad. J. Biochem., 51:772.

Feigin, R. D., 1977, Interaction of nutrition and infection: plans for future research, Am. J. Clin. Nutr., 30:1553.

Fernandez, N. A., Burgos, J. C., Plough, I. C., Roberts, L. J., and Asenjo, C. F., 1965, Nutritional status of people in isolated areas of Puerto Rico. Survey of Barrio Mavilla, Vega Alta, Puerto Rico, Am. J. Clin. Nutr., 17:305.

Bamji, M. S., Sharada, D., and Naidu, A. N., 1973, A comparison of the fluorometric and microbiological assays for estimating riboflavin content of blood and liver, Internatl. J. Vit. Nutr. Res., 43:351.

Bayoumi, R. A., and Rosalki, S. B., 1976, Evaluation of the methods of coenzyme activation of erythrocyte enzymes for detection of deficiency of vitamins $B_1$, $B_2$, and $B_6$, Clin. Chem. 22:327.

Beisel, W. R., Herman, Y. F., Sauberlich, H. E., Herman, R. H., Bortelloni, P. J., and Canham, J. E., 1972, Experimentally induced sandfly fever and vitamin metabolism in man, Am. J. Clin. Nutr., 25:1165.

Berry, W. T., and Darke, S. J., 1972, Letter: Nutrition of the elderly, Age and Aging, 1:177.

Bessey, O. A., Horwitt, M. K., and Love, R. H., 1956, Dietary deprivation of riboflavin and blood riboflavin levels in man, J. Nutr., 58:367.

Bessey, O. A., Lowry, O. H., and Love, R. H., 1949, The fluorometric measurement of the nucleotides of riboflavin and their concentration in tissues. J. Biol. Chem., 180:775.

Bessey, O. A., and Wolbach, S. B., 1939, Vascularization of the cornea of the rat in riboflavin deficiency with a note on corneal vascularization in vitamin A deficiency, J. Exper. Med., 69:1.

Beutler, E., 1969, Effect of flavin compounds on glutathione reductase activity: in vivo and in vitro studies, J. Clin. Invest., 48:1957.

Bovina, C., Landi, L., Pasquali, P., and Morchette, M., 1969, Biosynthesis of folate coenzymes in riboflavin-deficienct rats, J. Nutr., 99:320.

Brewster, M. A., Berry, D. H., and Murphey, M. N., 1974, Automated reaction rate analysis of erythrocyte glucose-6-phosphate dehydrogenase and glutathione reductase activities, Biochem. Med., 10:229.

Briggs, M., and Briggs, M., 1974, Oral contraceptives and vitamin nutrition, Lancet, 1:1234.

Briggs, M., 1979, Biochemical basis for the selection of oral contraceptives, Int. J. Gyn. Obstet., 16:509.

Brin, M., Dibble, M. V., Peel, A., McMullen, E., Bourquin, A., and Chen, N., 1965, Some preliminary findings on the nutritional status of the aged in Onondaga County, New York, Am. J. Clin. Nutr., 17:240.

Brocklehurst, J. C., Griffiths, L. L., Tayler, G. F., Marks, J., Scott, D. L., and Blackley, J., 1968, The clinical features of chronic vitamin deficiency, a therapeutic trial in geriatric hospital patients, Geront. Clin., 10:309.

Bro-Rasmussen, 1958, The riboflavin requirement of animals and man and associated metabolic relations. Part I: Technique of estimating requirement, and modifying circumstances. Part II: Relations of requirement to the metabolism of protein and energy, Nutr. Abst. Rev., 28:1,369.

Flatz, G., 1970, Enhanced binding of FAD to glutathione reductase in G6PD deficiency, Nature, 226:755.

Flatz, G., 1971a, Studies of enzymes connected with erythrocyte glutathione metabolism in a rural tropical population, Humangenetik, 11:221.

Flatz, G., 1971b, Population study of erythrocyte glutathione reductase activity. I. Stimulation of the enzyme by flavin adenine dinucleotide and by riboflavin supplementation, Humangenetik, 11:269.

Flatz, G., 1971c, Population study of erythrocyte glutathione reductase activity. II. Hematological data of subjects with low enzyme activity and stimulation characteristics in their families, Humangenetik, 11:278.

Fordyce, M. K., and Driskell, J. A., 1975, Effects of riboflavin repletion during different developmental phases on behavioral patterns, brain nucleic acid and protein contents and erythrocyte glutathione reductase activity of male rats, J. Nutr., 105:1150.

Forker, B. R., and Morgan, A. F., 1954, Effect of adrenocortical hormone on the riboflavin deficient rat, J. Biol. Chem., 209:303.

Foy, H., and Kondi, A., 1968, Comparison between erythroid aplasia in marasmus and kwashiorkor and the experimentally induced erythroid aplasia in baboons by riboflavin deficiency, Vit. Horm., 26:653.

Foy, H., and Mbaya, V., 1977, Riboflavin, Prog. Food Nutr. Sci., 2:357.

Foy, H., Kondi, A., and MacDougall, L., 1961, Pure red cell aplasia in marasmus and kwashiorkor treated with riboflavin, Br. Med. J., 1:937.

Foy, H., Kondi, A., and Mbaya, V., 1966, Serum vitamin $B_{12}$ and folate levels in normal and riboflavin-deficient baboons (Papio anibus), Br. J. Haemat., 12:239.

Foy, H., Kondi, A., and Verjee, Z. H. M., 1972, Relation of riboflavin deficiency to corticosteroid metabolism and red cell hypoplasia in baboons, J. Nutr., 102:571.

Foy, H., Kondi, A., Harriss, E. B., and Preston, J. K., 1968, Isotopic and cytological estimations of marrow erythroid activity in normal and riboflavin deficient baboons, Acta Haemat., 39:118.

Frenkel, E. P., Kitchens, R. L., Savage, H. E., Seibert, R. A., and Lane, M., 1979, Lack of effect of riboflavin deficiency on vitamin $B_{12}$-related metabolic pathways and fatty acid synthesis, Am. J. Clin. Nutr., 32:10.

Frischer, H., Bowman, J. E., Carson, P. E., Rieckmann, K. H., Willerson, Jr., D., and Colwell, E. J., 1973, Erythrocytic glutathione reductase, glucose-6-phosphate dehydrogenase, and 6-phosphogluconic dehydrogenase deficiencies in populations of the United States, South Vietnam, Iran and Ethiopia, J. Lab. Clin. Med., 81:603.

Fujita, A., and Matsuura, K., 1950, Determination of free and esterified riboflavin by the lumiflavin method, J. Bicohem., 37:445.

Gabay, S., and Harris, S. R., 1966, Studies of flavin adenine dinucleotide requiring enzymes and phenothiazines. II. Structural requirements for D-amino acid oxidase inhibition, Biochem. Pharmacol., 15:317.

Gabay, S., and Harris, S. R., 1967, Studies of flavin adenine dinucleotide requiring enzymes and phenothiazines. III. Inhibition kinetics with highly purified D-amino acid oxidase, Biochem. Pharmacol., 16:803.

Garber, M., Marquette, M. M., and Parsons, H. T., 1949, The availability of vitamins from yeasts. V. Differences in the influence of liver yeast on the absorption of pure thiamine hydrochloride, pure riboflavin and nitrogen by human subjects, and the effect of distribution of the vitamin doses, J. Nutr., 38:225.

Garry, P. J., and Owen, G. M., 1976, An automated FAD - dependent glutathione reductase assay for assessing riboflavin nutriture, Am. J. Clin. Nutr., 29:663.

Garthoff, L. H., Garthoff, S. K., Tobin, R. B., and Mehlman, M. A., 1973, The effect of riboflavin deficiency on key gluconeogenic enzyme activities in rat liver, Proc. Soc. Exptl. Biol. Med., 143:693.

Glatzle, D., Korner, W. F., Christeller, S., and Wiss, O., 1970, Method for the detection of a biochemical riboflavin deficiency stimulation of $NADPH_2$ - dependent glutathione reductase from human erythrocytes by FAD in vitro. Investigations on the vitamin $B_2$ status in healthy people and geriatric patients, Internatl. J. Vit. Nutr. Res., 40:166.

Glatzle, D., Vuilleumier, J. P., Weber, F., and Decker, K., 1974, Glutathione reductase test with whole blood, a convenient procedure for the assessment of the riboflavin status in humans, Experientia, 30:665.

Glatzle, D., Weber, F., and Wiss, O., 1968, Enzymatic test for the detection of a riboflavin deficiency. NADPH-dependent glutathione reductase of red blood cells and its activation by FAD in vitro, Experientia, 24:1122.

Glatzle, D., Weiser, H., Weber, F., and Wiss, O., 1973, Correlations between riboflavin supply, glutathione reductase activities and flavin levels in rats, Internatl. J. Vit. Nutr. Res., 43:187.

Goldsmith, G. A., 1956, Experimental niacin deficiency in man, J. Am. Dietet. Assoc., 32:312.

Goldsmith, G. A., 1959, "Nutritional Diagnosis", S. O. Waife and P. Gyorgy, eds., pp. 3-25, 108-113, Charles C. Thomas, Springfield.

Goldsmith, G. A., 1964, Nutrition, A comprehensive treatise, Vol. 2, pp. 109-198, Academic Press, New York.

Goldsmith, G. A., 1975, Vitamin B complex, thiamine, riboflavin, niacin, folic acid (folacin), vitamin $B_{12}$, biotin, Prog. Food Nutr. Sci., 1:559.

Gorus, F., and Schram, E., 1979, Applications of bio- and chemiluminescence in the clinical laboratory, Clin. Chem. 25:512.

Greene, H. L., 1972, Vitamins, in "Symposium on Total Parenteral Nutrition", pp. 78-91, Nashville, American Medical Association.

Gromisch, D. S., Lopez, R., Cole, H. S., and Cooperman, J. M., 1977, Light (phototherapy)-induced riboflavin deficiency in the neonate, J. Ped., 90:118.

Guggenheim, K., and Segal, S., 1977, Oral contraceptives and riboflavin nutriture, Internatl. J. Vit. Nutr. Res., 47:234.

Gyorgy, P., 1954, Early experiences with riboflavin, a retrospective, Nutr. Rev., 12:97.

Gyorgy, P., 1967, Reminiscences on the discovery and significance of some of the B vitamins, J. Nutr., 91(Suppl.):5.

Hamil, B. M., Coryell, M., Roderuck, C., Kaucher, M., Moyer, E. Z., Harris, M. E., and Williams, H. H., 1947, Thiamine, riboflavin, nicotinic acid, pantothenic acid and biotin in the urine of newborn infants, Am. J. Dis. Child., 74:434.

Hassan, F. M., and Thurnham, D. I., 1977, Effect of riboflavin deficiency on the metabolism of the red blood cell, Internatl. J. Vit. Nutr. Res., 47:349.

Hegsted, D. M., Gershoff, S. N., Trulson, M. F., and Jolly, D. H., 1956, Variation in riboflavin excretion, J. Nutr., 60:581.

Heller, S., Salkeld, R. M., and Korner, W. F., 1974, Riboflavin status in pregnancy, Am. J. Clin. Nutr., 27:1225.

Hines, C., 1978, Vitamins, absorption and malabsorption, Arch. Intern. Med., 138:619.

Hodges, R. E., and Krehl, W. A., 1965, Nutritional status of teenagers in Iowa, Am. J. Clin. Nutr., 17:200.

Honda, Y., 1968, Folate derivatives in the liver of riboflavin-deficient rats, Tohoku J. Exp. Med., 95:79.

Hoorn, R. K. J., Flikweert, J. P., and Westerink, D., 1975, Vitamin B-1, B-2 and B-6 deficiencies in geriatric patients, measured by coenzyme stimulation of enzyme activities, Clin. Chim. Acta, 61:151.

Hoppel, C. L., and Tandler, B., 1976, Relationship between hepatic mitochondrial oxidative metabolism and morphology during riboflavin deficiency and recovery in mice, J. Nutr., 106:73.

Hornbeck, C. L., and Bradley, D. W., 1974, Concentrations of FAD and glutathione as they affect values for erythrocyte glutathione reductase, Clin. Chem., 20:512.

Horvath, C., Szonyi, L., and Mold, K., 1976, Preventive effects of riboflavin and ATP on the teratogenic effects of the phenothiazine derivative T-82, Teratology, 14:167.

Horwitt, M. K., 1966, Nutritional requirements of man, with special reference to riboflavin, Am. J. Clin. Nutr., 18:458.

Horwitt, M. K., 1972, Riboflavin V. Occurrence in food, in "The Vitamins", W. H. Sebrell and R. S. Harris, eds., pp. 46-49, Academic Press, New York.

Horwitt, M. K., Harvey, C. C., Hills, O. W., and Liebert, E., 1950, Correlation of urinary excretion of riboflavin with dietary intake and symptoms of ariboflavinosis, J. Nutr., 41:247.

Horwitt, M. K., Hills, O. W., Harvey, C. C., Liebert, E., and Steinberg, D. L., 1949, Effects of dietary depletion of riboflavin, J. Nutr., 39:357.

Horwitt, M. K., and Witting, L. A., 1972, Riboflavin IX. Biochemical systems, in "The Vitamins", W. H. Sebrell and R. S. Harris, eds., pp. 53-70, Academic Press, New York.

Icen, A., 1967, Glutathione reductase of human erythrocytes. Purification and properties, Scand. J. Clin. Lab. Invest., 20(Suppl. 96):1.

Interdepartmental Committee on Nutrition for National Defense, Nutrition Survey Reports, National Institute of Health, Office of International Research, DHEW, Bethesda, Maryland.

Iyengar, L., 1973, Oral lesions in pregnancy, Lancet, 1:680.

Jelliffe, D. B., 1966, "The Assessment of the Nutritional Status of the Community", World Health Organization, Monograph No. 58, Geneva.

Jolliffe, N., Fein, H. D., and Rosenblum, L. A., 1939, Riboflavin deficiency in man, New Eng. J. Med., 221:921.

Jusko, W. J., and Levy, G., 1967, Absorption, metabolism and excretion of riboflavin-5-phosphate in man, J. Pharm. Sci., 56:58.

Jusko, W. J., and Levy, G., 1969, Plasma protein binding of riboflavin and riboflavin-5-phosphate in man, J. Pharm. Sci., 58:58.

Jusko, W. J., Levy, G., Yaffe, S. J., and Allen, J. E., 1971, Riboflavin absorption in children with biliary obstruction, Am. J. Dis. Child., 121:48.

Kaminetzky, H. A., Langer, A., Baker, H., Frank, O., Thomson, A. D., Munves, E. D., Opper, A., Behrle, F. C., and Glista, B., 1973, The effect of nutrition in teen-age gravidas on pregnancy and the status of the neonate. I. A nutritional profile, Am. J. Obstet. Gynecol., 115:639.

Kaminetzky, H. A., Baker, H., Frank, O., and Langer, A., 1974, The effects of intravenously administered water-soluble vitamins during labor in normovitaminemic and hypovitaminemic gravidas on maternal and neonatal blood vitamin levels at delivery, Am. J. Obstet. Gynecol., 120:697.

Karrer, P., Solomon, H., Schopp, K., and Benz, F., 1935, Synthetische flavine VII, Helv. Chim. Acta, 18:1143.

Kaufmann, N. A., and Guggenheim, K., 1977, The validity of biochemical assessment of thiamine, riboflavin and folacin nutriture, Internatl. J. Vit. Nutr. Res., 47:40.

Kay, R. G., Tasman-Jones, C., Rybus, J., Whiting, R., and Black, H., 1976, A syndrome of acute zinc deficiency during total parenteral alimentation in man, Ann. Surg., 183:331.

Kelsay, J. L., 1969, A compendium of nutritional status studies and dietary evaluation studies conducted in the United States, 1957-1967, J. Nutr., 99(Suppl. 1):123.

Kondi, A., Foy, H., and Mbaya, V., 1967, Erythroid aplasia in riboflavin-deficient baboons and its relation to marasmus and kwashiorkor, Br. J. Haemat., 13:967.

Kornberg, H. A., Langdon, R. S., and Cheldelin, V. H., 1948, Microbiological assay for riboflavin, Anal. Chem., 20:81.

Kostenbauder, H. B., and Sanvordeker, D. R., 1973, Riboflavin enhancement of bilirubin photocatabolism in vivo, Experientia, 29:282.

Koyanagi, T., and Oikawa, K., 1965, Effect of riboflavin on the metabolism of polyunsaturated fatty acids in the body of rat, Tohoku J. Exp. Med., 86:19.

Koziol, J., 1971, Fluorometric analyses of riboflavin and its coenzymes, in "Methods in Enzymology, Volume XVIII, Vitamins and Coenzymes, Part B", D. B. McCormick and L. D. Wright, eds., pp. 253-285, Academic Press, New York.

Krehl, W. A., Winters, R. W., 1950, Effect of cooking methods on retention of vitamins and minerals in vegetables, J. Am. Dietet. Assoc., 26:966.

Krishnaswamy, K., 1971, Erythrocyte glutamic oxaloacetic transaminase activity in patients with oral lesions, Internatl. J. Vit. Nutr. Res., 41:247.

Kuhn, R., Reinemund, K., Weygand, F., and Strobele, R., 1935, Uber die synthese der lactoflavins (Vitamin $B_2$), Ber. Deut. Chem. Ges., 68:1765.

Kumar, M., and Axelrod, A. E., 1978, Cellular antibody synthesis in thiamine, riboflavin, biotin, and folic acid-deficient rats, Proc. Soc. Exptl. Biol. Med., 157:421.

Lai, C. S., and Ransome, G. A., 1970, Burning feet syndrome case due to malabsorption and responding to riboflavin, Br. Med. J., 2:151.

Lakshmi, A. V., and Bamji, M. S., 1974, Tissue pyridoxal phosphate concentration and pyridoxamine phosphate oxidase activity in riboflavin deficiency in rats and man, Br. J. Nutr., 32:249.

Lakshmi, A. V., and Bamji, M. S., 1976, Regulation of blood pyridoxal phosphate in riboflavin deficiency in man, Nutr. Metab., 20:228.

Lane, M., 1971, Induced riboflavin deficiency in treatment of patients with lymphomas and polycythemia vera, Proc. Am. Assoc. Cancer Res., 12:85 (Abst.).

Lane, M., and Alfrey, C. P., 1965, The anemia of human riboflavin deficiency, Blood, 25:432.

Lane, M., Alfrey, C. P., Mengel, C. E., Doherty, M. D., and Doherty, J., 1964, The rapid induction of human riboflavin deficiency with galactoflavin, J. Clin. Invest., 43:357.

Lane, M., Smith, F. E., and Alfrey, C. P., 1975, Experimental dietary and antagonist-induced human riboflavin deficiency, in "Riboflavin", R. S. Rivlin, ed., pp. 245-277, Plenum Press, New York.

Langer, B. W., Jr., and Charoensiri, S., 1966, Growth response of Lactobacillus casei (ATCC 7469) to riboflavin, FMN and FAD, Proc. Soc. Exptl. Biol. Med., 122:151.

Leevy, C. M., Baker, H., TenHove, W., Frank, O., and Cherrick, G. R., 1965a, B-Complex vitamins in liver disease of the alcoholic, Am. J. Clin. Nutr., 16:339.

Leevy, C. M., Cardi, L., Frank, O., Gellene, R., and Baker, H., 1965b, Incidence and significance of hypovitaminemia in a randomly selected municipal hospital population, Am. J. Clin. Nutr., 17:259.

Leevy, C. M., Thompson, A. and Baker, H., 1970, Vitamins and liver injury, Am. J. Clin. Nutr., 23:493.

Leverton, R. M., 1964, Nutritional well being in the U.S.A., Nutr. Rev., 22:321.

Levy, G., and Jusko, W. J., 1966, Factors affecting the absorption of riboflavin in man, J. Pharm. Sci., 55:285.

Lopez, R., Cole, H. S., Montoya, F., and Cooperman, J. M., 1975, Riboflavin deficiency in a pediatric population of low socioeconomic status in New York City, J. Ped., 87:420.

Lopez, R., Gromisch, D. S., Cole, H. S., and Cooperman, J. M., 1978, Effect of erythrocyte glucose-6-phosphate dehydrogenase (G-6-PD) deficiency on light-induced riboflavin deficiency in the neonate, Proc. Soc. Exptl. Biol. Med., 157:41.

Lossy, F. T., Goldsmith, G. A., and Sarett, H. P., 1951, A study of test dose excretion of five B-complex vitamins in man, J. Nutr., 45:213.

Lowry, O. H., 1952, Biochemical evidence of nutritional status, Physiol. Rev., 32:431.

Lust, J. E., Hagerman, D. D., and Villee, C. A., 1954, The transport of riboflavin by human placenta, J. Clin. Invest., 33:38.

McCormick, D. B., 1961, Flavokinase activity of rat tissues and masking effect of phosphatases, Proc. Soc. Exptl. Biol. Med., 107:784.

McCormick, D. B., 1964, Specificity of flavin adenine dinucleotide pyrophosphorylase for flavin phosphates and nucleotide triphosphates, Biochem. Biophys. Res. Commun., 14:493.

McCormick, D. B., 1975, Metabolism of Riboflavin, in "Riboflavin", R. S. Rivlin, ed., pp. 153-198, Plenum Press, New York.

McGanity, W. J., Little, H. M., Fogelman, A., Jennings, L., Calhoun, E., and Dawson, E. B., 1969, Pregnancy in the adolescent. I. Preliminary summary of health status, Am. J. Obstet. Gyn., 103:773.

Mason, M., 1953, The metabolism of tryptophan in riboflavin deficient rats, J. Biol. Chem., 201:513.

May, J. M., 1961-1972, Studies in medical geography. The ecology of malnutrition, Vol. 1-11, Hafner Publishing Company, New York.

Mayerson, M., Feldman, S., and Gibaldi, M., 1969, Bile salt enhancement of riboflavin and flavin mononucleotide absorption in man, J. Nutr., 98:288.

Mayfield, H. L., and Hedrick, M. T., 1949, Thiamine and riboflavin retention in beef during roasting, canning and corning, J. Am. Dietet. Assoc., 25:1024.

Meighan, E. A., and MacKenzie, R. E., 1973, Flavin specificity of enzyme-substrate intermediates in the bacterial luminescence reaction. Structural requirements of flavin side-chain, Biochem., 12:1482.

Mellor, N. P., and Maass, A. R., 1972, An automated fluorometric method for the determination of riboflavin in human urine, "Technicon International Congress on Advances in Automated Analysis", June, p. 64, New York, (Abstract).

Melnick, D., Hochberg, M., and Oser, B. L., 1945, Physiological availability of the vitamins. I. The human bioassay technic, J. Nutr., 30:67.

Merrill, A. H., and McCormick, D. B., (Abst.), Fed. Proc.

Merrill, A. H., Addison, R., and McCormick, D. B., 1978, Induction of hepatic and intestinal flavokinase after oral administration of riboflavin to riboflavin-deficient rats, Proc. Soc. Exptl. Biol. Med., 158:572.

Merrill, A. H., Froehlich, J. A., and McCormick, D. B., 1979, Purification of riboflavin-binding proteins from bovine plasma and discovery of a pregnancy-specific riboflavin-binding protein, J. Biol. Chem., 254:9362.

Merrill, A. H., and McCormick, D. B., 1980, Chemistry and Physiology of Nutrients and Growth Regulators: Riboflavin, CRC Handbook Series on Nutrition and Food (In Press).

Mezey, E., 1978, Liver diseases and nutrition, Gastroenterology, 74:770.

Miller, Z., Poncet, I., and Takacs, E., 1962, Biochemical studies on experimental congenital malformations: Flavin nucleotides and folic acid in fetuses and livers from normal and riboflavin-deficient rats, J. Biol. Chem., 237:968.

Mitchell, G. W., and Hastings, J. W., 1969, The effect of flavin isomers and analogues upon the color of bacterial bioluminescence, J. Biol. Chem., 244:2572.

Mookerjea, S., and Hawkins, W. W., 1960, Some anabolic aspects of protein metabolism in riboflavin deficiency in the rats, Brit. J. Nutr., 14:231.

Mookerjea, S., and Jamdar, S. C., 1962, Liver transaminase activity in riboflavin-deficient rats, Canad. J. Biochem., 40:1065.

Morgan, A. G., Kelleher, J., Walker, B. E., Losowsky, M. S., Droller, H., and Middleton, R. S. W., 1975, A nutritional survey in the elderly, blood and urine vitamin levels, Internatl. J. Vit. Nutr. Res., 45:448.

Morrison, A. B., and Campbell, J. A., 1960a, Vitamin absorption studies. I. Factors influencing the excretion of oral test doses of thiamine and riboflavin by human subjects, J. Nutr.,

72:435.
Morrison, A. B., and Campbell, J. A., 1960b, The relationship between physiological availability of salicylates and riboflavin and in vitro disintegration time of enteric coated tablets, J. Am. Pharm. Assoc., 49:473.
Morrison, A. B., and Campbell, J. A., 1962, Physiological availability of riboflavin and thiamine in 'chewable' vitamin products, Am. J. Clin. Nutr., 10:212.
Morrison, A. B., Chapman, D. G., and Campbell, J. A., 1959, Further studies on the relation between in vitro disintegration time of tablets and the urinary excretion rates of riboflavin, J. Am. Pharm. Assoc., 48:634.
Morrison, A. B., Perusse, C. B., and Campbell, J. A., 1960, Physiologic availability and in vitro release of riboflavin in sustained release vitamin preparations, New Eng. J. Med., 263:115.
Morse, E. H., Merrow, S. B., and Clarke, R. F., 1965, Some biochemical findings in Burlington (Vt.) junior high school children, Am. J. Clin. Nutr., 17:211.
Moynahan, E. J., 1974, Acrodermatitis enteropathica: a lethal inherited human zinc-deficiency disorder, Lancet, 2:399.
Mueller, J. F., and Vilter, R. W., 1950, Pyridoxine deficiency in human beings induced with desoxypyridoxine, J. Clin. Invest., 29:193.
Muenzinger, K. F., Poe, E., and Poe, C. F., 1937, The effect of vitamin deficiency upon acquisition and retention of the maze habit in the white rat. II. Vitamin $B_2$, J. Comp. Psychol., 23:59.
Muller, E. M., and Bates, C. J., 1977, The effect of riboflavin deficiency on white cell glutathione reductase in rats, Internatl. J. Vit. Nutr. Res., 47:46.
Murthy, U. S., and Adiga, P. R., 1977a, Riboflavin-binding protein of hen's egg: Purification and radioimmunoassay, Indian J. Biochem. Biophys., 14:118.
Murthy, U. S., and Adiga, P. R., 1977b, Estrogen-induction of riboflavin-binding protein in immature chicks. Nature of the secretory protein, Biochem. J., 166:331.
Murthy, U. S., and Adiga, P. R., 1977c, Estrogen induction of riboflavin-binding protein in immature chicks. Modulation by thyroid status, Biochem. J., 166:647.
Murthy, U. S., and Adiga, P. R., 1978, Estrogen-induced synthesis of riboflavin binding proteins in immature chicks. Kinetics and hormonal specificity, Biochem. Biophys. Acta, 538:364.
Najjar, V. A., Johns, G. A., Medairy, G. C., Fleischmann, G., and Holt, Jr., L. E., 1944, The biosynthesis of riboflavin in man, J.A.M.A., 126:357.
Narisawa, K., Tamura, T., Tanno, K., Ohara, K., and Arakawa, T., 1968, Tetrahydrofolate-dependent enzyme activities of the rat liver in riboflavin deficiency, Tohoku J. Exp. Med., 94:417.

Newman, R. G., and Martin, S., 1971, The national nutrition survey in New York City, Trans. N.Y. Acad. Sci., 33:316.

Newman, L. J., Lopez, R., Cole, H. S., Boria, M. C., and Cooperman, J. M., 1978, Riboflavin, deficiency in women taking oral contraceptive agents, Am. J. Clin. Nutr., 31:247.

Nichoalds, G. E., Lawrence, J. D., and Sauberlich, H. E., 1974, Assessment of status of riboflavin nutriture by assay of glutathione reductase activity, Clin. Chem., 20:624

Norton, W. N., Daskal, I., Savage, H. E., Seibert, R. A., and Lane, M., 1976, Effects of riboflavin deficiency on the ultrastructure of rat sciatic nerve fibers, Am. J. of Path., 85:651.

Nutr. Rev., 1975, Central nervous system changes in deficiency of vitamin $B_6$ and other B-complex vitamins, 33:22.

Nutr. Rev., 1976, Riboflavin deficiency, galactose metabolism and cataract, 34:77.

Nutr. Rev., 1979, Riboflavin and thiamine binding proteins: their physiological significance and hormonal specificity, 37:261.

Okada, A., Takagi, Y., Itakura, T., Satani, M., Manabe, H., Iida, Y., Tanigaki, T., Iwasaki, M., and Kasahara, N., 1976, Skin lesions during intravenous hyperalimentation, zinc deficiency, Surg., 80:629.

Oldham, H., Johnston, F., Kleiger, D., and Hedderich - Arismendi, H., 1944, A study of the riboflavin and thiamine requirements of children of preschool age, J. Nutr., 27:435.

Oldham, H., Lounds, E., and Porter, T., 1947, Riboflavin excretion and test dose returns of young women during period of positive and negative nitrogen balance, J. Nutr., 34:69.

Oldham, H., Sheft, B. B., and Porter, T., 1950, Thiamine and riboflavin intakes and excretions during pregnancy, J. Nutr., 41:231.

O'Neal, R. M., Johnson, O. C., and Schaefer, A. E., 1970, Guidelines for classification and interpretation of group blood and urine data collected as part of the National Nutrition Survey, Ped. Res., 4:103.

Owen, G. M., Garry, P. J., Kram, K. M., Nelsen, C. E., and Montalvo, J. M., 1969, Nutritional status of Mississippi preschool children, Am. J. Clin. Nutr., 22:1444.

Pallauf, J., and Kirchgessner, M., 1973, Zur wirksamkeit erhohter Zulagen an vitamin $B_1$, $B_2$, $B_6$, $B_{12}$, pantothen-und nicotin saure bei zinkmangel, Internatl. Z. Vit. Ern. Forsch., 43:339.

Pascale, J. A., Mims, L. C., Greenberg, M. H., Gooden, D. S., and Chronister, E., 1976, Riboflavin and bilirubin response during phototherapy, Ped. Res., 10:854.

Pearson, W. N., 1962, Biochemical appraisal of nutritional status in man, Am. J. Clin. Nutr., 11:462.

Pearson, W. N., 1967a, Blood and urinary vitamin levels as potential indicies of body stores, Am. J. Clin. Nutr., 20:514.

Pearson, W. N., 1967b, Riboflavin, in "The Vitamins", Vol. 7, 2nd ed., P. Gyorgy and W. N. Pearson, eds., pp. 99-136, Academic Press, New York.

Pinto, J., Huang, Y. P., and Rivlin, R. S., 1979, Physiological significance of inhibition of riboflavin metabolism by therapuetic doses of chlorpromazine, Clin. Res., 27:444A, (Abstract).

Plough, I. C., and Bridgeporth, E. B., 1960, Relations of clinical and dietary findings in nutrition surveys, Pub. Health Rep., 75:699.

Plough, I. C., and Consolazio, F. C., 1959, The use of casual urine specimens in the evaluation of the excretion rates of thiamine, riboflavin and N'-methylnicothinamide, J. Nutr., 69:365.

Pollack, H., and Bookman, J. J., 1951, Riboflavin excretion as a function of protein metabolism in the normal, catabolic and diabetic human being, J. Lab. Clin. Med., 38:561.

Prager, M. D., Hill, J. M., Speer, R. J., and Goerner, M., 1958, Whole blood riboflavin levels in healthy individuals and in patients manifesting various blood dyscrasias, J. Lab. Clin. Med., 52:206.

Pruzansky, J., and Axelrod, A. E., 1955, Antibody production to diphtheria toxoid in vitamin deficiency states, Proc. Soc. Exptl. Biol. Med., 89:323.

Ramachandran, M., and Iyer, G. Y. N., 1974, Erythrocyte glutathione reductase in iron deficiency anemia, Clin. Chim. Acta., 52:225.

Rasmussen, K. M., Barsa, P. M., and McCormick, D. B., 1979, Pyridoxamine (pyridoxine) 5'-phosphate oxidase activity in rat tissues during development of riboflavin or pyridoxine deficiency, Proc. Soc. Exptl. Biol. Med., 161:527.

Recommended Dietary Allowances, 1980 (In Press), Ninth Revised Edition, Food and Nutrition Board, National Academy of Sciences - National Research Council, Washington, D. C.

Register, U. D., and Sonnenberg, L. M., 1973, The vegetarian diet, J. Am. Diet. Assoc., 62:253.

Rivlin, R. S., 1970a, Regulation of flavoprotein enzymes in hypothyroidism and in riboflavin deficiency, Adv. Enzym. Regulation, 8:239.

Rivlin, R. S., 1970b, Riboflavin metabolism, New Engl., J. Med., 283:463.

Rivlin, R. S., 1975a, Riboflavin and Cancer, in "Riboflavin", R. S. Rivlin, ed., pp. 369-391, Plenum Press, New York.

Rivlin, R. S., 1975b, Hormonal regulation of riboflavin metabolism, in "Riboflavin", R. S. Rivlin, ed., pp. 393-426, Plenum Press, New York.

Rivlin, R. S., 1979, Hormones, drugs and riboflavin, Nutr. Rev., 37:241.

Rivlin, R. S., Menendez, C., and Langdon, R. G., 1968, Biochemical similarities between hypothyroidism and riboflavin deficiency, Endocrinology, 83:461.

Roderuck, C., Coryell, M. N., Williams, H. H., and Marcy, I. G., 1946, Metabolism of women during the reproductive cycle. IX. The utilization of riboflavin during lactation, J. Nutr., 32:267.

Roe, D. A., 1971, Drug-induced deficiency of B vitamins, N. Y. State J. Med., 71:2770.

Roe, D. A., McCormick, D. B., Lin, R., 1972, Effects of riboflavin on boric acid toxicity, J. Pharm. Sci., 61:1081.

Rose, D. P., and McGinty, F., 1968, The influence of adrenocortical hormones and vitamins upon tryptophan metabolism in man, Clin. Sci., 35:1.

Rosenberg, I. H., Solomons, N. W., and Schneider, R. E., 1977, Malabsorption associated with diarrhea and intestinal infections, Am. J. Clin. Nutr., 30:1248.

Rosenthal, S. R., 1978, Malnutrition and the developing fetus: an investigation of the effects of riboflavin, B-12, and folate deficiencies during pregnancy both on the mother and fetus, Mt. Sinai J. Med., 45:581.

Rosenthal, W. S., Adhurn, N. F., Lopez, R., and Cooperman, J. M., 1973, Riboflavin deficiency in complicated chronic alcoholism, Am. J. Clin. Nutr., 26:858.

Rubaltelli, F. F., Allegri, G., Costa, C., and DeAntoni, A., 1974, Urinary excretion of tryptophan metabolites during phototherapy, J. Ped., 85:865.

Saad, S., and Shehata, M., 1978, The effect of combined administration of diphenylhydantoin and certain oral hypoglycemic drugs on the hepatic content of some vitamins of the B-complex group, Materia Med. Polona, 10:122.

Sanpitak, N., and Chayutimonkul, L., 1974, Oral contraceptives and riboflavin nutrition, Lancet, 1:836.

Sanvordeker, D. R., and Korstenbauder, H. B., 1974, Mechanism for riboflavin enhancement of bilirubin photodecomposition in vitro, J. Pharm. Sci., 63:404.

Sarette, H. P., and Perizweig, W. A., 1943, The effect of protein and B-vitamin levels of the diet upon the tissue content and balance of riboflavin and nicotinic acid in rats, J. Nutr., 25:173.

Sauberlich, H. E., Dowdy, R., and Skala, J. H., 1974, Riboflavin, in "Laboratory Tests for the Assessment of Nutritional Status", pp. 30-37, CRC Press, Cleveland.

Sauberlich, H. E., Judd, J. H., Nichoalds, G. E., Broquist, H. P., and Darby, W. J., 1972, Application of the erythrocyte glutathione reductase assay in evaluating riboflavin nutritional status in a high school student population, Am. J. Clin. Nutr., 25:756.

Schorah, C. J., and Messenger, D., 1975, Flavin nucleotides, glutathione reductase and assessment of riboflavin status, Internatl. J. Vit. Nutr. Res., 45:39.

Schrier, S. L., Kellermeyer, R. W., Carson, P. E., Ickes, C. E., and Alving, A. S., 1958, The hemolytic effect of primaquine. IX. Enzymatic abnormalities in primaquine - sensitive erythrocyte, J. Lab. Clin. Med., 52:109.

Scott, E. M., Duncan, I. W., and Ekstrond, V., 1963, Purification and properties of glutathione reductase of human erythrocytes, J. Biol. Chem., 238:3928.

Sebrell, W. H., and Butler, R. E., 1938, Riboflavin deficiency in man, Public Health Reports, 53:2282.

Sharada, D., Bamji, M. S., 1972, Erythrocyte glutathione reductase activity and riboflavin concentration in experimental deficiency of some water soluble vitamins, Internatl. J. Vit. Nutr. Res., 42:43.

Shaw, J. H., and Phillips, P. H., 1941, The pathology of riboflavin deficiency in the rat, J. Nutr., 22:345.

Singer, T. P., Salach, J., Hemmerich, P., and Ehrenberg, A., 1971, Flavin peptides, in "Methods in Enzymology, Vol. XVIII, Vitamins and Coenzymes, Part B", D. B. McCormick and L. D. Wright, eds., pp. 416-427, Academic Press, New York.

Singher, H. O., Kensler, C. J., Levy, H., Poore, E., Rhoads, C. P., and Unna, K., 1944, Interrelationship between thiamine and riboflavin in the liver, J. Biol. Chem., 154:69.

Slater, E. C., and Morell, D. B., 1946, The fluorometric determination of riboflavin in urine, Biochem. J., 40:652.

Sleisenger, M. H., and Brandborg, L. L., 1977, Malabsorption, in "Major Problems in Internal Medicine", Vol. XIII, L. H. Smith, ed., W. B. Saunders Co., Philadelphia.

Smithells, R. W., Sheppard, S., and Schorah, C. J., 1976, Vitamin deficiencies and neural tube defects, Arch. Dis. in Child., 51:944.

Smith, J. M., Lu, S. D. C., Hare, A., Dick, E., and Daniels, M., 1959, J. Nutr., 69:85.

Snell, E. E., and Strong, F. M., 1939, Microbiological assay for riboflavin, Ind. Eng. Chem. (Anal. Ed.), 11:346.

Snyderman, S. E., Ketron, K. C., Burch, H. B., Lowry, O. H., Bessey, O. A., Guy, L. P., and Holt, L. E., Jr., 1949, The minimum riboflavin requirement of the infant, J. Nutr., 39:219.

Speck, W. T., Chen, C. C., and Rosenkranz, H. S., 1975, In vitro studies on effects of light and riboflavin on DNA and He La Cells, Ped. Res., 9:150.

Spies, T. D., Bean, W. B., and Ashe, W. F., 1939, Recent advances in the treatment of pellagra and associated deficiencies, Ann. Intern. Med., 12:1830.

Srivastava, S. K., and Beutler, E., 1970, Increased susceptibility of riboflavin deficient rats to galactose cataract, Experientia, 26:250.

Staal, G. E. J., and Veeger, C., 1969, The reaction mechanism of glutathione reductase from human erythrocytes, Biochem. Biophys. Acta, 185:49.

Staal, G. E. J., Visser, J., and Veeger, C., 1969, Purification and properties of glutathione reductase of human erythrocytes, Biochem. Biophys. Acta, 185:39.

Stanley, P. E., 1971, Determination of submicromole levels of NADH and FMN using bacterial luciferase and the liquid scintillation spectrophotometer, Anal. Biochem., 39:441.

Steier, M., Lopez, R., and Cooperman, J. M., 1976, Riboflavin deficiency in infants and children with heart disease, Am. Heart. J., 92:139.

Sterner, R. T., and Price, W. R., 1973, Restricted riboflavin, within-subject behavioral effects in human, Am. J. Clin. Nutr., 26:150.

Stickle, G., 1975, Pregnancy in adolescents: scope of the problem, Contemp. Ob. Gyn., 5:1975.

Street, H. R., Cowgrill, G. R., and Zimmerman, H. M., 1941, Further observations of riboflavin deficiency in the dog, J. Nutr., 22:7.

Strehler, B. L., and Cormier, M. J., 1954, Kinetic aspects of the bacterial luciferin-luciferase reaction in vitro, Arch. Biochem. Biophys., 53:138.

Strong, F. M., and Carpenter, L. E., 1942, Preparation of samples for microbiological determination of riboflavin, Ind. Eng. Chem. (Anal. Ed.), 14:909.

Sugioka, G., Porta, E. A., Corey, P. N., and Hartroft, W. S., 1969, The liver of rats fed riboflavin-deficient diets at two levels of protein, Am. J. Path., 54:1.

Suvarnakich, K., Mann, G. V., and Stare, F. J., 1952, Riboflavin in human serum, J. Nutr., 47:105.

Sydenstricker, V. P., Sebrell, W. H., Cleckley, H. M., and Kruse, H. D., 1940, The ocular manifestations of ariboflavinosis, J.A.M.A., 114:2437.

Tan, E. L., and Trachewsky, D., 1975, Effect of aldosterone on flavin coenzyme biosynthesis in the kidney, J. Steroid Biochem., 6:1471.

Tandler, B., Erlandson, R. A., and Wynder, E. L., 1968, Riboflavin and mouse hepatic cell structure and function. I. Ultrastructural alterations in simple deficiency, Am. J. Path., 52:69.

Tanphaichitr, V., Kulapongse, S., and Komindr, S., Assessment of nutritional status in adult hospitalized patients, (In press).

Tatt, L. K., Tan, I. K., and Seet, A. M., 1975, A new colorimetric method for the determination of NADH/NADPH dependent glutathione reductase in erythrocytes and in plasma, Clin. Chim. Acta, 58:101.

Tautt, J., 1976, Changes in the levels of certain vitamins from B group under conditions of vitamin deficiency, Acta Physiol. Pol., 27:1.

Ten-State Nutrition Survey, 1968-1970, U.S. Department of Health, Education and Welfare, Health Services and Mental Health Administration, Center for Disease Control, Atlanta, Georgia 30333, DHEW Publication No. (HSM) 72-8132.

Thackray, G. B., Sharman, I. M., Hyams, D. E., 1972, Riboflavin status of the elderly, Proc. Nutr. Soc., 31:89A.

Thurnham, D. I., 1972, Influence of glucose-6-phosphate dehydrogenase deficiency on the glutathione reductase test for ariboflavinosis, Annals Trop. Med. Parasitol., 66:505.

Thurnham, D. I., Migasena, P., and Pavapootanon, N., 1970, The ultramicro red-cell glutathione reductase assay for riboflavin status: its use in field studies in Thailand, Mikro. Chim. Acta (Wien), 5:988.

Thurnham, D. I., Migasena, P., Vudhivia, N., and Supawan, V., 1972, Riboflavin supplementation in a resettlement village in north-east Thailand, Br. J. Nutr., 28:91.

Thomas, J., and Ellis, F. R., 1977, The health of vegans during pregnancy, Proc. Nutr. Soc., 36:46A.

Tillotson, M. S., and Baker, E. M., 1972, An enzymatic measurement of the riboflavin status in man, Am. J. Clin. Nutr., 25:425.

Trachewsky, D., 1978, Aldosterone stimulation of riboflavin incorporation into rat renal flavin coenzymes and the effect of inhibition by riboflavin analogues on sodium reabsorption, J. Clin. Invest., 62:1325.

Tucker, R. G., Mickelsen, O., and Keys, A., 1960, The influence of sleep, work, diuresis, heat, acute starvation, thiamine intake and bed rest on human riboflavin excretion, J. Nutr., 72:251.

Tuszynsky, S., Myszkowska, K., Woziniak, W., Tautt, J., and Lewandowska, K., 1957, Influence of sulphaquanidine on the degree and order of inhibition of intestinal synthesis of vitamin $B_1$, riboflavin and nicotinamide in rats, Acta. Physiol. Polon., 8:727 (Nutr. Abst. Rev., 1958, 28:1065).

Verjee, Z. H. M., 1975, Tryptophan metabolism in baboons, effect of riboflavin and pyridoxine deficiency, Acta Vit. Enzymol., 29:198.

Vilter, R. W., Mueller, J. F., Helen, S., Glazer, H. S., Jarrold, T., Abraham, J., Thompson, C., Virginia, R. F., and Hawkins, V. R., 1953, The effect of vitamin B-6 deficiency induced by deoxypyridoxine in human beings, J. Lab. Clin. Med., 42:335.

Vir, S. C., and Love, A. H. G., 1977, Riboflavin status of institutionalized and non-institutionalized aged, Internatl. J. Vit. Nutr. Res., 47:336.

Vo-Khactu, K. P., Sims, R. L., Clayburgh, R. H., and Sandstead, H. H., 1976, Effect of simultaneous thiamin and riboflavin deficiencies on the determination of transketolase and glutathione reductase, J. Lab. Clin. Med., 87:741.

Wada, H., and Snell, E. E., 1961, The enzymatic oxidation of pyridoxine and pyridoxamine phosphate, J. Biol. Chem., 236:2089.
Wagner-Jauregg, T., 1954, Vitamins Vol. III, W. H. Sebrell, Jr., and R. S. Harris, eds., pp. 301-332, Acacemic Press, New York.
Wagner-Jauregg, T., 1972, Riboflavin II. Chemistry, in "The Vitamins", W. H. Sebrell, Jr., and R. S. Harris, eds., pp. 3-43, Academic Press, New York.
Warkany, J., 1975, Riboflavin deficiency and congenital malformation, in "Riboflavin", R. S. Rivlin, ed., pp. 279-302, Plenum Press, New York.
Warkany, J., and Nelson, R. C., 1942, Congenital malformations induced in rats by maternal nutritional deficiency, J. Nutr., 23:321.
Warkany, J., and Schraffenberger, E., 1943, Congenital malformations induced in rats by maternal nutritional deficiency. V. Effects of a purified diet lacking riboflavin, Proc. Soc. Exptl. Biol. Med., 54:92.
Warkany, J., and Schraffenberger, E., 1944, Congenital malformation induced in rats by maternal nutritional deficiency. VI. The preventive factor, J. Nutr., 27:447.
Weber, F., Glatzle, E., and Wiss, O., 1973, Symposium on recent advances in the assessment of vitamin status in the man. The assessment of riboflavin status, Proc. Nutr. Soc., 32:237.
West, D. W., and Owen, E. C., 1969, The urinary excretion of metabolites of riboflavin by man, Brit. J. Nutr., 23:889.
Whitehead, T. P., Kricka, L. J., Carter, T. J. N., and Thorpe, G. H. G., 1979, Analytical luminescence: its potential in the clinical laboratory, Clin. Chem., 25:1531.
Williams, R. R., and Cheldelin, V. H., 1942, Destruction of riboflavin by light, Science, 96:22.
World Health Organization, 1965, Nutrition in pregnancy and lactation, Tech. Rep. Ser. No. 302.
World Health Organization, 1967, Requirements of vitamin A, thiamine, riboflavin and niacin, Tech. Rep. Ser. No. 362.
World Health Organization, 1974, "Handbook of Human Nutritional Requirements", W.H.O. Monograph Series 61, Geneva.
Yagi, K., 1951, Micro-determination of riboflavin, flavin mononucleotide and flavin adenine dinucleotide by filter paper chromatography, J. Biochem., 38:161.
Yagi, K., 1962, Chemical determination of flavins, Methods Biochem. Anal., 10:319.
Yagi, K., and Okuda, J., 1958, Phosphorylation of riboflavin by transferase action, Nature, 181:1663.
Yagi, K., Nagatsu-Ishibashi, I., and Oshashi, A., 1966, Migration of injected $^{14}C$-labelled riboflavin into rat tissues, J. Biochem. (Tokyo), 59:313.

Yang, C. S., Charalampoo, A., and McCormick, D. B., 1964, Microbiological and enzymatic assays of riboflavin analogues, J. Nutr., 84:167.

Yang, C. S., and McCormick, D. B., 1967, Degradation and excretion of riboflavin in the rat, J. Nutr., 93:445.

Yawata, Y., and Tanaka, K. R., 1971, Effect of metabolic stress on activation of glutathione reductase by FAD in human red cells, Experientia, 27:1214.

Zannoni, V. G., and Sato, P. H., 1976, The effect of certain vitamin deficiencies on hepatic drug metabolism, Fed. Proc., 35:2464.

Ziegler, J. A., 1944, Photochemical destruction of vitamin $B_2$ in milk, J. Am. Chem. Soc., 66:1039.

# ROLE OF VITAMIN C IN HEALTH AND DISEASE

Herbert K. Naito, Ph.D.

Head, Lipid Lipoprotein Laboratories, Division of Laboratory Medicine and Division of Research, The Cleveland Clinic Foundation and Clinical Professor Department of Chemistry, Cleveland State University

## INTRODUCTION

The biosynthesis, physiological functions and catabolism of L-ascorbic acid (Vitamin C) in man remain topics of great interest and challenge for the clinical biochemist. L-ascorbic acid is a water-soluble vitamin, which chemically is a lactone (internal ester of a hydroxycarboxylic acid) and is characterized by the enediol group which makes it a strongly reducing compound. Ascorbic acid itself is the most active reducing agent known to occur naturally in living tissues. The empirical formula of this vitamin is $C_6H_8O_6$. The Levo-ascorbic acid is an active antiscorbutic substance, while the dextro-ascorbic acid is not (Figure 1). One of the isomers, D-isoascorbic acid or erythorbic acid, is produced commercially for use as a food additive. It has the reducing power of L-ascorbic acid but cannot support growth. L-ascorbic acid is readily and reversibly oxidized to dehydro-L-ascorbic acid which retains vitamin C activity (Figure 2). This compound can be further oxidized to diketo-2-gluconic acid, in a non-reversible reaction; the compound has no reported biological activity. Dehydration and decarboxylation can lead to the formation of furfural, which can polymerize to form brown pigments or combine with amino acids in the Strecker degradation. The failure of man, non-human primates, guinea pigs and a few other species of animals to synthesize their own vitamin C from glucose is due to a deficiency of an enzyme in the liver which converts L-gluconic acid to ascorbic acid. Thus, for these animals vitamin C is an essential vitamin that is required in the daily diet. While scurvy is not as common today in the U.S.A. as compared to a century ago, vitamin C deficiency is one of the more common vitamin deficiencies, particularly in school-age children (Ten State Nutrition Survey, 1972).

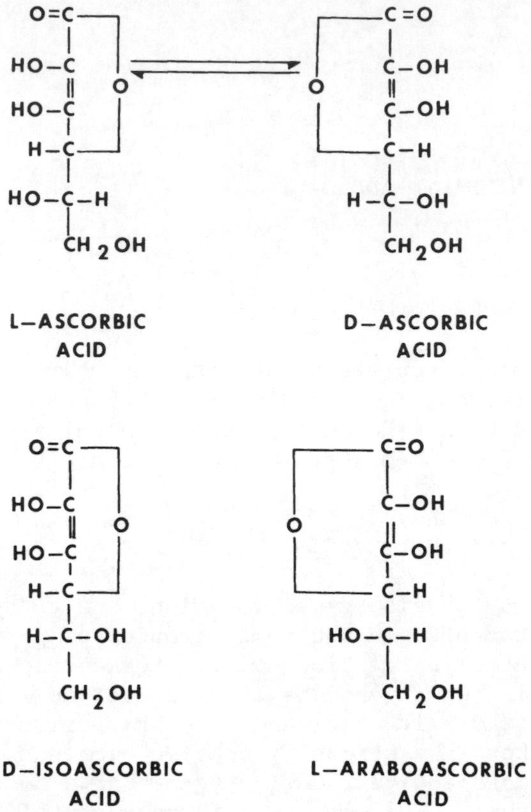

L-ASCORBIC ACID

D-ASCORBIC ACID

D-ISOASCORBIC ACID

L-ARABOASCORBIC ACID

Fig. 1. Structural formulas of ascorbic acid and its stereoisomers.

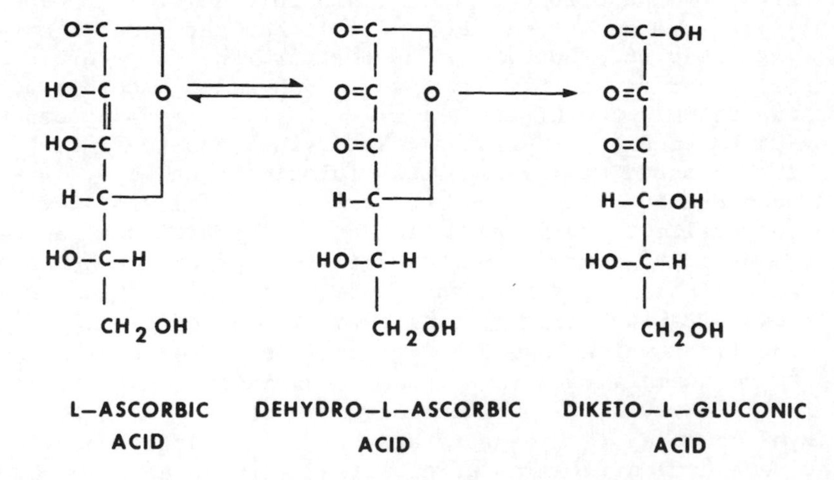


Fig. 2. Oxidation of L-ascorbic acid.

SYNTHESIS

Vitamin C is a common and essential nutritional element of all animals. Except for the invertebrates, fishes, Indian fruit-eating bat, guinea pig, non-human primates and man, vitamin C can be synthesized by plants and most animals (Figure 3). The emergence of the biosynthetic ability of amphibians suggests that a greater need of the vitamin was somehow linked with the evolution of vertebrates from aquatic to the terrestrial environment. It is likely that the evolution of all lines of vertebrates generally required loss of some specialization as new specialization came about. It is difficult to determine why vitamin C is made in the kidney and not the liver of amphibians. It may be considered that the early amphibians started synthesis in an organ where ascorbic acid could be produced at a high rate. *In vitro* results show that in the toads and frogs the activity of the kidney enzyme (L-glucono oxidase) is much higher than that of the mammalian liver enzyme (Chatterjee et al., 1975). The amphibians' capacity to synthesize

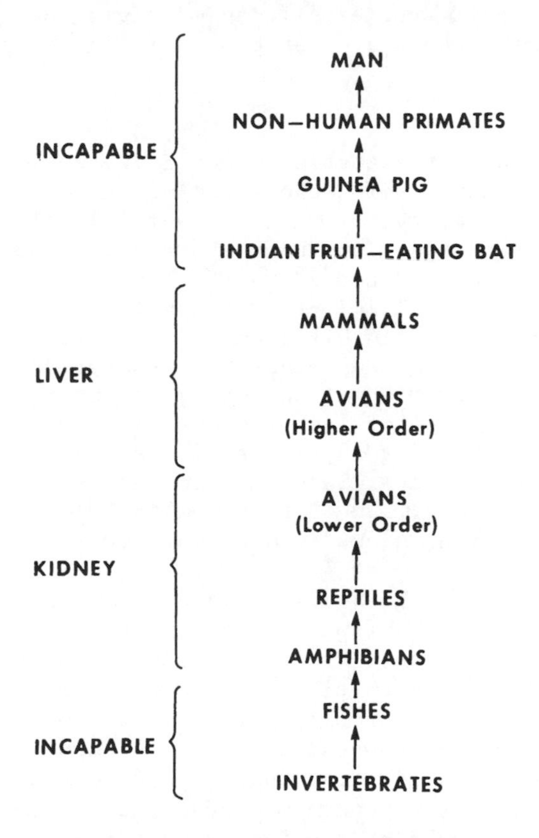


Fig. 3. Schematic diagram of vitamin C - synthesizing abilities of various species of animals in relation to their phylogeny (adapted from Chatterjee, 1973).

vitamin C in the kidney continued in that of the reptiles, but was then transferred to the liver of mammals. This change in site of synthesis appeared to have taken place when the vertebrates were evolving thermoregulatory mechanisms and changed from poikilothermic to homeothermic species. The kidney activities had been altered to accommodate the necessities of life on dry land with increased physiological demands such as regulation of calcium, phosphate, urea and other analytes. Primitive birds retained the ascorbate biosynthetic activity in the kidney, while the passeriform birds (i.e., house crow and myna) can be considered transitional, on the borderline in transferring the biosynthetic capacity from the kidney to the liver. Thereafter, the ascorbate synthetic capacity was taken over by the liver of the more evolved passeriform birds, while a number of other highly evolved passerers are incapable of producing vitamin C. The failure of the guinea pig, Indian fruit-eating bat, non-human primates and man to synthesize ascorbic acid is due to a common defect, namely, the absence of the terminal enzyme L-glucono oxidase (Burns, 1957). Figure 4 illustrates the pathway of ascorbic acid synthesis in animals via the glucuronic acid pathway, the main substrate for biosynthesis being D-glucose.

## CATABOLISM

The metabolic fate of ascorbic acid and its derivatives in animals depends on a number of factors including animal species, route of ingestion, quantity of material, nutritional status, and, of course, the compound itself. There appears to be a central core of metabolism of ascorbic acid that is observed with low dietary intakes and that is similar from species to species. Excess ascorbic acid is handled quite differently in different species. Thus, in man there is a limited intestinal absorption followed by an efficient urinary excretion of the excess ascorbic acid. In the guinea pig, some 60-70% of ingested ascorbic acid is rapidly catabolized to $CO_2$. What, if any, part of the central core metabolism of ascorbic acid is related to the nutritional requirement for vitamin C is not known. The catabolism to $CO_2$ does not appear to be a required function, since it does not occur in man (Figure 4).

This metabolic path seems well defined in animals through xylonate and lyxonate, but definite proof for the subsequent steps and isolation studies on the enzyme involved are not available (Tolbert et al., 1975). In particular, the question of whether the catabolic pathway proceeds through xylitol or through the L-ribulose to D-xyulose step for the conversion of L to D sugars is unresolved. In this process ascorbate is first converted to dehydroascorbic acid by a variety of enzymic and nonenzymic processes in animals. No specific ascorbic acid oxidase has been reported in animals. Enzymic delactonization of dehydro-L-ascorbic acid has been studied in animal tissue (Kagawa et al., 1961; Kagawa and Takiguchi, 1962). The enzyme is present in ox, rabbit, rat, and guinea pig livers, but is

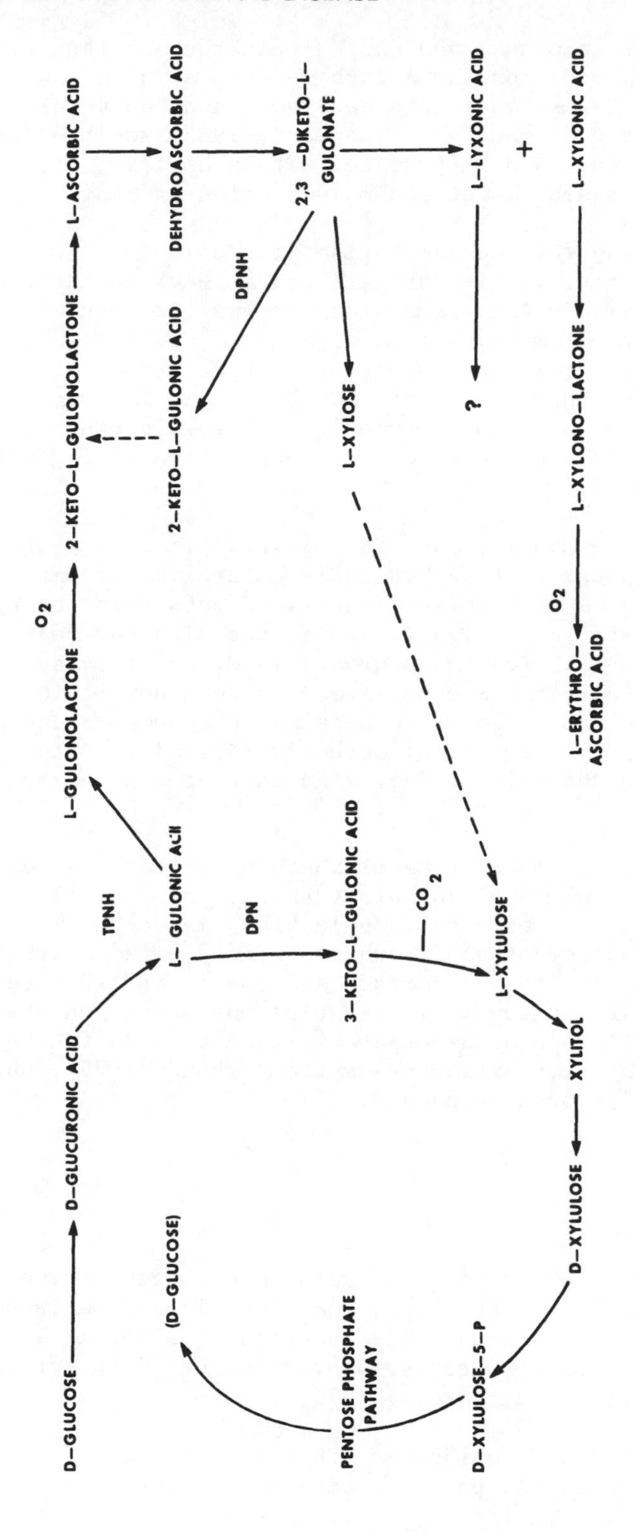


Fig. 4. Pathway of ascorbic acid synthesis.

essentially absent in monkey and man, a result consistent with the absence of $^{14}CO_2$ production for ascorbic-1-$^{14}$C acid in man (Baker et al., 1966). 2,3-Diketogulonate decarboxylase has been studied *in vivo* and *in vitro* (Kagawa and Takaguchi, 1962) and the products shown to be L-xylonate and L-lyxonate. It is a widely distributed enzyme, probably present in all animals, including man.

Another pathway for the catabolism of ascorbate is by C(2)-C(3) carbon chain cleavage, to give oxalate and a 4-carbon intermediate. This pathway appears in all animals and represents about 4-10% of the essential catabolism of ascorbate in man. Banay and Dimant (1962) used double labeling to show that the 2 carbons of oxalate come from the $C_1$ and $C_2$ carbons of ascorbate. It is commonly suggested that oxalate is produced from 2,3-diketogulonate, because oxalate is formed in the spontaneous decomposition of ascorbic acid solutions.

The 4-carbon metabolites of ascorbic acid from C(2)-C(3) cleavage do not appear to be extensively catabolized in man, because very little of the label from ascorbate-4-$^3$H gets into the body water pool (Tolbert et al., 1967). In guinea pigs the body water pool is extensively labeled by ascorbate-4-$^3$H, probably by the catabolism of a 5-carbon compound formed by decarboxylation of 2,3-diketogulonate. Even so, threonate and threonolacetone are also present in guinea pig urine, probably derived from the oxalate path for ascorbate catabolism, including that of man (Baker et al., 1971).

Ascorbate-3-sulfate has recently been shown to be a metabolite of ascorbic acid in urine of animals (Baker et al., 1971). The rapid clearance of ascorbate sulfate in blood is one of the possible reasons for the absence of antiscorbutic effect with ascorbate sulfate in the guinea pig. Ascorbate sulfate is ineffective in preventing or curing scurvy in guinea pigs whether given orally or by injection. A role of ascorbate sulfate as a sulfating agent has been demonstrated in biological systems (Bond, 1975), but further work in this area is needed.

## PHYSIOLOGY

### Collagen Metabolism

Direct involvement of ascorbic acid in collagen synthesis is well known and represents, perhaps, the most clearly defined biochemical role of this vitamin. The function of ascorbic acid in collagen synthesis has been reviewed by a number of investigators (Barnes and Kodicek, 1972; Barnes, 1975).

Studies in isolated collagen-synthesizing systems have demonstrated that ascorbic acid participates in the synthesis of collagen hydroxyproline and hydroxylysine, both of which are formed by the hydroxylation of particular prolyl and lysyl residues previously incorporated into peptide linkage during the process of ribosomal collagen protein synthesis. The precise mode of action of the vitamin in these hydroxylations has yet to be elucidated.

The need for ascorbic acid as a reductant *in vitro* in these reactions is not highly specific. Because of this lack of specificity, the possibility existed that compounds other than vitamin C might perform this function *in vivo* and that impaired collagen synthesis in scurvy might imply a role for ascorbic acid in collagen metabolism other than one in hydroxylation. There appears to be no evidence, however, for the participation of other reductants in the hydroxylation mechanism *in vivo*. Studies in scorbutic guinea pigs have indicated that hydroxylation at least of peptidyl proline is impaired *in vivo* in ascorbic acid deficiency, and it is possible that this is the primary lesion in collagen synthesis in scurvy.

Nevertheless, hydroxylation of collagen proline *in vivo* is only slightly impaired in scorbutic animals. There is no evidence for the occurrence of a substantially underhydroxylated moiety. It is believed that a slight additional reduction beyond an initial 5-10%, which does not appear to impair collagen function, causes a lack of formation of the triple helical structure of collagen with consequent degradation of the unassociated α-chains. Accumulation of the latter within the cell may also cause a feedback inhibition of further collagen protein synthesis. The continuous absence of accumulation of unhydroxylated material in the tissues of scorbutic guinea pigs may be attributable to reduced collagen protein synthesis, which is regarded as a secondary feature of impaired hydroxylation.

## Lipid Metabolism

In the last few years, there has been considerable interest in vitamin C and its role in lipid metabolism and atherosclerosis. In this regard vitamin C seems to be involved in at least two systems: the maintenance of vascular wall integrity and the metabolism of cholesterol to bile acids.

The etiology of atherogenesis has not been fully elucidated. However, there appear to be several possible mechanisms for the initial development and progression of the arterial vascular disease; one possible common denominator being a change in vascular wall permeability (Gore and Stefanovic, 1967). Since collagen is a major component in the arterial blood vessel, it is not unreasonable to suspect that a dysmetabolism of collagen may later alter the arterial wall structure and ultimately affect vascular wall permeability, predisposing the blood vessel to the development of atherosclerotic lesion.

The guinea pig, like man, non-human primates, certain birds and the Indian fruit-eating bat, but unlike most other species, cannot synthesize vitamin C and is therefore dependent upon exogenous sources for meeting its daily requirements. It has been reported that there is an inverse relationship between liver ascorbic acid concentration and hepatic cholesterol concentration in guinea pigs (Ginter and Ondreicka, 1971; Ginter et al., 1967). Cholesterol accumulation in the liver during low ascorbate intake did not appear to originate from increased synthesis or absorption during times when liver ascorbate was low. However, decreased liver ascorbate appeared to decrease the conversion of cholesterol to bile acids for excretion via the bile (Ginter et al., 1971). Vitamin C may be a necessary cofactor for the hydroxylation of the cholesterol nucleus to form bile acids. It was concluded that this decreased rate of cholesterol catabolism and excretion leads to hypercholesterolemia.

There is a wide divergence of opinion concerning the significance of vitamin C levels in relation to atherosclerosis in human beings. Willis (1953) noted that there frequently was a deficiency of ascorbic acid in the arterial wall of hospitalized patients, and that old age seemed to accentuate the deficiency. He also noted that the lowest levels were in areas most prone to development of atherosclerosis (Willis and Fishman, 1955). Sokoloff et al. (1967) reported that of 20 atherosclerotic, hypercholesterolemic patients, the serum cholesterol decreased 30 percent and the β-lipoprotein decreased 25 percent after a 30-day course of 0.5 g of ascorbic acid three times a day. Normal subjects showed a significant decrease in serum cholesterol levels when 1.0 g ascorbic acid was added to theri diet, while those with atherosclerosis receiving clofibrate for treatment of hyperlipidemia or anticoagulant therapy tended to show increased cholesterol levels with high vitamin C intake (Spittle, 1972). Young, normal subjects, when given 1.0 g ascorbic acid per day for 6 weeks in addition to their usual dietary intake, failed to show any significant change in serum cholesterol level (Anderson et al., 1972). Ginter et al. (1970) found decreased plasma cholesterol levels after ascorbic acid therapy in hypercholesterolemic subjects (>200 mg/dl) with initial low blood ascorbic acid concentrations. In another study Ginter (1975) reported that 1 g of ascorbic acid daily for a period of 3 months given to subjects 50-75 years of age with a mean serum cholesterol concentration of <200 mg/dl had no effect on plasma cholesterol. Peterson et al. (1975) administered 4 g of ascorbic acid daily for 2 months to hypercholesterolemic patients and found no effect on serum lipids.

Since there are conflicting reports on the effects of ascorbate on serum cholesterol levels, we examined the relation of varying intake of ascorbate on serum lipid levels in the guinea pig and in human beings.

First of all, it should be recognized that some of the confusion in the literature appears to be related to the fact that several studies focused upon acute avitaminosis C rather than chronic hypovitaminosis C. Furthermore, the studies used different means of ascorbate administration (via drinking water, gastric intubation, or intraperitoneal injection). Another factor that may have influenced the results is the vehicle that was used in the vitamin C solution. Some investigators used ascorbate in water while others used sucrose to increase the palatability of the ascorbic acid. In the present study, a commercial source was used (Ce-Vi Sol from Mead Johnson and Co., Evansville, IN), in which ascorbic acid was diluted in a glycerine solution to minimize the oxidation of L-ascorbate to dehydroascorbate, a reaction that occurs extremely rapidly. It is possible that those who provided ascorbate in the drinking water did not obtain the effect of ascorbate, but rather that of dehydroascorbate. Furthermore, when vitamin C is provided in drinking water bottles, it is not possible to predict the exact amount of ascorbate intake, because there is much spillage when the guinea pigs drink. There is also the problem of varying water intake by different animals.

In the first experiment, an acute vitamin C deficiency study was made by feeding 10 adult male guinea pigs (English short-hair strain) a chow diet (Reid-Briggs guinea pig diet, ICN Pharmaceutical Inc., Cleveland, OH) completely devoid of vitamin C (Table 1).

Table 1. Composition of the Reid Briggs Guinea Pig Diet[a,b]

| Nutrients | Percent (By Weight) |
|---|---|
| Vitamin-free casein | 30.0 |
| Starch, corn | 20.0 |
| Alphacel | 15.0 |
| Sucrose | 10.3 |
| Glucose | 7.8 |
| Corn Oil | 7.3 |
| Salt Mixture | 6.0 |
| Potassium acetate | 2.5 |
| Vitamin fortification mixture[c] | 0.6 |
| Magnesium oxide | 0.5 |

[a] Reid and Briggs (1953).
[b] Semi-synthetic diet was made by ICN Pharmaceutical Inc., Cleveland, OH.
[c] Vitamin C fortification normally consisting of 0.2 percent (by weight) was omitted when the scorbutogenic diet was made.

Ten age-matched control male guinea pigs fed a guinea pig chow containing vitamin C were used to obtain baseline values. Ascorbate content in tissues and sera was measured by the method described by Roe and Kuthner (1943). All lipid analyses were done on fasting samples. A detailed description of the methods is described elsewhere (Naito and Gerrity, 1979). The results are presented in Table 2. After 2 weeks on the experimental diet,

Table 2. Effects of Acute Avitaminosis C ($\overline{X} \pm$ S.D.)[a]

| Analytes | Initial Day 0[b] (Baseline Values) | Final Day 14[b] | Significance |
|---|---|---|---|
| Total weight change (grams) | 925 ± 68 | -47.3 ± 33.0 | P<0.05 |
| Serum Ascorbate (mg/dl) | 0.64 ± 0.41 | 0.052 ± 0.04 | P<0.01 |
| Urine Ascorbate (mg/dl) | 1.37 ± 0.79 | 0.86 ± 0.55 | N.S. |
| Liver Ascorbate (mg/100 g wet wt.) | 12.37 ± 6.11 | 1.15 ± 0.38 | P<0.01 |
| Adrenal Ascorbate (mg/100 g wet wt.) | 86.77 ± 29.22 | 8.05 ± 2.45 | P<0.01 |
| Spleen Ascorbate (mg/100 g wet wt.) | 36.11 ± 11.8 | 4.17 ± 0.6 | P<0.01 |
| Serum Cholesterol (mg/dl) | 44.7 ± 13.3 | 58.0 ± 4.4 | N.S. |
| Liver Cholesterol (mg/100 wet wt.) | 302.6 ± 62.8 | 328.8 ± 26.5 | N.S. |
| Serum Triglycerides (mg/dl) | 61.8 ± 26.7 | 146.6 ± 50.3 | P<0.1 |

[a]Animals were fed the Reid-Briggs guinea pig scorbutogenic diet for 14 days.
[b]N=10 animals per group.
[c]Students t-test

signs of acute scurvy were demonstrated by microhemorrhages of the small blood vessels of the tibia-femur joint. After 2 weeks on the scorbutogenic diet, there was a significant reduction in body weight. The level of vitamin C in most organs reached scorbutogenic levels. However, the acute avitaminosis C condition did not appear to be associated with changes in serum lipid levels. Because the animals demonstrated signs of metabolic distress (i.e., stopped eating, loss of body weight, and lethargy), it appears that these types of acute studies may yield misleading lipid results. This was confirmed after examination of data that was collected by repeating the experiment; it was found that the avitaminosis C condition resulted in a 21% elevation of the serum cholesterol concentration rather than no change as observed in the first study (unpublished report).

The following study was designed to examine the chronic effects of varying levels of vitamin C (Ce-Vi Sol) intake on serum and liver cholesterol concentration in adult male guinea pigs fed either a chow diet or a 2.0% cholesterol (by weight) supplemented chow diet. The experimental design is presented in Table 3.

Table 3. Experimental Design -- Effects of Varying Levels of Vitamin C Intake With and Without Cholesterol Supplementation[a,b]

| Ascorbate Intake mg/animal/day | Group | Diet |
|---|---|---|
| 0.5 | Hypovitaminosis C | Vitamin C Free, no cholesterol |
| 0.5 | Hypovitaminosis C | Vitamin C Free, cholesterol[c] |
| 5.0 | "Experimental Control" | Vitamin C Free, no cholesterol |
| 5.0 | "Experimental Control" | Vitamin C Free cholesterol |
| 50.0 | Hypervitaminosis C | Vitamin C Free, no cholesterol |
| 50.0 | Hypervitaminosis C | Vitamin C Free, cholesterol |
| Normal | Control | Chow |

[a]Adult Male Guinea Pigs (N=6 animals/group).
[b]Animals were fed the Reid-Briggs guinea pig chow diet with 0.2 g% (w/w) vitamin C fortification.
[c]cholesterol = 2.0% (by weight) supplementation.

Since it has been reported that 5-10 mg/animal/day is the recommended requirement for vitamin C for adult guinea pigs (Banerjee and Singh, 1958; Ginter and Ondreicka, 1971), that daily dose was arbitrarily adopted for the present studies. They are referred to here as "experimental" controls. The 0.5 mg/animal/day group, then, was considered hypovitaminosis C group, and the 50 mg/animal/day was considered hypervitaminosis C group. A fourth group (control) was fed the standard guinea pig diet (Reid-Briggs), which contained about 0.2 g percent of vitamin C (by weight). The First three groups were fed the Reid-Briggs scorbutogenic diet (Table 3) and were given daily vitamin C via gastric intubation for 6 weeks, after which time the study was terminated. After removing 2.5 ml of blood (for lipid studies) via the jugular vein, an equal amount of Evans blue dye was infused and allowed to circulate for 2 hours while the animals were under amytol anesthesia. The animals were then sacrificed and the tissues removed, frozen with liquid nitrogen, and stored until analyzed. The aorta was quickly removed, rinsed in buffered 1.3% formaldehyde solution, and cut longitudinally, and the intima was examined for Evans blue uptake.

The effects of different levels of vitamin C intake on body weight, serum and liver ascorbate, and cholesterol concentrations of guinea pigs fed a low-cholesterol diet are shown in Tables 4 and 5. All animals gained body weight except the hypovitaminosis C group, which lost weight. The liver ascorbate concentration correlated positively with the increasing dietary intake of vitamin C, while the serum ascorbate levels did not. It is interesting that the 50 mg vitamin C group had a hepatic ascorbate concentration similar to that of the control group. Through feed-intake studies, it was estimated that 40-50 mg of vitamin C was consumed per day by the control animals on the Reid-Briggs chow diet containing 0.2% vitamin C (by weight). The presumed euvitaminosis C group, 5 mg/day, had a significantly lower liver vitamin C content than the control group. The serum cholesterol level was lowest in the 50 mg/day group and highest in the 0.5 mg/day group, suggesting an inverse relationship bewteen vitamin C intake and serum cholesterol level. The hepatic cholesterol was higher in all three experimental groups when compared to the control group, but the 0.5 mg/day group had the most accumulation of hepatic cholesterol.

In another study, the Reid-Briggs chow diet supplemented with 2% cholesterol (by weight) was fed to three groups of guinea pigs provided with different levels of vitamin C for 6 weeks. The control group from the previous study was also used for this study. The 50 mg/dl group maintained body weight, whereas the 5.0 and 0.5 mg/day groups lost weight (Table 5). It appears that with a high cholesterol diet, the consumption of larger amounts of vitamin C may be necessary to prevent the loss of body weight. Like the group fed the low-cholesterol diet, those groups of guinea pigs fed the high-cholesterol diet and given varying daily amounts of vitamin C via stomach intubation showed a positive correlation with the intake and hepatic ascorbate levels, but not with the serum ascorbate concentration (Table 5). Thus, it appears that while the serum ascorbate level is not a good index

Table 4. Effect of Different Levels Of Vitamin C Intake on Serum and Liver Lipid Concentration of Guinea Pigs Fed Chow Diet ($\overline{X} \pm$ S.E.M.)

| Ascorbate Intake (mg/animal/day)[a] | Body Weight (g) | Ascorbate | | Serum Lipids (mg/dl) | | | Liver Lipids (mg/100 g tissue)[e] | | |
|---|---|---|---|---|---|---|---|---|---|
| | | Serum (mg/dl) | Liver (mg/100 g tissue)[c] | Chol[d] | TG[e] | PL[f] | Chol | TG | PL |
| Normal[b] | 661<br>±33.2 | 0.35<br>±0.02 | 9.18<br>±1.0 | 44.8<br>±5.0 | 77.2<br>±19.1 | 68.8<br>±6.6 | 275.3<br>±4.0 | 4227.9<br>±591.4 | 3824.2<br>±101.6 |
| 0.5 | 498<br>±36 | 0.028<br>±0.006 | 1.001<br>±0.187 | 110.7<br>±13.2 | 168<br>±58.1 | 88.0<br>±14.3 | 570.1<br>±95.3 | 1887.3<br>±444.4 | 3710.2<br>±130.5 |
| 5.0 | 590<br>±44 | 1.068<br>±0.01 | 5.467<br>±0.176 | 73.8<br>±15.7 | 87.3<br>±14.9 | 22.8<br>±10.4 | 366.7<br>±39.6 | 3133.2<br>±1175.4 | 3063.2<br>±54.6 |
| 50.0 | 535<br>±27 | 0.1475<br>±0.04 | 8.649<br>±1.720 | 61.0<br>±3.5 | 102.0<br>±35.0 | 47.6<br>±1.7 | 372.7<br>±31.7 | 1426.5<br>±334.6 | 3738.4<br>±153.4 |

[a] Animals were fed Reid-Briggs scorbutogenic diet and given Vitamin C via tubal intubation.
[b] Animals were fed Reid-Briggs guinea pig chow diet with 0.2 g% (by weight) Vitamin C fortification.
[c] Wet weight
[d] Chol = Total cholesterol
[e] TG = Triglycerides
[f] PL = Total phospholipids

Table 5. Effect of Different Levels of Ascorbate Intake on Serum and Liver Lipids of Guinea Pigs Fed a Scorbutogenic Diet with 2.0 Percent Cholesterol-Supplementation ($\overline{X} \pm$ S.E.M.)

| Ascorbate Intake (mg/animal per day)[a] | Final Body Weight (g) | Δ in Body Weight (g) | Ascorbate | | Serum Cholesterol (mg/dl) | Liver Cholesterol (mg/100 g tissue)[c] |
|---|---|---|---|---|---|---|
| | | | Serum (mg/dl) | Liver (mg/100 g tissue)[c] | | |
| 0.5 | 381 ±24 | (-96) | 0.052 ±0.016 | 1.232 ±0.419 | 463.8 ±57.1 | 4803.5 ±326.2 |
| 5.0 | 414 ±5 | (-48) | 0.045 ±0.369 | 4.389 ±1.041 | 322.7 ±40.8 | 4539.9 ±450.5 |
| 50.0 | 544 ±28 | (-7) | 1.392 ±0.078 | 8.774 ±1.391 | 229.3 ±29.7 | 3371.7 ±736.9 |
| Control[b] | 661 ±33 | (+183) | 0.350 ±0.020 | 9.180 ±1.000 | 44.8 ±5.0 | 275.3 ±4.0 |

[a] Animals were fed Reid-Briggs scorbutogenic diet with 2.0 percent cholesterol supplementation and given vitamin C via tubal intubation.

[b] Animals were fed Reid-Briggs diet with 0.2 g percent (by weight) vitamin C fortification and no cholesterol supplementation. Calculated intake = 40-50 mg/animal/day.

[c] Wet weight.

for vitamin C intake, hepatic ascorbate levels are. Again the 50 mg/day group had a hepatic vitamin C content comparable to that of the controls. Feeding the 2% cholesterol diet resulted in an increase in serum cholesterol levels in all three experimental groups as compared to animals on a chow diet alone (Table 5). The data suggested that an inverse relationship existed between vitamin C intake and serum cholesterol level. However, even the intake of 50 mg/day was not sufficient to normalize the serum cholesterol levels. Similarly, the high vitamin C intake had little effect on the accumulation of hepatic cholesterol. The Evans blue dye study indicated that the guinea pigs fed the no-cholesterol diet showed no evidence of dye uptake. However, about 30% of the animals in the 0.5 and 5.0 mg ascorbate per day groups fed 2% cholesterol showed evidence of Evans blue uptake, as exhibited by areas of streaking in the thoracic aorta. The 50 mg/day group and the control animals did not demonstrate any Evans blue uptake by the intima. It can be concluded from this study that:

a. Liver ascorbate level correlated with the amount of oral intake of vitamin C, while the serum ascorbate level did not.
b. Five mg vitamin C per animal per day was not an adequate amount for maintaining normal health and body weight in this strain of guinea pigs; 50 mg per animal per day may have represented a more adequate intake.
c. Cholesterol feeding caused a loss of body weight, particularly in the animals with insufficient vitamin C intake.
d. Vitamin C insufficiency caused an increase in serum and hepatic cholesterol concentrations over a 6-week period in both cholesterol-fed guinea pigs and those fed no cholesterol.
e. About 1/3 of the animals in the 0.5 and 5.0 mg ascorbate per day groups fed the cholesterol diet showed Evans blue uptake in the thoracic aorta. The guinea pigs receiving no cholesterol showed no indication of such changes. The 50 mg ascorbate per day group fed cholesterol also showed no uptake of Evans blue.
f. This study suggested that vitamin C, in some way, does influence cholesterol metabolism in guinea pigs. Further studies are necessary to precisely define the role of vitamin C in cholesterol metabolism and arterial wall integrity.

It appears that a bolus dose of 50 mg vitamin C via stomach intubation provided a sufficient amount of ascorbate to maintain hepatic vitamin C levels comparable to those of control animals who had vitamin C fortified in their diet. By bolus dose, it is difficult to determine how much vitamin C was actually absorbed by the gastrointestinal tract. The level of 8-9 mg ascorbate per 100 g liver probably represents the saturated level. The question arises whether two 25 mg doses or five 10 mg doses per day lead to greater coefficient of absorption of vitamin C, thus leading to more effective action of ascorbate on cholesterol metabolism.

Since the 50 mg/day dose was neither effective in inhibiting the rise of nor lowering the serum cholesterol levels in the cholesterol-fed pigs, it would be interesting to determine whether the pharmacological or mega dosages, i.e., 250 or 500 mg/day vitamin C, can be effective. However, because our attempts to administer 250 mg/day to the animals were associated with diarrhea, it appears to be physiologically unpractical to experiment with higher dosages.

Results of our study with the guinea pigs suggest that insufficient vitamin C intake causes a rise in serum cholesterol concentrations. Since scurvy or hypovitaminosis C in man is rare today, it is difficult to extrapolate the usefulness of these animal studies to man. It is probably more important to know whether large dosages can act as hypocholesterolemic agents in man, especially when previous studies are in conflict (Willis and Fishman, 1955; Sokoloff et al., 1967; Spittle, 1972; Anderson

| Time Period | Period 0 | | | Period 1 | | | Period 2 | | | Period 3 | | |
|---|---|---|---|---|---|---|---|---|---|---|---|---|
| | 1st Week | 2-3 Weeks Later | 4th Week (Time Zero) | End of 10th Week | | | End of 16th Week | | | End of 22nd Week | | |
| Treatment | No Vitamin C or Placebo (Maintain Regular Diet) | | | On Vitamin C | | | Off Vitamin C (Placebo) | | | On Vitamin C | | |
| | | | | 2nd | 4th | 6th | 2nd | 4th | 6th | 2nd | 4th | 6th |
| Physical exam | | | X | | | X | | | X | | | X |
| Dietary history | Briefing | | X | | | X | | | X | | | X |
| Laboratory tests (need blood and urine) | | | | | | | | | | | | |
| Cholesterol | From chart | X | X | X | X | X | X | X | X | X | X | X |
| Triglyceride | From chart | X | X | X | X | X | X | X | X | X | X | X |
| Phospholipid | | X | X | X | X | X | X | X | X | X | X | X |
| Lipoprotein electrophoresis | From chart | X | X | X | X | X | X | X | X | X | X | X |
| Total protein | From chart | X | X | | | X | | | X | | | X |
| Protein electrophoresis | From chart | X | X | | | X | | | X | | | X |
| SMA-12/60 | From chart | X | X | | | X | | | X | | | X |
| SMA-6/60 (kidney profile) | From chart | X | X | | | X | | | X | | | X |
| Thyroxine | From chart | X | X | | | X | | | X | | | X |
| Ascorbate | X | X | X | X | X | X | X | X | X | X | X | X |

n = 6 subjects/group
Male only

Fig. 5. Experimental design of vitamin C study on Type IIa hyperlipoproteinemic subjects: Group A.

| Time Period | Period 0 | | | Period 1 | | | Period 2 | | | Period 3 | | |
|---|---|---|---|---|---|---|---|---|---|---|---|---|
| | 1st Week | 2-3 Weeks Later | 4th Week (Time Zero) | End of 10th Week | | | End of 16th Week | | | End of 22nd Week | | |
| Treatment | No Vitamin C or Placebo (Maintain Regular Diet) | | | Off Vitamin C (Placebo) | | | On Vitamin C | | | Off Vitamin C (Placebo) | | |
| | | | | 2nd | 4th | 6th | 2nd | 4th | 6th | 2nd | 4th | 6th |
| Physical exam | | | X | | | X | | | X | | | X |
| Dietary history | Briefing | | X | | | X | | | X | | | X |
| Laboratory tests (blood) | | | | | | | | | | | | |
| Cholesterol | From chart | X | X | X | X | X | X | X | X | X | X | X |
| Triglyceride | From chart | X | X | X | X | X | X | X | X | X | X | X |
| Phospholipid | | X | X | X | X | X | X | X | X | X | X | X |
| Lipoprotein electrophoresis | From chart | X | X | X | X | X | X | X | X | X | X | X |
| Total protein | From chart | X | X | | | X | | | X | | | X |
| Protein electrophoresis | From chart | X | X | | | X | | | X | | | X |
| SMA-12/60 | From chart | X | X | | | X | | | X | | | X |
| SMA-6/60 (kidney profile) | From chart | | X | | | X | | | X | | | X |
| Thyroxine | From chart | X | X | | | X | | | X | | | X |
| Ascorbate | X | X | X | X | X | X | X | X | X | X | X | X |

n = 5 subjects/group
Male only

Fig. 6. Experimental design of Vitamin C study on Type IIb hyperlipoproteinemic subjects: Group B

et al., 1972; Ginter et al., 1970; Ginter, 1975; Peterson et al., 1975) as discussed earlier.

The following study was designed to determine whether one gram (500 mg in the morning and 500 mg in the evening) of vitamin C intake per day would lower serum lipids in persons with hyper-β-lipoproteinemia (types IIa and IIb). The double blind, double-crossover experimental design is shown in Figures 5 and 6. The ten male participants (mean age = 45 years; all had cholesterol concentration >300 mg/dl) were divided into two groups, one of which (A) started with the vitamin C period first (two 250 mg tablets in the morning and two 250 mg tablets in the evening), followed by the placebo period (calcium lactate) and ending with the vitamin C period. Group A consisted of type IIa hyperlipoproteinemic subjects, while group B consisted of participants with type IIb hyperlipoproteinemia. The second group (B) started with the placebo period first, followed by the vitamin C period and ending with the placebo period. Each period but the baseline period (4 weeks) lasted 6 weeks. A baseline period when blood samples were taken about 2 weeks apart was set up at the initiation of the experiment. During each of the three experimental periods, fasting blood samples were collected at 2-week intervals. The mean levels the parameters measured of the three bleedings for each period were used to express the effect or lack of effect of vitamin C. Since the participants were free-living, a dietary history was recorded by a dietitian by 3-day detailed record keeping each week. The composition of the dietary intake was calculated using tables from Feeley et al. (1972), Watt and Merrell (1963) and Church and Church (1977). The results are presented in Tables 6 and 7. The dietary intake study indicated that the caloric intake, percent of calories as carbohydrates, proteins, fat, alcohol, as well as the ascorbate (besides the vitamin C tablets) and cholesterol intake during each of the four periods (baseline periods 1, 2 and 3 were not statistically different from each other in both groups A and B. The low cholesterol intake is due to the low-cholesterol diet that the subjects were ingesting. The effect of 1 g of vitamin C per day on serum lipids is shown in Table 8. Despite a 2.5 to 3-fold elevation in serum ascorbate levels during the vitamin C periods (compared to baseline), the serum cholesterol, triglyceride and phospholipid concentrations did not change in either group A or B. To ensure that all the participants consumed the placebo and vitamin C each day during the study, a pill count was made every 2 weeks when they came in for bleeding and physical examination. Compliance on the pills was 98% in group A and 99% in group B during the entire study. Because high dosages of vitamin C intake may result in oxylate stone formation in the kidneys (Barness, 1975) and electrolyte imbalances (Lewin, 1974), 18 different analytes were monitored on the Technicon SMA 12/60 and 6/60 (1973). The results (Tables 9-12) indicated that little change in the concentration of the various analytes occurred during the different experimental periods. In addition, there was no change in body weight, blood pressure, or serum thyroxine concentration (in either group) during the different periods. This study indicates that in adult male individuals with hyper-β-lipoproteinemia, large intake of vitamin C does not appear to be effective in lowering serum lipids. This does not negate the possibility that during insufficient vitamin C intake derangement of lipid and lipoprotein metabolism can occur.

## Hematopoiesis

In addition to the above physiological functions, vitamin C plays a number of roles that relate to hematopoiesis. Although anemia is not a prominent component of clinical scurvy, it does occur in some instances for a variety of reasons. Ascorbic acid is a potent reducing agent and as such it enhances absorption of iron and inhibits absorption of copper from the digestive tract (Hodges, 1976). It also facilitates the transfer of iron from transferrin to ferritin. Ascorbic acid aids in formation of active compounds from tetrahydrofolates by linkage of the $N^5$ or $N^{10}$ position to formyl, hydroxymethyl, methyl or formimino groups (Stokes et al., 1975). Massive doses of ascorbic acid, however,

Table 6. Group A: Daily Food-Intake Data During the Various Study Periods

| Diet Components | Baseline Period | Ascorbic Acid Period | Placebo Period | Ascorbic Acid Period |
|---|---|---|---|---|
| Total Calories | 1993.2 ± 389.81 | 2191.9 ± 408.16 | 2083.6 ± 49.04 | 2939.3 ± 787.81 |
| % Calories from Carbohydrate | 40.6 ± 5.69 | 42.3 ± 3.98 | 37.5 ± 1.84 | 37.6 ± 7.68 |
| % Calories from Protein | 20.0 ± 3.74 | 19.6 ± 2.33 | 22.0 ± 5.19 | 15.4 ± 3.44 |
| % Calories from Fat | 34.2 ± 7.15 | 31.8 ± 3.64 | 34.9 ± 8.63 | 35.4 ± 3.50 |
| % Calories from Alcohol | 5.2 ± 4.23 | 6.3 ± 3.29 | 5.6 ± 8.69 | 11.6 ± 8.10 |
| Ascorbic Acid (mg) | 138.2 ± 73.83 | 114.5 ± 30.29 | 86.5 ± 14.53 | 77.7 ± 27.08 |
| Cholesterol (mg/dl | 328.4 ± 145.60 | 327.6 ± 164.55 | 366.4 ± 138.00 | 386.9 ± 89.27 |

Table 7. Group B: Daily Food - Intake Data During the Study Periods

| Diet Components | Baseline Period | Ascorbic Acid Period | Placebo Period | Ascrobic Acid Period |
|---|---|---|---|---|
| Total Calories | 1980.0 ± 392.11 | 1861.8 ± 277.37 | 2058.4 ± 267.74 | 1855.5 ± 277.48 |
| % Calories from Carbohydrate | 39.2 ± 15.52 | 35.0 ± 14.94 | 33.6 ± 11.30 | 34.5 ± 10.29 |
| % Calories from Protein | 19.9 ± 3.42 | 20.3 ± 3.55 | 18.9 ± 4.15 | 19.6 ± 3.81 |
| % Calories from Fat | 32.4 ± 7.74 | 32.2 ± 7.27 | 35.4 ± 5.15 | 35.0 ± 3.50 |
| % Calories from Alcohol | 9.5 ± 9.43 | 13.2 ± 14.35 | 12.8 ± 14.77 | 11.4 ± 11.38 |
| Ascorbic Acid (mg) | 120.0 ± 39.54 | 105.6 ± 20.33 | 130.0 ± 91.06 | 108.4 ± 46.52 |
| Cholesterol (mg/dl) | 293.5 ± 97.86 | 258.9 ± 79.72 | 294.6 ± 62.95 | 305.0 ± 63.44 |

Table 8. Serum Vitamin C and Lipid Levels of Group A and B Subjects (mg/dl, $\overline{X} \pm$ S.E.M.)

| | Group A | | | | Group B | | | |
|---|---|---|---|---|---|---|---|---|
| | Baseline | Period 1 | Period 2 | Period 3 | Baseline | Period 1 | Period 2 | Period 3 |
| TC | 314.1 ± 10.8 | 309.9 ± 4.7 | 308.1 ± 5.9 | 299.9 ± 6.7 | 352.6 ± 13.0 | 353.9 ± 19.9 | 340.7 ± 15.9 | 338.3 ± 11.2 |
| TG | 93.4 ± 2.8 | 112.2 ± 5.3 | 103.1 ± 4.6 | 100.8 ± 3.6 | 211.6 ± 25.1 | 223.8 ± 25.9 | 212.1 ± 21.4 | 209.9 ± 23.3 |
| PL | 278.2 ± 9.3 | 271.1 ± 6.2 | 291.1 ± 10.1 | 273.8 ± 10.3 | 317.0 ± 10.1 | 303.5 ± 11.5 | 293.8 ± 11.7 | 313.5 ± 8.1 |
| Serum Ascorbate | 1.11 ± 0.14 | 2.05 ± 0.27 | 1.96 ± 0.12 | 1.96 ± 0.23 | 0.47 ± 0.14 | 1.80 ± 0.20 | 1.88 ± 0.35 | 1.83 ± 0.19 |

TC = Total Cholesterol
TG = Triglycerides
PL = Phospholipids
Group A = Type IIa hyperlipoproteinemic subjects
Group B = Type IIb hyperlipoproteinemic subjects

Table 9. Group A: SMA 12/60® Results of Various Serum Analytes of the Subjects During the Vitamin C Study ($\bar{X} \pm$ S.D.)

| | T.P. (g/dl) | Alb. (g/dl) | $Ca^{++}$ (mg/dl) | Inor Phos. (mg/dl) | Chol. (mg/dl) | Glu. (mg/dl) | Uric Acid (mg/dl) | T. Bili. (mg/dl) | Alk. Phos. (U/L) | CPK (U/L) | LDH (U/L) | SGOT (U/L) |
|---|---|---|---|---|---|---|---|---|---|---|---|---|
| Baseline | 7.36 ±.07 | 4.54 ±.11 | 9.8 ±.2 | 3.30 ±.07 | 298 ±11 | 102 ±4 | 7.04 ±.26 | 0.78 ±.23 | 60 ±7 | 128 ±9 | 168 ±12 | 28.6 ±2.9 |
| Period 1 | 7.34 ±.08 | 4.48 ±.05 | 9.5 ±.08 | 3.04 ±.19 | 301 ±39 | 96 ±5 | 6.7 ±.22 | 0.72 ±.22 | 58 ±7 | 112 ±10 | 162 ±8 | 3.0 ±2.7 |
| Period 2 | 7.34 ±.03 | 4.42 ±.12 | 9.48 ±.19 | 3.16 ±.09 | 297 ±20 | 99 ±6 | 7.14 ±.22 | 0.84 ±.29 | 58 ±10 | 130 ±22 | 165 ±11 | 28.0 ±2.5 |
| Period 3 | 7.4 ±.03 | 4.48 ±.09 | 9.58 ±.15 | 3.10 ±.18 | 297 ±16 | 101 ±4 | 6.66 ±.36 | 0.58 ±.09 | 57 ±8 | 121 ±12 | 164 ±10 | 27.5 ±3.2 |

Table 10. Group A: SMA 6/60® Results of Various Serum Analytes and of Thyroxine Levels of the Subjects During the Vitamin C Study ($\overline{X} \pm$ S.D.)

| | $Na^+$ (mEq/L) | $K^+$ (mEq/L) | $Cl^-$ (mEq/L) | $CO_2$ (mEq/L) | BUN (mg/dl) | Creat. (mg/dl) | $T_4$[a] (%) |
|---|---|---|---|---|---|---|---|
| Baseline | 141 ±.5 | 4.56 ±.16 | 100 ±.8 | 28 ±1 | 20 ±3.2 | 1.24 ±.07 | 0.94 ±.02 |
| Period 1 | 143 ±.5 | 4 ±.27 | 106 ±5 | 30 ±.9 | 21 ±3.3 | 1.35 ±.06 | 0.91 ±.01 |
| Period 2 | 142 ±1 | 4.15 ±.17 | 101 ±2 | 28 ±.5 | 22 ±3.5 | 1.33 ±.1 | 0.92 ±.04 |
| Period 3 | 142 ±.8 | 4.36 ±.1 | 102 ±1 | 29 ±.9 | 21 ±3.4 | 1.24 ±.04 | 0.93 ±.02 |

[a] As Free Thyroxine Index

Table 11. Group B: SMA 12/60® Results of Various Serum Analytes of the Subjects During the Vitamin C Study ($\overline{X} \pm$ S.D.)

| | T.P. (g/dl) | Alb. (g/dl) | $Ca^{++}$ (mg/dl) | Inor. Phos. (mg/dl) | Chol. (mg/dl) | Glu. (mg/dl) | Uric Acid (mg/dl) | T. Bili. (mg/dl) | Alk. Phos. (U/L) | CPK (U/L) | LDH (U/L) | SGOT (U/L) |
|---|---|---|---|---|---|---|---|---|---|---|---|---|
| Baseline | 7.34 ±.17 | 4.55 ±.13 | 9.82 ±.11 | 3.46 ±.05 | 296 ±11 | 104 ±4 | 6.64 ±67 | 1.06 ±14 | 68 ±5 | 122 ±36 | 163 ±8 | 27 ±2.5 |
| Period 1 | 7.18 ±.22 | 4.44 ±.81 | 9.5 ±.13 | 3.12 ±.14 | 304 ±5 | 108 ±6 | 6.88 ±.61 | 1.02 ±.24 | 69 ±3 | 145 ±24 | 166 ±12 | 26 ±3.3 |
| Period 2 | 7.36 ±.11 | 4.48 ±.15 | 9.82 ±.25 | 3.28 ±.19 | 300 ±13 | 106 ±8 | 6.64 ±.62 | 1.12 ±.26 | 69 ±3 | 155 ±46 | 165 ±11 | 30 ±2.7 |
| Period 3 | 7.26 ±.15 | 4.5 ±.13 | 9.58 ±.14 | 3.3 ±.09 | 296 ±15 | 106 ±7 | 6.96 ±.41 | .84 ±.18 | 67 ±3 | 122 ±14 | 167 ±7 | 37 ±4.4 |

Table 12. Group B: SMA 6/60® Results of Various Serum Analytes and of Thyroxine Levels of the Subjects During the Vitamin C Study ($\bar{X} \pm$ S.D.)

| | $Na^+$ (mEq/L) | $K^+$ (mEq/L) | $Cl^-$ (mEq/L) | $CO_2$ (mEq/L) | BUN (mg/dl) | Creat. (mg/dl) | $T_4$[a] (%) |
|---|---|---|---|---|---|---|---|
| Baseline | 142 ±.8 | 4.6 ±.09 | 103 ±.4 | 28 ±1.1 | 17 ±1.0 | 1.07 ±3.9 | 0.88 ±.02 |
| Period 1 | 142 ±.6 | 4.68 ±.34 | 102 ±2 | 30 ±1.4 | 19 ±1.9 | 1.30 ±5.1 | 0.91 ±.02 |
| Period 2 | 143 ±1 | 4.78 ±.47 | 103 ±1 | 29 ±.8 | 19 ±2.9 | 1.08 ±6.0 | 0.89 ±.02 |
| Period 3 | 141 ±.7 | 4.58 ±.24 | 101 ±1 | 29 ±1.0 | 18 ±1.8 | 1.17 ±4.6 | 0.92 ±.03 |

[a]As Free Thyroxine Index

may have an adverse effect upon the availability of vitamin $B_{12}$ (Herbert and Jacob, 1974). Whether this proves to be of clinical significance remains to be determined.

By inference if not by established fact, vitamin C is thought to have a number of other functions related to hematology. Blood clotting activity is obviously impaired in scurvy as manifested by petechial hemorrhages and ecchymoses, bleeding gums and hemarthroses. Despite claims that one or another of the blood clotting factors may be impaired by lack of vitamin C other investigators are unable to demonstrate any specific abnormalities of blood clotting in scorbutic subjects although the patients may have had obvious bleeding. It is true, however, that large amounts of ascorbic acid can reverse the anticoagulant effects of either heparin *in vitro* or dicumarol *in vivo* (Signell and Flessa, 1970).

One explanation for the bleeding that occurs in patients with scurvy is lack of "intercellular substances" leading to capillary leakage of blood. Biopsies of skin and subcutaneous tissues from men with scurvy examined by electron-microscopy failed to provide support for this theory (Hodges et al., 1969). Furthermore Wolback's original article describing intercellular substances probably related to collagen fibers that lie between fibroblasts, not "intercellular cement", although he tended to agree with previous workers who coined this term (Wolback and Howe, 1926).

## Sparing Effect on Other Vitamins

Ascorbic acid is known to have a sparing effect on several other vitamins including those of the B complex group, vitamin A and vitamin E. This may be related primarily to its role as an antioxidant.

## Common Cold

Vitamin C has been a popular medication for preventing the common cold and reducing the severity of cold symptoms for many years. Pauling's book on the subject has led to even further use of vitamin C, in even larger doses (Pauling, 1970). Most researchers today find that megadosages of vitamin C appear to have little beneficial effect in terms of cold prevention in normal, healthy individuals eating a balanced meal. Lewis et al. (1975), for example, studied a population of 2,500 employees at the National Institutes of Health and found that 3,000 mg of vitamin C intake did not reduce the incidence of the common cold when compared to the group receiving the placebo. This is in agreement with the double-blind trial findings of Coulehan et al. (1974), Anderson et al. (1972) and Wilson and Loh (1973) that vitamin C does not have a prophylactic effect with doses ranging from 500-2,000 mg per day. Likewise, vitamin C in 1.5 to 3 gram supplements has failed to prevent colds induced by intranasal rhinovirus innoculation (Sasaki et al., 1973; Schwartz et al., 1973). This does not eliminate the possibility that inadequate intake of vitamin C may make a person more susceptible to acquiring the common cold or upper respiratory infections or syndromes. It should also be pointed out that an adequate ascorbate intake for a normal, healthy individual is not an adequate intake for chronically ill patients or females taking oral contraceptive pills or during pregnancy.

The human requirement for vitamin C is increased during pregnancy. It appears to be increased in persons ingesting oral contraceptive steroids. The rationale for increased requirements is based primarily on the findings of a downward trend in plasma ascorbic acid concentrations in successive trimesters of pregnancy and decreased plasma, white cell, and platelet ascorbic acid concentrations in women taking oral contraceptives (Rivers and Devine, 1975).

The 1974 recommended dietary allowances (RDA) for vitamin C is 60 mg/day for pregnant women and 80 mg/day for lactating women. According to the data of Rivers and Devine (1975), this suggested RDA during pregnancy and lactation may be markedly underestimated.

## Stress

During the past decade vitamin C has become the most heavily consumed vitamin, following revivals of the belief in its prophylactic action on various diseases. In stress, vitamin C once held a position of preeminence. Not only is ascorbate abundant in the adrenal cortex, but its concentration fails promptly after administration of adrenal corticotropic hormone (ACTH). It is not known, however, if ascorbate is a necessary component of steroidogenesis. On the contrary, accumulated evidence summarized in man and experimental animals appears to militate against the facilitating role of ascorbate in adrenal steroidogenesis (Kitabchi and West, 1975). This is confusing in light of the fact that vitamin C is found to be involved in the formation of norepinephrine from dopamine and in the conversion of tryptophan to 5-hydroxyptophan which is the first step in the synthesis of serotonin (Thoa and Booker, 1963; Sulkin and Sulkin, 1967; Bhagat et al., 1966; Thoa et al., 1966; Abboud et al., 1970; Cooper, 1961). Since two of the important neurotransmitters are dependent upon ascorbic acid, it is possible that vitamin C may have an important role in the observed fatigue, weakness and vasomotor instability in scorbutic patients. Fuller et al. (1971) have shown that vitamin C deficiency in guinea pigs causes greater susceptibility to endotoxin shock. This was expressed by a high mortality rate in scorbutic animals within the first hour following sublethal dose of endotoxin administration. Morphologic changes in the animals which died from the shock were most severe in the lungs and myocardium.

On the other hand, supplementation of vitamin C intake may not necessarily aid a person in stress. Rats forced to swim in ice water were reported to survive longer if they were given supplements of ascorbic acid. Subsequent studies generally failed to confirm the "anti-stress" actions of ascorbic acid in man. Army troops transported to the top of a mountain where they were subjected not only to high altitude hypoxia, but also to cold and sleeplessness, nonetheless metabolized radioactively labeled ascorbic acid at a normal rate. Other studies showed that athletic performance did not improve following massive doses of vitamin C (Ryer et al., 1954a, 1954b).

Prior to World War II, vitamin C in moderate doses was recommended as a means to temporarily increase the "high-altitude resistance" of pilots, airplane passengers and mountaineers (Petersen, 1941). However, Pfannenstiel (1938) and Doerholt (1938) also reported that larger doses of vitamin C actually diminish the high-altitude resistance. Thus, rabbits that had been given large amounts of sodium ascorbate intravenously (100 mg each day for four days, an additional 200 mg 30 minutes before the experiment) developed convulsions typical of hypoxia with unprecedented violence in low-pressure chambers under conditions that were well tolerated by untreated animals. The same effect was in turn demonstrated with human volunteers.

While it is fairly easy to make generalizations on the role of vitamin C during stress, the dynamics of the body's metabolism is too complicated to evaluate the effects on a singular basis. First of all, the duration, frequency and type of stress are independent components that result in varying effects of stress on vitamin C requirements and metabolism. Secondly, the effects of "stress" should be viewed as a systemic rather than a focal effect. A case and point could be made with vitamin C supplementation for stressed laying hens. The biochemical role of ascorbic acid in avian systems has been for the most part neglected. It is thought that ascorbic acid is metabolized in the same manner as it is in mammalian species. The big difference in ascorbic acid metabolism among mammalian species is seen in the excretion products. Rats and guinea pigs oxidize the ascorbic acid molecule to respiratory $CO_2$, while man excretes urinary oxalates.

It is well-known that stress, as manifested by unfavorable environmental temperature, disease and crowding, causes a decline in egg quality as measured by shell thickness, egg weight and interior protein (Warren and Schnepel, 1940; Wilhelm, 1940).

Dietary supplementation of ascorbic acid to laying hens has produced variable and inconsistent results on hen performance.

Thornton and Moreng (1958, 1959) were the first to show a possible beneficial effect of exogenous ascorbic acid on egg quality. Interior egg quality and shell thickness appeared to be improved. They noticed that the response to ascorbic acid was greater during the summer months. Perek and Kendler (1962, 1963) found ascorbic acid supplementation beneficial on egg production and egg weight, but non-significant in increasing shell thickness. Heywang and Kemmerer (1955), Hunt and Aitken (1962) and Arscott et al. (1962) did not find a beneficial effect in shell thickness or in general egg shell quality with ascorbic acid supplementation.

The effects of ascorbic acid on egg quality may be influenced by the types of diets employed. Thornton (1960a, b) reported an interaction between ascorbic acid and the calcium level on shell thickness, with responses appearing at the 2.5 and 3.0 percent levels in contrast to no response at the 2 percent calcium level. In addition, more consistent improvements have been obtained with rations containing 13 as compared with 17 percent protein.

Damage to the structure of the egg shell caused by climatic stress has received little attention. Mather et al. (1962) concluded from their work on the influence of environmental temperature on the microscopic structure of the egg that the quantity of matrix in the spongiosa was not influenced to the same degree as shell thickness. El-Boushy et al. (1968) studied the effects of heat stress and ascorbic acid supplementation on thickness, quality, structure and ultra-structure of the egg shell. Climatic stress significantly decreased the shell thickness and all its layers except the membranes. The favorable effect of ascorbic acid addition proved to be highly significant only for the thickness of the shell as a whole.

The use of controlled environmental temperature on laying hen performance has been studied. Ahmad et al. (1967) found that ascorbic acid supplementation was helpful in maintaining interior egg quality in White Leghorns as environmental temperatures increased from 21 °C to 29.4 °C and 35 °C. Egg shell thickness was also maintained during the increasing temperature. Lyle and Moreng (1968) also found that egg shell thickness was maintained with ascorbic acid supplementation when temperature was increased to 29 °C.

Nockels et al. (1968) presented data that withdrawal of ascorbic acid after prolonged supplementation depressed interior egg quality in eggs from hens which had been maintained at either optimum or 30.2 °C. Herrick and Nockels (1969) observed a significant increase in interior egg quality due to ascorbic acid supplementation. Eggs stored for two weeks maintained the higher egg quality when compared with the controls.

The influence of rapid changing environmental temperatures on the body temperature control of hens was studied by Thornton (1962). The effect of dietary ascorbic acid on body temperatures was recorded. Body temperature of both control and supplemented hens acclimated at 80 $^{o}$F was increased when the ambient temperature was increased; but the increase was significantly greater in control individuals. A decrease in ambient temperature resulted in a greater fall in body temperature of the controls. These data indicated that ascorbic acid was of aid for maintaining body temperature. Ahmad et al. (1967), Lyle and Moreng (1968), Subaschandran and Balloun (1968) also presented evidence that ascorbic acid supplementation helps maintain body temperature during times of heat stress.

Squibb et al. (1955) studied the effects of the three diseases, coryza, cholera and Newcastle disease, on the level of serum ascorbic acid. Serum ascorbic acid decreased in birds infected with coryza, while it increased in the birds infected with cholera and Newcastle disease. This work presents evidence that ascorbic acid metabolism in chickens is disturbed by stress caused by disease.

Blood ascorbic acid levels in the chicken can be influenced by thyroid regulators (Thornton and Deeb, 1961). Their results showed that dietary iodinated casein caused an increase in blood ascorbic acid while thiouracil resulted in a decreased ascorbic acid level in the blood. It follows that the rate of ascorbic acid synthesis can be influenced by changes in the metabolic rate of the chicken.

Rumsey (1969) imposed a low level stress in hens to evaluate the "anti-stress" properties of ascorbic acid. Laying hens were subjected to simulated stress conditions by injection of ACTH or a synthetic corticosteroid, Dexamethasone. Both exogenous ACTH and Dexamethasone had a depressing effect on kidney ascorbic acid biosynthesis as well as blood ascorbic acid levels. Exogenous ascorbic acid also depressed kidney biosynthesis. Egg weight, egg shell thickness, and egg shell protein were not affected by the stress conditions imposed or by ascorbic acid supplementation.

In summary, it is clear that many factors may influence the ability to regulate ascorbic acid metabolism in avian species. At the present time, supplementation of ascorbic acid is not practiced in poultry feeding, but the evidence indicates that much may be learned by studying the physiological responses of chickens under various conditions when their diets are supplemented with ascorbic acid.

## THE RECOMMENDED DIETARY ALLOWANCES FOR VITAMIN C

Recommended Dietary Allowances are used as the basis for decisions about the adequacy of food supplies, so it is important that the basis for establishing them and their limitations be clearly understood.

Recommended Dietary Allowances are defined as "the levels of intake of essential nutrients considered, in the judgment of the Food and Nutrition Board on the basis of available scientific knowledge, to be adequate to meet the known nutritional needs of practically all healthy persons" (Food and Nutrition Board, 1974). That is, they are recommendations for the quantities of a group of specific chemical compounds that should be consumed by each member of a population in order to provide reasonable assurance that the physiologic needs of all will be met. This is a public health concept. The RDA are nutritional standards. They are recommendations that can be used as the basis for practical decisions. They are standards designed to be used in planning the food supply of a population in order to ensure that the food procured will be nutritionally adequate to maintain the health of that population. They are standards also against which the nutritional adequacy of the available food supply can be assessed in order to identify possible nutrient shortages that could create public health problems.

How are these standards established? The starting point is the quantitative information about human requirement. Two types of information are used for the most part: (1) information from studies on human subjects in which the amount of nutrient required to maintain satisfactory growth and body weight, to prevent depletion of the nutrient from the body, to maintain some specific body function or to prevent the development of unique signs or symptoms has been established; (2) information from dietary surveys indicating how much of the nutrient is consumed by populations that are generally healthy and show none of the signs that are associated with an inadequate intake of the nutrient. From the first of these a figure for the average requirement of the population studied can be obtained; from the second, an estimate of the upper limit of human needs is obtained, but this may not be very precise.

But RDA are not average individual requirements, they must obviously exceed average requirements if they are to cover the needs of those with high requirements. Since individual requirements cannot be predicted, a statistical approach must be used in estimating the amount by which the average requirement must be increased to accomplish this. When information is insufficient to permit a statistical prediction, selection of an appropriate estimate involves judgment.

From information obtained in studies of human subjects, an estimate of variability among the individuals studied can be calculated. Assuming that requirements follow a normal distribution pattern, 97.5% of the population should have requirements below the mean plus twice the coefficient of variation. This amount, the average plus twice the coefficient of variation, provides an estimate for a standard that should cover the needs of most of the population, provided the nutrient is utilized with high efficiency. However, if efficiency of utilization of the ingested nutrient is known to be low, owing for example, to incomplete absorption, the recommended level of intake must be increased further to allow for this. The value obtained after allowing for incomplete absorption should meet or exceed the needs of the vast majority of individuals.

It is important to emphasize that RDA do not take into consideration the losses of nutrients during the processing and preparation of foods, nor are the RDA designed to cover increased needs resulting from severe stress, disease, or trauma above the usual minor infections and stresses of everyday life. The processing losses must be allowed for separately in planning a food supply, and the increased needs represent clinical problems that must be given individual consideration.

Nutritionists agree among themselves about the amounts of each essential nutrient that should be consumed daily. This generalization, however, does not apply to vitamin C or ascorbic acid which continues to be controversial. This controversy centers around two major topics. The first is how much ascorbic acid should be recommended "to meet the known nutritional needs of practically all healthy persons" and the second is what are the effects of giving very large amounts of ascorbic acid?

As a result of this controversy, most persons (and a great many physicians) are confused about the daily allowance of vitamin C needed to support optimal health. In attempting to clarify this, two basic concepts should be acknowledged. First, on the basis of available scientific evidence, the Committee on Dietary Allowances of the National Academy of Sciences National Research Council establishes Recommended Dietary Allowances which in the judgment of its members provide adequate intakes for practically all healthy persons. Secondly, there is a vast difference between "physiologic" amounts of any nutrient and "pharmacologic" doses which may be given for various medically indicated reasons. The lowest range, representing the amounts needed or recommended to prevent signs or symptoms of deficiency in practically all healthy persons can be termed the "physiologic dose" (the RDA level, 45 mg of vitamin C/day). Somewhat larger amounts (100-2000 mg/day), generally in the range of ten times the physiologic dose, may be used to treat an illness or condition quite unrelated to the generally recognized manifestations of deficiency of that nutrient. This level of intake can be termed "pharmacologic dose". Very large intake of an essential nutrient, for example 100 times (or more) the physiologic dose, may induce undesirable or toxic signs and symptoms, and can be called "toxic dose" (2000-4000 mg/day). It is not uncommon for pharmacologic doses to cause deleterious side reactions (Barnes, 1975).

With the understanding of the above concepts, it is safe to recommend the following levels for vitamin C intake. As little as 10 mg ascorbic acid will prevent scurvy. This level may be regarded as a minimum requirement but it does not ensure fully satisfactory tissue levels. Each day the adult male removes about 30 mg ascorbic acid from body stores. The recommended allowance has been set at 45 mg for males and females over 11 years, 35 mg for infants, 40 mg for children, and 60 mg for pregnancy.

During infections such as tuberculosis, rheumatic fever, and pneumonia and severe stress such as burn injuries the ascorbic acid recommended allowance should be increased.

Fruits and vegetables are excellent sources of vitamin C. Table 13 illustrates the vitamin C content in some selected foods. Like other water soluble vitamins, ascorbate is labile to heat and oxygen. Thus, fresh fruits and vegetables, including juices, are excellent dietary sources in meeting RDA.

Since vegetable cells contains an enzyme called ascorbic oxidase, cutting vegetables finely causes a greater quantity of release of this enzyme which causes more breakdown of vitamin C. The rate of ascorbic oxidase activity increases as the temperature is raised; thus, gradual heating of vegetables destroys ascorbate. Boiling causes a breakdown of the enzyme. The loss of vitamin C due to ascorbic oxidase can be minimized by immersing the vegetables directly in boiling water which causes immediate destruction of the enzyme. Vitamin C is more readily destroyed than other vitamins. During cooking, ascorbate loss occurs as a result of (1) cutting the vegetables finely; (2) using an excess of water which leaches the vitamin C out of the cells; (3) gradually heating the vegetables rather than putting them in boiling water; (4) overcooking; (5) adding sodium bicarbonate in order to preserve the color of the vegetables as alkalinity destroys ascorbate.

## TOXIC EFFECTS OF LARGE VITAMIN C INTAKE

### Gastrointestinal Disturbances

Gastrointestinal disturbances are perhaps the most consistent abnormalities noted following the ingestion of large quantities of ascorbic acid. Nausea, abdominal cramps, and diarrhea are

Table 13. Vitamin C Content in Some Selected Foods

| Item | Vitamin C (mg/100 g edible portion) |
|---|---|
| Acerola (Barbados Cherry) | 1,300 |
| Apples (raw) | 4 |
| Apricots (raw) | 10 |
| Avocados (raw) | 14 |
| Bananas (raw) | 10 |
| Beans, Lima (cooked) | 17 |
| Beans, snap (cooked) | 12 |
| Beef | 0 |
| Broccoli (cooked) | 90 |
| Cabbage (cooked) | 24 |
| Carrots (raw) | 8 |
| Collards (cooked) | 46 |
| Corn, sweet (cooked) | 9 |
| Currants, Black European (raw) | 200 |
| Eggs | 0 |
| Eggplant (cooked) | 3 |
| Fish | 0 |
| Grapefruit, pulp | 38 |
| Grapefruit, juice unsweetened | 34 |
| Lemon, Juice | 45 |
| Lettuce (raw) | 15 |
| Milk, cow | 1 |
| Muskmelons (raw) | 33 |
| Oranges (raw) | 55 |
| Oranges, Juice (canned) | 40 |
| Peaches (raw) | 7 |
| Pears (raw) | 4 |
| Peas, green (cooked) | 20 |
| Pineapple (raw) | 17 |
| Pineapple (juice) | 9 |
| Pork | 0 |
| Poultry | 0 |
| Potato (cooked) | 20 |
| Spinach (raw) | 51 |
| Spinach (cooked) | 28 |
| Summer squash (cooked) | 10 |
| Strawberries (raw) | 59 |
| Tomato, ripe (raw) | 23 |
| Tomato juice | 16 |
| Wheat flours | 0 |

[a]from Watt and Merrill (1963)

frequently mentioned. These may be due to the ascorbic acid itself or to sensitization reactions such as hives, angioneurotic edema, and skin rashes, which have also been reported in some of these patients. These effects may be ameliorated or eliminated by taking the ascorbic acid as a buffered salt or after meals.

### Renal Stone Formation

One of the feared consequences of excessive ascorbate intake is the possible development of renal stones or nephrocalcinosis. Stone formation may be accelerated with the administration of large doses of ascorbate. In those who are prone to renal stone formation, stones may form more easily. This includes persons with oxalosis, hyperuricemia, and cystinuria.

### Oxaluria-Oxalosis

Studies on the sources of oxalic acid indicate that in the normal subject approximately 50% of the daily excretion of oxalic acid, about 50 mg, is derived from dietary ascorbate. The source of the other 50 mg varies. According to Lewin (1973), the total intake of vitamin C should be kept below 4 grams daily because of the increasingly unacceptable levels of oxalate ions that are likely to accompany the high ascorbate intake.

## VITAMIN C DEFICIENCY

### Scurvy

Scurvy is produced by prolonged deficiency of vitamin C. Human volunteers deprived of vitamin C showed no clinical manifestations for 17 weeks. The first signs noted are hyperkeratotic hair follicles from 17 to 21 weeks, followed by perifollicular hemorrhages at 26-34 weeks (Medical Research Council, 1948). Since, in ordinary circumstances, an infant or adult is not completely deprived of vitamin C, it can be assumed that at least 6-9 months would elapse before clinical scurvy appears even with gross deficiency. The basic pathological change in scurvy is defective formation of collagen. The lining of blood vessels is defective because the cells are not cemented together. The bony matrix or framework is poorly formed for the same reasons.

### Subclinical Scurvy

Before the development of clinical signs of scurvy, vague general symptoms such as lassitude, fatigue, weakness, irritability, a tendency to recurrent infections and aching of bones may be noted (Hodges, 1969).

Clinical Scurvy

Clinical symptoms appear when the body pool of vitamin C decreases to about 300 mg (Hodges, 1969). Clinical scurvy is manifest by changes in: (1) blood vessels; (2) skin; (3) bones; (4) teeth and gums; (5) anemia; and (6) general symptoms.

Blood vessels. The blood vessels become fragile and porous due to defective formation of collagen in the lining of the wall. This leads to spongy gums, hemorrhages into the skin, seen as petechiae and ecchymoses, and hemorrhages into the alimentary and urinary tract. Conjunctival hemorrhages occur by 74-95 days in experimental human scurvy. Muscular and subperiosteal hemorrhages may also occur.

Skin. The skin becomes rough and dry. Early hyperkeratotic changes are seen in the hair follicles, and are most marked on legs and buttocks. The occurrence of petechial hemorrhages around hair follicles and larger hemorrhages, ecchymoses, have already been referred to.

Bones. During normal growth, cartilage cells at the growing end of bones proliferate. These are invaded by osteoblasts to form the bony matrix, also known as osteoid tissue. Ossification occurs when calcium salts are deposited in this matrix, forming bone.

In the absence of vitamin C the formation of bony matrix (osteoid tissue) is defective. Instead of osteoblasts, fibroblasts develop and bone formation is retarded. This gives rise to osteoporosis and spontaneous fractures. The periosteum is not attached firmly to the bone and subperiosteal hemorrhages occur, usually at the lower end of the femur and upper end of the humerus. These bony changes are found in infants and children with scurvy.

Teeth and Gums. Deficiency of vitamin C in children leads to defective formation of the teeth. The dentine becomes porous, the alveolar bone is absorbed and the teeth may fall out. The gums become spongy and bleed easily if slight pressure is applied. Hypertrophy of the gums tends to bury the teeth and infection and ulceration of the unhealthy gums is common. These changes in the gums may not be seen in an edentulous subject.

Anemia. Microcytic hypochromic anemia is a common feature in scurvy. The anemia is partly due to deficient absorption of iron, with a low intake of vitamin C and partly due to blood loss from hemorrhages. Deficiency of vitamin C also disturbs folic acid metabolism and a megaloblastic anemia may develop.

General Manifestations. Scurvy is manifested by pyrexia, rapid pulse and susceptibility to infection. Wound healing is delayed.

Aspirin is more likely to produce gastric mucosal bleeding in a vitamin C-deficient subject.

METHODS OF ASSESSING VITAMIN C ADEQUACY

A person's nutritional status, including adequacy of vitamin C, may be determined on the basis of clinical signs of a dietary deprivation, biochemical measurements, or dietary intake findings. Information on the dietary intakes of ascorbic acid can be useful in the general assessment and prediction of the nutritional status of a population, but usually such data cannot provide the desired direct knowledge as to the vitamin C nutritional status of the individuals. Even the presence of clinical signs associated with a vitamin C deficiency may not always be a reliable indicator of a deficiency of the nutrient, since such signs may be the result of local infections or improper oral hygiene. Therefore, for the most part, biochemical measurements represent the most objective assessment of the vitamin C nutritional status of an individual.

Healthy adult men have an average body pool of 1.5 grams of ascorbic acid. In experimental vitamin C deficiency, scurvy developed when the body pool of ascorbic acid had been depleted to a level of 300 mg or less (Baker et al., 1971). Occurred in the plasma and whole blood ascorbate levels. The fall in plasma ascorbic acid appears to be more pronounced than the reduction in the ascorbate body pool. Consequently, a low plasma ascorbate level may not necessarily indicate scurvy, but scurvy will invariably ensue if the absence of ascorbic acid in the serum persists. Moreover, continued serum ascorbic acid levels of less than 0.10 mg/dl would probably eventually lead to signs of scurvy. Such was observed in the study of Hodges et al. (1971) where the first signs of scurvy appeared when the serum ascorbic acid levels ranged from 0.13 to 0.24 mg/dl. There was also a definite relationship between the body reserves of ascorbic acid and whole-blood ascorbate levels. Signs of scurvy were observed when the whole-blood ascorbate level fell below 0.3 mg/dl. At this point, the body pool of ascorbic acid had fallen from an initial normal level of approximately 1500 mg to a pool size ranging from 96 to 490 mg.

As noted previously, serum (or plasma) ascorbate levels may not always fully reflect vitamin C intake or the state of the body ascorbate reserves. Nevertheless, within a relatively limited range, serum levels of ascorbic acid show a linear relationship with the intake of vitamin C. With a given intake of the vitamin, the serum ascorbic acid concentration will plateau at a given level, around 1.4 mg/dl. We have observed this when subjects were given two doses of 500 mg of vitamin C, one in the morning and one in the evening. When serum vitamin C levels fall below 0.2 mg/dl, signs of scurvy are commonly observed. With an intake of 45 g of vitamin C per day, representing the 1974 Recommended Dietary Allowance, a serum ascorbate level of approximately 0.6-0.8 mg/dl could be expected. Higher serum ascorbate levels may be obtained with greater intakes of vitamin C. However, the maximal serum ascorbic acid level appears to be at about 1.4 mg/dl, at which point renal clearance of the vitamin rises sharply. The current practice of some people to ingest high doses of vitamin C can result in the temporary attainment of higher serum ascorbic acid concentrations.

The Interdepartmental Committee on Nutrition for National Defense (ICNND) has considered serum ascorbic acid levels of less than 0.10 mg/dl "deficient" and levels from 0.10 to 0.19 mg/dl as "low", while levels of 0.20 mg/dl and above are termed "acceptable" (Manual for Nutrition Surveys, 1963). In the nutrition survey of Canada (Nutrition Canada National Survey, 1973) subjects with serum vitamin C levels of less than 0.2 mg/dl were considered at "high risk". Levels of 0.6 mg/dl to 0.4 mg/dl were considered necessary for subjects 19 years of age or younger and for subjects 20 years and above, respectively, to be considered at "low risk" with respect to vitamin C nutriture. No differences were considered with respect to the sex of the subjects.

The results of the studies of Hodges et al. (1971) indicated that signs of scurvy may be encountered in subjects with serum ascorbate levels below 0.20 mg/dl and would support the guidelines employed in the Canadian nutrition survey. Based on the above data, a revision of the guidelines interpreting serum vitamin C data has been suggested by Sauberlich (1975) as seen in Table 14.

Leukocyte ascorbic acid concentrations have been considered to be more closely related to tissue stores of the vitamin than

Table 14. Guidelines for the Interpretation of Vitamin C Levels (mg/dl)[a]

| Measurement | Acceptable | Low | Deficient |
|---|---|---|---|
| Whole Blood Vitamin C | ≥0.50 | 0.30-0.49 | <0.30 |
| Serum Vitamin C | ≥20 | 0.10-0.19 | <0.10 |
| Leukocyte Vitamin C | ≥15 | 8-15 | 0-7 |

[a]Adapted from Sauberlich, 1975.

serum concentration (Sauberlich, 1975, Sauberlich et al., 1973, 1974). Others have suggested that leukocyte ascorbic acid concentrations do not reliably reflect tissue status and that both serum and leukocyte ascorbate concentrations should be used to do this (Loh, 1972, Keith and Pelletier, 1974). In general, serum vitamin C levels tend to respond more readily than leukocytes to recent dietary intake of ascorbic acid. However, low serum ascorbate concentrations do indicate low or inadequate intakes of vitamin C with probably only partial reserves present. Nevertheless, under controlled intakes of ascorbic acid, there is a relationship between ascorbate intake and ascorbate levels in both the leukocytes and serum. With an intake of 45 mg of vitamin C per day (RDA), leukocytes will contain approximately 20 mg/dl. Subjects with leukocyte ascorbate levels of less than 8 mg/dl are considered to be at "high risk". Concentrations may fall to zero in clinical cases of scurvy (Sauberlich, 1975).

Although numerous procedures and modifications have been reported for the determination of vitamin C in white blood cells, the methods must still be considered technically difficult and require relatively large blood samples. Consequently, present procedures are not practical for routine use in nutrition surveys. With adequate laboratory facilities, measurement of leukocyte ascorbate levels can be a useful diagnostic technique for clinical cases of vitamin C deficiency. To some extent measurement of serum ascorbate levels have been favored over measurement in white blood cells because of the greater percentage increase in serum ascorbate concentrations in response to graded intakes of vitamin C. Such factors, along with the relative ease of measurement, have led to the common use of serum ascorbate determinations for assessing vitamin C nutritional status. This approach was used in national nutrition surveys of Canada and the United States.

Urinary vitamin C excretion data have received only limited use in determining the nutritional status of this vitamin. As a result, guides for the interpretation of urinary vitamin C levels in terms of nutritional status have not been established. Analytical procedures for measuring ascorbic acid in urine are more difficult and less reliable than those used for measuring the vitamin in serum. Nevertheless, in the scorbutic patient, the urinary excretion of vitamin C would be expected to be essentially zero and hence, could provide supportive diagnostic information.

On occasion, ascorbic acid saturation tests are used to obtain information as to tissue ascorbate deficit in individual patients (Sauberlich et al., 1973, 1974; Dutra DeOliveira et al., 1959; Lowry, 1952). For nutrition surveys such tests are not practical and are probably of little use. At best, the saturation or loading test must be conducted with considerable care and the results interpreted with caution. However, for clinical cases the test may be of conclusive value in excluding scurvy as a diagnosis and of considerable usefulness in obtaining the diagnosis of scurvy.

The various analytical procedures used to measure ascorbic acid are summarized by Sauberlich (1975). The method of Roe and Kuether (1943) has generally been considered the "classical technique" for vitamin C quantitation.

## SUMMARY

In summary, it is obvious that considerable interest and research has developed in the area of ascorbic acid metabolism. The biochemist has contributed much information concerning the identification and detailed study of enzyme systems and related nutrients on metabolites that control the synthesis, functions and end products of vitamin C in living organisms. Further studies are needed on the role of antihistamine effect of ascorbic acid, vitamin C and bile acid metabolism, vitamin C and the common cold, ascorbic acid and bioenergetics, vitamin C and prevention of carcinoma, ascorbate and ascorbate-$SO_4$ metabolism.

Furthermore, reports on the effects of massive intakes of vitamin C have raised many questions and violently different opinions. Most of the claims for dramatic benefit from very high intakes have been based on resistance to or therapy toward the common cold and on quantities very far above the amount apparently needed for normal reproduction, lactation, early growth, and longevity. Very few physicians with extensive experience in this type of research recommend the practice of using large dosages, such as 1 to 20 g per day.

Hence, we have a situation where many experienced nutrition scientists trained as biochemists or physicians and in other specialized branches of science are not in agreement on either of two major issues:

a. Should the quantities recommended for daily consumption of vitamin C be those adopted in the table of Recommended Dietary Allowances by the National Academy of Sciences? Differences in opinion on this quantity are not very great -- seldom more than a factor of twice the values now used.
b. What should be the recommended intake of vitamin C with special reference to prevention or therapy of the common cold or other disease risks to which the public is exposed?

The Food and Nutrition Board has adopted a consensus of views among well-trained clinical biochemists and physicians in arriving at the recommended daily intake, which ranges from 35 to 80 mg per day, depending on age and including an extra allowance for lactation. There is an increasing concern that the recommendation for growing children and adults should be higher than the present stated values of 40 to 45 mg per day, perhaps in the range of 70 to 75 mg, and 100 mg during lactation.

The area of most intense controversy currently is in regard to the quantity that should be recommended for consumption by the public without further consideration when there is an indication of an oncoming cold. Quantities in the range of 5 to 10 g or more per day are sometimes recommended by laymen and some scientists. A few enthusiastic laymen and some physicians suggest practices such as a regular intake in the range of 1 g or more each day. Claims are made that these very large intakes are harmless. Others with excellent research training and much experience in medicine, public health, and biochemistry believe that such large intakes have little if any significant value and entail significant hazards to health, such as development of kidney and bladder stones, and alterations in carbohydrate metabolism. Thus far there has been little evidence to indicate a mechanism to account for the claims made in supporting the high dosages, and very little research has been reported to compare the value of vitamin C with other potential drugs such as antihistamines. Thus, until these questions are clarified by further studies, a conservative approach should be taken in terms of recommending large amounts of vitamin C.

REFERENCES

Abboud, F. M., Hood, J., Hodges, R.E., and Mayer, H. E., 1970, Antonomic reflexes and vascular reactivity in experimental scurvy in man, J. Clin. Invest. 49:298-307.

Ahmad, M. M., Moreng, R. E., and Muller, H. D., 1967, Bread responses in body temperature and ascorbic acid, Poultry Sci. 46:6-15.

Anderson, T. W., Reid, D. B. W., Beaton, G. H., 1972, Vitamin C and the common cold: A double-blind trial, Can. Med. Assoc. J. 107:503.

Arscott, G. H., Rachapaetayakom, P., Bernier, P. E., and Adams, F.W., 1962, Influence of ascorbic acid, calcium and phosphorus on specific gravity of eggs, Poultry Sci. 41:485.

Baker, E., Hammer, D., March, S., Tolbert, B. and Canham, J., 1971, Ascorbate sulfate: A urinary metabolite of ascorbic acid in man, Science 173:826.

Baker, E. M., Hodges, R. E., Hood, J., Sauberlich, H.E., March, S. C., and Canham, J. E., 1971, Metabolism of $^{14}$C- and $^{3}$H-labeled L-ascorbic acid in human scurvy, Amer. J. Clin. Nutr. 24:444.

Banay, M. and Dimant, E., 1962, On the metabolism of L-ascorbic acid in the scorbutic guinea pig, Biochim. Biophys. Acta 59:313.

Banerjee, S. and Singh, H. D., 1958, Cholesterol metabolism in scorbutic guinea pigs, J. Biol. Chem. 233:336.

Barnes, M. J., 1975, Function of ascorbic acid in collagen metabolism, Ann. N. Y. Acad. Sci. 258:264.

Barnes, M. J. and Kodicek, E., 1972, Biological hydroxylation and ascorbic acid with special regard to collagen metabolism, Vit. Horm. 30:1.

Barness, L. A., 1975, Safety considerations with high ascorbic acid dosage, Ann. N. Y. Acad. Sci. 258:523.

Bhagat, B., West, W. L., and Robinson, I. M., 1966, Sensitivity to norepinephrine of isolated atria from scorbutic guinea pigs, Biochem. Pharmacol. 15:1637-1639.

Bond, A.D., 1975, Ascorbic-2-sulfate metabolism by human fibroblasts, Ann. N. Y. Acad. Sci. 258:307.

Burns, J. J., 1957, Missing step in man, monkey and guinea pig required for the biosynthesis of L-ascorbic acid, Nature 180:553.

Chatterjee, I. B., Majumder, A. K., Nandi, B. K., and Subramanian, N., 1975, Synthesis and some major functions of vitamin C in animals, Ann. N. Y. Acad. Sci. 258:24.

Chaudhuri, C. R. and Chatterjee, I. B., 1969, L-ascorbic acid synthesis in birds: Phylogenetic trend, Science 166:435.

Church, C. F. and Church, N. H. (eds.), 1977, "Bowes and Church's Food Values or Portions Commonly Used," 12th ed., J. B. Lippincott, Philadelphia, Pennsylvania.

Cooper, J. R., 1961, The role of ascorbic acid in the oxidation of tryptophan to 5-hydroxytryptophan, Ann. N. Y. Acad. Sci. 92:208-211.

Coulehan, J. L., Reisinger, K. S., Rogers, K. D., and Bradley, D. W., 1974, Vitamin C prophylaxis in a boarding school, New Engl. J. Med. 290:6.

Doerholt, G., 1938, Tierexperimentelle untersuchungen über den einfluss des vitamin C auf die höhenfestigkeit, Luftfahrtmediz. Abhandl. 2(3/4):240.

Feeley, R. M., Criner, P. E., and Watt, B. K., 1972, Cholesterol content of foods, J. Am. Dietet. Assoc. 61:134.

Food and Nutrition Board, 1974, Recommended Dietary Allowances, 8th ed., National Academy of Sciences, Washington, D.C.

Fuller, R. N., Henson, E. C., Shannon, E. L., Collins, A. D., and Brunson, J. G., 1971, Vitamin C deficiency and susceptibility to endotoxin shock in guinea pigs, Arch. Path. 92:239.

Ginter, E., 1975, Ascorbic acid in cholesterol and bile acid metabolism, Ann. N. Y. Acad. Sci. 258:410.

Ginter, E. and Ondreicka, R., 1971, Liver cholesterol esters composition in guinea pigs with chronic ascorbic acid deficiency, Nutr. Metab. 13:321.

Ginter, E., Kajaba, I., and Nizner, O., 1970, The effect of ascorbic acid on cholesterolemia in healthy subjects with seasonal deficit on vitamin C, Nutr. Metab. 12:76.

Ginter, E., Cerven, J., Nemec, R., and Mikus, L., 1971, Lowered cholesterol catabolism in guinea pigs with chronic ascorbic acid deficiency, Am. J. Clin. Nutr. 24:1238.

Ginter, E., Bobek, P., Kopec, Z., Ovecka, M., and Cerey, K., 1967, Metabolic disorders in guinea pigs with chronic vitamin C hyposaturation, Versuchskierk 9:228.

Gore, I., and Stefanovic, V., 1967, The relation of permeability of rabbit aorta to dietary induced lipid accumulation, Fed. Proc. 26:431.

Herbert, V., and Jacob, E., 1974, Destruction of vitamin $B_{12}$ by ascorbic acid, J. Amer. Med. Assoc. 230:241.

Herrick, R. B., and Nockels, C. F., 1969, Effects of a high level of dietary ascorbic acid on egg quality, Poultry Sci. 48:1518-1519.

Heywang, B. W., and Kemmerer, A. R., 1955, The effects of procaine penicillin and ascorbic acid on egg weight and shell thickness during hot weather, Poultry Sci. 34:1032.

Hodges, R. E., 1976, Ascorbic acid, in "Recent Knowledge in Nutrition," The Nutrition Foundation, Inc., Washington, D.C.

Hodges, R. E., Baker, E. M., Hood, J., Sauberlich, H. E., and March, S. C., 1969, Experimental scurvy in man, Amer. J. Clin. Nutr. 22:535.

Hodges, R. E., Hood, J., Canham, J. E., Sauberlich, H. E., and Baker, E. M., 1971, Clinical manifestations of ascorbic acid deficiency in man, Amer. J. Clin. Nutri. 24:432.

Hunt, J. R., and Aitken, J. R., 1962, Studies on the influence of ascorbic acid on shell quality, Poultry Sci. 41:219.

Kagawa, Y., 1962, Enzymatic studies on ascorbic acid catabolism in animals, I. Catabolism of 2,3-diketo-L-gulonic acid, J. Biochem. 51:134.

Kagawa, Y., Takaguchi, H., 1962, Enzymatic studies on ascorbic acid catabolism in animals, II. Delactonization of dehydro-L-ascorbic, J. Biochem. 51:197.

Kagawa, Y., Takaguchi, H., and Shimazono, N., 1961, Enzymatic delactonization of dehydro-L-ascorbate in animal tissues, Biochim. Biophys. Acta 51:413.

Keith, M. O., Pelletier, O., 1974, Ascorbic acid concentrations in leukocytes and selected organs of guinea pigs in response to increasing ascorbic acid intake, Amer. J. Clin. Nutr. 27:368.

Kitabchi, A. E. and West, W. H., 1975, Effect of steroidogenesis on ascorbic acid content and uptake in isolated adrenal cells, Ann. N. Y. Acad. Sci. 258:422.

Lewin, S., 1973, Evaluation of potential effects of high intake of ascorbic acid, Comp. Biochem. Physiol. 46:427.

Lewis, T. L., Karlowski, T. R., Kapikian, A. Z., Lynch, J. M., Shaffer, G. W., and George, D. A., 1975, A controlled clinical trial of ascorbic acid for the common cold, Am. N. Y. Acad. Sci. 258:505.

Loh, H. S., 1972, The relationship between dietary ascorbic acid intake and buffy coat and plasma ascorbic acid concentrations at different ages, Int. J. Vit. Res. 42:80.

Lowry, O. H., 1952, Biochemical evidence of nutritional status, Physiol. Rev. 32:431.

Lyle, G. R., and Moreng, R. E., 1968, Elevated environmental temperature and duration of post exposure ascorbic acid administration, Poultry Sci. 46:410-416.

Manual for Nutrition Surveys, 1963, (2nd ed.), Interdepartmental Committee on Nutrition for National Defense, Superintendent of Documents, U. S. Government Printing Office, Washington, D.C.

Mather, F. B., Epling, G. P., and Thornton, P. A., 1962, The microscopic structure of the egg shell matrix as influenced by shell thickness and environmental temperature, Poultry Sci. 4:191-200.

Medical Research Council, 1948, Vitamin C requirement of human adults - Experimental study of vitamin C deprivation in man, A preliminary report by the Vitamin C Subcommittee of the Accessory Food Factors Committee, Lancet i:853.

Naito, H. K., 1981, Nutritional Modification for the prevention and treatment of hyperlipidemia and hyperlipoproteinemia, in "Cardiovascular Disease and Nutrition," Spectrum Publications, New York.

Naito, H. K., and Gerrity, R. G., 1979, Unusual resistance of the ground squirrel to the development of dietary-induced hypercholesterolemia and atherosclerosis, Exptl. Mol. Path. 31:452.

Nockels, C. F., Herrick, R. B., and Shutze, J. V., 1968, Effects of ascorbic acid withdrawal on interior egg quality, Poultry Sci. 47:1702.

Pauling, L., 1970, "Vitamin C and the Common Cold," W. H. Freeman and Co., San Francisco, California.

Pedersen, J. M., 1941, Ascorbic acid and resistance to low oxygen tension, Nature 148:84.

Perek, M. and Kendler, J., 1962, Vitamin C supplementation to hens' diets in a hot climate, Poultry Sci. 41:677-678.

Perek, M. and Kendler, J., 1963, Ascorbic acid as a dietary supplement for White Leghorn hens under condition of climatic stress, British Poultry Sci. 4:191-200.

Peterson, V. E., Ceapo, P. A., Weininger, J., Ginsbers, G., and Olefsky, J., 1975, Quantification of plasma cholesterol and triglyceride levels in hypercholesterolemic subjects receiving ascorbic acid supplements, Am. J. Clin. Nutr. 28:584.

Pfannenstiel, W., 1938, Tierversuche über die vitaminbeeinflussbarkeit der Höhenfestigkeit, Luftfahrtmediaz. Abhandl. 2(3/4):234.

Rivers, J. M., and Devine, M. M., 1975, Relationships of ascorbic acid to pregnancy, and oral contraceptive steroids, Am. N. Y. Acad. Sci. 258:465.

Roe, J. H., and Kuether, C. A., 1943, The determination of ascorbic acid in whole blood and urine through the 2,4-Dinitrophenylhydrazine derivative of dehydroascorbic acid, J. Biol. Chem. 147:399.

Roy, R. N., and Guha, B. C., 1958, Species difference in regard to the biosynthesis of ascorbic acid, Nature 182:319-320.

Rumsey, G. L., 1969, Effects of stress upon egg shell quality in the laying hen, Proceedings, Cornell Nutrition Conference, pp. 103-109.

Ryer, III, R., Grossman, M. I., Friedemann, T. E., Best, W. R., Consolazio, C. F., Kuhl, W. J., Insull, Jr., W., and Hatch, F. T., 1954a, The effect of vitamin supplementation on soldiers residing in a cold environment, Part I, Physical performance and response to cold exposure, J. Clin. Nutrition 2:97.

Ryer, III, R., Grossman, M. I., Friedemann, T. E., Best, W. R., Consolazio, C. F., Kuhl, W. J., Insull, Jr., W., and Hatch, F. T., 1954b, The effect of vitamin supplementation on soldiers residing in a cold environment, Part II, Psychological, biochemical and other measurements, J. Clin. Nutrition 2:179.

Sasaki, Y., Togo, Y., Wagner, Jr., H. N., Hornick, R. B., Schwartz, A. R., and Procter, D. F., 1973, Mucociliary function during experimentally induced rhinovirus infection in man, Ann. Otol. 82:203.

Sauberlich, H. E., Skala, J. H. and Dowdy, R. P., 1973, Laboratory Tests for the assessment of nutritional status, in "CRC Critical Reviews in Clinical Laboratory Sciences," CRC Press, Inc., Cleveland, Ohio.

Sauberlich, H. E., Skala, J. H. and Dowdy, R. P., 1974, Laboratory Tests for the assessment of nutritional status, in "CRC Critical Reviews in Clinical Laboratory Sciences," CRC Press, Inc., Cleveland, Ohio.

Signell, L. T., and Flessa, H. C., 1970, Drug interactions with anticoagulants, J. Amer. Med. Assoc. 214:2035.

Sokoloff, B., Hori, M., Saelhof, C., McConnell, B., and Imai, T., 1967, Effect of ascorbic acid on certain blood fat metabolism factors in animals and man, J. Nutr. 91:107.

Spittle, C. R., 1972, Atherosclerosis and vitamin C, Lancet 2:1280.

Squibb, R. L., Braham, J. E., Guzmamand, M. and Scrimshaw, N., 1955, Blood serum, total proteins, riboflavin, ascorbic acid, carotenoids and vitamin A of New Hampshire chickens infected with coryza, cholera or Newcastle disease, Poultry Sci. 34:1054-1058.

Stokes, P. L., Melikian, V., Leeming, R. L., Graham-Portman, H., Blair, J. A., and Cooke, W. T., 1975, Folate metabolism in scurvy, Amer. J. Clin. Nutr. 28:126.

Subaschandran, D. V., and Balloun, S. L., 1968, Acetyl-p-aminophenol and vitamin C in heat stressed birds, Poultry Sci. 46:1073.

Sulkin, D. F., and Sulkin, N. M., 1967, An electron microscopic study of antonomic ganglion cells of guinea pigs during ascorbic acid deficiency and partial inanition, Lab. Invest. 16:142-152.

Technicon AutoAnalyzer 12/60 and 6/60, 1974, Technicon Instruments Corp., Tarrytown, New York.

Ten State Nutrition Survey, 1968-1970. IV, 1972, U. S. Department of Health, Education, and Welfare, Center for Disease Control, Atlanta, Georgia, Publ. No. (HSM) 72-8132.

Thoa, N. B., and Booker, W. M., 1963, Cardiovascular Dynamics in normal and scorbutic guinea pigs infused with dopamine and norepinephrine, Fed. Proc. 22:448.

Thoa, N. B., Wurtman, R. J., and Axelrod, J., 1966, A deficient binding mechanism for norepinephrine in hearts of scorbutic guinea pigs, Proc. Soc. Exp. Biol. Med. 121:267-270.

Thornton, P. A., 1960a, The effects of dietary calcium level on the efficiency of ascorbic acid in maintenance of egg shell thickness at increased environmental temperatures, Poultry Sci. 40:1401-1406.

Thornton, P. A., 1960b, Tyrosine, protein and ascorbic acid effects on egg shell thickness from chickens subjected to heat stress, Poultry Sci. 40:1832-1835.

Thornton, P. A., 1962, The effects of environmental temperature on body temperature and oxygen uptake by the chicken, Poultry Sci. 41:1053-1060.

Thornton, P. A., and Deeb, S. S., 1961, The influence of thyroid regulators on blood ascorbic acid levels in the chicken, Poultry Sci. 41:1063-1067.

Thornton, P. A., and Moreng, R. E., 1958, The effects of ascorbic acid on egg quality factors, Poultry Sci, 37:691-698.

Thornton, P. A., and Moreng, R. E., 1959, Further evidence on the value of ascorbic acid for maintenance of shell quality in warm environmental temperature, Poultry Sci. 38:594-599.

Tolbert, B. M., Chen, A., Bell, E., and Baker, E. M., 1967, Metabolism of ascorbic-4-$^{3}$H acid in man, J. Clin. Nutr. 20:250.

Warren, D. C., and Schnepel, R. L., 1940, The effect of air temperature on egg shell thickness in the fowl, Poultry Sci. 19:67-72.

Watt, B. K., and Merrill, A. L., 1963, Composition of Foods: Agriculture Handbook, No. 8, U. S. Government Printing Office, Washington, D. C.

Wilhelm, L. A., 1940, Some factors affecting variations in egg shell quality, Poultry Sci. 19:246-253.

Willis, G. C., 1953, An experimental study of the intimal ground substance in atherosclerosis, Can. Med. Assoc. J. 69:17.

Willis, G. C., and Fishman, S., 1955, Ascorbic acid content of human arterial tissue, Can. Med. Assoc. J. 72:500.

Wilson, C. W. M. and Loh, H. S., 1973, Common cold and vitamin C, Lancet 1:638.

Wolback, S. B. and Howe, P. R., 1926, Intercellular substances in experimental scorbutus, Arch. Path. 1:1.

ACKNOWLEDGEMENTS

This study was supported by Grant No. CRP-400 from the Cleveland Clinic Foundation, Grant No. HL-6835 from NHLBI, and Grant No. 8557 from the Bleeksma Fund of the Cleveland Clinic Foundation. I am grateful to Dr. Helen B. Brown, Dr. Victor G. deWolfe, Ms. Karen Wilcoxen and Ms. Ingrid Raulinaitis for their assistance on this project. I would like to express my gratitude to Helen Brewster for typing this manuscript.

# VITAMIN A AND RETINOL BINDING PROTEIN ALTERATIONS IN DISEASE

Raymond J. Shamberger, Ph.D.
Department of Biochemistry
The Cleveland Clinic Foundation
9500 Euclid Avenue
Cleveland, Ohio 44106

## VITAMIN A

### Historical Developments

A lipid-soluble compound which is essential for life and present in egg yolk, was first described by Stepp (1909). A similar compound was later found in butterfat, egg yolk, and cod liver oil and was named "fat soluble A" by McCollum and Davies (1913, 1915) to distinguish it from essential water-soluble nutrients, which they called "water soluble B". The "fat-soluble A" factor was capable of preventing xerophthalmia (McCollum and Simmonds, 1917) and night blindness (Fridericia and Holm, 1925). Drummond (1920) named the active lipid "vitamin" A.

A growth-promoting substance in plant extracts was found by Steenbock et al. (1921). After Karrer et al., (1930) elucidated the structure of β-carotene and that of retinol (Karrer et al., 1931, 1933), the provitamin role of β-carotene became obvious.

Holmes and Corbett (1937) succeeded in crystallizing vitamin A from fish liver. Wald (1935 a, b, c, 1936 a, b) isolated the chromophore from bleached retinas. Morton (1944) and Morton and Goodwin (1944) demonstrated that the chromophore was retinal. Arens and Van Drop (1946) and Isler et al., (1947) succeeded in achieving the chemical synthesis of pure vitamin A. Shortly thereafter, the total synthesis of β-carotene was also reported (Karrer and Eugster, 1950).

Nomenclature and Chemistry

The nomenclature follows the rules on Biochemical Nomenclature (IUPAC, 1960; IUPAC-IUB, 1966). The parent compounds retinol, retinal and retinoic acid (Figure 1) are the most frequently involved in vitamin A metabolism. The synthetic

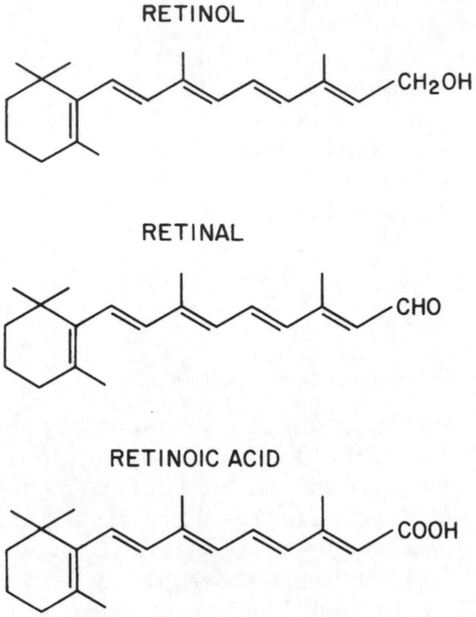

RETINOL

$CH_2OH$

RETINAL

CHO

RETINOIC ACID

COOH

Fig. 1. The structures of retinol, retinal and retinoic acid.

chemistry of vitamin A and derivatives have been extensively reviewed (Isler et al., 1967, 1970; Schwieter and Isler, 1967). A review of synthetic processes for obtaining β-carotene and carotenoids has appeared (Mayer and Isler, 1971). Vitamin A is a generic term used for all compounds, other than carotenoids, that exhibit qualitatively the biological activity of retinol. The term "retinoids" in recent years has been accepted as a general term that includes both the natural forms of vitamin A and synthetic analogs. Vitamin A is necessary for the growth, health and life of higher animals. Animals without vitamin A will cease to grow, and in time die. Vitamin A is necessary for vision, reproduction, maintenance of differentiated epithelia and for mucus secretion. One International Unit of vitamin A is defined as 0.3 μg of all-trans retinol. Generally, 1 μg of retinol is assumed to be biologically equivalent to about 6 μg of β-carotene, or about 12 μg of mixed dietary carotenoids.

## Absorption

Vitamin A is obtained in the diet primarily as long-chain fatty acid esters of retinol. The intestine hydrolyzes dietary retinyl esters, and the resulting retinol is then absorbed into the mucosal cell. The newly absorbed retinol in the mucosol cell or retinol newly synthesized from carotene is re-esterified with long-chain, mainly saturated, fatty acids and incorporated into lymph chylomicrons (Huang and Goodman, 1965). The chylomicrons are absorbed in the lymph and enter the circulation when the lymphatic vessels join the blood stream. Most of the chylomicron triglyceride is removed from the blood stream by extrahepatic tissues. The remaining chylomicrons are smaller (Redgrave, 1970) and contain virtually all of the chylomicron retinyl esters which are almost entirely removed from the circulation by the liver (Goodman et al., 1965; Redgrave, 1970). In the liver, hydrolysis and re-esterification occur, and the resulting retinyl esters (mainly retinyl palmitate) are stored within hepatocytes associated with lipid droplets.

## Biosynthesis

β-carotene is converted to vitamin A primarily in the intestinal mucosa. Two enzymes are involved in the biosynthetic processes: first, β-carotene-15, 15'-dioxygenase, and second, retinaldehyde reductase. β-carotene-15, 15'-dioxygenase catalyzes the cleavage of β-carotene at the central double bond to yield two molecules of retinaldehyde (Goodman and Olson, 1969). The cleavage reaction appears to occur by a dioxygenase mechanism (Goodman et al., 1966; Goodman and Olson, 1969).

The reduction of the newly formed retinaldehyde to retinol is catalyzed by another soluble mucosal enzyme, retinaldehyde reductase (Fidge and Goodman, 1968). This enzyme has a molecular weight in the range of 60,000-80,000, and has a requirement of NADH or NADPH. Retinaldehyde reductase appears to be a relatively nonspecific aldehyde reductase for several short- or medium-chain aliphatic aldehydes.

## Functions

Night blindness results from vitamin A deficiency, because the vitamin is required to regenerate the chemical which allows us to see in dim light. Rhodopsin is the complex of opsin (a protein present in the rods of the eye) and the retinal form of vitamin A. The simplified process is shown in Figure 2. A supply of vitamin A is needed to regenerate the complex, because some retinal is degraded in the process of rhodopsin degradation caused by bright light. Normal vision cannot be quickly obtained after rhodopsin bleaching by bright light, such as that from an oncoming car, unless there is an ample supply of vitamin A.

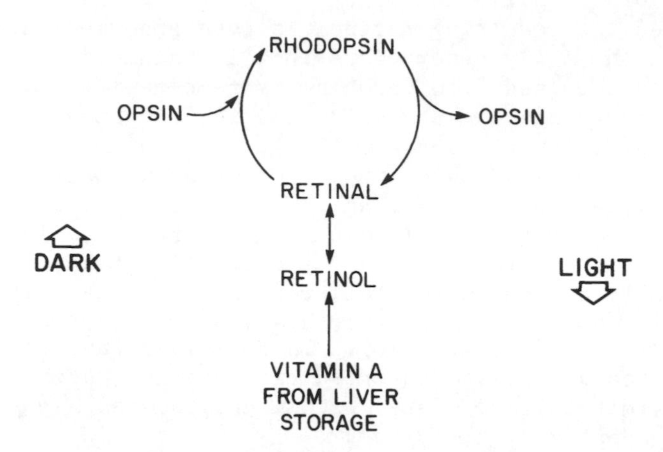


Figure 2. Vitamin A function in the visual cycle.

Only retinol and retinal can be used for maintenance of the visual cycle, whereas retinol, retinal and retinoic acid function in maintaining growth. Rats fed retinoic acid maintained their growth rate, but soon developed eye lesions.

Vitamin A is also required for the maintenance of epithelial tissue; that is, cells found in the outer layers of the skin as well as lining the gastrointestinal, respiratory and urogenital tracts. Vitamin A deficiency results in a transformation of epithelial cells from soft, moist tissue to cells which are hard and dry, or keratinized. This effect has been related to the protective effect of vitamin A in maintaining the integrity of epithelial tissue against infective organisms, particularly of the respiratory tract. The biochemical effect of vitamin A in this function may directly effect the synthesis of mucopolysaccharides and protein. DeLuca et al., (1975) found that labeled mannose incorporation into liver glycoproteins in both mildly and severely vitamin A deficient hamsters decreased by about 70% compared to pair-fed control animals.

Some research suggests that vitamin A allows the maturation of osteoblasts to osteoclasts, a conversion necessary for bone remodeling and therefore for continued bone growth. Mellanby (19744) likened vitamin A to "the director of building operations in the bone" and concluded that it might be expected that in vitamin A deficiency that the osteoblasts and osteoclasts would either cease to work or work in a completely disorderly way. This result was indeed obtained in fowl (Howell and Thompson, 1967 a,b). Avitaminosis A affected the nervous system indirectly by causing an overgrowth of the periosteum. Similar findings were obtained in fowl in which deficiency of vitamin A was precipitated suddenly by removal of retinoic acid from the diet.

Likewise, in tooth development vitamin A is necessary for the normal differentiation and function of the ameloblasts, cells of ectodermal origin which are responsible for the formation of dental enamel.

Evans and Bishop (1922) found that female rats fed diets low in vitamin A suffered from prolongation of estrus and failed to enter diestrus; a vaginal smear showed the presence of cornified cells. The rats also failed to ovulate. Administration of butterfat or cod liver oil readily reversed these changes within 1-3 days. One hundred days after placement on a low vitamin A diet, only 22% of rat matings resulted in the birth of living young. Implantation of the sperm into the egg generally did not occur, indicating that the sperm did not reach the egg because of the keratinized surface of the vaginal epithelium (Evans, 1928).

Evans (1932) and Mason (1932, 1933) showed that rats fed diets containing adequate amounts of vitamin E but deficient in vitamin A cause degeneration of the germinal epithelium in the testes and absence of spermatozoa in the epididymal fluid. Vitamin A can reverse the changes resulting from its deficiency. Howell et al., (1963) found that depletion of retinol in rats caused a basic lesion at a state in spermatocytogenesis before the meiotic division of the spermatocytes to spermatids. Similar results were obtained in the guinea pig (Howell et al., 1967).

Vitamin A also has a known effect on lysosomes and their enzymatic contents. Hypervitaminosis A reduces the stability of rat liver lysosomes and increases the release of acid hydrolases from these subcellular particles (Anonymous, 1966).

## RETINOL BINDING PROTEINS

### Role in Transport of Vitamin A

In the 1960's vitamin A was shown to be transported in plasma as the alcohol, retinol, bound to a specific transport protein, retinol-binding protein (RBP) (Kanai et al., 1968). Human RBP is a single polypeptide chain with a molecular weight of about 21,000 with a single binding site for one molecule of retinol. In plasma, most of RBP normally circulates as the retinol-RBP complex. The usual concentration of RBP in plasma is about 40-50 μg/ml.

RBP interacts strongly with prealbumin, and normally circulates as a 1:1 molar RBP-prealbumin complex. Evidence suggests that specific receptors for serum RBP are present on the surface of certain vitamin A-requiring target cells (Heller, 1975; Rask and Peterson, 1976). RBP may deliver retinol to target tissues specifically at these locations. The retinol then presumably enters the target cell, where it may become associated with the intracellular cytosol binding protein for retinol (CRBP). The complex of retinol with CRBP may be the form in which retinol is transmitted to specific sites within the cell, and may be involved directly in the subsequent metabolism and/or biochemical functions or retinol within the cell. The intracellular binding proteins differ from plasma RBP with regard to molecular weight, immunoreactivity, binding affinity for serum prealbumin, and ultraviolet and fluorescence spectral characteristics.

Clinical Studies of RBP

Even though RBP has been studied in many diseases to date, no disease has been found where RBP was totally absent or has had abnormal immunological properties. In patients with liver disease, the plasma concentration of vitamin A, RBP and prealbumin were all found to be markedly decreased (Smith and Goodman, 1971). These low plasma concentrations of RBP and prealbumin may reflect a reduced rate of synthesis of the proteins by the diseased liver. Patients with liver disease and low plasma RBP concentrations (below 20 μg/ml) were found to have impaired dark adaptation (Vahlquist et al., 1978). This study suggests that these patients had peripheral vitamin A deficiency symptoms secondary to their inability to mobilize vitamin A from the liver. In these patients, vitamin A therapy did not affect either the reduced dark adaptation ability or the low plasma RBP concentrations. In another study (Russell et al., 1978) vitamin A therapy of patients with alcohol-associated cirrhosis and marginal vitamin A status appeared to stimulate RBP release from the liver, and led to improvement in dark adaptation.

In patients with chronic renal disease, the plasma concentrations of RBP and of vitamin A were greatly elevated, while that of prealbumin remained normal (Smith and Goodman, 1971). The elevated concentrations of RBP reflect the fact that the kidney is normally the main catabolic site for RBP, so that RBP metabolism is impaired in chronic renal disease. RBP is small enough to be filtered by the renal glomeruli, whereas prealbumin and the RBP-prealbumin complex are not. Very little RBP is normally present in the free state. Its glomerular filtration and renal metabolism are sufficiently large to constitute RBP's major catabolic route. The important role of the RBP-prealbumin complex in sparing RBP from glomerular filtration was shown by turnover studies demonstrating a much higher fractional turnover rate for free RBP than for RBP bound to prealbumin (Vahlquist et al., 1973). Very little RBP appears in the urine normally, as nearly all of the filtered RBP is resorbed and degraded by the renal tubules. In contrast, patients with tubular proteinuria clear RBP from plasma at a normal rate, but excrete large amounts of RBP in the urine.

## VITAMIN A AND RETINOL BINDING PROTEINS AND CANCER

### Tumor Cell RBP's

Some tumors contain both CRBP and CRABP (cytosol binding protein for retinoic acid), some only CRABP and in some none of these proteins have been detected (Ong and Chytil, 1976). A positive correlation between the responsiveness to vitamin A treatment and the presence of CRABP has been observed.

Specific antibodies against cellular retinol-binding protein, raised in rabbit, were detected by sucrose gradient centrifugation and gel filtration using tritium-labeled, CRBP prepared by reductive methylation. By means of a radioimmunoassay, pure CRBP from liver and testis as well as CRBP present in a crude extract of liver were compared. All preparations showed identical immunoreactivity, suggesting CRBP is not tissue-specific (Ong and Chytil, 1979).

### Vitamin A and Cancer

Rowe and Garlin (1959) investigated the effect of vitamin A deficient diet on oral carcinogenesis induced by 7,12-dimethyl-1,2-benzantracene (DMBA) in golden hamsters. A 5% solution of carcinogen was applied twice weekly for 13 weeks. Neoplastic malignant lesions in hamsters given the vitamin A deficient diet were about 30% greater than the supplemented hamsters. Levij and Polliak (1968) found that topical vitamin A as retinyl palmitate applied to hamster and cheek pouches with DMBA for 12 weeks caused an increase in the size of squamous cell carcinomas. Chu and Malmgreen (1965) compared the incidence of neoplastic lesions of the alimentary tract of Syrian golden hamsters after oral feeding of DMBA alone or in combination with retinyl palmitate and benzopyrene alone or in combination with retinyl palmitate. Vitamin A had marked inhibitory effect on stomach carcinomas.

Saffiotti et al., (1967), Cone and Nettesheim (1973) and Smith et al., (1975 a,b) found that vitamin A prevented benzopyrene induced respiratory squamous cell carcinoma in animals.

Mouse skin papillomas can also be prevented by oral administration of 100 IU of vitamin A per gram of food (Davies, 1967). Shamberger (1971) has also reduced DMBA-croton oil induced mouse skin papillomas by applying vitamin A with the carcinogen to the skin. Bollag (1971) has induced mouse skin tumors with DMBA and croton oil. When papillomas reached at least 4 mm in diameter vitamin A (either as retinyl palmitate or as retinoic acid) was given orally or intraperitoneally. After 2 weeks of treatment with retinoic acid, 60-80% regression of the papilloma volume was estimated. Similar results were obtained with oral administration of retinoic acid or retinyl palmitate. Bollag (1972) has also given mice retinoic acid (200 mg/kg every 14 days) orally during the promotion phase of carcinogensis. The volume of papillomas was greatly reduced and the appearance of both papillomas and carcinomas as delayed. Bollag and Ott (1971) used retinoic acid as a chemotherapeutic agent to regress human skin basal cell carcinomas.. Verma et al., (1978) has observed that retinoids inhibit 12-0-tetradecanoylphorbol-13-acetate induced ornithine decarboxylase activity in mouse epidermis. There is some evidence that this phenotypic change is an essential component of the mechanism of skin tumor promotion.

Smith et al., (1972) have transplanted fetal lung tissues mixed with 20-methylcholanthrene (MCA) into the thigh muscle of Balb/c mice. The animals were placed on deficient, normal and excess retinyl palmitate diets. Carcinomas arose in the adequate or deficient groups but no carcinomas were observed in the group fed high vitamin A levels.

Prostate glands of mice were grown in tissue culture for 7-9 days. Methylcholanthrene caused hyperplasia of the alveolar epithelium, with subsequent squamous metaplasia or parakeratosis (Lasnitzki and Goodman, 1974). β-Retinol, β-retinoic acid, α-retinoic acid and cyclopentenyl analogue inhibited the effects of methylcholanthrene.

Vitamin A enhances the anti-tumor effect of cyclophosphamide in mammary gland adenocarcinoma in rats and mice (Anton and Brandes, 1968) and of 1,3-bis (2-chloro-ethyl-1-nitrosourea; (BCNU) (Cohen and Carbone, 1972). Synthetic retinoids can prevent DMBA-induced mammary cancer (Moon et al., 1976).

Hypervitaminosis A prevented the appearance of bladder squamous metaplasia and squamous cell neoplasia in rats fed (N-4-(5-Nitro-2 furyl)-2-thiazolyl formamide (Cohen et al., 1976). Feeding of 13-cis-retinoic acid after completion of carcinogen treatment diminished the number and severity of tumors induced in the bladders of male Fischer rats by 12 oral doses of N-butyl-N-(4-hydroxy-butyl) nitrosamine (Grubbs et al., 1979). Similar results have been observed in C57BL/6 mice (Becci et al., 1978).

The growth of Shope rabbit papilloma has been inhibited by hypervitaminosis A (McMichael, 1965). Vitamin A has also reduced the tumor size of mice innoculated with a murine sarcoma virus of the Maloney strain (Levine et al., 1975). Retinoic acid has shown an inhibitory effect on the growth of untransformed, transformed and tumor cells in vitro (Lotan and Nicolson, 1977). Some carcinogenesis test systems have not been effected by vitamin A.

## Vitamin A Toxicity and Cancer Prevention

The toxic effects of vitamin A may limit its effective use in human cancer prevention. Sporn et al., (1976) have summarized the effectiveness of retinoids. These derivatives of vitamin A have the same anticarcinogenic effect as vitamin A but are not as toxic as vitamin A. For example, retinyl methyl ether is 20 to 100 times less toxic than retinol or retinoic acid in their comparative toxicity to tracheal cartilage in organ culture. Because of their lesser toxicity vitamin A anticancer experiments in animals mostly utilize retinoids in their experimental design.

## Need for Further Research

A large clinical trial would be useful to test retinoids as a general cancer preventative on a massive scale in a general population. In addition, retinoids might also be used in conjunction with the regular chemotherapeutic cancer drugs which are given to cancer patients. Ways to measure retinol binding protein are needed. Concentrations of retinol binding protein and its relationship to various disease states, particularly cancer, should be determined.

## REFERENCES


Anonymous. 1966. Vitamin A and lysosomes. Nutrition Reviews 24:240.

Anton, E. and Brandes, D., 1968, Lysosomes in mice mammary tumors treated with cyclophosphamide. Distribution related to the course of the disease. Cancer 21:483.

Arens, J.F. and Van Drop, D.A., 1946, Synthesis of some compounds possessing vitamin A activity. Nature (London) 157:190.

Becci, P.J., Thompson, H.J., Grubbs, C.J., Squire, R.A., Brown, C.C., Sporn, M.B. and Moon, R.C., 1978, Inhibitory effects of 13-cis-retinoic acid on urinary bladder carcinogenesis induced in C57BL/6 mice by N-butyl-N-(4-hydroxybutyl)-nitrosamine. Cancer Res. 38:4463.

Bollag, W., 1971, Therapy of chemically induced skin tumors of mice with vitamin A palmitate and vitamin A acid. Experientia 37:90.

Bollag, W. and Ott, F., 1971, Vitamin A-Sawie in der Tumortherapie: Lokalbehandlung von Prakanzerosen und Basalzellkarzinomen der Haut mit Vitamin A-Sawre. Schweiz. Med. Wochenschr. 101:17.

Bollag, W., 1972, Prophylaxis of chemically induced benign and malignant epithelial tumors by vitamin A acid (retinoic acid). Eur. J. Cancer 8:689.

Chu, E.W. and Malmgren, R.A., 1965, An inhibitory effect of vitamin A on the induction of tumors of forestomach and cervix in the Syrian golden hamster by carcinogenic polycyclic hydrocarbons. Cancer Res. 25:884.

Cohen, M.H. and Carbone, P.P., 1972, Enhancement of the antitumor effects of 1,3-bis-(2-chloroethyl)-1-nitrosourea and cyclophosphamide by vitamin A. J. Nat. Cancer Inst. 48:921.

Cohen, S.M., Wittenberg, J.F. and Bryan, G.T., 1976, Effects of avitaminosis A and hypervitaminosis A on urinary bladder carcinogenicity of N- 4-(5-Nitro-2-furyl)-2-thiazolyl formamide. Cancer Res. 36:2334.

Cone, M.V. and Nettesheim, P., 1973, Effects of vitamin A on 3-methylcholanthrene-induced squamous metaplasias and early tumors in the respiratory tract of rats. J. Natl. Cancer Inst. 50:1599.

Davies, R.E., 1967, Effect of vitamin A on 7,12-dimethylbenzanthracene-induced papilloma in rhino mouse skin. Cancer Res. 27:237.

DeLuca, L.M., Silverman-Jones, C.S. and Barr, R.M., 1975, Biosynthetic studies on mannolipids and mannoproteins of normal and vitamin A-depleted hamster livers. Biochem. Biophys. Acta 409:342.

Drummond, J.C., 1920, The nomenclature of the so called accessory food factors (vitamins). Biochem. J. 14:660.

Cohen, S.M., Wittenberg, J.F. and Bryan, G.T., 1976, Effects of avitaminosis A and hypervitaminosis A on urinary bladder carcinogenicity of N- 4-(5-Nitro-2-furyl)-2-thiazolyl formamide. Cancer Res. 36:2334.

Cone, M.V. and Nettesheim, P., 1973, Effects of vitamin A on 3-methylcholanthrene-induced squamous metaplasias and early tumors in the respiratory tract of rats. J. Natl. Cancer Inst. 50:1599.

Davies, R.E., 1967, Effect of vitamin A on 7,12-dimethylbenzanthracene-induced papilloma in rhino mouse skin. Cancer Res. 27:237.

DeLuca, L.M., Silverman-Jones, C.S. and Barr, R.M., 1975, Biosynthetic studies on mannolipids and mannoproteins of normal and vitamin A-depleted hamster livers. Biochem. Biophys. Acta 409:342.

Drummond, J.C., 1920, The nomenclature of the so called accessory food factors (vitamins). Biochem. J. 14:660.

Evans, H.M. and Bishop, K.S., 1922, On an invariable and characteristic disturbance of reproductive function in

animals reared on a diet poor in fat soluble vitamin A. Anat. Rec. 23:17.
Evans, H.M., 1928, The effects of inadequate vitamin A on the sexual physiology of the female. J. Biol. Chem. 77:651.
Evans, H.M., 1932, Testicular degeneration due to inadequate vitamin A in cases where E is adequate. Am. J. Physiol. 99:477.
Fidge, N.H. and Goodman, D.S., 1968, The enzymatic reduction of retinal to retinol in rat intestine. J. Biol. Chem. 243:4372.
Fridericia, L.S. and Holm, E., 1925, Relation between night blindness and malnutrition-influence of deficiency of fat-soluble A vitamin in the diet on the visual purple in the eyes of rats. Am. J. Physiol. 73:63.
Goodman, D.S., Huang, H.S. and Shiratori, T., 1965, Tissue distribution and metabolism of newly absorbed vitamin A in the rat. J. Lipid Res. 6:390.
Goodman, D.S., Huang, H.S. and Shiratori, T., 1966, Mechanism of the biosynthesis of vitamin A from β-carotene. J. Biol. Chem. 241:1929.
Goodman, D.S. and Olson, J.S., 1969, In: Methods in Enzymology. Vol. XV. Steroids and Terpenoids. (R.B. Claxton, ed.) New York, Academic.
Grubbs, C.J., Moon, R.A., Squire, R.A., Farrow, G.M., Stinson, S.F., Goodman, D.G., Brown, C.C., Spron, B.M., 1979, 13-cis-Retinoic acid: Inhibition of bladder carcinogenesis induced in rats by N-butyl-N-(4-hydroxylbutyl) nitrosamine. Science 198:743.
Heller, J., 1975, Interactions of plasma retinol-binding protein with its receptor. Specific binding of bovine and human retinol-binding protein to pigment epithelium cells from bovine eyes. J. Biol. Chem. 250:3613.
Holmes, H.N. and Corbett, R.E., 1937, The isolation of crystalline vitamin A. J. Am. Chem. Soc. 59:2042.
Howell, J.M., Thompson, J.N. and Pitt, G.A.J., 1963, Histology of the lesions produced in the reproductive tract of animals fed a diet deficient in vitamin A alcohol but containing vitamin A acid. I. The male rat. J. Reprod. Fertil. 5:159.
Howell, J.M. and Thompson, J.N., 1967a, Lesions associated with the ataxia in vitamin A deficient chicks. Br. J. Nutr. 21:741.
Howell, J.M. and Thompson, J.N., 1967b, Observations on the lesions in vitamin A deficient adult fowls with particular reference to changes in bone and central nervous system. Br. J. Exp. Pathol. 48:450.
Howell, J.M., Thompson, J.N. and Pitt, G.A.J., 1967, Changes in the tissues of guinea-pigs fed on a diet free from vitamin A but containing methyl retinoate. Br. J. Nutr. 21:37.

Huang, H.S. and Goodman, D.S., 1965, Vitamin A and carotenoids. I. Intestinal absorption and metabolism of 14C-labeled vitamin A alcohol and β-carotene in the rat. J. Biol. Chem. 240:2839.

Isler, O., Huber, W., Ronco, A. and Koffler, M., 1947, Synthese des Vitamin A. Helv. Chim. Acta 30:1911.

Isler, O., Klaui, H. and Solms, U., 1967, Vitamin A. III. Industrial preparation and production. In the Vitamins (W.H. Sebrell, Jr. and R.S. Harris, eds.) p. 101. Academic Press, New York.

Isler, O., Solms, U. and Wursch, J., 1970, In: Fat Soluble Vitamins (R.A. Morton, ed.) p. 99. Pergamon Press, Oxford.

IUPAC. 1960, Commission on the nomenclature of biological chemistry. J. Am. Chem. Soc. 82:5581.

IUPAC-IUB. 1966, Commission on biochemical nomenclature, tenative rules. J. Biol. Chem. 241:2987.

Kanai, M., Raz, A. and Goodman, D.S., 1968, Retinol-binding protein: The transport protein for Vitamin A in human plasma. J. Clin. Invest. 47:2025.

Karrer, P., Helfenstein, A., Wehrli, H. and Wehstein, A., 1930, Pflanzenfarbstoffe. XXV. Ueber die Konstitution des Lycopins and Corotins. Helv. Chim. Acta 13:1084.

Karrer, P., Morf, R. and Schopp, K., 1931, Zur Kenntis des Vitamins-A aus Fischtranen. Helv. Chim. Acta 14:1036.

Lasnitzki, I. and Goodman, D.S., 1974, Inhibition of the effects of methylcholanthrene on mouse prostate in organ culture by vitamin A and its analogs. Cancer Res. 34:1564.

Levij, I.S. and Polliack, A., 1968, Potentiating effect of vitamin A on 9,10 dimethyl 1-2 benzanthracene carcinogenesis in the hamster cheek pouch. Cancer 22:300.

Levine, N.S., Salisbury, R.E., Seifter, E., Walker, H.L., Mason, A.D., Jr., and Pruitt, B.A., Jr., 1975, Effect of vitamin A on tumor development in burned, unburned and glucocorticoid-treated mice inoculated with an oncogenic virus. Experientia 15:1309.

Lotan, R. and Nicolson, G.L., 1977, Inhibitory effects of retinoic acid or retinyl acetate on the growth of untransformed, transformed, and tumor cells in vitro, 59:1717.

Mason, K.E., 1932, Differences in testes injury and repair after vitamin A deficiency, vitamin E deficiency and inanition. Am. J. Anat. 51:153.

Mason, K.E., 1933, Differences in testes injury and repair after vitamin A deficiency and inanition. Am. J. Anat. 52:153.

Mayer, H. and Isler, O., 1971, Total synthesis of carotenoids. In: Carotenoids (O. Isler, ed.) p. 328. Binkhauser. Verlog, Basel.

McCollum, E.V. and Davies, M., 1913, The necessity of certain lipids on the diet during growth. J. Biol. Chem. 15:167.

McCollum, E.V. and Davies, M., 1915. The nature of the dietary

deficiency of rice. J. Biol. Chem. 23:181.
McCollum, E.V. and Simmonds, N., 1917, A biological analysis of pellagra producing diet. II. The minimum requirements of the two unidentified dietary factors for maintenance as contrasted with growth. J. Biol. Chem. 32:181.
McMichael, H., 1965, Inhibition of growth of Shope rabbit papilloma by hypervitaminosis A. Cancer Res. 25:947.
Mellanby, E., 1944, Nutrition in relation to bone growth and the nervous system. Proc. R. Soc. London Ser. B. 132:28.
Moon, R.C., Grubbs, C.J. and Spron, M.B., 1976, Inhibition of 7, 12-dimethylbenzanthracene-induced mammary carcinogenesis by retinyl acetate. Cancer Res 36:2626.
Morton, R.A., 1944, Chemical aspects of the visual process. Nature (London) 153:69.
Morton, R.A. and Goodwin, T.W., 1944, Preparation of retinene "in vitro". Nature (London) 153:405.
Ong, E.D. and Chytil, F., 1976, Presence of retinol and retinoic acid binding proteins in experimental tumors. Cancer Lett. 2:25.
Ong, D.E. and Chytil, F., 1979, Immunochemical comparison of vitamin A binding proteins in rat. J. Biol. Chem. 254:8733.
Rask, L. and Peterson, P.A., 1976, In vitro uptake of vitamin A from the retinol-binding plasma protein to mucosal epithelial cells from monkey's small intestine. J. Biol. Chem. 251: 6360.
Redgrave, T.G., 1970, Formation of cholesteryl ester-rich particulate lipid during metabolism of chylomicrons, J. Clin. Invest 49:465.
Rowe, N.H. and Gorlin, R.J., 1959, The effect of vitamin A deficiency upon experimental oral carcinogenesis. J. Dent. Res. 38:72.
Russell, R.M., Morrison, S.A., Smith, F.R., Oaks, E.V. and Carney, E.A., 1978, Vitamin A reversal of abnormal dark adaptation in cirrhosis. Study of effects on the plasma retinol transport system. Ann. Int. Med. 88:622.
Saffiotti, U., Montesano, R., Sella-Kumar, A.R., and Borg, S.A., 1967, Experimental cancer of the lung: Inhibition by vitamin A of the induction of tracheobronchial squamous metaplasia and squamous cell tumors. Cancer 20:857.
Schwieter, U. and Isler, O., 1967, Vitamin A. II. Chemistry, In: The vitamins, Vol. I (W.H. Sebrell, Jr. and R.S. Harris, eds.) Academic Press, New York.
Shamberger, R.J., 1971, Inhibitory effect of vitamin A on carinogenesis. J. Nat. Cancer Inst. 48:1491.
Smith, F.R. and Goodman, D.S., 1971, The effect of disease of the liver, thyroid, kidneys on the transport of vitamin A in human plasma. J. Clin. Invest. 50:2426.
Smith, W.E., Yazdi, E. and Miller, L., 1972, Carcinogenesis in pulmonary epithelia in mice on different levels of vitamin A. Environ. Res. 5:152.

Smith, D.M., Rogers, A.E., Herndon, B.J. and Newberne, P.M., 1975a, Vitamin A (retinyl acetate) and benzopyrene-induced respiratory tract carcinogenesis in hamsters fed a commercial diet. Cancer Res. 35:11.

Smith, D.M., Rogers, A.E. and Newberne, P.M., 1975b, Vitamin A and benzopyrene carcinogenesis in the respiratory tract of hamsters fed a semi-synthetic diet. Cancer Res 35:1485.

Sporn, M.B., Dunlop, N.M., Newton, D.L. and Henderson, W.R., 1976, Relationship between structure and activity of retinoids. Nature 263:110.

Steenbock, H., Sell, M.T., Nelson, E.M. and Buell, M.V., 1921, The fat-soluble vitamin. Proc. Am. Soc. Biol. Chem. J. Biol. Chem. 46: Proc. XXXII.

Stepp, W., 1909, Versuche uber Futterung mit lipoid freier Ernahrung. Biochem. Ztschr. 22:452.

Vahlquist, A., Peterson, P.A. and Wibell, L., 1973, Metabolism of the vitamin A transporting protein complex. I. Turnover studies in normal persons and in patients with chronic renal failure. Eur. J. Clin. Invest. 3:352.

Verma, A.K., Rice, H.M., Shapas, B.G. and Boutwell, R.K., 1978, Inhibition of 12-O-tetradecanoylphorbol-13 acetate-induced ornithine decarboxylase activity in mouse epidermis by vitamin A analogs (retinoids). Cancer Res. 38:793.

Wald, G., 1935a, Vitamin A in eye tissue. J. Gen. Physiol. 18:905.

Wald, G., 1935 b, Carotenoids and visual cycle. J. Gen. Physiol. 19:351.

Wald, G., 1935c, Pigments of the bull frog retina. Nature (London) 136:832.

Wald, G., 1936c, Pigments in the retina. I. J. Gen. Physiol. 19:781.

Wald, G., 1936b, Pigments in the retina. II. J. Gen. Physiol. 20:45.

# VITAMIN D - ITS EXCESSIVE USE IN THE U.S.A.

C. Bruce Taylor, Associate Chief of Staff for Research and Development and Shi-Kaung Peng, Staff Pathologist of Laboratory Service, Veterans Administration Medical Center and Albany Medical College of Union University Albany, New York 12208

## INTRODUCTION

Sir Edward Mellanby (1920) first demonstrated that cod-liver oil could prevent rickets. Mellanby (1920) named this antirachitic material "fat soluble vitamin D." In retrospect it is tragic that this extremely potent sterol was classified as a vitamin; its reclassification as a potent, carefully controlled hormone should be seriously considered (Loomis, 1967). There is a mass of published information on the toxic effects of excesses of vitamin D on the arterial system in man (Forfar et al., 1956; Taussig, 1966; Seelig, 1969; Taylor et al., 1972 and Scientific reviews (GRAS) food ingredients - vitamin D, NTIS, U.S. Department of Commerce, 1974) and animals (Hass et al., 1958; Bajwa et al., 1971; Scientific reviews (GRAS) food ingredients - vitamin D, NTIS, U.S. Department of Commerce, 1974; Liu et al., 1979 and Peng et al., 1978).

The saga of the use and abuse of vitamin D since the first demonstration of its antirachitic properties in 1920 is most interesting. It appears that a combination of an initial lack of understanding of the actions and metabolism of vitamin D plus a general expectation that vitamin D would have a favorable therapeutic effect on many skeletal diseases led to a rather serious, extensive and somewhat protracted overuse of this agent. The report by Seelig (1969) covers the history of our development of knowledge of vitamin D very thoroughly.

## REVIEW OF PERTINENT LITERATURE

### History of Development of Knowledge of Vitamin D Requirements

It was observed in 1953 and 1954 that approximately 100 new cases of infantile hypercalcemia were being diagnosed annually in Great Britain. This observation led to a survey of the vitamin D enrichment practices in Great Britain (Lightwood et al., 1956; Lightwood et al., 1957 and Forfar, 1956). This survey disclosed that the then liberal enrichment of national dried milk, enrichment of infant cereals plus the daily ingestion of one teaspoonful of national cod-liver oil compound containing 700-800 I.U. would result in an ingestion of about 4,000 I.U. of vitamin D per day. Interestingly, a study in Glasgow (Graham, 1959) revealed that 26 infants and children had hypercalcemia while ingesting about 2,000 I.U. per day and 8 receiving about 1,000 I.U. of vitamin D per day showed hypercalcemia. It should also be pointed out that since essentially all British infants were ingesting 3,000 to 4,000 I.U. of vitamin D per day there were many individuals who suffered no ill effects. These studies suggested a rather wide range of tolerance to ingestion of variable excessive amounts of vitamin D. Following this study vitamin D was removed from infant cereal and from national dried milk; the amount of vitamin D in national cod-liver oil was also reduced significantly. Since daily intake of vitamin D in Great Britain has been reduced to about 400 I.U. the occurrence of hypercalcemia in infants and children has been essentially eliminated. Jeans (1950) reported a recommended daily intake of 400 I.U. of vitamin D (for newborns, infants, growing children to 19 years of age and pregnant and lactating females) which is generally accepted in the Western World.

More recently the Committee on Nutrition, American Academy of Pediatrics has published an excellent study on the prophylactic requirements and the toxicity of vitamin D (Fomon, 1963). This committee wisely recommends minimal but adequate daily intakes of vitamin D. A daily intake of 400 I.U. of vitamin D for premature and term infants, children and adolescents up to age 19 is recommended. No vitamin D is recommended for adult males and non-pregnant adult females. The committee further advised that pregnant females during the second half of gestation and lactating females should receive 400 I.U. of vitamin D. It should be mentioned that Loomis (1967) states that unpigmented skin of man, when exposed to solar radiation, converts a significant amount of 7-dehydrocholesterol to vitamin $D_3$; he indicated that about 20 to 25 square centimeters of minimally pigmented skin (an area of skin present on one cheek or the back of one hand) when exposed to 3 hours of sunshine can convert a sufficient quantity of 7-dehydrocholesterol to 400 I.U. of vitamin $D_3$. Individuals with more pigment in their skin such as Blacks and Indians have a reduced capacity to synthesize vitamin D in skin exposed to sunlight. The author states that persons with more pigment in their skin, such as Blacks and Indians, who live in inland areas far from the equator have a greater propensity to develop rickets if they have no oral or subcutaneous sources of vitamin D. Persons with pigmented skin living on the sea coast (ex. Eskimos) are not susceptible to rickets because of the rich source of vitamin D in fish liver and other tissues of fish.

## Vitamin D Concentration in Foods

Fomon et al. (1963) include a table in their paper (Table IIA, page 519) which lists amounts of vitamin D added to various foods. They list 25 common food products eaten frequently) which contain very significant quantities of vitamin D (80 to 1,000 I.U. vitamin D per serving). During the past 16 years the addition of vitamin D to frequently ingested foods has become much more common. Kummerow (1979) presents data indicating that the average daily per capita intake of vitamin D from vitamin D enriched foods is 2435 I.U. If an individual is also taking one vitamin capsule containing 400 I.U. of vitamin D each day his daily vitamin D intake is 2835 I.U.; this level of daily vitamin D dosage is comparable to the toxic levels of vitamin D intakes reported by the British in 1953 and 1954. (Lightwood et al., 1956; Lightwood et al., 1957 and Forfar et al., 1956).

Another important source of dietary vitamin D is that present in meats and fish. Kummerow et al. (1976) and Kummerow (1979) present startling data on the excess amounts of vitamin D added to animal feeds with a resultant retention of vitamin D in various edible portions of these animals (Examples:- when layer chickens were fed a commercial ration which assayed at 1,600 I.U. of vitamin D per pound their skin contained 4,086 I.U. vitamin D/pound, breast muscle - 363 I.U. and liver 1,725 I.U. of vitamin D. When swine were fed 100,000 I.U. of vitamin D/pound of commercial ration for 6 weeks, their lean muscle contained 5,765 I.U. and liver 2,270 I.U. of vitamin D per pound). Fish frequently have very rich concentrations of vitamin D which are not of exogenous origin; their vitamin D is endogenously synthesized (Example: vitamin D content of rock fish - 2,700 I.U. of vitamin D per pound). Kummerow (1979) states that the average per capita intake from vitamin D enriched foods is 2,435 I.U. which is six times the National Research Council requirements. Further, many persons take an additional 400 I.U. of vitamin D in a daily vitamin capsule.

## Absorption, Turnover and Storage of Vitamin D

The metabolism and turnover of vitamin D is important and should be discussed (Avioli et al., 1967) as well as the accumulation of vitamin D in tissues resulting from excess vitamin D intake (Kummerow et al., 1976). Avioli et al. (1967) report that vitamin D excretion is principally via the enterohepatic circulation; its excretion is comparable to that of cholesterol. The biological half-life of labeled vitamin $D_3$ is nearly 5 days (112.5 hours). Since vitamin D is a fat soluble vitamin, once it gains entrance to the organism (via injection or intestinal absorption) its elimination is a slow process. If given in excess vitamin D cannot be excreted rapidly by the kidneys like the water soluble vitamins. Excess intake of vitamin D leads to profound increases in its content in tissues and serum; Kummerow et al. (1976) have shown that feeding 100,000 I.U. of vitamin $D_3$ per pound of ration to weanling pigs for 6 weeks results in a more than 15 fold increase in vitamin $D_3$ content in skeletal muscle and a 12 fold increase in serum vitamin $D_3$ concentration. They also present data showing a 40 fold increase in skeletal muscle of vitamin $D_3$ content of humans given massive doses of vitamin $D_3$ for osteomalacia. They reported no data on accumulation of vitamin $D_3$ in arteries, kidney or heart. These tissues are most sensitive to excessive accumulation of vitamin D and will be discussed in the next section.

## Toxicity of Excess Vitamin D

Excessive intake of vitamin D can adversely affect a number of tissues (Seelig, 1969; Hass et al., 1958 and Liu et al., 1979). The heart and kidneys have an intermediate sensitivity to vitamin D. When there is an excess intake of vitamin D, morphologic and functional changes involving the heart and kidneys become apparent rather early (Seelig, 1969; Liu et al., 1979); both organs show abnormal accumulation of calcium which destroys functional cardiac and renal tissue. Liu et al. (1979) observed severe diffuse calcific cardiomyopathy with cardiac failure after 23 weeks of excess vitamin D intake plus nicotinism in rhesus monkeys. Seelig (1969) presents strong evidence that excess intake of vitamin D is the cause of the supravalvular aortic stenosis syndrome; she also presents evidence of renal damage due to hypervitaminosis D.

There is an abundance of experimental evidence in many species of animals indicating that arterial tissues are most susceptible to hypervitaminosis D. (Hass et al., 1958; Bajwa et al., 1971; Scientific Reviews (GRAS) food ingredients-vitamin D, NTIS, U.S. Department of Commerce, 1974; Liu et al., 1979 and Peng et al., 1978). Evidence implicating mild, protracted hypervitaminosis D in human arteriosclerosis is also quite convincing (Forfar et al., 1956; Taussig, 1966; Meyer & Stelzig, 1968; Seelig, 1969 and Taylor et al., 1972). The arteries' elastic tissue is the principal target tissue of excess vitamin D; elastic membranes and adjacent ground substance accumulate calcium and are severely damaged or disrupted. Smooth muscle cells are less susceptible but at a later stage undergo necrosis (Hass et al., 1958; Hass et al., 1960 and Liu et al., 1979). Peng et al. (1978) reported a very significant study in which minimal daily excesses of vitamin $D_3$ produced arterial damage in squirrel monkeys. The established daily vitamin $D_3$ requirement for squirrel monkeys is 100 I.U. (Lehner et al., 1967); it is of interest that vitamin $D_2$ is essentially inactive in squirrel monkeys. Daily ingestion of 500 I.U. of vitamin $D_3$ plus 0.5 percent cholesterol in the diet produced early but significant arteriosclerosis within 10 months (Peng et al., 1978). Daily intakes of 1,000 I.U. of vitamin $D_3$ without the addition of dietary cholesterol for 10 to 18 months also produced severe arteriosclerotic changes (Peng et al., 1978).

When one reflects upon the fact that the average person in the U.S. ingests 6 to 7 times the maximal, recommended daily amount, 400 I.U. vitamin D (for infants, children and pregnant females - adult males and non-pregnant females require no exogenous vitamin D) over many decades, it is shocking to contemplate the effect of ingestion of 5 to 10 times the daily required dose of vitamin D on the arteries of squirrel monkeys. If significant arterial damage and repair can be induced in squirrel monkeys by feeding 5 to 10 times the recommended daily intake for 10 to 18 months it is frightening to contemplate what the daily ingestion of 6 to 7 times the required daily intake of vitamin D (for infants and pregnant women - 400 I.U.) might do to the arterial tree of humans over a period of 50 or more years. Consideration of the above information suggests that "operation over-kill" in the area of vitamin D ingestion in this country may be the most significant risk factor in the genesis of arteriosclerosis; cholesterol intake and levels of serum cholesterol may well be only important secondary factors in the pathogenesis of atherosclerosis.

It should be re-stated that vitamin D is a fat soluble vitamin. After it has been absorbed or ingested its elimination is a slow process principally via the enterohepatic circulation. All persons interested in nutrition as related to cardiovascular disease should always be mindful that vitamin D is not a water soluble vitamin and that excess intakes of this vitamin cannot be rapidly excreted through the kidneys.

Finally, it should be mentioned that the traditional "end-stage" complication of human arteriosclerosis, arterial thrombosis, has been produced in rabbits (Hass et al., 1960) and monkeys (Liu et al., 1979) by administration of vitamin D, nicotine and cholesterol to these animals.

## SUMMARY AND CONCLUSIONS

Vitamin D was first identified in 1920. Numerous clinical and experimental studies have been reported. In 1950 a recommended daily intake of 400 I.U. for premature and term infants, children and adolescents up to 19 years and pregnant and lactating females was reported and accepted; no exogenous vitamin D was recommended for adult males and non-pregnant females.

Many processors of foods add vitamin D to a wide variety of foods. It has been estimated that the average person in the United States ingests 2,435 I.U. of vitamin daily which is added to many foods; also, many persons take 400 I.U. of vitamin D in a daily vitamin capsule. There is also a great propensity for those feeding hogs, chickens and beef to add massive quantities of vitamin D to animal feeds. This results in accumulation of shocking amounts of vitamin D in meat products consumed by persons in this country. Data was presented indicating that the average person on a conventional diet consumes more than 6 to 7 times the recommended daily intake of 400 I.U. of vitamin D. The excessive consumption of vitamin D represents another example of "operation over-kill" in the U.S.A. It seems that those making policies for vitamin D intake in this country are not remembering that vitamin D is a fat soluble vitamin with a very slow excretion rate. When excess vitamin D is absorbed into the body it cannot be excreted rapidly by the kidneys like the water soluble vitamins. The half turnover time is about five days. Excess amounts of vitamin D are stored in various tissues, e.g. cardiac muscle, skeletal muscle, kidneys and arterial walls.

It has been shown that a daily intake of five to ten times the required dose of vitamin D can produce significant arteriosclerosis in squirrel monkeys within ten months. It seems likely that man's daily intake, which is seven times greater than that required for a growing child, may well be the major cause of his arteriosclerosis; the average adult in this country may be taking this excessive vitamin D for 50 to 60 years. Cholesterol intake and elevated serum cholesterol levels are probably secondary contributing factors in the development of arteriosclerosis.

It is suggested that all individuals involved in the planning, production and distribution of food and vitamins in the United States develop and enlightened consciousness regarding vitamin D intake and the grave complications related to prolonged, moderate overdosage of this vitamin. It is also suggested that all persons in this country check labels on their food packages for their vitamin D content in order to avoid chronic hypervitaminosis D.

Nutritionists should seriously consider establishing a national committee to evaluate vitamin D additives in foods; hopefully, the rather promiscuous habit of adding significant amounts of vitamin D to many foods could be corrected. The Food and Drug Administration might also increase its involvement in this area in order to make certain that many individuals in the U.S. are not unknowingly inducing mild, chronic hypervitaminosis D by daily ingestion of excess vitamin D.

## REFERENCES


Avioli, L.V., Lee, S.W., McDonald, J.E., Lund, J. and De Luca, H.F., 1967, Metabolism of vitamin $D_3$ - $^3H$ in human subjects: Distribution in blood, bile, feces and urine, J. Clin. Invest., 46:983.

Bajwa, G.S., Morrison, L.M. and Ershoff, B.H., 1971, Induction of aortic and coronary athero-arteriosclerosis in rats fed a hypervitaminosis D, cholesterol-containing diet, Proc. Soc. Exptl. Biol. Med., 138:975.

Fomon, S.J., Chairman, Committee on Nutrition, Am. Academy of Pediatrics, 1963, The prophylactic requirement and the toxicity of vitamin D, Pediatrics (Springfield) 31:512.

Forfar, J.O., Balf, C.L., Maxwell, G.M. and Tompsett, S.L., 1956, Idiopathic hypercalcemia of infancy, Clinical and metabolic studies with special reference to the aetiological role of vitamin D, Lancet, 1:981.

Graham, S., 1959, Idiopathic hypercalcemia, Postgrad. Med., 25:67.

Hass, G.M., Trueheart, R.E., Taylor, C.B. and Stumpe, M., 1958, An experimental histologic study of hypervitaminosis D, Amer. J. Pathol., 36:295.

Hass, G.M., Trueheart, R.E. and Hemmens, A., 1960, Production of calcific atheroarteriosclerosis and thromboarteritis with nicotine, vitamin D and dietary cholesterol, Am. J. Pathol., 49:739.

Jeans, P.C., 1950, Vitamin D, JAMA, 143:177.

Kummerow, F.A., Cho, C.H.S., Huang, W.Y.T., Imai, H., Kamio, A., Deutsch, M.J. and Hooper, W.M., 1976, Additive risk factors in atherosclerosis, Am. J. Clin. Nutr., 29:579.

Kummerow, F.A., 1979, Nutrition imbalance and angiotoxins as dietary risk factors in coronary heart disease, Am. J. Clin. Nutr., 32:58.

Lehner, N.D.M., Bullock, B.C., Clarkson, T.B. and Lofland, H.B., 1967, Biologic activities of vitamin $D_2$ and $D_3$ for growing squirrel monkeys, Lab. Anim. Care, 17:483.

Lightwood, R. (Chairman), Sheldon, W., Harris, C. and Stapleton, T., (Committee for Study of Hypercalcemia in Infants and Vitamin D), 1956, Hypercalcemia in infants and vitamin D, Brit. Med. J., 2:149.

Lightwood, R. and Stapleton, T., 1957, National policies for the prevention of rickets, Ann. Pediat., 188:1957.

Liu, L.B., Taylor, C.B., Peng, S.K. and Mikkelson, B., 1979, Experimental arteriosclerosis in Rhesus monkeys induced by multiple risk factors: cholesterol, vitamin D and nicotine, Arterial Wall, 5:25.

Loomis, W.F., 1967, Skin-pigment regulation of vitamin D biosynthesis in man. Variation in solar ultraviolet at different latitudes may have caused racial differentiation in man, Science, 157:501.

Mellanby, E., 1920, Accessory food factors (vitamins) in the feeding of infants, Lancet, 1:856.

Meyer, W.W. and Stelzig, H.H., 1968, Calcification patterns of the internal elastic membrane, Calc. Tiss. Res., 3:266.

Peng, S.K., Taylor, C.B., Tham, P. and Mikkelson, B., 1978, Role of mild excesses of vitamin $D_3$ in arteriosclerosis. A study in squirrel monkeys, Arterial Wall, 4:229.

Scientific literature reviews on generally recognized as safe (GRAS) food ingredients - vitamin D, 1974, Washington, D.C., Natl. Tech. Information Serv., U.S. Dept. of Commerce, July.

Seelig, M.S., 1969, Vitamin D and cardiovascular, renal and brain damage in infancy and childhood, Annals N.Y. Acad. of Sciences, 147:537.

Taussig, H.B., 1966, Possible injury to cardiovascular system from vitamin D, Ann. Intern. Med., 65:1195.

Taylor, C.B., Hass, G.M., Ho, K.J. and Liu, L.B., 1972, Risk factors in the pathogenesis of arteriosclerotic heart disease and generalized atherosclerosis, Ann. Clin. Lab. Sc., 2:239.

# THE CLINICAL IMPLICATIONS OF LIPID ANTIOXIDANT NUTRITION

Jeffrey Bland Ph.D.

Assoc. Prof., Dept. of Chemistry
University of Puget Sounds
Dir. Bellevue-Redmond Medical Labs.
Tacoma, Washington 98416

## BACKGROUND

Few areas in human nutrition are so charged with claims and counterclaims as is that of the antioxidant nutrients, and vitamin E specifically. The very name of vitamin E, tocopherol, reflects the considerable controversy which surrounds it, in that it is derived from the Greek word meaning "bringing forth in childbirth." The fertility effect, however, since its early demonstration in rats by Evans and Bishop (1923), has not been demonstrated in humans. Vitamin E was discovered when Evans and Bishop (1923) isolated from vegetable oils a substance that helped rats produce robust offspring. Unfortunately, this observation was picked up by many individuals and promoted in the lay literature as a human "sex" vitamin. This clouded the scene with regard to the biological importance of vitamin E and has led to much misunderstanding concerning both the prophylactic and potential therapeutic benefit of vitamin E and other members of the lipid antioxidant family for the human being. The controversy surrounding it is summed up in a statement made by Tappel (1973): "While few compelling uses have been found for vitamin E, the more research which is done on this substance, the more intriguing it appears. Thus, there is a nagging suspicion that there is a very important use for the vitamin and we are just not smart enough to see it."

In this review the biological effects of vitamin E and others of the lipid antioxidant family will be explored in hopes of better differentiating between the supportable or therapeutic uses and those which are based solely upon testimony.

The early history of vitamin E in the period of 1922 to 1932 was fraught with considerable controversy as to what it was and how it actually played a significant role in animal physiology (Mason, 1977). Sterility in rats was found in high-protein, low-fat diets by Osborne, Mendel and Mattill. Evans and Bishop proposed the presence of an antisterility factor in 1922, and fetal death and resorption in rats as a result of deficiency of this factor was reported by Mattill in 1924. Testicular degeneration was then found in the animals treated with a fat-deficient diet, and the essence of this substance termed vitamin E, as being related to these fertility changes, was propounded in the monograph by Evans and Burr, entitled "The Antisterility Vitamin: Fat-soluble E" in 1927. It was not, however, until 1931 that Olcott and Mattill recognized the potential antioxidant effect of vitamin E.

In 1936, Evans et al. isolated a pure fraction named alpha-tocopherol, and in 1937 Emerson et al. isolated the beta and gamma tocopherol forms while in that same year Olcott and Emerson discussed the mechanism of action of vitamin E, utilizing an anti-oxidant model. The structural formula of alpha-tocopherol was published by Fernholz in 1938, as shown in Figure 1. During this period it was found that vitamin E deficiency produced a nutritionally-induced muscular dystrophy in rabbits, similar to forms which had been observed in the guinea pig some seven years earlier. Later it was found that the muscle wasting in chicks, termed "exudative diathesis" was also related to vitamin E deficiency.

Dam's laboratory provided evidence of the antiencephalo-malasic property of alpha-tocopherol in chicks (1938), and, MacKenzie et al. (1940) found that a lack of alpha-tocopherol alone was responsible for the myopathic changes previously described in the rabbit. The two decades following 1920 witnessed the isolation of a fat-soluble constituent of food of importance in optimizing physiologic function in many animal species, but which had not yet been identified with any human deficiency disease. A major question which remained from this period was, and still is, whether the tocopherols as a family have functions other than those as biologic antioxidants in animals. It has become clear, however, that biologic antioxidation is a major pathway by which the tocopherol family and other lipid soluble antioxidants do, in fact, elicit their function.

We now recognize many human pathologic conditions as being associated with increased oxidation, particularly of the free-radical oxidation type (Dormandy, 1978). Generally speaking, free-radical oxidation is a result of an oxidant material interacting with a biological component, such as a cell membrane lipid or a protein, to produce a free-radical chain-carrying species which

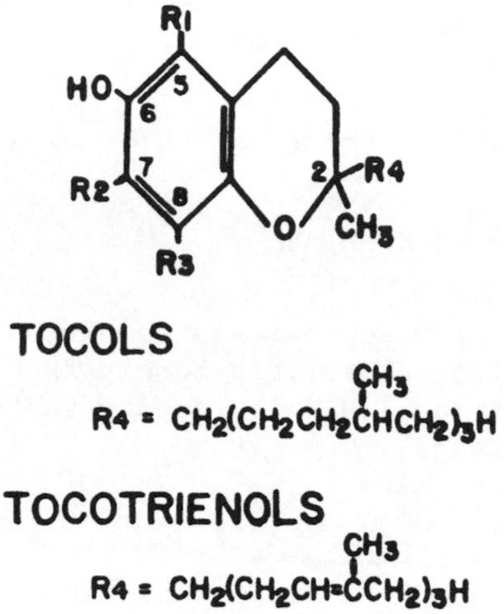

| Tocol | Tocotrienol | Methyl Positions |
|---|---|---|
| α - (alpha) | ζ - (zeta) | 5, 7, 8 |
| β - (beta) | ε - (epsilon) | 5, 8 |
| γ - (gamma) | η - (eta) | 7, 8 |
| δ - (delta) | 8-methyl-tocotrienol | 8 |

Figure 1. Chemical structures of naturally occurring tocopherols

$$A - B \xrightarrow[h\nu]{\Delta \text{ or}} A. + B. \quad (1)$$

$$M^{+} + ROOH \longrightarrow M^{+2} + .OR + {}^{-}OH \quad (2)$$

$$H_2O \xrightarrow{h\nu} H. + .OH \quad (3)$$

Figure 2. Generation of free radicals in biological systems by (1) thermal or photolytic homolysis, (2) one electron redox reactions, (3) high energy radiation and photolysis of water.

can then produce considerable cell or tissue degeneration. Lipid peroxides, for instance, are formed by a series of oxidation reactions of unsaturated lipid material and have been identified as important degradative biomolecules involved in the cellular aging process. In oxygen toxicity, air pollution, oxidative damage to cells, and a host of other degenerative biochemical transformations, the formation of lipid peroxides involves a reaction of biochemical oxidants with polyunsaturated lipids (Bland, 1976), as is diagrammed in Figure 2.

## VITAMIN E AS A FREE-RADICAL TRAP

The free-radical intermediates, as shown in Figure 2, occur in almost all biologic systems. Cells and tissues are generally protected against oxidizing free radicals by complexing of antioxidant mechanisms, some of which are intimately involved with the fat-soluble vitamin families, such as the tocopherols.

In disease, these mechanisms may fail to protect or conversely the mechanisms may fail and cause disease. The primary products of free-radical oxidation undergo rapid and spontaneous fragmentation. Many of these fragments are also highly active in biologic systems. Some have considerable survival time and may induce long-term degenerative effects. In the excellent review by Dormandy (1978), he discusses the importance of free-radical oxidation and antioxidants in human health and disease and says: "Because of the numerousness and instability of free-radical oxidation products, their characterization and isolation are fraught with difficulties, but there is no longer any doubt that many are extraordinarily powerful and have an astonishing range of effects upon biologic systems. Some are cytotoxic, some have striking bacteriostatic properties, some affect platelet aggregation, some have prostaglandin-like activity on neuromuscular transmission, and some of the processes which might be influenced or governed by these products *in vivo* have considerable survival value." The insidious character of these free-radical-induced lipid peroxidations suggests that the mechanism of formation involves a chain-carrying propagation process as seen in Figure 3 (Bland, 1976). Conceivably, then, only a few free radicals need necessarily be produced within a biochemical system to induce considerable lipid peroxidation and associated damage.

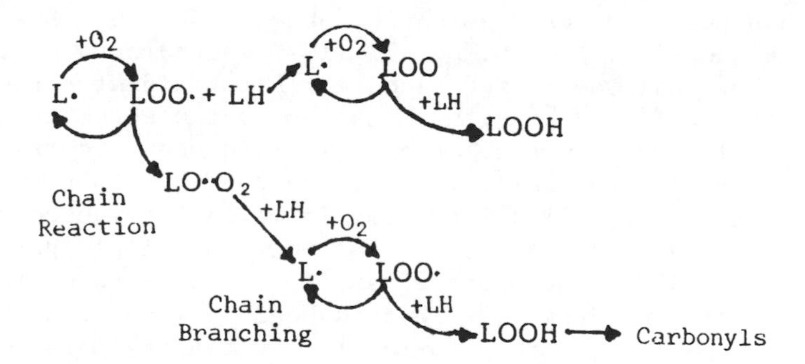


Figure 3. Mechanism of free radical reaction leading to fatty acid hydroperoxides.

Recently, Cohen and Cedarbaum (1979) have demonstrated that reduction of oxygen during microsomal electron-transfer reactions can ultimately produce hydroxyl radicals, which can participate as active chain-initiating species in production of the peroxides. Other free-radical initiators found in biological systems include hydrogen peroxide, superoxide, anion radical, and singlet oxygen. Formation of lipid peroxides is most commonly associated with regions of the biochemical systems which are abundant in polyunsaturated fatty acid phospholipids. Cellular membranes, including mitochondrial and microsomal membranes, are composed of fatty acids with two, four, five, and six double bonds, which have been shown to be rapidly peroxidized via free-radical mechanisms with increasing ease in the absence of protection by the nutrient antioxidant agents. The autooxidation of fats is a common phenomenon. In food stuffs it is known as the process which causes rancidity. It has also been shown to occur in vitro in adipose tissue and in vivo in human red blood cells under conditions of hyperoxia. It is enhanced in the presence of iron, a potent generator of free radicals, and of ascorbic acid. Co-oxidation of unsaturated fatty acids and ascorbic acid occurs as a coupled reaction with ferrous iron. The forms of vitamin E, such as alpha-tocopherol, are methyl-substituted phenolic derivatives, which have a chroman ring, accounting for their reactivity with free radicals, and a fatty acid side chain making them soluble in lipids (Figure 1).

Because of its capacity to trap free radicals, as a one-electron reducing agent, vitamin E prevents the oxidation reactions described above. As can be seen from Figure 4, as alpha tocopherol serves as a selective oxidant trap (thereby reducing the free-radical chain-carrying process), it leads to the ultimate oxidation product tocoquinone. Evidence is accumulating to demonstrate that vitamin E presents itself nicely in vivo as a selective lipid antioxidant in fat-rich organelles, such as cellular membranes. For instance, patients afflicted with beta-thalassemia major show virtually every kind of erythrocyte membrane deformation abnormality that has been described. In addition, Stocks, et al. (1972) have found an increasing susceptibility of thalassemic red blood cells to oxidant stress. Rachmilewitz et al. (1976) have found that beta-thalassemics show a nearly two-fold increase in total red cell lipids and a correspondingly larger amount of lipid per cell, which is susceptible to peroxidation. After oxidant stress they find increases in malonaldehyde concentrations which can only exist as a result of increased lipid peroxidation (Trostler et al. 1979). Graziano et al. (1976) have found recently that the iron-chelating drug 2,3-dihydroxybenzoic acid is an effective inhibitor of red cell membrane peroxidation in thalassemics. Hemolysis is almost always found to be preceded by lipid peroxidation. Barker and Brenn (1975) have found that peroxidation of red cell membrane of vitamin E deficient rats proceeds in a similar manner to the peroxidation of mycellar phospholipids in vitro, and they suggest that peroxidation exposes both polar and nonpolar lipid sites in the red cell membrane, resulting in its weakening. Kahane and Rachemilewitz (1976) have recently found that alterations due to increased lipid peroxidation in thalassemics' erythrocyte membranes can be prevented by the ingestion of larger amounts of vitamin E, which is consistent with the antioxidant model for the tocopherols.

Recently Bowie et al. (1979) have reported that the crisis of sickle cell anemia is associated with increased lipid peroxidation of the red cell membrane leading to increased malonaldehyde concentrations of membrane lipids. The sickle cells are much more susceptible to oxidatively-induced hemolysis when exposed to hydrogen peroxide than normal red cells. Vitamin E, when present in the incubation medium, was found to suppress the formation of malonaldehyde in the sickled red cell when exposed to peroxide. Interestingly, intracellular calcium, which is known to become elevated during the formation of irreversibly sickled cells, also was decreased at high vitamin E levels.

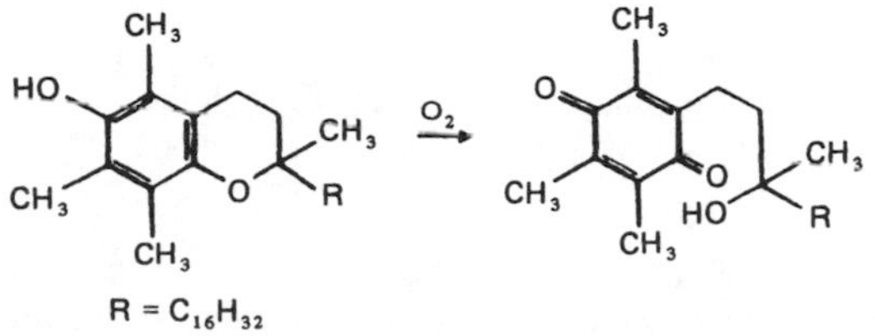

Figure 4. Oxidation of α-tocopherol

Vitamin E does not participate in this process of antioxidation by itself. It works in conjunction with other nutrient-derived antioxidants, as can be seen in Figure 5. The interaction of vitamin E with the selenium-activated enzyme, glutathione peroxidase, and sulfur amino acids renders the lipid-rich regions of tissue systems much less susceptible to peroxidative damage induced by free radical reactions.

## VITAMIN E DEFICIENCY AND ANEMIAS

Vitamin E deficiency in infants is known to lead to hemolytic anemia, with the typical syndrome appearing at four to six weeks of

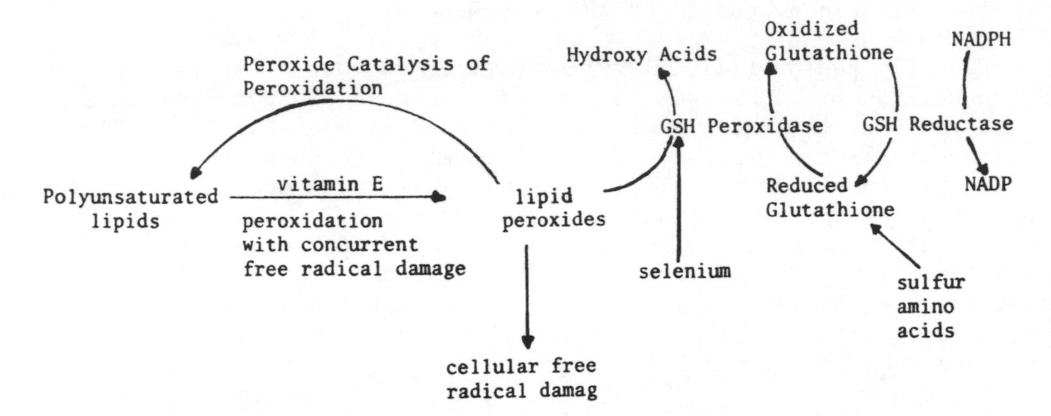


Figure 5. Interactions among selenium, vitamin E, and sulfur amino acids in inhibition of lipid peroxidation damage.

age in infants of birth weights less than 1500 grams, and is characterized by reticulocytosis, thrombocytosis and low hemoglobin values. Blood studies have revealed a variable red cell morphology, with decreased plasma concentrations of vitamin E and very marked increased sensitivity of the erythrocytes to hemolysis induced by exposure to dilute solutions of hydrogen peroxide. This demonstrates that these infants are unable to withstand oxidative stress and thereby have increased tendencies toward lipid peroxidation with the attendant loss of membrane integrity and function, resulting in anemia. A number of factors have been found to contribute to the vitamin E deficient status of the small premature infant. Even full-term infants with birth weights of 3500 grams or more are relatively vitamin E deficient at birth, having plasma concentration of 0.2 to 0.6 mg/dl (normal adult 0.75 to 0.92 mg/dl), and body stores of about 20 mg. Breast-fed infants quickly achieve adult values, but infants fed artificial formula may take months to reach adult values if their diet is not adequately supplemented. It is interesting to note that iron supplements, which are commonly given to these low-birth-weight babies, increase the rate of free-radical peroxidation, and can actually exacerbate this vitamin E deficiency in premature infants (Oski and Barness, 1967). The studies of Melhorn and Gross

(1969) demonstrated that administration of exogenous iron at a dose of 8 mg/kg/day to vitamin E deficient premature infants produced a more severe anemia than non-iron supplementation. Their conclusion was that iron precipitated hemolysis in the vitamin E deficient state by increasing free radical peroxidation. The work of Oski (1977) has demonstrated that these premature infants, when given iron in conjunction with d-alpha tocopherol administered intramuscularly in a dose of 200 to 800 mg, responded by a rise in hemoglobin from 7.6 to 9.8 g/dl, and a reduction in the hydrogen peroxide hemolysis test from 80 percent to 8 percent. In a subsequent study it was shown that daily administration of 25 mg of vitamin E by mouth, starting in the first week of life will prevent the hemolytic anemia associated with vitamin E deficiency in the low-birth-weight premature infant. The optimum dose was found to be 125 mg/kg of dl-alpha-tocopherol in a water dispersible vehicle, administered intramuscularly in four divided doses on days one, two, seven and eight of life.

The antioxidant action of vitamin E may have important implications for the integrity of other cells in addition to the red cells that are susceptible to oxidant-induced injury. Of particular importance to the human infant is a possible role of vitamin E deficiency in the genesis of retrolental fibroplasia and bronchopulmonary dysplasia. Johnson and associates (1976) have shown that even when oxygen is carefully administered, controlled or omitted, retrolental fibroplasia may occur in premature infants because the retinal vessels, which usually develop during intrauterine life at oxygen tensions of 50 to 55mm/Hg, are exposed to abnormally high tensions before retinal vascularization is complete. The result may be vasoconstriction and obliteration of retinal arteriols and capillaries. Johnson et al. (1976) conducted a clinical trial in which alternate infants with birth weights of less than 2000 g were started on vitamin E or a placebo within four to 24 hours of birth. The study was designed to determine if the vitamin E supplemented infants who had higher serum levels of vitamin E had less incidence of retrolental fibroplasia, or regression of the symptoms as determined by indirect ophthalmoscopy. Follow-up of 48 infants at one year of age revealed that six of eight with retrolental fibroplasia treated with vitamin E had normalized as compared with seven of 18 who received placebo injections, and one of 12 who were not treated.

Recently, Ehrenkranz et al. (1978) have reported that respiratory distress syndrome in neonates who develop bronchopulmonary dysplasia can be significantly ameliorated by the administration of vitamin E. Twenty infants received vitamin E administered intramuscularly during the acute phase of the syndrome and twenty infants served as controls. Administration of vitamin E significantly increased the serum vitamin E concentration. None of the nine vitamin E treated patients had changes characteristic of bronchopulmonary dysplasia, and all survived, whereas six of the 13 controls had x-ray changes consistent with the condition and four died. The conclusion of the authors was that administation of vitamin E during the acute phase of the respiratory distress syndrome appears to modify significantly the development of dysplasia.

Consistent with this view that infants, and more specifically premature infants of low birth weight, are susceptible to increased lipid peroxidation as a result of antioxidant insufficiencies is a report from Money (1978). He reported the low liver contents of vitamin E, selenium, and vitamin A in infants with sudden infant death syndrome as compared to that of control children. The results of this study indicate that low levels of vitamin E and organoselenium compounds may be important in the development of certain forms of sudden infant death syndrome and should be given further attention. Further, ferrous sulphate, an iron supplement which is known to increase peroxidation, could involve a special risk.

## VITAMIN E INSUFFICIENCY IN THE ADULT

One of the lengthiest and most extensive of the studies in adults with regard to vitamin E nutriture was that of Horwitt and coworkers (1960, 1962, 1978). Under the sponsorship of the Food and Nutrition Board of the National Academy of Sciences, this long-term project was begun in 1953 to determine the need for vitamin E by adult male subjects. Low plasma tocopherol levels and increased susceptibility of red blood cells to hemolysis in the presence of hydrogen peroxide were used as criteria of insufficient vitamin E. The basal diet supplied about three milligrams of d-alpha-tocopherol daily, and depletion occurred slowly, but was found to be accelerated by increasing the daily consumption of polyunsaturated fatty acids, principally linoleic acid. This is presumably a result of the increased need of antioxidants to protect the reactive sites of unsaturation in these fatty acids against peroxidation. During the course of this eight-year study, several changes were made, each one accentuating the decrease in plasma tocopherol. Horwitt (1956) observed that subjects maintained on low tocopherol diets for long periods of time showed small, but significant, decreases in erythrocyte lifetime and that subsequent administration of 300 mg d-alpha tocopheryl acetate daily resulted in small, but significant, increases in reticulocyte count. These results were consistent with those of Marvin et al. (1960), who showed a substantial decrease in erythrocyte survival time in monkeys which were severely deficient in vitamin E.

Malabsorption syndromes were long suspected as causes of vitamin E deficiency. In some cases of cystic fibrosis, chronic pancreatitis, biliary atresia, and celiac disease, findings of extensive deposition of lipid peroxidized material, such as ceroid, and muscle necrosis were taken as indirect evidence of vitamin E deficiency (Herting, 1966). Increasing numbers of reports have demonstrated the direct correlation between the classic signs of vitamin E deficiency (including creatinuria, ceroid deposition, muscle weakness, increased serum creatine phosphokinase, and subnormal levels of alpha tocopherol in blood or adipose tissue) in cases of cystic fibrosis, biliary atresia, nontropical sprue, xanthomatous biliary cirrhosis, chronic pancreatitis, and a host of other conditions generally characterized by steatorrhea. Thus, there appears to be little question that malabsorption syndromes lead to vitamin E deficiencies in the human.

In a recent report by Tsai et al. (1975) the effect of megavitamin E supplementation in humans was examined. Two hundred and two subjects were randomly assigned to treatment groups, one receiving 600 I.U. of dl-alpha tocopheryl acetate daily and the other a placebo tablet. The experiment was double blind and proceeded for a period of four weeks. Results of this study showed that the megavitamin E supplementation did not have an effect on work performance, sexuality, and general well being, and did not cause muscular weakness or gastrointestinal disturbances, based upon subjective evaluations of the treated group. Vitamin E supplementation at this level did not affect the prothrombin time, total blood leucocyte count, or serum creatine phosphokinase activity of the subjects. Serum triglyceride levels were significantly elevated in vitamin E supplemented females, but not in males. Megavitamin E supplementation significantly decreased serum tri- and tetra-iodothyronine concentrations in males and in females not using steroid oral contraceptive agents. In general, then, the study indicated that under experimental conditions of the study, vitamin E supplementation at fairly large levels does not have significant beneficial or undesirable effects on general health conditions.

Recently, evidence has accumulated which suggests that higher doses of vitamin E than the National Academy of Sciences Recommended Dietary Allowance of 15 I.U. per day may be advantageous in certain circumstances in promoting enhanced antioxidant activity when an individual is exposed to oxidant stress. Working with human red blood cells, our group found that when 600 I.U. of vitamin E are administered orally on a daily basis an enhanced protection against photodynamic-induced oxidative damage to the erythrocyte (achieved by exposing the red blood cell to oxygen and light simultaneously) can be achieved *in vivo* as compared to the same group serving as its own control on a non-vitamin E supplemented regime (Bland et al. 1975). Figure 6 shows the significant difference between the vitamin E supplemented group and the same group with no vitamin E supplementation. This work *in vivo* is consistent with the work of

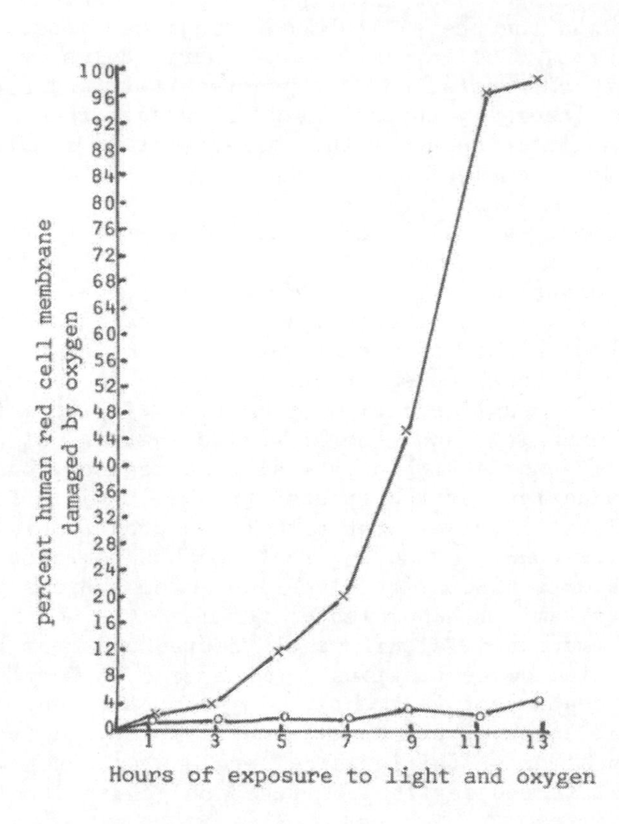


Figure 6. Effect of α-tocopherol on human red blood cell sensitivity to photo-induced membrane damage.
x - controls, ingesting normal diet unsupplemented
o - same subjects, ingesting normal diet supplemented with 600 IU α-tocopherol daily
from Bland et al. (1975).

Goldstein and Harber (1972), which demonstrated that alpha-tocopherol was found to be an effective agent in providing partial protection against photohemolysis when erythrocytes from erythropoietic protoporphyric donors were treated in vitro with an emulsion of vitamin E and then exposed to long wave-length ultraviolet radiation in the presence of oxygen. As was demonstrated by Grams et al. (1972), vitamin E is readily oxidized in the presence of light, oxygen and a sensitizing dye, such as hemoglobin. The model which has been formulated to account for the ability of vitamin E to prevent oxidative hemolysis of red blood cells involves the inhibition of conversion via photooxygenation of membrane-bound cholesterol to its hydroperoxide (as shown in Figure 7), or possibly other unsaturated fatty acids to their hydroperoxides, which could then participate in free radical chain-carrying destruction of the cell membrane. Our work has shown that vitamin E treated donors have red cell membranes with higher vitamin E concentration and an enhanced ability to prevent the conversion of membrane-bound cholesterol to its hydroperoxide. This hydroperoxide leads to membrane deformation and increased membrane lysis (Lamola et al. 1973). Incorporation of cholesterol hydroperoxide into the erythrocyte membrane has been demonstrated by Lamola et al. (1973) to create a much more fragile erythrocyte membrane, as shown in Figure 8. The erythrocyte system is uniquely disposed to oxidative hemolysis, in that the hemoglobin in the red cell is a very efficient photosensitizer of oxygen to its excited species, singlet oxygen, which can then participate vigorously in peroxidation of unsaturated residues within the cell membrane bilayer. In fact, Lubin et al. (1972) report evidence for involvement of hemoglobin-bound oxygen in hemolysis of vitamin E deficient erythrocytes. To determine the active oxidant species in the erythrocyte of the vitamin E insufficient case, we examined the appearance of cholesterol hydroperoxide in the erythrocyte membrane, utilizing a $C_{14}$ cholesterol-labeled membrane in the presence and absence of various levels of vitamin E. In Figure 9, it can be seen that in the absence of vitamin E, the rate of production of hydroperoxide within the cellular membrane was considerably greater than in the cells which had been treated with 0.56 mM tocopherol (Bland et al. 1978). Merkel and Kearns (1972) have shown that one of the potential oxidant species in this process, singlet oxygen (an activated oxidation species some 22 Kcals per mole above the ground state energy of oxygen), is intimated to be involved in many of these photodynamic, light-induced oxidation processes, and has a ten-fold longer lifetime in deuterium oxide than in normal water. This enhanced lifetime has been used as a test for the presence of singlet oxygen in photooxidations. As can be seen in Figure 10, the rate of photoperoxidation is greatly enhanced in the presence of a selective singlet oxygen trapping agent, isobenzofuran. These data then clearly implicate the involvement of singlet oxygen in the erythrocyte photooxidation process and demonstrate that tocopherols

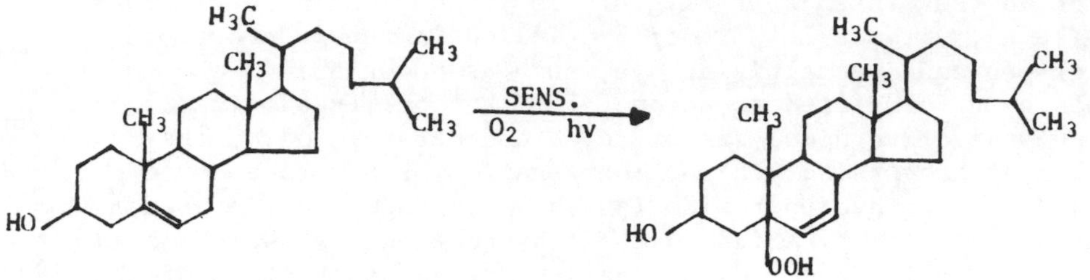

Figure 7. Photooxidation of cholesterol and conversion to its 5-hydroperoxide.

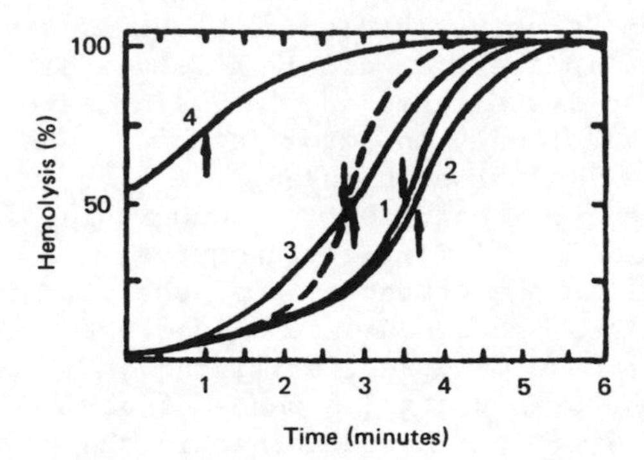


Figure 8. Cholesterol hydroperoxide effects upon the stability of the erythrocyte membrane. The solid curves show hemolysis of cholesterol hydroperoxide containing RBC after incubation with the hydroperoxide for 10 minutes (curve 1), 20 minutes (curve 2), 50 minutes (curve 3) or 80 minutes (curve 4) at 37°C. The arrows represent the time taken in each for 50 percent hemolysis to occur.

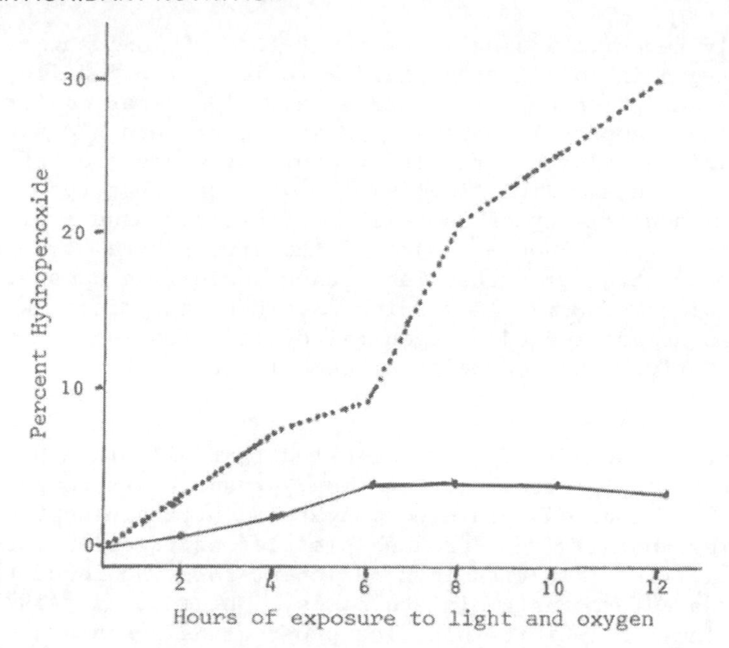


Figure 9. $^{14}$C-Cholesterol-labeled erythrocyte membrane photooxidation (Bland et al. 1978).

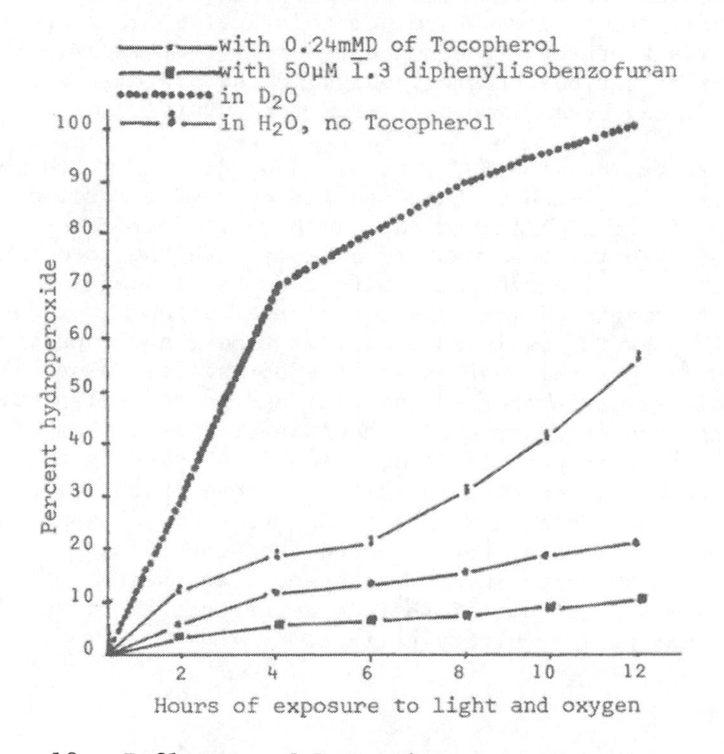


Figure 10. Influence of $D_2O$ and isobenzofuran on erythrocyte membrane photoxidations (Bland et al. 1978).

must selectively react with singlet oxygen before the oxidant reacts with singlet oxygen in solution to yield a variety of oxidation products. A recent paper by Matsushita et al. (1977) has confirmed the fact that the tocopherols have very significant quenching effects upon photosensitized oxidation resulting from the production of singlet oxygen. Prince and Little (1973) have found that tocopherol reduces the radiosensitivity of mammalian erythrocytes and again suggests that its effect upon exposure of the erythrocyte to high energy radiation is that of protection against oxidative damage. Other types of blood abnormalities which have been associated with vitamin E use in humans include congenital dyserythropoietic anemia of the type II variety and reduction of osmotic fragility in beta-thalassemia major.

The work of Steiner (1977) has suggested that vitamin E has an inhibitory effect upon platelet aggregation, presumably through the inhibition of the endogenous platelet conversion of arachidonic acid into prostaglandin. The increased platelet aggregation observed by Machlin et al. (1975) in vitamin E deficient rats was found to be related to increased prostaglandin synthesis. White et al. (1977) suggest that vitamin E inhibits platelet prostaglandin synthesis by blocking conversion of arachidonic acid into a free radical state necessary for the insertion of arachidonic acid into the cyclo-oxygenase or lipooxygenase pathways. This raises then the potentiality that vitamin E may have hematological effects other than just its antioxidant effect. The suggested role that vitamin E plays in the synthesis of various prostaglandins in the platelet and elsewhere in the body may be an important area for future research.

Recently, Meyers et al. (1977) have found that the anti-tumor antibiotic adriamycin, which is known to induce severe cardiac toxicity by way of lipid peroxidation, can have its cardiac toxicity reduced by prior treatment of animals with the free radical scavenger tocopherol. The treatment with tocopherol does not, however, alter the magnitude or duration of the adriamycin-induced suppression of DNA synthesis in P388 ascites tumor, nor does it diminish the anti-tumor responsiveness of P388 ascites tumor. This remarkable result suggests that adriamycin, and potentially other cytotoxic agents, has at least two mechanisms of tissue damage: one which involves lipid peroxidation, which is blocked by tocopherol and may result in cardiac toxicity, and one which involves binding to DNA, is not antagonized by tocopherol and is responsible for tumor response. It might be suggested from these data that most all drugs which are known to induce increased free radical perioxidation damage may have some of their side effects ameliorated by pretreatment of the patient with vitamin E.

## RELATIONSHIP OF VITAMIN E DIETARY LEVELS TO SERUM VITAMIN E CONCENTRATIONS

Farrell and Bieri (1975), in assessing the possible toxic and/or beneficial effects of vitamin E supplementation, examined a group of 28 adults who were voluntarily ingesting 100 to 800 I.U. tocopherol per day for an average of three years. No evidence of toxicity was apparent on reviewing past medical histories with the subjects. Plasma tocopherol was found to be elevated significantly, from 650 μg/dl (control group mean) to 1340 μg/dl (experimental group mean). Plasma alpha tocopherol concentrations were found not to correlate with total daily dose in this group, but did relate to plasma triglyceride and cholesterol concentrations. Laboratory screening for potential toxic side effects of vitamin E supplementation by performance of 20 standard clinical blood tests failed to reveal any disturbance in liver, kidney, muscle, thyroid, erythrocytes, leucocytes, coagulation parameters, or blood glucose concentrations. It was concluded from this study that megavitamin E supplements in this group produced no apparent toxic side effects, and that subjective claims for beneficial effects were highly variable. Herting and Drury (1965) compared plasma tocopherol concentrations to erythrocyte hemolysis data in normal subjects, stating that plasma alpha-tocopherol levels less than 0.5 mg/dl suggested vitamin E insufficiency. Average plasma tocopherol levels were found to range between 0.358 to 0.507 mg/dl, suggesting that there was considerable incidence of vitamin E insufficiency in the subjects studied in Pittsburgh, Pennsylvania and New York.

The question as to the absorption of dietary vitamin E, in that it is a fat-soluble vitamin and would be expected to be more transported through the lymphatic system than the portal blood system into the liver, has been examined by several investigators. Blomstrand and Forsgren (1968) administered radio-labeled tocopheryl acetate to human subjects whose thoracic ducts were cannulated. Serial samples of lymph were collected and vitamin E concentration was determined. Their results show that tocopherols were absorbed via the lymphatic pathway in humans, that tocopherol enters the thoracic duct lymph intact, and that tocopheryl acetate is split and recovered as tocopherol, presumably via the action of lipase or esterase enzymes in pancreatic juice. It was found that in patients with biliary obstruction the lymph contained only minute quantities of vitamin E and therefore gallbladder problems, pancreatitis, or pancreatic insufficiency may all be related to inefficient absorption of vitamin E. In a study by Gallo-Torres et al. (1971), it was found that the absorption of vitamin E varied with the fat content of the diet. Serum concentrations of vitamin E were found to be generally logarithmically related to dietary intake, with ten-fold higher ingestion leading to doubled serum concentrations of vitamin E. It was also found that a higher percentage of dietary saturated triglycerides promoted a correspondingly higher absorption of vitamin E, whereas an increase in linoleic acid, or other polyunsaturated fatty acids, depressed the absorption of vitamin E. In their work it was found that 80 to 90 percent of the vitamin E absorbed into the lymphatic system was in the form of unesterified tocopherol, regardless of the dietary lipid in which the vitamin was emulsified, or the form in which vitamin E was administered, be it either the acetate or succinate form. This same association between depressed vitamin E absorption and greater unsaturated fatty acid content of the diet was reported by Muralidhara and Hollander (1977) who found that vitamin E is absorbed by a passive diffusion process into the lymph and is dependent upon micelle formation, which is depressed in the presence of polyunsaturated fatty acids. Bieri et al. (1977) have examined the kinetics of uptake of vitamin E and have found that it exchanges rapidly between plasma and red cells, with an equilibrium being reached in six to eight hours. There are, however, several variables which can influence the distribution of tocopherol between red cells and plasma. Farrell and Bieri (1975) found that the plasma concentration of alpha-tocopherol correlates strongly with the plasma concentration of total lipids, as beta-lipoprotein, cholesterol, triglycerides, and phospholipids. They showed that as total lipids in the plasma increase beyond 15 mg/ml, the concentration of vitamin E in red cells goes down appreciably. This suggests that individuals suffering from lipoprotein abnormalities may have reduced red cell vitamin E concentrations and enhanced red cell hemolysis, as a result of free radical peroxidation. In order to determine the influence that feeding of increasing amounts of polyunsaturated fatty acids and other saturated fats had upon serum vitamin E concentration, a study was undertaken by Bieri et al. (1978) in rats. Alpha-tocopherol was kept at a constant level between 30 to 50 mg/kg in different dietary experiments. After eight to ten weeks, tissues were analyzed for tocopherol and total polyunsaturated fatty acids (PUFA). There was a tendency for plasma alpha-tocopherol to decrease with increasing dietary PUFA. Tissue polyunsaturated fatty acid increased as dietary PUFA increased. The ratio of alpha-tocopherol to PUFA in several tissues declined more with the first increase in dietary PUFA than with the second increase. It was concluded that dietary PUFA has a variable but small effect on lowering the tissue content of alpha-tocopherol, and that the change in vitamin E status, as judged by the tissue ratio of alpha tocopherol to PUFA, is primarily a result of increased tissue deposition of PUFA. This would suggest then that high polyunsaturated fat diets utilized in the management of lipoprotein abnormalities, may be resulting in the production of low tocopherol-to-PUFA concentrations within cellular membranes, thereby rendering them susceptible to greater lipid peroxidation and free radical pathology. It appears that vitamin E should be ingested in larger amounts when the unsaturated lipid content of the diet is increased.

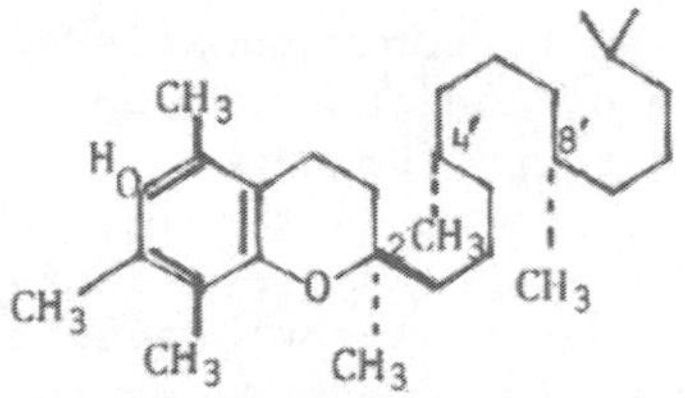

Fig. 11. Configuration of d-α-tocopherol (2D, 4'D, 8'D-α-tocopherol).

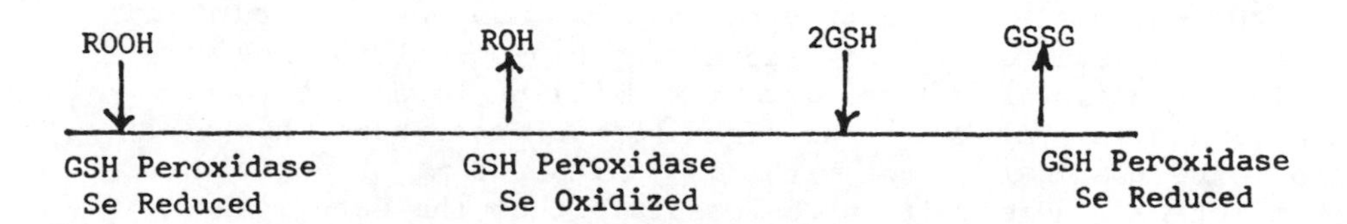


Figure 12. Reaction of glutathione peroxidase, glutathione (GSH), and glutathione disulfide (GSSG).

FORMS OF VITAMIN E AND THEIR BIOLOGIC ACTIVITY

Vitamin E is known to be a vitamin which has several members in its family, including alpha, beta, gamma, and delta forms of tocopherol, as shown in Figure 1. Vitamin E is also known to have three chiral centers, and therefore possesses eight potential stereoisomers, only one of which is the naturally derived form of the tocopherols. Synthetic production of alpha-tocopherol will provide a racemic modification containing 8 stereoisomers. The α-tocopherol mixture that was formerly considered synthetic, shown in Figure 11, was called ambo, and can be seen to have only a configurational change at carbon atom number two. The presently available synthetic dl-α-tocopherol is totally racemic at all 3 chiral centers and therefore its biological activity has been found by Ames (1979) to be lower than has been suggested in the National Formulary. This relationship is demonstrated in Table I. Dam and Sondergaard (1964) found from a comparison of the natural versus the synthetic tocopheryl acetates against encephalomalacia that the potencies were 168 for the d form, 100 for the dl form, and 44 for the unnatural ℓ form of the vitamin. It is important, however, to recall that the potencies of these compounds as related to their activities in international units are based upon a specific biological assay criterion. With regard to vitamin E, the relative activities have been measured by one of the following methods, including resorption, gestation in the rat, dialuric acid hemolysis in the rat, respiratory decline in rat liver, encephalomalacia in the chick, and exudative diathesis in the chick. As we learn more about the human activities of vitamin E, it may possibly turn out that the standard methods of determining biological potency for vitamin E, utilizing the rat or chick, will be found not to be suitable assays. Although the alpha form of tocopherol has been found to be the most potent with 1.38 units per mg of activity in rat and chick studies, it may be possible that the beta, gamma, or delta forms of vitamin E play significant roles in human-derived antioxidant functions. It is, therefore, possible that we have gone to high therapeutic potencies of the alpha form of vitamin E commercially at the expense of losing some of the important biological activities of the additional forms of vitamin E found in the natural mixtures, as beta, gamma, and delta.

THE MEASUREMENT OF SERUM VITAMIN E CONCENTRATIONS

In the late 1940s the syndrome of malabsorption and deficiency of tocopherol were linked (Pappenheimer, 1946; Ansanelli and Lane, 1957). Isolated reports initially associated nontropical sprue with histologic evidence of avitaminosis E, characterized by ceroid pigment in smooth muscle of the alimentary tract. Subsequently, Nitowsky and coworkers (1956a, 1956b) found low blood tocopherol levels in children with cystic fibrosis, congenital biliary atresia

Table I. Determination of Relative Potencies of Several Forms of α-tocopherol

| Form of α-tocopherol | Standard | No. of bioassays | Log relative potency | Relative potency |
|---|---|---|---|---|
| *RRR*-α-tocopherol acetate | *2-ambo*-α-tocopheryl acetate | 28 | 0.219 | 1.66** |
| *RRR*-α-tocopherol | *2-ambo*-α-tocopherol | 18 | 0.118 | 1.31 |
| *all-rac*-α-tocopheryl acetate | *2-ambo*-α-tocopheryl acetate | 11 | -0.080 | 0.83* |
| *all-rac*-α-tocopheryl acetate | *RRR*-α-tocopheryl acetate | 19 | -0.282 | 0.52** |

* $p \leq 0.05$ or ** $p < 0.01$

and, to a lesser extent, cirrhosis of the liver. In a more recent study Braunstein (1961) found that tocopherol deficiency, as determined by lowered serum tocopherol concentrations, was associated with the presence of chronic pancreatitis.

In most early studies the serum tocopherol concentrations were measured colorimetrically by the Emmerie-Engel method after elimination of the interference by carotenoids and similar substances by catalytic hydrogenation (Swick and Bauman, 1952; Quaife and Biehler, 1945). This procedure, and variants which eliminated the hydrogenation by a background correction for interfering carotenoids, was both time-consuming and subject to problems in maintaining reproducibility. More recently, fluorimetric techniques have been developed for the determination of serum or plasma tocopherol (Taylor et al.1976; Hashim and Schuttringer, 1966). The method involves saponification of plasma by KOH in the presence of ascorbic acid. The vitamin E is extracted into spectroscopic-grade hexane and the fluorescence intensity in the hexane phase measured at 298nm excitation and 322nm emission at slit widths of 8 and 10nm respectively. It has been found, however, that contamination of the sample can occur as a result of exposure to the rubber stopper of the Vacutainer tube (Sinclair and Slattery, 1978). Whereas vitamin E values between 1.40 and 1.50 mg/liter were found for all plasma samples from plastic tubes without Vacutainer rubber stoppers, those with stoppers were found to be 11.0, 8.7, and 41.5 mg/liter tocopherol for samples left in contact with red, green and purple stoppers respectively. Clearly, these results indicate precaution when blood is collected for sensitive fluorimetric assay of vitamin E.

The most recent method which may become the method of choice for serum vitamin E determination is that of high pressure liquid chromatographic procedures. This technique allows not only the measurement of total tocopherol concentration, but also the fractionation of the various forms of tocopherol, be it alpha, beta, gamma, or delta. The method of DeLeenheer and coworkers (1978, 1979) has allowed for the routine separation of tocopherols in about 30 minutes utilizing a reverse phase column. This technique is both fast and free of most interferences and has the additional advantage of being able to look at the concentrations of each of the tocopherol positional isomers. It is important to recall that in the future it may become important to fractionate the tocopherols, because each has a different biological potency with regard to specific antioxidant or other physiological effects. Using this technique, concentrations of the various tocopherols in serum vary between 6.6 and 15 mg/liter for alpha, 0.0 to 0.2 mg/liter for beta, and 0.7 to 2.7 mg/liter for gamma. Utilizing high pressure-high performance liquid chromatography, the determination of serum vitamin E concentration should be readily obtainable in the clinical laboratory and provide important information relating to the antioxidant sufficiency of the patient.

## LIVER NECROSIS, VITAMIN E AND SELENIUM DEFICIENCIES

In 1961 Schwarz recognized the Factor Three as a separate dietary agent, and this product has ultimately been shown to be an organic selenium compound. It is effective at very small dose levels and is replaceable in the diet by small amounts of inorganic selenium compounds. The factor protects against various deficiency disease, which hitherto have been found to be solely attributable to a deficiency of vitamin E. Factor Three has been shown to prevent dietary liver necrosis in the rat and multiple necrotic degeneration of the heart, liver, muscle, and kidney in the mouse, as well as fatal exudative diathesis in the chick and in the turkey (Schwarz and Foltz, 1958). Factor Three, which is derived in reasonably high concentration from brewers yeast, has ultimately been suggested to be a mixture of organoselenium compounds which seem to act as precursors to the active antioxidant enzyme glutathione peroxidase. This relationship between selenium and the metalloenzyme glutathione peroxidase represents the first reported example of the essentiality in humans of the inorganic element selenium. Selenium, one of the most potent nutritional agents known, affords protection when added to the diet in the ten-parts-per-billion range, and the National Research Council suggested daily dietary intake is from 90 to 120 μg. The element is in the active site of the enzyme glutathione peroxidase, which is essential for the maintenance of normal functions and structural integrity of erythrocytes, liver, kidney, testes, and other organs. It has subsequently been found by Schwarz and Pathak (1975) that the organoselenium compounds, including the selenodicarboxylic acids, are much less toxic with higher biologic potencies than the inorganic selenite or selenate forms. Hoekstra (1975) has found that simultaneous dietary deficiencies of vitamin E and selenium in weanling rats result in fatal hepatic necrosis. They found that glutathione peroxidase activity was reduced by 97 percent in selenium and vitamin E deficiency in the liver and the oxidation products of lipid peroxidation, including malonaldehyde, were increased by seven fold, as compared with rats fed the diet supplemented with selenium and vitamin E. The decrease in glutathione peroxidase was completely prevented by selenium administration, but vitamin E had a small effect. More recently, Lane et al. (1979) have found the effects of various levels of selenium and vitamin E on glutathione peroxidase activity in the intestine and liver to be related to whether the rats were fed a corn/soy bean diet or torula yeast diet. Rats fed the corn/soy bean diets had greater glutathione peroxidase activity in the small intestine, colon and liver tissues, catalase activity and selenium in the liver, and body weight gains were higher than those fed the selenium deficient-torula yeast diets. Tocopherol supplementation had no significant effect on glutathione peroxidase activity in rats fed torula yeast or corn/soy bean diets supplemented with selenium, suggesting that selenium does, in fact, play the selective role in activation of the peroxidase enzyme. Tappel (1974) has demonstrated that, of the many known interrelationships between vitamin E and selenium which have been investigated, the influence of selenium upon the glutathione peroxidase enzyme and its subsequent effect upon the general chain-breaking antioxidant properties of vitamin E seems to explain the synergistic relationship between these nutrients. Direct correlation has been found by Flohe (1971) and Little and O'Brien (1968) between the glutathione perioxidase enzymes from the tissues studied and their specificity for the reduction of hydroperoxides through the intermediary of glutathione as a reductant. The activity of glutathione peroxidase seems to be directly related to the level of dietary selenium up to a point of saturation. The influence of glutathione peroxidase is as shown in Figure 12, where the enzyme is seen to reactivate glutathione disulfide, a membrane-bound product of free radical oxidation, to glutathione in order for an additional antioxidation to occur (Fischer and Whanger 1977). Vitamin E and glutathione peroxidase work together synergistically to provide a dual mechanism for protection against lipid peroxidation (Christophersen, 1969; Schwarz, 1965). The vitamin E, which is lipid soluble and sits within regions such as the cellular membrane, awaits the production of an incipient free radical, or hydroperoxide. This hydroperoxide can be chain radical terminated by abstraction of a hydrogen atom from tocopherol to produce tocopherol quinone, thereby quenching the chain radical process. If, however, the vitamin E is ineffective in dealing with this free radical, the glutatione system can render the free radical oxidant species into a reduced form which inactivates it chemically, thereby oxidizing glutathione to glutathione disulfide. In order for the cycle to continue, glutathione disulfide is then regenerated back to glutathione by glutathione peroxidase, and the resultant glutathione is now ready to serve as an additional reducing agent for another free radical. It is, therefore, clear that vitamin E in conjunction with organoselenium compounds and the essential amino acid cystine or methionine as a precursor to glutathione, is essential for the production of an optimal antioxidant arsenal which protects against _in vivo_ lipid peroxidation.

## SUMMARY

It can be seen from this discussion vitamin E is a powerful biological antioxidant, which works in conjunction with other essential nutritional constituents, such as selenium and the sulfur-containing amino acids, to affect protection against free radical peroxidation and pathology _in vivo_. Considerable evidence has accumulated in many animal systems that vitamin E deficiency produces discrete pathology; however, as of yet there is only a small amount of literature dealing with human chronic vitamin E insufficiencies and their relationship to histopathologic changes. The suggested use of vitamin E as a therapeutic agent in certain human problems, such as respiratory distress in the neonate, and prevention of retrolental fibroplasia in the premature infant, as well as the potential use in the autoimmune disease of the adult human, suggest that both clinicians and fundamental bioscientists have yet considerable more work to do on both the biological mechanisms and the clinical implications of therapeutic doses of vitamin E. The data at this time, taken *in toto*, would suggest that vitamin E is not an agent which has many of the direct therapeutic effects in the human that are proven in animals, but that it may be a significant factor whose deficiency is related to the progression of many of the lipid peroxidation-induced degenerative diseases, such as atherosclerosis, connective tissue diseases, liver necrosis, the myopathies, and potentially even some forms of cancer, as it relates to carcinogenesis through lipid peroxidation. At this point in time, these associations are working hypotheses, however, and not discrete proven causal relationships.

As a recent review by Chow (1979) has pointed out, cellular antioxidant mechanisms vary from tissue to tissue. Many metabolic reactions and dietary components play an important role in cellular antioxidant defense, many of these may yet remain to be studied. Further study at the cellular level with humans will be necessary to have a better understanding of the full clinical importance of this area.

The best operative philosophy as it relates to the clinical application of vitamin E is taken from Leonardo de Vinci, who is reputed to have generalized, "The supreme misfortune that can befall any human is for him to embrace a theory, mistaking it for a fact." The spectrum of activities of the antioxidant nutrients appears at this point to be quite wide, yet the full unequivocal demonstration of their effects remains for future work and proof.

REFERENCES

Ames, S., 1979, Bipotencies in rats of several forms of α-tocopherol, *J. Nutr.*, 109:2198.

Ansanelli, G., and Lane, N., 1957, Ceroid pigmentation of the small intestine, *Ann. Surg.*, 146:117.

Barker, M. O., and Brin, M., 1975, Vitamin and phospholipid peroxidation, *Arch. Biochem. Biophys.*, 169:506.

Bieri, J. G., Evarts, R. P., and Thorp, S., 1977, Factors affecting the exchange of tocopherol between red cells and plasma, *Am. J. Clin. Nutr.*, 30:686.

Bieri, J. G., Thorp, S. L., and Tolliver, T. J., 1978, Effect of dietary polyunsaturated fatty acids on tissue vitamin E status, *J. Nutr.*, 108:392.

Bland, J., 1976, Biochemical effects of excited state molecular oxygen, J. Chem. Ed., 53:274.
Bland, J., Madden, P., and Herbert, E.J., 1975, Effect of α-tocopherol on the rate of photohemolysis of human erythrocytes, Physiol. Chem. Physics, 7:69.
Bland, J., Canfield, W., Kennedy, T., Vincent, J., and Wells, R., 1978, Effect of tocopherol on photooxidation rate in human erythrocyte membrane in vitro, Physiol. Chem. Physics, 10:145.
Blomstrand, R., and Forsgren, L., 1968, Labeled tocopherols in man, Intern. Z. Forschung, 38:328.
Bowie, L., Carreathers, S. A., and Wright, A. G., 1979, "Lipid Membrane Peroxidation, Vitamin E, and the Generation of Irreversibly Sickled Cells," Abstracts of 31st National Meeting of American Association for Clinical Chemistry, New Orleans (July, 1979).
Braunstein, H., 1961, Tocopherol deficiency in adults with chronic pancreatitis, Gastroenterology, 40:224.
Chow, C. K., 1979, Nutritional influence on cellular antioxidant defense systems, Am. J. Clin. Nutr., 32:1066.
Christophersen, B. O., 1969, Reduction of linolenic acid hydroperoxide by glutathione peroxidase, Biochem. Biophys. Acta, 176:463.
Cohen, G., and Cederbaum, A. I., 1979, Chemical evidence for the production of hydroxyl radicals during microsomal electron transport, Science, 204:66.
Dam, H., and Glavind, J., 1939, Vitamin E and exudative diathesis, Skand. Arch. Physiol., 82:229.
Dam, H., and Sondergaard, E., 1964, Comparison of isomers of vitamin E, Z. Ernahrungswissenschaft, 5:73.
Dam, H. J., Glavind, J., and Hagens, E., 1938, Vitamin E and encephalomalacia in chicks, Nature, 142:1157.
De Leenheer, A. P., De Bevere, V. O., and Claeys, A., 1978, Determination of serum α-tocopherol by high performance liquid chromatography, Clin. Chem., 24:585.
De Leenheer, A. P., Veerle, O. R., and Daeys, A., 1979, Measurement of α-, β-, γ-tocopherol in serum by liquid chromatography, Clin. Chem., 25:425.
Dormandy, T. L., 1978, Free radical oxidation and antioxidants, Lancet, I:647.
Ehrenkranz, R. A., Donta, B. W., Albow, R. C., and Warshaw, J. B., 1978, Amelioration of bronchopulmonary dysplasia after vitamin E administration, New Engl. J. Med., 229:564.
Evans, H. M., and Bishop, K. S., 1923, Fat soluble antifertility factor, J. Metab. Res., 2:223.
Farrell, P. M., and Bieri, J. G., 1975, Megavitamin E supplementation in man, Am. J. Clin. Nutr., 28:1381.
Fischer, W. C., and Whanger, P. D., 1977, Effects of selenium deficiency on vitamin E deficiency in rats, J. Nutr. Sci. Vitaminol., 23:273.

Flohe, L., 1971, Glutathione peroxidase: enzymology and biological aspects, Klin. Wochschr., 49:669.

Gallo-Torres, H. E., Weber, F., and Wiss, O., 1971, The effect of different dietary lipids on lymphatic appearance of vitamin E, Int. J. Vit. Nutr. Res., 41:504.

Goldstein, B. D., and Harber, L. C., 1972, EPP red cell hemolysis and vitamin E, J. Clin. Invest., 51:892.

Grams, G. W., Eskins, K., and Inglett, G. E., 1972, Photooxidation of vitamin E, J. Am. Chem. Soc., 94:866.

Graziano, J. A., Miller, D. R., and Cerami, A., 1976, Erythrocyte membrane lipid peroxidation during the sickling process, Br. J. Haematol., 32:351.

Hashim, S. A., and Schuttringer, G. R., 1966, Rapid determination of tocopherol in microquantities of plasma, Am. J. Clin. Nutr., 19:137.

Herting, D. C., 1966, Perspective on vitamin E, Am. J. Clin. Nutr., 19:210.

Herting, D. C., and Drury, E., 1965, Plasma tocopherol levels in man, Am. J. Clin. Nutr., 17:351.

Hoekstra, W. G., 1975, Biochemical role of selenium and relationship to vitamin E, Fed. Proc., 34:2083.

Horwitt, M. K., 1960, Vitamin E and lipid metabolism in man, Am. J. Clin. Nutr., 8:451.

Horwitt, M. K., 1962, Interrelations between vitamin E and polyunsaturated fatty acids in adult men, Vitamins and Hormones, 20:541.

Horwitt, M. K., Harvey, C. C., Duncan, G. D., and Wilson, W. C., 1956, Effects of limited tocopherol intake in man with relationship to erythrocyte hemolysis, Am. J. Clin. Nutr., 4:408.

Johnson, L. H., Schaffer, D. B., Rubinstein, D., Crawford, C. S., and Bogg, T. R., 1976, The role of vitamin E in retrolental fibroplasia, Fed. Proc., 1966:425 (A748).

Kahane, I., and Rachemilewitz, E. A., 1976, Alterations in red blood cell membrane and the effect of vitamin E on osmotic fragility in β-thalassemia major, Isr. J. Med. Sci., 12:11.

Lamola, A. A., Yamane, T., and Trozzolo, A. M., 1973, Erythrocyte membrane cholesterol photooxidation, Science, 179:1131.

Lane, H. W., Shirley, R. L., and Cerda, J. J., 1979, Glutathione peroxidase activity in intestinal and liver tissue of rats fed various levels of selenium, sulfur and α-tocopherol, J. Nutr., 109:444.

Little, C., and O'Brien, P. J., 1968, An intracellular GSH-peroxidase with lipid peroxide substance, Biochem. Phiophys. Acta, 31:145.

Lubin, B., Fromm, M., and Oski, F., 1972, The role of hemoglobin oxygen in the hemolysis of vitamin E deficient human erythrocytes, Pediatr. Res., 6:371.

Machlin, L., 1975, Vitamin E and prostaglandins, Proc. Soc. Exp. Biol. Med., 149:275.

Mackenzie, C. G., Levine, M. D., and McCollum, E. V., 1940, Myopathic changes in the rabbit as related to vitamin E deficiency, J. Nutr., 20:399.

Marvin, H. N., Denning, J. S., and Day, P. L., 1960, Erythrocyte survival in vitamin E deficient monkeys, Proc. Soc. Exper. Biol. Med., 105:473.

Mason, K., 1977, The first two decades of vitamin E, Fed. Proc., 36:1906.

Matsushita, S., Terao, J., and Yamauchi, R., 1977, Photosensitized oxidation of unsaturated fatty acid esters and quenching effects of tocopherols on singlet oxygen, Int. Symposium on Tocopherol, Oxygen and Biomembranes (Japan, 1977).

Melhorn, D. K., and Gross, S., 1969, Relationships between dextran and vitamin E in iron deficiency anemia in children, J. Lab. Clin. Med., 74:789.

Merkel, P. B., and Kearns, D. R., 1972, Deuterium effects on singlet oxygen lifetimes in solution, J. Am. Chem. Soc., 94:1030.

Money, D. F. L., 1978, Vitamin E, selenium and vitamin A content of livers from sudden infant death cases, N. Zealand J. Sci., 21:41.

Muralidhara, K. S. and Hollander, D., 1977, Intestinal absorption of α-tocopherol. The influence of luminal constituents on the absorptive process, J. Lab. Clin. Med., 90:88.

Myers, C. E., McGuire, W. P., Liss, R. H., Ifrim, I., and Young, R. C., 1977, Adriamycin: The role of lipid peroxidation in cardiac toxidity and tumor response, Science, 197:165.

Nitowsky, H. M., Cornblath, M., and Gordon, H. H., 1956a, Studies of tocopherol deficiency in infants and children II. Plasma tocopherol and erythrocyte hemolysis by hydrogen peroxide, J. Dis. Child., 92:164.

Nitowsky, H. M., Gordon, H. H., and Tildon, J. J., 1956b, Studies of tocopherol deficiency in infants and children IV. The effect of alpha-tocopherol on creatinuria in patients with cystic fibrosis of the pancreas and biliary atresia, Bull. Johns Hopkins Hospt., 98:361.

Oski, F. A., 1977, Metabolism and physiological roles of vitamin E, Hosp. Prac., 1977 (10):79.

Oski, F. A., and Barness, I. A., 1967, Vitamin E deficiency: a previously unrecognized cause of hemolytic anemia in the premature infant, J. Pediatr., 70:211.

Pappenheimer, A. M., 1946, Ceroid pigment in human tissues, Amer. J. Path., 22:395.

Prince, E. W., and Little, J. B., 1973, Effect of dietary fatty acids and tocopherol on the radiosensitivity of mammalian erythrocytes, Rad. Res., 53:59.

Quaife, M. L., and Biehler, R., 1945, A simplified hydrogenation technique for the determination of blood plasma tocopherols, J. Biol. Chem., 159:663.

Rachmilewitz, E. A., Lubin, B. H., and Shohet, S. B., 1976, Lipid content of thalassemic erythrocytes, Blood, 47:495.

Schwarz, K., 1961, Factor three: A nutritional essential, Proc. Exptl. Biol. Med., 78:852.

Schwarz, K., 1965, Role of vitamin E, selenium and related factors in experimental nutritional liver disease, Fed. Proc., 24:58.

Schwarz, K., and Foltz, C. M., 1958, Factor 3 activity of selenium compounds, J. Biol. Chem. 233:245.

Schwarz, K., and Pathak, K. D., 1975, Biological essentiality of selenium and the development of biologically active selenium compounds, Chemica Scripta, 8A:85.

Sinclair, A. J., and Slattery, W., 1978, Blood-collecting tube as a contamination source in vitamin E fluorimetry, Clin. Chem., 24:2073.

Steiner, M., 1977, Vitamin E: An inhibitor of platelet aggregation, International Symposium on Tocopherol, Oxygen and Biomembranes, (Japan, 1977).

Stocks, J., Offerman, E. L., and Modell, C. B., 1972, Membrane dysfunction in beta-thalassemia major, Br. J. Haematol., 23:713.

Swick, R., and Bauman, L., 1952, Chemical assay for tocopherol in animal materials, Anal Chem., 24:758.

Tappel, A. L., 1973, Vitamin E, Nutrition Today, 1973:4.

Tappel, A. L., 1974, Selenium-glutathione peroxidase and vitamin E, J. Am. Clin. Nutr., 27:960.

Taylor, S. L., Lamden, M. P., and Tappel, A. L., 1976, Sensitive fluorimetric method for tissue tocopherol analysis, Lipids, 11:530.

Trostler, N., Brady, P. S., Romsos, D. R., and Levelle, G. A., 1979, Influence of dietary vitamin E on malonaldehyde levels in liver, J. Nutr., 109:345.

Tsai, A. C., Kelly, J. J., Peng, B., and Cook, N., 1978, Study of the effect of megavitamin E supplementation in man, Am. J. Clin. Nutr., 31:831.

White, G., 1977, Vitamin E inhibition of platelet prostaglandin biosynthesis, Fed. Proc., 36:350.

# VITAMIN $B_{12}$ AND FOLIC ACID:

# CLINICAL AND PATHOPHYSIOLOGICAL CONSIDERATIONS

Ruth C. Steinkamp, M.D.

Former Director, Bureau of Cancer and Special Services
Arkansas State Department of Health, Little Rock
Arkansas

## INTRODUCTION

The last decade has brought new understanding to the pathophysiological mechanisms and biochemical interrelationships of vitamin $B_{12}$ and folic acid and their role in DNA synthesis. The evolution of facts concerned with these vitamins continues to unfold in the meticulous fashion begun over 50 years ago (Kass, 1978) when it was recognized that pernicious anemia was a deficiency disease. Even such a momentous contribution at that time could not foretell the present promise to understand basic cellular processes.

Both vitamin $B_{12}$ and folic acid deficiency result in an anemia designated as megalobastic, characterized by blood and bone marrow changes as a result of DNA synthesis impairment.

The first account of a patient with an intractable anemia is generally credited to Combe of Edinburgh in 1824 (Chanarin, 1969; Kass, 1976b). Addison's description concisely presented the clinical findings. Biermer, in 1872, described 15 cases of "progressive pernicious anemia" subsequently analyzed to include six pregnancy-related anemias, two with steatorrhea, one with chlorosis, and six with obscure cause, of whom one may have had pernicious anemia (Kass, 1976b).

In the 100 years since Addison's initial description, the succession of basic discoveries, adapted from Kass (1976b), include the following: recognition of related disorders of the stomach by Flint in 1860; gastric achlorhydria by Fenwick in 1870; bone marrow changes by Pepper in 1875; peripheral blood cell characteristics by Quincke in 1876-77, Hayem in 1877, Eichhorst in 1878, and Laache in 1883; increased numbers of erythroid precursors in the bone marrow by Cohnheim, 1877; designation of megaloblasts as abnormal erythroid precursors by Ehrlich in 1880; neurological abnormalities by Leichtenstein in 1884, and by Lichtheim in 1887; kinetic relationship by Levine and Ladd in 1921; effectiveness of treatment with liver in dogs by Whipple and Robscheit-Robbins in 1925 and in man by Minot and Murphy (1926); bone marrow changes with liver therapy by Peabody in 1927; production of a potent liver extract by Cohn et al. in 1928; determination of an intrinsic factor in stomach by Castle (1929); isolation of vitamin $B_{12}$, the antipernicious anemia principle, by Rickes et al. (1948) and by Smith in 1948 and purification by Smith and Parker (1948); demonstration that extrinsic factor and vitamin $B_{12}$ were the same (Berk et al., 1948); x-ray diffraction of the molecule and structure determination by Hodgkin et al. from 1950-56; discovery of coenzyme $B_{12}$ (5'-deoxyadenosylcobalamin) by Barker et al. (1958); and synthesis of vitamin $B_{12}$ by Woodward and Eschenmoser in 1972.

Discoveries about folic acid have been no less remarkable. In the early thirties, Wills (1931, 1932, 1933), while working in South India, recognized that the "pernicious anemia of pregnancy", first described by Osler (1919), and "tropical anemia" were essentially the same. They differed from pernicious anemia in not responding to purified liver extract but did respond to "marmite", an autolyzed yeast substance. After the purification of pteroylglutamic acid in 1943 (Stokstad, 1943), crystallization from liver (Pfiffner et al., 1943), and synthesis and structural identification (Angier, et al., 1945), it became apparent that the "Wills factor" and pteroylglutamic acid were the same. Furthermore, the vitamin M shown by Day et al. (1945) to correct a deficiency anemia in monkeys, the vitamin $B_c$ which corrected the deficiency in chickens, and the "norite eluate factor" essential for growth of *Lactobacillus casei* (Snell and Peterson, 1940) were the same as pteroylglutamic acid. Mitchell et al. (1941) termed the *Lactobacillus casei* factor, "folic acid", concentrating it from four tons of spinach.

The apparent difference in physiological response to vitamin $B_{12}$ and to folic acid emerged with the work of Vilter et al. (1950) by observations that pernicious anemia patients were not able to maintain their hematological response and the neurological symptoms exacerbated with folic acid treatment. The basic interrelationships of folic acid and vitamin $B_{12}$, best explained by the methyltetrahydrofolate trap hypothesis, see Biochemical Considerations, (Herbert and Zalusky, 1962), are not completely understood. Other problems relating to drug interrelationships, to neurological and psychiatric manifestations, to immune mechanisms, and to genetic findings have recently promoted intensive research. Development of and use of new techniques to study biochemical and cellular mechanisms offers promise to understand DNA synthesis and its involvement in neoplastic change of cells.

## CHARACTERISTICS OF MEGALOBLASTIC ANEMIAS

### General

Classically, the megaloblastic anemias are characterized by pernicious anemia, better termed Addisonian pernicious anemia, as the fatal disease which Addison and others described no longer inexorably results in death. Of insidious onset and progression, the pancytopenia (anemia, leucopenia and thrombocytopenia) associated with macrocytosis results, in probably 95% of cases, from nutritional imbalance, and thus is reversible.

Herbert (1970a) distinguished between vitamin-deficient and those anemias responsive to vitamins. Lack of vitamin $B_{12}$ and folic acid produce nutrient deficient megaloblastic anemias. The importance of their distinction is paramount, for anemias resulting from vitamin $B_{12}$ deficiency respond, for a time, to folic acid but the neurological problems fail to respond. The determination of the etiologic factor or factors is basic to successful outcome. The possible causes for nutrient deficiency may be categorized in one or more of five situations: inadequate ingestion, inadequate absorption, inadequate utilization, increased excretion, and increased requirement (Herbert, 1973). Table I presents such a classification for vitamin $B_{12}$ and folic acid deficiencies. Clinical nutrition has been defined by de Oliviera (1979) as the medical specialty dealing with the physiopathology, diagnosis, and treatment of nutritional diseases and related problems.

### Occurrence

Owing to the development and widespread use of electronically automated peripheral blood counting devices, identification of macrocytic erythrocytes (increased mean corpuscular volume) and thus subsequent recognition of megaloblastosis, with or without anemia, has led to a greater awareness of macrocytic red cells associated with various disease processes (McPhedran et al., 1973), or life styles (Helman, 1973).

The prevalence of nutritional anemias as a world wide problem has been examined recently by Baker and DeMaeyer (1979). Collaborative studies were undertaken by the World Health Organization in seven countries with responsible laboratories utilizing standardized techniques and reference assay materials. Women in the third trimester of pregnancy were chosen as the most likely to show a deficiency state as a result of their increased requirements. Generally, the women were from lower socio-economic groups in hospital or clinic settings. Hemoglobin levels were below

Table I. Etiologic Classification of Megaloblastic Anemias in Man (modified from Herbert (1979), Rosenberg (1976), and Stebbins and Bertino (1976))

| Vitamin $B_{12}$ | Folic Acid |
|---|---|
| **Inadequate Ingestion** | |
| 1. Diet lacks sources (microorganisms and animal foods) | 1. Diet lacks adequate sources of unprocessed, fresh or slightly cooked foods |
| a. Vegans | a. Chronic Alcoholics |
| b. Chronic Alcoholics | b. Drug Addicts? |
| c. Drug Addicts? | c. Poverty |
| d. Poverty | d. Food Faddism |
| e. Food Faddism | e. Religious Commitment |
| f. Religious commitment | f. Tropical or non-tropical megaloblastic anemia |
| **Inadequate Absorption** | |
| 1. Gastric Intrinsic Factor Secretion absent or inadequate | 1. No comparable factor to vitamin $B_{12}$? Presence of folate binders? |
| a. Addisonian Pernicious Anemia | |
| 1) Not associated with antigen-antibody stimulus | |
| a) Hereditary absence of IF | |
| b) Congenital defect of IF molecule | |
| c) Adult Pernicious Anemia | |

2) Associated with antigen-antibody stimulus
   a) Juvenile Pernicious Anemia
   b) Hereditarily determined gastric atrophy, about one-half of adult Pernicious Anemia
   c) Acquired gastric atrophy
   d) Gastric damage associated with endocrine disorders and autoimmune diseases
   e) IF Blocking antibody
   f) IF-Vitamin $B_{12}$ complex binding antibody
3) Gastrectomy, total or subtotal
4) Lesions destructive to gastric mucosa
   a) Chemical corrosives
   b) Disease processes

2. Ileal Associated Factors Resulting in Malabsorption

   a. Enteropathy (less common)

      1) Gluten-induced
      2) Celiac Disease
      3) Idiopathic Steatorrhea
      4) Non-tropical Sprue
      5) Tropical Sprue
      6) Regional Enteritis
      7) Strictures
      8) Surgically induced by resection or anastamosis
      9) Malignancy, granuloma, or other small bowel disease

2. Jejunal or Ileal Associated Factors Resulting in Malabsorption
   a. Enteropathy (more common)
      1) Gluten-induced
      2) Celiac Disease
      3) Idiopathic Steatorrhea
      4) Non-tropical Sprue
      5) Tropical Sprue
      6) Regional Enteritis
      7) Strictures
      8) Surgically induced by resection or anastamosis
      9) Malignancy, granuloma, or other small bowel disease, scleroderma, amyloid, diabetic enteropathy, Whipple's Disease

b. Drugs
   1) Para-amino-salicylic acid
   2) Colchicine
   3) Neomycin
   4) Ethanol
   5) Metformin
   6) Oral contraceptive agents?
   7) Calcium-chelating agents

c. ph >6
   1) Pancreatic exocrine secretion deficient
   2) Zollinger-Ellison Syndrome

d. Disorders of Intestinal Receptors for IF-$B_{12}$ Complex and/or Transport Across the Intestinal Cell
   1) Congenital (Imerslund-Gräsbeck Syndrome)
   2) Acquired
      a) Absent
      b) Nonfunctioning

b. Drugs
   1) Dilantin
   2) Primidone
   3) Barbiturates
   4) Ethanol
   5) Metformin
   6) Oral contraceptive agents
   7) Nitrofurantoin
   8) Phenformin
   9) Tetracycline
   10) Glutethimide
   11) Cycloserine

c. Malabsorption with other Disorders
   1) Systemic Bacterial infection
   2) Dermatitis herpetiformis
   3) Cardiac Failure
   4) Zinc Deficiency?

d. Disorders of Intestinal Receptors?

e. Congenital and acquired nonconjugase defects

f. Inadequate biliary or intestinal conjugase

g. Conjugase inhibitors in food

3. Competition for vitamin $B_{12}$ by parasite or bacteria
   a. Blind Loop Syndrome
      1) Small bowel diverticuli
      2) Small bowel anastamosis
   b. Fish Tapeworm
4. Radiation Therapy?

3. Competition for Folates by Bacteria
   a. Blind Loop Syndrome
      1) Small bowel diverticuli
      2) Small bowel anastamosis
4. Radiation Therapy?

## Inadequate Utilization

### Vitamin $B_{12}$

1. Vitamin $B_{12}$ Antagonists
   a. Substituted vitamin $B_{12}$ amides and Anilides
   b. Cobaloximes
2. Congenital or Acquired Enzyme Deficiency or Deletion
   a. Methylmalonyl-CoA Mutase
   b. Methyltetrahydrofolate-homocysteine methyl-transferase
   c. $B_{12a}$ reductase
   d. $B_{12r}$ reductase
   e. Deoxyadenosyl transferase
   f. Other enzyme deficiency or deletion

### Folic Acid

1. Folate Antagonists (dihydrofolate inhibitors)
   a. Methotrexate
   b. Pyrimethamine
   c. Triamterene
   d. Pentamidine
   e. Trimethoprim
2. Congenital or Acquired Enzyme Deficiency or Deletion
   a. Formimino transferase
   b. Dihydrofolate reductase
   c. Methyl tetrahydro-folate transmethylase
   d. 5-10-methenyltetra-hydrofolate cyclohydro-lase
   e. 5-10-methylene tetra-hydrofolate reductase
   f. Cystathionine synthetase?
   g. Other enzyme de-ficiency or deletion

3. Congenital or Acquired Deficiency, Abnormality or Deletion of Transcobalamins (Vitamin $B_{12}$ Binding Proteins)
   a. Abnormal TC I and/or TC III glycoprotein with myeloproliferative disease
   b. Increased TC II with liver disease
   c. Abnormal TC with hepatoma, cirrhosis
   d. Congenital lack of TC II

4. Liver Disease

5. Protein-calorie malnutrition?

6. Malignancy

7. Renal Disease

8. Thiocyanate Intoxication

3. Congenital or Acquired Abnormality of folate binders?
   a. Increased binding by lithium
   b. Increased binding in chronic myelocytic leukemia

4. Liver Disease

5. Vitamin $B_{12}$ Deficiency

6. Ethanol

7. Excess dietary glycine and methionine

8. Diphenylhydantoin and other anticonvulsants

9. Ascorbic acid deficiency

10. Isoniazid

11. Homocystinurias

<u>Increased Requirements</u>

Vitamin $B_{12}$

1. Physiological
   a. Pregnancy
   b. Infancy

2. Metabolic
   a. Hyperthyroidism?
   b. Chronic Temperature elevation?

3. Increased Hematopoiesis?

Folic Acid

1. Physiological
   a. Pregnancy
   b. Infancy

2. Metabolic
   a. Hyperthyroidism?
   b. Chronic Temperature elevation?

3. Increased Hematopoiesis
   a. Myeloproliferative Disorders
   b. Hemolytic Processes
   c. Chronic Blood Loss

| | |
|---|---|
| 4. Malignancy? | 4. Destruction by Dietary Oxidants? |
| 5. Megadoses of Vitamin C? | 5. Lesch-Nyhan Syndrome |

<u>Increased Excretion</u>

| | |
|---|---|
| 1. Lack of, or inadequate TC II | 1. Vitamin $B_{12}$ deficiency? |
| 2. Liver Disease | 2. Liver Disease? |
| 3. Renal Disease? | 3. Kidney Dialysis |
| | 4. Chronic exfoliative dermatitis and psoriasis |

<u>Inhibition of or Interferences with DNA Synthesis</u>

1. Interference with purine synthesis or metabolism

    a. Purine Antagonists
        1) 6-mercaptopurine
        2) Thioguanine
        3) Azathioprine

    b. Congenital enzymatic defect in purine synthesis
        1) Lesch-Nyhan Syndome

2. Interference with pyrimidine synthesis

    a. Pyrimidine Antagonists
        1) 5-Fluorouracil
        2) 5-Fluorodeoxyuridine
        3) Hydroxyurea
        4) Cytosine Arabinoside
        5) 6-Azauridine

3. Interference with DNA synthesis?

    a. L-Asparaginase
    b. Azulphidine
    c. Arsenic
    d. Benzene

4. Inhibition of Ribonucleotide Reductase

   a. Cytosine Arabinoside
   b. Hydroxyurea
   c. Iron deficiency
   d. Procarbazine

5. Inhibition of Protein Synthesis

   a. L-Asparaginase

<u>Unknown Mechanisms in Diseases Sometimes</u>

<u>Associated with Megaloblastic Anemia</u>

1. Pyridoxine-responsive anemias

2. Thiamin-responsive anemias

3. Myeloproliferative diseases

   a. Polycythemia vera
   b. DiGuglielmo's anemia

4. Miliary Tuberculosis

5. Aplastic anemia

11 g/dl in from 21.8% (Poland) to 82% (Burma). There was no definition of the causes of the anemia. Correlations were found between hemoglobin concentrations and serum iron or the per cent saturation of transferrin, and with hemoglobin concentrations and serum folate values. No correlation was found with serum vitamin $B_{12}$ levels.

The National Nutrition Survey in California (1971) studied a random sample drawn from census enumeration districts which had a high per cent of low-income families. Serum folate was below 3 ng/ml (deficient level) in from 5.7% to 20.5% of all ages of subjects. The proportion with unacceptable values (<6 ng/ml) was greater for serum folate, up to 71.5%, than for any of the other indices of nutritional status used. By age groups, hematocrit levels were deficient and low in from 5.6% to 23.5% of all groups. For all subjects, the serum folate was deficient and low in 64.3% and the hematocrit in 17.8%.

## Clinical Manifestations

The classical accounts of clinical findings in "idiopathic" anemia continue to hold for some patients in late presentation of pernicious anemia. Symptoms and findings include: weakness, lethargy, anorexia without cachexia, dyspepsia, smooth tongue, glossitis, constipation, peripheral nerve loss with paresthesias, loss of vibratory sense, forgetfulness, visual loss, mild icterus, hyperpigmentation which is noticeable even in pigmented races, extreme pallor, and evidence of cardiac decompensation. All of which insidiously progress to the point of complete debility and finally death in untreated patients. Variation of symptomatology and severity depend on severity of the anemia, its duration, and cause.

Clinical symptomatology for other causes of megaloblastic anemia may relate to the underlying mechanism, as in sprue and the malabsorption syndromes where persistent diarrhea and weight loss become prominent. Likewise in the increasingly identified, but rare, hereditable syndromes, the early onset in infancy with failure to thrive may be a clinical clue. More common are the megaloblastic anemias associated with chronic alcohol intake, with pregnancy, and with other factors such as chronically administered medications, e.g., anticonvulsants, contraceptive agents, or antineoplastic drugs. In these situations, the symptoms of the underlying process may bring the patient to medical care and intervention before the megaloblastic process has become full-blown.

## Morphological Findings

Reviews of cellular morphology in megaloblastic anemias adequately provide detailed descriptions (Kass, 1976b). The peculiar characteristics of morphology and staining are attributable to the DNA impairment. All cells are larger than normal; macro-ovalocytic red cells, macrocytic granulocytes with hypersegmented nuclei, and giant platelets. As the erythrocytic cells develop, the cytoplasmic portion appears more mature than the nucleus (nuclear-cytoplasmic dissociation), the nuclear chromatin retaining a coarse granularity with failure of clumping (Kass, 1976a). Kass (1968) has pointed out the "clock-face" appearance of the large nucleus in all stages of erythropoietic precursors. Adherence of separate parts of chromatin to the nuclear membrane is attributed to result in the dial formation.

Failure to mature also occurs in the white blood cell series. The cytoplasm appears more basophilic and less mature; the nucleus becomes pinched off and forms excess segments; the cell is larger than normal. The thrombocytes also display large size and bizarre shapes.

Characteristic in the peripheral blood is the macro-ovalocytic red blood cell, along with hypersegmented neutrophils, low reticulocyte count, leucopenia, and thrombocytopenia. In patients with iron deficiency complicating a megaloblastic process, a dimorphic situation may appear. Here, there is less evidence of megaloblastic change in erythroid precursors, and two populations of red cells may be seen; the macro-ovalocytes and the microcytic hypochromic cells typical of iron deficiency. After therapy with iron alone, the megaloblastic changes become evident. Prior to therapy, the hypersegmented neutrophils or macrocytic metamyelocytes in the bone marrow may be diagnostically helpful.

Morphological changes occur in other tissues of the body and may cause diagnostic dilemmas. For example, van Niekirk (1966) demonstrated that cytological preparations of the uterine cervix and vaginal tract in women with folic acid deficiency showed nuclear and nucleoli changes suggestive of carcinoma. Similarly, mucosal cells of the stomach (Gardner, 1956), and of the intestinal tract (Bianchi et al., 1970; Davidson and Townley, 1977) show megaloblastic changes which respond to therapy. The smoothness of the tongue has been attributed to megaloblastosis of cells of the papillary surface (Herbert, 1979).

Ineffective Cellular Production and Increased Cellular Destruction

Finch et al. (1956) have shown that the total erythropoietic activity in pernicious anemia is about three times normal. However, the rate of destruction of erythrocytes was equal to or greater than production. Cellular evidence for intramedullary destruction was elucidated by Myhre (1964a, 1964b). Maturation time of megaloblasts, a failure of maturation to orthochromic erythroblasts, and a low incorporation of $^{59}Fe$ into hemoglobin before therapy normalized after two or more days following vitamin $B_{12}$ treatment. Other works (Nathan and Gardner, 1962; Astaldi, 1960) have also found maturation time prolonged in untreated pernicious anemia. Failure of DNA synthesis in the $G_2$ phase was postulated as the mechanism leading to erythroid precursor destruction and ineffective erythropoiesis (Wickramasinghe et al., 1967, 1968).

Wickramasinghe and Pratt (1970) have shown ineffective granulopoiesis by autoradiographic and spectroscopic study of granulopoietic cells. A prolonged S period, with up to 30% of giant metamyelocytes showing DNA synthesis retardation, and a premitotic arrest in $G_2$ were found.

Kinetics of platelet metabolism in pernicious anemia, studied by Harker and Finch (1969), demonstrated that despite increased intramedullary megakaryocytic mass in pernicious anemia, the pletelet production and turnover were decreased. Cellular evidence in bone marrow preparations is present in the degenerating megakaryocytes. Harker and Finch (1969) suggested the term, "ineffective thrombopoiesis", be reserved for situations in which daily platelet production per nuclear megakaryocyte unit is less than half that expected. Not only are the platelets quantitatively abnormal but also qualitatively defective. Levine (1973) showed no platelet aggregation response to collagen or to epinephrine prior to therapy in three patients with pernicious anemia. Serial studies on one patient were done. Following therapy with vitamin $B_{12}$, platelet aggregation increased above normal and platelet factor 3 availability, as induced by addition of either ADP or epinephrine, suddenly increased on day nine after therapy. Retrospective study of their 21 cases showed thrombotic episodes in two patients. The latter is not uncommon and is probably a result of increased functional capacity of young platelets and of patient predilection. Ingeborg and Stofferson (1979) confirmed evidence of abnormal platelet function before therapy in pernicious anemia. Although only three of ten patients had a prolonged bleeding time, six had disturbed platelet function. After therapy, one patient developed thrombosis, consistent with the 10% noted by Levine (1973).

That unbalanced growth of megaloblasts is the result of impaired synthesis of DNA, together with premature cell death or abnormal cell division with resultant dysplasia, has been suggested as a "useful model" for the mechanism for megaloblastic transformation by Beck (1977a). Support for this hypothesis was developed: 1) megaloblasts have an even larger cytoplasm than nucleus and hence a larger amount of RNA with relatively unimpaired RNA synthesis; 2) megaloblasts have a normal or slightly increased amount of DNA per cell, accounting for the cytoplasmic basophilia which results from the arrest in the S (synthesis of DNA) phase in order to complete the doubling of DNA before division; and 3) thymidine and deoxyuridine are incorporated into the DNA of megaloblasts, with thymidine having the capability to correct the megaloblastosis of folic acid deficiency but not of $B_{12}$ deficiency. These factors reinforce the explanation of biochemically defective DNA synthesis and explain the microscopic findings of an imbalance between the young, finely divided, chromatin network and the mature cytoplasm. Lewis and Verwilghen (1973) have presented a review of the various dyserythropoietic anemias and mechanisms of dyserythropoiesis.

Biochemical Considerations

Plasma bilirubin concentrations are moderately increased reflective of the hemolytic component in megaloblastic anemias and result in mild clinical icterus (London and West, 1950). The serum iron and high saturation of transferrin may be the result of the ineffective erythropoiesis; on therapy there is a rapid decrease in serum iron levels with a decrease in the saturation of transferrin. Recent studies indicate that iron may have a definite role in DNA synthesis (Wickramasinghe and Longland, 1974). Absorption of the intrinsic factor-vitamin $B_{12}$ complex by intestinal mucosa involves a bivalent cationic bond, $Ca^{++}$, with release of vitamin $B_{12}$ in the intestinal cell or on the cell surface upon conversion of the bond (Cooper and Castle, 1960). The sharp decrease of potassium occurring after treatment of megaloblastic anemias may result in cardiac complications and because of this, Lawson et al. (1972) have suggested careful monitoring of potassium during therapy. They found 15 cases of sudden death in a series of 108 patients treated for megaloblastic anemia with hematocrits below 25%. The low potassium could have produced a cardiac arrhythmia, but the causes of death were uncertain.

Zinc has been shown to increase hepatic content of folic acid (Williams et al., 1973) in zinc deficiency. Silink et al. (1975) found zinc an essential part of bovine hepatic conjugase, γ-glutamyl hydroxylase. Tamura et al. (1978) found zinc deficiency led to impaired absorption of dietary folate in six adult men and proposed that intestinal conjugase could be a zinc metalloenzyme. Prasad et al. (1963) found no evidence of folic and deficiency in a population with chronic zinc deficiency. The syndrome of zinc deficiency, including hypogonadism, dwarfism and iron deficiency anemia, was associated with hyperpigmentation and hepatosplenomegaly, two associated findings in pernicious anemia.

Hyperpigmentation is found in many disease states (Greipp, 1978) including endocrinopathies, nutritional deficiencies (pernicious anemia, folic acid deficiency, chronic liver disease), porphyria, autoimmune diseases, after therapy with drugs interfering with sulfhydryl metabolism (busulfan and arsenicals), and with radiation therapy. The biochemical changes resulting in hyperpigmentation are related to the decreased intracellular glutathionine levels which release melanin synthesis from feedback inhibition of tyrosinase (Gilliam and Cox, 1973). Excess tyrosine levels are found in pernicious anemia (Levine, 1967). As the precursor of melanin, tyrosine is converted by tyrosinase to DOPA which is metabolized to dopa-quinone. A series of non-enzymatic oxidations then produces melanin (Greipp, 1978). Hyperpigmentation in folate deficiency may be explained by elevated biopterin levels as biopterin is involved in the hydroxylation of phenylalanine, necessary in melanin production. Ogbuawa et al. (1978) noted, in three black women with pernicious anemia, that the diffuse palmar, buccal, acral and vulvar hyperpigmentation present at diagnosis regressed with therapy.

## VITAMIN $B_{12}$ DEFICIENCY

### Occurrence

The highest frequency of pernicious anemia has been recognized for many years among the Northern European peoples and their descendants. Age related, its occurrence increases proportionately (Herbert, 1979). Incidences according to age groups were one per million for ages six months to one year, one per 10,000 for ages one to 10 years, one per 5,000 for ages 30 to 40 years, with progressive increases to one per 200 for the sixth decade of life. Study of population surveys of total cases in the United Kingdom and in Sweden and Denmark showed a range of 100 to 130 cases per 100,000 (Chanarin, 1969).

Kaplan and Rigler (1945) studied autopsy data of 23,231 patients over 45 years of age to determine 293 with pernicious anemia, a frequency of 1.3%.

Chanarin analyzed (1969) the changing age distribution of pernicious anemia in London and Boston over the first half of the twentieth century and found the mean age at diagnosis was 60.5 years where 50 years before, the peak incidence was in the 40 to 50 year age group.

Relatively uncommon in Jews and Orientals, pernicious anemia has also been less frequently seen in blacks than in whites. A greater prevalance than previously thought to exist in blacks in both the United Sates (Hart and McCurdy, 1971; Solanki et al., 1978; Carmel and Johnson, 1978) and in South Africa (Solanki et al., 1978) has been found. It appears to occur at a younger age than in whites (Hart and McCurdy, 1971; Solanki et al., 1978; Carmel and Johnson, 1978) and more frequently in women (Hart and McCurdy, 1971; Carmel and Johnson, 1978; Solanki et al., 1978; Bash and Rosner, 1979). Seventeen per cent of women and 23 per cent of men were under 40 years of age in the Washington, D. C.

group, as compared with 43% for women and 41% for men in the blacks described from South Africa (Solanki et al., 1978). The ages of those from Los Angeles and Detroit studied by Carmel and Johnson (1978), were similar to the Washington, D. C. group. A higher prevalence for intrinsic factor circulating antibody was found in both groups: 23 of 24 tested by Carmel and Johnson (1978) and 85% of 20 black women tested by Solanki et al. (1978). Metz et al. (1961) reported on three young Bantu women, ages 23, 28, and 34 years with Addisonian pernicious anemia associated with gastric mucosal atrophy. Genetic or environmental roles may play a part, but they are unknown at present.

Vitamin $B_{12}$ deficiency associated with specific absorptive defects, e.g., gastrectomy, small bowel disorders, or others, may not be common owing to medically recognized risks, preventive vitamin therapy, and vitamin $B_{12}$ reserves which are adequate for at least three years in most people. Pernicious anemia has been observed frequently in association with leukemia (Zarafonetis et al., 1957; Blackburn et al., 1968; Corcino et al., 1971), with gastric cancer (Kaplan and Rigler, 1945; Blackburn et al., 1968), with polycythemia vera or erythremic myelosis (Zarafonetis et al., 1957; Riddell and Davidson, 1968; Engel and Stickney, 1972), with myelosclerosis (Hoffbrand et al., 1968), and with a variety of neoplasms (Larsson, 1962; Lowe, 1976; Fraser, 1969; Burnier et al., 1976; Arvanitakis et al., 1979).

Dietary deficiencies of vitamin $B_{12}$ may be increasing. As a result of economic circumstances, food faddism, religious commitment, or other reasons, totally vegetarian diets with no milk, eggs, meat, or other animal products are being consumed by more people.

The recognized inborn errors of metabolism of vitamin $B_{12}$ are quite rare, but the recent advances in diagnosis may reveal more cases.

## Biochemical Considerations

Reviews of the chemical and biological aspects of vitamin $B_{12}$ may supply references for the major developments in the distinguished and remarkable series of studies which have employed the full range of the basic sciences, clinical medicine and biology (Castle, 1961; Huennekens, 1966; Stadtman, 1971; Herbert, 1975; Beck, 1977b; Kass, 1976b).

The structure of vitamin $B_{12}$ bears a porphyrin-like ring, the corrin portion, linked to a central cobalt atom. This is the only known instance of cobalt in a vital metabolic system in man. Lying nearly perpendicular to the planar ring is a 5,6-dimethylbenzimidazolyl nucleotide, a unique compound not previously known in nature. The cyanocobalamin molecule contains a cyano-group which lies above the pyrrole rings which, with one exception, are connected by methylene carbon bridges. Methyl groups or acetamide and propionamide residues almost completely saturate the pyrrole rings.

Corrinoid compounds are numerous. Nomenclature has been systematized (Smith, 1965; IUPAC-IUB, 1966; IUPAC-IUB, 1974).

Two compounds, both alkyl cobalt derivaties, serve as coenzymes: 5'-deoxyadenosyl cobalamin (Barker et al., 1958; Barker, 1967), and methyl cobalamin (Guest et al., 1962).

Deoxyadenosyl cobalamin mediates the methylmalonyl-CoA mutase reaction to succinyl-CoA and is reduced in vitamin $B_{12}$ deficiency (Gompertz, 1971; Silber and Moldow, 1970).

The block in enzyme reaction leads to an accumulation of metabolic intermediates normally converted to succinyl-CoA and metabolized in the citric acid cycle. Kass (1976b) has found cytochemical evidence for increased levels of methylmalonyl-CoA in megaloblasts of untreated pernicious anemia. Other biochemical evidence of methylmalonyl-CoA block are the methylmalonic aciduria, characteristic in untreated pernicious anemia (Cox and White, 1962), and propionic aciduria (Cox et al., 1968). Decreased activity of methylmalonyl-CoA mutase may also be involved in development of neurological manifestations of vitamin $B_{12}$ deficiency. Frenkel (1973) has demonstrated abnormalities of fatty acid synthesis by propionate metabolism in nerve tissue of pernicious anemia patients. Total fatty acid synthesis was decreased from that of normal nerve tissue. In addition it was suggested that the abnormal fatty acids found may be pathogenetic. As early as 1937, (Williams et al., 1937) a relation between abnormal lipid metabolism in vitamin $B_{12}$ deficiency and myelin degeneration was postulated. In pernicious anemia plasma the neutral fat levels were increased while cholesterol esters and phospholipid were decreased. After therapy these returned to normal.

In the second corrinoid coenzyme, a methyl group is the ligand of cobalt (Beck, 1977B). Methyl cobalamin is the prosthetic group (Loughlin et al., 1964) for the $N^5$-methyltetrahydrofolate-homocysteine methyltransferase enzyme and is responsible for conversion of homocysteine to methionine. $N^5$-methyltetrahydrofolate is the major methyl donor in the biosynthesis of methionine (Guest et al., 1962; Guest et al., 1964; Larrabee et al., 1961; Larrabee et al., 1963).

The methyltetrahydrofolate trap hypothesis is at present the most accepted theory for the mechanism of the $N^5$-methyltetrahydrofolate transferase enzyme. In vitamin $B_{12}$ deficiency, there is an accumulation of methyltetrahydrofolate and a depletion of other folates necessary for the synthesis of purines and pyrimidines required in the synthesis of DNA (Herbert and Zalusky, 1962; Buchanan, 1964; Buchanan et al., 1964; Johns and Bertino, 1965; Nixon and Bertino, 1970, 1972; Hoffbrand, 1975; Das and Herbert, 1976; Herbert and Das, 1976; Hoffbrand et al., 1977).

Only by cobalamin-dependent methyl transferase reaction can methyltetrahydrofolate be converted to other folate derivatives. As a consequence of reduced methyl transferase activity, methyltetrahydrofolate increases in the body with resultant decreases in tissue levels of tetrahydrofolate, formyltetrahydrofolate, and methylenetetrahydrofolate, and thereby the rate of reactions dependent on these folate coenzymes is slowed. As a consequence, biosynthesis of purines and thymine, and therefore of DNA, is impaired.

Hoffbrand et al. (1977) expanded and confirmed these interrelationships. Folate polyglutamates are the naturally occurring folate coenzymes required for folate mediated reactions in amino acid metabolism and purine and pyrimidine synthesis. As the circulating form of plasma folate is 5-methyltetrahydrofolate, a monoglutamate, cells must build the active tetrahydrofolate polyglutamate coenzymes from 5-methyltetrahydrofolate.

Sauer and Wilmanns (1977) point out that while cobalamin dependent methionine synthesis is the primary metabolic fault in pernicious anemia, mechanisms other than the methyl folate trap theory do occur and require other explanations. Among these are direct cobalamin dependency of folate polyglutamate synthesis, of folate transfer through the cell membrane, and of thymidylate synthetase enzyme protein synthesis. Clinically, there are other objections. Children with defective vitamin $B_{12}$ metabolism and congential methylmalonic aciduria do not have megaloblastic transformation indicative of impaired DNA. The explanation for the latter may be afforded by the activity of vitamin $B_{12}$ in the synthesis of thymidylate synthetase.

Haurani (1973) studied an unusual case of megaloblastic anemia (pernicious anemia) in whom the serum vitamin $B_{12}$ was 2000 to 3000 pg/ml during relapse although the thymidylate synthetase activity of cultured phytohemagglutinin-stimulated lymphocytes was low. Additional pernicious anemia patients, as well as folic acid deficient individuals and normals, were studied. Only in pernicious anemia in relapse were the enzyme values low or absent. This important work demonstrated for the first time that vitamin $B_{12}$ deficiency results in a decreased synthesis of thymidylate synthetase and an accumulation of unused 5-methyltetrahydrofolate. As folate acts as a coenzyme for thymidylate synthetase in DNA synthesis, Haurani noted that vitamin $B_{12}$ could not be the cause of the megaloblastosis.

Chanarin et al. (1974) and Chanarin and Perry (1977) have pointed out difficulties in accepting the methyl folate trap theory as the total explanation for vitamin $B_{12}$-folic acid interrelationships. Other reactions in which a cobalamin acts as a coenzyme are those mediated by: glutamate mutase, dioldehydrase, ribonucleotide reductase, glycerol dehydrase, ethanolamine deaminase, ribonucleotide triphosphate reductase, and β-lysine isomerase which utilize 5'adenosyl cobalamin; and in synthesis of methane and of acetate which utilize methyl cobalamin (Beck, 1977b; Herbert and Das, 1976). Beck (1977b) points out that the validity of system dependency on a vitamin $B_{12}$ coenzyme is based on the demonstrated requirement in a cell-free system. Only adenosyl- and methyl- cobalamin have been so designated.

Kass (1976b) has observed cytochemical evidence of histone metabolism abnormality in pernicious anemia and folic acid deficiency. The histones are thought to bind DNA into compact masses resei ling a clock face on the interior surface of the nuclear membrane of megaloblasts (Kass, 1968). Additional evidence is based on the preponderance of lysine-rich and lack of arginine-rich histones in pernicious anemia megaloblasts.

Aberrant chromosomal patterns of bone marrow cells in untreated pernicious anemia demonstrate unusual metaphase configurations, chromosomal gaps, and constrictions (Kiossoglou et al., 1965; Powsner and Berman, 1965; Menzies et al., 1966).

## Nutritional Requirements of Vitamin $B_{12}$

Vitamin $B_{12}$ is unique in that only foods of animal origin supply the dietary needs. The ultimate source of vitamin $B_{12}$ in animals depends upon microbial synthesis.

The "average" daily diet in the U.S.A. contains between 3 μg and 30 μg vitamin $B_{12}$ depending on the proportion of animal foodstuffs of which between 1 μg and 5 μg are absorbed (WHO, 1972; Heyssel et al., 1966). The average daily requirement for adult men and nonpregnant women has been estimated between 0.1 μg to 1 μg (Herbert, 1968; WHO, 1972). The 0.1 μg/day requirement is based on data in which daily intramuscular injections of 0.1 μg vitamin $B_{12}$ given to pernicious anemia patients resulted in a slow return to normal serum vitamin $B_{12}$ levels following a suboptimal reticulocyte response. Optimal responses occurred with daily injections of 1 μg (West and Reisner, 1949; Sullivan and Herbert, 1965).

Studies using whole body turnover of radioactive tagged vitamin $B_{12}$ in normal individuals who were in a steady state provided a range of 0.5 to 2.5 μg per day (Gräsbeck, 1960; Heinrich, 1964; Heyssel et al., 1966; Reizenstein, et al., 1966). Darby et al. (1958) had shown earlier in a patient with pernicious anemia in relapse the range of utilization of vitamin $B_{12}$ was 0.5 to 2.0 μg/day.

The most recent edition of the NRC recommended daily dietary allowances (1974) has set the daily level of vitamin $B_{12}$ intake "for maintenance of good nutrition of practically all people in the U.S.A." at 3 μg for both sexes over the age of 10 years with an additional 1 μg per day for pregnant or lactating women. Allowances set for infants are based on approximations of vitamin $B_{12}$ in human milk and amount to 0.3 μg daily. Interpolation of values from 0.3 μg at one year of age to 3 μg at 10 years of age have been made.

Young and Scrimshaw (1979) discuss genetic and biologic variables which affect human nutrient requirements. Individual requirements for vitamin $B_{12}$ are affected by these variables. Beck (1977b) considered allowances in relation to obligatory loss approximating 0.1 per cent of the total body pool, as shown by Heyssel et al. (1966), and suggests the minimal daily requirement is adequate to maintain nutritional balance only in subjects with low body stores.

A diet usual in Western countries is believed to be adequate to gradually replenish depleted body stores.

## Absorption of Vitamin $B_{12}$

Absorption of vitamin $B_{12}$ occurs by two mechanisms: simple diffusion, accounting for less than three per cent of the usual range of 1 μg to 5 μg absorbed, and by complexing with intrinsic factor (IF). The latter is the significant absorptive mechanism whereby vitamin $B_{12}$ derivatives, released from food by peptic digestion (Doscherholmen and Swaim, 1973), are attached to specific binding sites of gastric IF (Castle, 1929; Castle and Townsend, 1929; Castle et al., 1930; Gräsbeck, 1959) secreted by parietal cells (Hoedemaker et al., 1964; Taylor, 1965; Jacob and Glass, 1969; Jacob and Glass, 1971). The IF-$B_{12}$ complex is required for absorption by specific absorptive sites on the ileal mucosal cells (Booth and Mollin, 1959). The ileal receptors in man are capable of binding 0.5 μg to 1.0 μg vitamin $B_{12}$ bound to IF (Hooper et al., 1973). Katz and Cooper (1974) have determined the rate-limiting factor in binding of the IF-$B_{12}$ complex is the number of ileal receptors or other intestinal process. A specific site on the IF molecule binds with the ileal receptor in the presence of $Ca^{++}$ (Herbert, 1959) and at neutral pH (Herbert and Castle, 1961; Mackenzie and Donaldson, 1972; Hooper et al., 1973).

Anti-IF antibodies of two types have been identified. The "blocking" type prevents binding of vitamin $B_{12}$ by IF. The "binding" type combines with the IF-$B_{12}$ complex or with the free IF without impairing its ability to bind vitamin $B_{12}$ (Kass, 1976b) and prevents attachment to ileal receptors.

Parietal cell antibodies have also been identified in pernicious anemia. The evidence for autoimmune mechanisms and the state of knowledge has been reviewed by Taylor (1976) and Kass (1976b).

The vitamin, released from the IF complex, passes into the cell or is incorporated into the cell by pinocytosis (Beck, 1977b) and then dissociates in the cell. From the cell it enters the portal circulation and combines with the transport proteins (Allen, 1975) to be delivered to liver, bone marrow and other tissues.

The enterohepatic circulation of vitamin $B_{12}$ is important in the body economy. Reizenstein (1959) found 3 μg to 9 μg vitamin $B_{12}$ per 24 hours were recirculated in the bile. All but approximately 1 μg was reabsorbed. As a result of the biliary reabsorption, 10 to 12 years are required for a person on a strict vegetarian intake to become devoid of vitamin $B_{12}$ (Heyssel et al., 1966) and to evidence clinical deficiency. For a person lacking in IF but with normal $B_{12}$ stores, one to four years after withholding parenteral vitamin $B_{12}$ supplementation would be required. In patients with ileal dysfunction or other cause for malabsorption, vitamin $B_{12}$ deficiency occurs more rapidly as a result of non-reabsorption of biliary vitamin $B_{12}$.

Inadequate secretion of IF may accompany lesions which destroy the gastric mucosa; ingestion of corrosives, linitis plastica, neoplastic lesions; lesions which accompany gastric atrophy; and other conditions such as iron deficiency, multiple sclerosis, subtotal or total gastrectomy, and certain endocrine disorders.

Inadequate absorption of the IF-$B_{12}$ complex may result from any condition interfering with ileal function: tropical sprue, non-tropical sprue (idiopathic steatorrhea, gluten-induced enteropathy), regional enteritis, ileal resection, neoplasias, granulomas, ileal strictures and anastamoses.

Pancreatic insufficiency results in malabsorption of vitamin $B_{12}$, which is correctable by pancreatic extracts and trypsin (Toskes and Deren, 1972; Toskes et al., 1971; Toskes et al., 1973). Low intestinal pH of the small bowel which accompanies pancreatic disease interferes with absorption of vitamin $B_{12}$. Sodium bicarbonate therapy enhances absorption in this situation by raising intestinal pH.

Malabsorption is commonly found in untreated pernicious anemia patients, which is reversible with treatment. In three of four pernicious anemia patients studied by Brody et al. (1966) the malabsorption was associated with sprue, with jejunal diverticuli, or with side to side entero-enterostomy. In the fourth patient no associated abnormality was found. In additional studies on these patients (Carmel and Herbert, 1967), the malabsorption was corrected after $B_{12}$ therapy continuing for one to ten months, suggesting that the malabsorption could be related to ileal damage associated with vitamin $B_{12}$ deficiency. Lindenbaum et al. (1974) studied intestinal function in 28 untreated pernicious anemia patients. D-xylose absorption was impaired in 29%, fat in 9%, and vitamin $B_{12}$ given with intrinsic factor in 75%. Two patients had high titers to serum IF antibodies. The malabsorption (xylose) continued from 2 weeks to 41 months after vitamin repletion. The work supports the hypothesis of Herbert (1972c) that a "vicious cycle" may occur with secondary malabsorption aggravating the pernicious anemia.

Mathan et al. (1974) have found that cobamides with unaltered corrin ring and an intact benzimidazole moiety competitively inhibit the binding of cyanocobalamin to IF. In addition, they reported the nature of the cobamide bound to IF had little effect on the binding of IF to the ileal mucosal receptor.

Mechanisms of action in the vitamin $B_{12}$ absorptive process by IF, so brilliantly elucidated by Castle, and the conditions responsible for the lesions, continue to challenge researchers. While an immune mechanism is strongly suspected (Taylor, 1976) in absorption impairment, additional exogenous as well as endogenous factors may be contributory.

## Transport of Vitamin $B_{12}$ - The Binding Proteins

Since Ternberg and Eakin (1949) identified a non-dialyzable substance from gastric juice and hog intrinsic factor which bound vitamin $B_{12}$, considerable knowledge has developed of the nature and activity of protein-binding of the vitamin. Extensive reviews have brought forward the clincial and scientific importance of this group of proteins (Gräsbeck, 1969; Hall, 1971; Gräsbeck, 1975; Allen, 1975).

All human body fluids that have been studied contain proteins that bind vitamin $B_{12}$ (Gräsbeck, 1969). Only found in gastric juice and stomach extract, IF is a glycoprotein which avidly binds the vitamin. Other vitamin $B_{12}$ binding proteins have undergone brisk investigations primarily as a result of improved conventional techniques (Gräsbeck et al., 1966; Visuri and Gräsbeck, 1973) and development of sophisticated new techniques, such as affinity chromatography (Allen and Majerus, 1972a,b,c; Allen and Mehlman, 1973;

Christensen et al., 1973), isoelectric focusing (Stenman, 1971), and a method to evaluate structural variants of Transcobalamin (TC) II (Frater-Schröder et al., 1979).

A new nomenclature of the three types of binding proteins has been suggested by Gräsbeck (1969, 1975). These are 1) IF, 2) R-type binding proteins or cobalophilin, and 3) TC II. "R-type" binding proteins attained the name by "rapid" electrophoretic mobility for the $B_{12}$ binding protein in gastric juice. Immunologically identical binding proteins in other tissues, including saliva, tears, granulocytes and other tissue, were subsequently found but lacked rapid mobility. Cobalophilin has been suggested to replace the term "R-type binding proteins".

The three classes of vitamin $B_{12}$ transport proteins have similarities in that they contain single polypeptide chains of 340 - 375 amino acids with each chain containing a single $B_{12}$ binding site. However, they do not cross react with each other immunologically (Allen, 1975). They differ in amino acid and carbohydrate composition, in molecular weight, and in their interactions with the vitamin $B_{12}$ molecule. That these proteins are coded by different structural genes is supported by documented cases with congenital deficiencies or structural abnormalities (Hall, 1973).

TC II is the transport protein which carries the exogenous vitamin $B_{12}$ to all tissues and facilitates entry into the cells (Finkler and Hall, 1967; Allen, 1975; Frater-Schröder et al., 1979). TC II is found in semen and CSF as well as plasma. TC II is thought to be necessary in CSF for the delivery of vitamin $B_{12}$ to cells of the nervous tissue (Friedman et al., 1977). Children with an inherited defect of TC II may develop normally *in utero* as the human placenta has a high affinity binding site for TC II. An earlier study (Boger et al., 1957) demonstrated higher vitamin $B_{12}$ concentration in cord blood than in maternal blood. However, children with TC II deficiency subsequently develop signs of $B_{12}$ deficiency within a few months after birth. Familial studies support the finding of Hakami (1971) in two siblings that the abnormality is related to a recessive mode of transmission. A significant study of transport of therapeutic vitamin $B_{12}$ in congenital TC II deficiency has been reported (Hall et al., 1979).

About eight hours after an oral dose of radioactive vitamin $B_{12}$, radioactivity begins to transfer from TC II to TC I (Doscherholmen et al., 1959; Cooper and White, 1968). According to prior nomenclature, so-named TC I and TC III are now considered in the cobalophilin group in which the endogenous vitamin $B_{12}$ is carried (TC I) and from which bound vitamin $B_{12}$ is slowly transferred to tissues. TC III has been shown to be derived from granulocytes by Burger et al. (1975).

Abnormalities of all vitamin $B_{12}$ binding proteins have been found. Hall (1973) noted inherited IF deficiency resulting in vitamin $B_{12}$ malabsorption. A biologically inert IF was described in a 13 year old boy whose parents were first cousins (Katz et al., 1972). The parents and sister of the patient were heterozygous for the abnormal IF (Katz et al., 1974).

Cobalophilin (TC I) levels are elevated in chronic myeloproliferative disorders, particulary chronic myelocytic leukemia (Carmel, 1975), in polycythemia vera (Hall and Finkler, 1969), in metastatic cancer (Carmel, 1975), and in adolescent hepatoma (Waxman and Gilbert, 1973). Jacob et al. (1977) have identified in a patient with bronchogenic carcinoma an abnormal macromolecular complex which bound 90% of the endogenous vitamin $B_{12}$. Distinct from cobalophilin, they suggested the complex was an immunoglobulin of polyclonal origin. Allen (1975) suggested that increased synthesis of R-type proteins, as well as prolongation of plasma survival as a result of decreased liver clearance, may be the mechanisms for elevations in myeloproliferative and neoplastic diseases.

Gräsbeck (1975) predicted additional causes of vitamin $B_{12}$ deficiency may be found as a result of autoantibodies directed against cell membrane receptors and mitochondria.

## Clinical Manifestations of Vitamin $B_{12}$ Deficiency

Other than the general findings of a megaloblastic anemia as described in Clinical Manifestations, vitamin $B_{12}$ deficiency has distinctive characteristics. Increasingly, it is diagnosed as a result of newer knowledge indicating that deficiency occurs in persons other than the typical middle or older-aged individual of Scandinavian or Northern European descent. These include persons with inherited abnormalities of absorption, transportation, and with increased vitamin $B_{12}$ requirements as well as persons with environmentally conditioned deficiency. Some of these are: vegans (strict vegetable diet), alcoholics (more often folic acid deficiency occurs), and accompanying other diseases such as autoimmune diseases, hyper- and hypothryoidism, hypergammaglobulinemia (Burnier et al., 1976), hypogammaglobulinemia (van Dommelen, et al., 1963), and endocrine disorders. Vitamin $B_{12}$ is also found to be deficient after gastrectomy (Deller and Witts, 1962; Deller et al., 1962; Hines et al., 1967), in the blind loop syndrome (Dellipiani et al., 1968; Booth et al., 1968; Donaldson, 1968; Brandt et al., 1977), in cystic fibrosis (Deren et al., 1973; Rucker and Harrison, 1973), with pancreatic insufficiency (Von der Lippe, 1977), with massive small bowel diverticulosis (Badenoch et al., 1955), and with bacterial overgrowth in the small bowel (Brandt et al., 1977). The latter has demonstrated that gut bacteria, in particular *E. coli*, convert vitamin $B_{12}$ to non assimilable analogues altered on the nucleotide portion of the molecule. The analogues competitively inhibit the binding of vitamin $B_{12}$ to IF.

Formerly, infestation by the fish tapeworm (Diphyllobothrum latum) was an important etiology for vitamim $B_{12}$ deficiency in Scandinavian and Far Eastern countries. However, with gradual dying out of the fish as a result of water pollution and less raw fish consumption, this is no longer so frequently seen. Nyberg (1960) demonstrated the possible inhibition of absorption of IF-$B_{12}$ complex in this clinical situation. More recent studies indicate the defect is a result of competition for vitamin $B_{12}$ by the worm (Beck, 1977b).

Malabsorption of vitamin $B_{12}$ is also caused by certain drugs (Waxman et al., 1970b; Herbert, 1972a). Para-amino-salicylic acid used in treatment of tuberculosis, colchicine for gout, neomycin for certain infections, and ethanol are the most important causes for drug induced malabsorption. In addition, the biguanide group of oral hypoglycemic agents produce vitamin $B_{12}$ malabsorption but are no longer available as the FDA restricted their use as a result of danger from lactic acidosis production. To what extent, if any, the oral contraceptive agents produce vitamin $B_{12}$ malabsorption is presently unknown. Herbert and Jacob (1974) also cautioned that intake of 500 mg or more of Vitamin C may result in vitamin $B_{12}$ destruction.

Environmentally induced vitamin $B_{12}$ deficiency may become more prominent as a result of the increasing use of macrobiotic diets for reason of religious or other commitment. Economics may also play a part in persons taking a strictly vegetarian diet. Hines (1966) reported severe anemia responsive to vitamin $B_{12}$ in a 41 year old vegan (no animal product food eaten) of 7 years duration following a six year period of a lacto-vegetarian diet (diet included milk products). He had also been a blood donor, giving a pint of blood every three to four months. Therapy with 1 μg vitamin $B_{12}$ intramuscularly per day produced a reticulocytosis of 14.5% as well as a marked reduction in serum iron and a return to normal serum lactic dehydrogenase activity.

In another case report, a 6 month old, exclusively breast fed child of a 26 year old vegan mother, presented in coma, with severe megaloblastic anemia, and hyperpigmentation (Higginbotham et al., 1978). Biochemical findings included methylmalonic aciduria, homocystinuria, cystathionuria, glycinuria, methylcitric aciduria, 3-hydroxypropionic aciduria and formic aciduria. The child also had hepatomegaly, gross abnormalities of the EEG, and monilial dermatitis. By one month after the start of vitamin $B_{12}$ therapy, all abnormalities had disappeared. In addition, striking increase in growth occurred with a gain of 5 cm in length and 3 cm in head circumference.

Clinical and laboratory findings specific to vitamin $B_{12}$ deficiency are neurologic abnormalities, decreased serum vitamin $B_{12}$ levels, methylmalonic aciduria, reticulocytosis after vitamin $B_{12}$ therapy with no response to physiologic doses of folic acid (100-400 μg/day), and abnormal response in deoxyuridine suppression (Killman, 1964,; Herbert et al., 1973). The latter probably offers the greatest clinical potential of any recent development for differentiation of megaloblastic anemias.

Neurologic abnormalities characteristically develop insidiously and symmetrically. First, paresthesias develop in the distal parts of the extremities, with associated loss of vibratory sense and proprioception. In the untreated patient these progress to spastic ataxia and the manifestations of subacute combined system disease of the dorsal and lateral columns of the spinal cord. Accompanying cerebral manifestations of irritability, somnolence, forgetfulness, EEG changes (Samson et al., 1952), and overt psychosis often revert dramatically with vitamin $B_{12}$ before hematologic response is apparent. Optic atrophy may occur. The neurologic lesion is a degeneration of the myelin sheath. Pathogenesis of the degenerative changes is incompletely known. It has been suggested that one possible reason is that adenosyl cobalamin (coenzyme $B_{12}$) is required for fat metabolism in propionic acid utilization by conversion of methylmalonic acid to succinate (Herbert, 1979), or possibly vitamin $B_{12}$ may facilitate reduction of the S-S form of coenzyme A to the active S-H form.

Once developed, the subacute combined system symptoms usually respond incompletely to therapy. The ability of large amounts of folic acid to mask the blood and bone marrow megaloblastosis, with a suboptimal but significant reticulocytosis, and at the same time to exacerbate the neurologic deficits (Vilter et al., 1950), is incompletely explained but is critical to the proper care of the patient (Carmel, 1979). Because of this danger FDA regulations prohibit the inclusion of more than physiologic levels of folic acid in generally available multivitamin preparations.

Inhibition of cell growth by anesthetic gases, considered in the past to be a narcotic effect, has been shown to interfere with DNA production by depressing incorporation of thymidine and to cause a delay in entry of cells into the DNA-synthetic phase (Sturrock and Nunn, 1974; Nunn et al., 1976). Sturrock and Nunn (1975) have also demonstrated multiple mitoses and multinucleate cells in mammalian cells exposed to nitrous oxide and halothane. Trudell et al. (1979) suggested that nitrous oxide is carcinogenic by its reduction action of aromatic nitro compounds, as determined on mammalian tissue and gastrointestinal microflora both _in vitro_ and _in vivo_.

Pregnant rats exposed to 45 to 50% nitrous oxide for 2, 4, or 6 days on day 8 underwent resorption of fetal products in 19%, 25%, and 57%, respectively (Fink et al., 1967). Lane et al. (1979) have confirmed the teratogenic nature of nitrous oxide.

Interference with cell division by various anesthetics was demonstrated as long ago as 1914 (Lillie, 1914).

For a limited time in the 1950's, nitrous oxide anesthesia was used to control neurologic symptoms in tetanus infections, but instances of bone marrow failure, with severe leucopenia, thrombocytopenia and bone marrow megaloblastosis, resulted in abandonment of that treatment (Lassen et al., 1956; Sando and Lawrence, 1958). Because of the bone marrow and hematologic effects, it was also studied as a treatment of acute leukemia (Eastwood et al., 1963) and subsequently abandoned as the leucopenic effect was not sustained. In two patients, both of whom died, the polymorphonuclear cell count fell faster than the lymphocytes, and the bone marrow became hypoplastic.

A recent report (Layzer, 1978) documented severe myeloneuropathy in six patients (five dentists and one hospital technician) who had prolonged nitrous oxide exposure. The myeloneuropathy was similar to that seen in subacute combined system disease with numbness and paresthesias in extremities and depressed reflexes. There appears to be a high rate of abuse of nitrous oxide, 20% of medical and dental students reported by Rosenberg et al., (1979). Another report cited three deaths from halothane abuse in hospital personnel (Spencer et al., 1976).

Amess et al. (1978) found eight patients receiving 50% nitrous oxide and 50% oxygen for 24 hours during and following cardiac bypass surgery who developed megalobastic bone marrow findings. Deoxyuridine suppression tests were abnormal at the end of ventilation with the anesthetic gas. Five patients undergoing cardiac bypass surgery, but without nitrous oxide, had normoblastic marrows and normal deoxyuridine suppression tests. Of nine additional patients who received nitrous oxide only during surgery, three had abnormal deoxyuridine suppression tests at 24 hours. The deoxyuridine abnormality was identical in that found in vitamin $B_{12}$ deficiency, but the serum vitamin $B_{12}$ levels were normal. The effect appeared after less than six hours exposure to nitrous oxide in one patient. Recovery occurred on stopping the nitrous oxide. Treatment with hydroxocobalamine before and after exposure did not prevent the abnormality.

In another study, Deacon et al. (1978) exposed rats to 50% nitrous oxide: 50% oxygen and measured liver methionine synthetase, excretion of methylmalonic acid and deoxyuridine suppression.

Methionine synthetase, which requires methyl cobalamin as a coenzyme in the conversion of homocysteine to methionine, was virtually absent in the livers after six hours exposure. Methylmalonic acid excretion was not affected by 24 hour exposure. The deoxyuridine suppression test became abnormal shortly after inactivation of the methionine synthetase.

An editorial (1978) has evaluated these findings with nitrous oxide and those with other anesthetic gases. They made the interesting speculation that chemotherapy given during anesthesia may preferentially affect tumor and spare the host since it has been reported that transplanted tumor cells were less inhibited in their proliferation than were normal marrow cells.

A recent abstract (Thom, 1979) indicates, from work with microbiological systems subjected to anesthetic gases and helium, that helium, by antagonizing narcosis, enhanced growth of organisms except in the presence of nitrous oxide. In this case, growth was inhibited.

While the mechanism of action of nitrous oxide in producing bone marrow failure, DNA synthesis disruption, and neuropathy may relate to impaired vitamin $B_{12}$ metabolism, continued work is needed to establish these mechanisms.

The complex nature of the many unsolved questions concerning neurological findings in folate and vitamin $B_{12}$ deficiency have been reviewed (Reynolds, 1976). Reynolds postulated that the hematological and neurological expressions of vitamin $B_{12}$ deficiency have a similar biochemical basis, and that a block in folate metabolism is responsible for the neurological consequences.

Diagnosis of vitamin $B_{12}$ deficiency rests on the demonstration of the typical clinical features of megaloblastic anemia, achlorhydria after histamine stimulation, and response to physiologic doses of parenteral vitamin $B_{12}$ (Minot and Castle, 1935; Sullivan and Herbert, 1965; Sullivan, 1970b; Beck, 1977b; Hall, 1973; Toskes and Deren, 1973; Glass, 1974; Herbert, 1979). The latter may be an incomplete response in the presence of active infection, renal disease, or tumor.

Achlorhydria is present if the pH of the gastric juice is never less than 3.5 and decreases less than one pH unit after histamine stimulation (Beck, 1977b). Hypochlorhydria is present should the pH unit decrease by more than one, even though the initial pH is above 3.5. While Addisonian pernicious anemia is precluded in the absence of achlorhydria, certain vitamin $B_{12}$ responsive conditions, especially congenitally acquired absence of IF, or other rare conditions of absorptive failure may occur in the presence of free gastric acid.

Additional diagnostic studies are necessary for the more commonly encountered patients presenting with atypical or less severe findings than in late Addisonian pernicious anemia. Indications for additional tests are the uncertainty of the cause of megaloblastic anemia and/or presence of neurologic findings in the absence of anemia. Special tests include measurement of the serum vitamin $B_{12}$, absorption of vitamin B12, assay of IF, measurement of abnormal metabolite products, and the deoxyuridine suppression test (Killmann, 1964).

Serum vitamin $B_{12}$ measurements employing microbiological assay by _Euglena gracilis_ or _Lactobacillus leichmanii_ are generally less than 100 pg/ml (normal, 315 ± 125 pg/ml) (Kass, 1976b) Radiodilution assay methods have been developed but appear less reliable than the microbiological techniques (Whitehead and Cooper, 1977; Donaldson, 1978; Cooper and Whitehead, 1978; Kolhouse et al., 1978), because cobalamin analogues normally present in serum may act as inhibitors of cobalamin or may be inert biologically. Kolhouse et al. (1978) suggest this variability may be a clue to further investigation to determine why some patients with vitamin $B_{12}$ deficiency may develop either hematologic or neurologic abnormality whereas in other patients both occur. Important clinical implications arise in the use of radiodilution assay methods. It has been estimated about 20% of the true cobalamin deficiencies are missed with this type of assay since it is extensively used in clinical settings in the form of specialized commercial kits. Kolhouse et al. (1978) feel the error rate may be overestimated as the trained clinician or hematologist should and would obtain additional tests should the clinical and other laboratory findings be in discrepancy.

The Schilling test (Schilling, 1953) evaluates absorption of an oral dose of radioactive vitamin $B_{12}$ by measurement of urinary excretion of the radiolabeled vitamin. In the absence of intrinsic factor or other cause for malabsorption, excretion is less than 5% in the first 24 hours after a "flushing" parenteral dose of nonradioactive vitamin $B_{12}$. If the test is positive, it is repeated utilizing an oral dose of IF with the radioactive vitamin $B_{12}$. A patient with pernicious anemia usually will then excrete more than 5% of the test dose of labelled vitamin. Certain precautions in usage and interpretation are needed. Reliable urine collection is required. The test may be invalid in the presence of vomiting, diarrhea, or renal disease. The large parenteral flushing dose precludes observation of response to physiologic vitamin $B_{12}$ doses. Impaired absorption as seen in sprue, other gastrointestinal disorders (Mahmud et al., 1971; Klipstein, 1972), or other conditions may result in abnormally low excretion both before and after IF.

Herbert (1972b) discussed modifications of the Schilling test technique to elucidate possible malabsorption by measuring amounts of radioactivity in blood, urine, and feces following 1) oral vitamin $B_{12}$ test dose, 2) oral dose plus IF, and, if these are both abnormal, 3) repeating the test after an appropriate trial of therapy, such as an anthelminthic or antibiotic in the case of fish tapeworm infestation or blind loop syndrome, respectively, or after treatment with sodium bicarbonate if pancreatic insufficiency is present. Herbert emphasized the importance of retesting for absorption following appropriate therapy to distinguish between primary or secondary malabsorption.

IF measurements were originally bioassay techniques based on reticulocytosis in pernicious anemia patients (Minot and Castle, 1935). These prolonged and tedious methods depended on pernicious anemia patient resources. An in vitro technique developed by Sullivan et al. (1963) utilized guinea pig ileal homogenate as a source of receptors for IF. A rapid, less complicated technique utilizing protein-coated charcoal separation was developed by Gottlieb et al. (1965). Subsequently, Begley and Trachtenberg (1979) reported an in vitro assay for IF which eliminated the need for elaborate forms of protein separation and for a source of antibody. IF is decreased in partial gastrectomy (Ardeman and Chanarin, 1966) and may be variable in pernicious anemia (Castle et al., 1931).

Additional laboratory findings which may be found in association with untreated vitamin $B_{12}$ deficiency are elevated serum folate in 20-35% of patients, elevated serum lactic dehydrogenase, formiminoglutamic aciduria, methylmalonic aciduria, hyperbilirubinemia, increased fecal stercobilin, elevated plasma iron with increased saturation of transferrin, accelerated plasma iron clearance, and impaired utilization of iron by red blood cells. Of the above biochemical abnormalities, the most specific for vitamin $B_{12}$ deficiency is the excretion of methylmalonic acid; normal excretion is less than 10 mg/day (Gutteridge and Wright, 1970; Frenkel and Kitchens, 1977). In about 34% of patients with pernicious anemia, blocking antibodies may be present; about 19% have both blocking and binding antibodies for IF (Sullivan, 1970a).

Finally, the most recent and probably the most important for differential diagnosis is the deoxyuridine suppression test introduced by Killmann in 1964. It is particularly useful in recognizing vitamin $B_{12}$ or folic acid deficiency, specifically differentiating the two, in non-anemic patients with peripheral blood macrocytosis and equivocal bone marrow morphology together with normal serum vitamin $B_{12}$ or red cell folate values (Herbert et al., 1973; Wickramasinghe and Longland, 1974). The test is based on the fact that preincubation of normal bone marrow with an appropriate concentration of deoxyuridine severely suppresses the subsequent incorporation of tritiated thymidine into DNA. The suppression is subnormal with vitamin $B_{12}$ or with folate deficiency and is correctable *in vitro* by the addition of the appropriate deficient vitamin.

In patients with megaloblastic changes as a result of neoplastic, chemotherapeutic, or other agents which interfere with DNA synthesis, the deoxyuridine suppression test is abnormal. Except for these, an abnormal test indicates vitamin $B_{12}$ or folic acid deficiency.

Immune Mechanisms in Vitamin $B_{12}$ Deficiency

Although studied for more than twenty years, basic mechanisms to account for immunologic aberrations associated with pernicious anemia continue to elude researchers.

Starting with the approach of Taylor and Morton (1958) in IF immunized rabbits, Schwartz (1958) then demonstrated IF-inhibiting substances in sera of pernicious anemia patients treated with oral hog IF-$B_{12}$ preparations. It had been shown earlier that immunosuppressive agents, ACTH and corticosteroids, could increase reticulocyte counts or change the bone marrow to produce mature red cells (Thorn et al., 1950; Wintrobe et al., 1951; Doig et al., 1957). Ardeman and Chanarin (1965) showed that steroids improved the vitamin $B_{12}$ absorption in pernicious anemia patients, irrespective of the presence of antibody to IF. In some, there was a return of free gastric juice hydrochloric acid and of IF. No improvement in absorption occurred in gastric resection patients. An additional pernicious anemia patient with recovery of gastric mucosal secretion of IF and of acid was reported by Jeffries (1965).

Antibodies to IF and to parietal cells have been demonstrated in over 50% of pernicious anemia patients (Irvine, 1963; Irvine, 1965). The association of pernicious anemia with autoimmunity was reviewed by Chanarin (1972). Where the prevalence of pernicious anemia in Great Britain is one to two per 1000 population, pernicious anemia in association with other diseases is 31 per 1000 for Grave's disease and 108 per 1000 for primary myxedema. Six of 118 with Addison's disease, and seven of 74 with idiopathic hypoparathyroidism also had pernicious anemia. Ardeman et al. (1966) suggested that patients with thyroid disorders should be screened for pernicious anemia. Other than the possible immune mechanisms associated with thyroid disease, increased requirements for vitamin $B_{12}$ have been reported in uncontrolled thyrotoxicosis (Alperin et al., 1970).

Schade et al. (1966, 1967) demonstrated IF antibodies to be of two types: precipitating antibody; antibody which combines with IF before or after addition of vitamin $B_{12}$ and which inhibits vitamin $B_{12}$ absorption *in vivo* and *in vitro*, and blocking antibody; antibody which blocks the ability of IF to complex with the vitamin to form the IF-$B_{12}$ complex. Also termed Type I, or "binding" antibody, the precipitating antibodies interfere with ileal mucosal absorption (Glass, 1974), and Type II, or "blocking" antibodies, interfere with coupling of vitamin $B_{12}$ to IF (Garrido-Pinson et al., 1966; Schade et al., 1966; Schade et al., 1967).

IF antibodies have also been demonstrated in the absence of pernicious anemia. In 11 of 23 patients with both thyroid disease and IF serum antibodies who were followed from three to seven years, none developed pernicious anemia (Rose et al., 1970). Both types of IF antibodies have been found in saliva, serum, and gastric juice. It has been suggested that IF antibodies produced locally by gastric mucosal cells are of greater importance than serum antibodies.

In separate studies, infants found to have no gastric IF at birth, as a result of maternal transfer of antibody *in utero*, subsequently developed IF as the antibody disappeared (Bar-Shany and Herbert, 1967; Goldberg et al., 1967). Charache et al. (1968) reported an additional infant with normal IF at birth although subjected to maternal anti-IF antibody *in utero*.

Parietal cell antibody was reported in 84% of pernicious anemia patients and in 36% of their healthy relatives (Jeffries and Sleisinger, 1965; Fisher and Taylor, 1965).

The heterogeneous nature of the IgG light chain auto-antibodies associated with pernicious anemia, similar tc that of normal gamma globulin, was described by Bernier and Hines (1967).

An extensive study of 24 HLA specificities on the A and B loci by platelet-complement fixation and microlymphocytotoxic techniques was done on 66 Addisonian pernicious anemia patients by Horton and Oliver (1976). No association was found for HLA A3, HLA B7, or the haplotype A3, B7 with the presence of serum antibodies to vitamin $B_{12}$-binding sites of IF. A small increase in frequency of the antigen HLA B8 (38%) was found for those patients who also had serum thyroid antibodies over that found in thyroid-antibody-negative patients (13%). The Horton and Oliver (1976) study was unable to confirm earlier work that HLA B7 was associated with pernicious anemia (Zittoun et al., 1975).

Antibodies against TCs have been found in pernicious anemia patients with high serum levels of vitamin $B_{12}$. Carmel and Schurafa (1977) found binding (Type I) antibodies in such a patient.

Marcoullis et al. (1979) recently reported the first instance of the simultaneous occurrence of blocking and binding antibodies to IF, TC II, and TC I and other R-type vitamin $B_{12}$-binding proteins in a patient with treated pernicious anemia. Following demonstration that the dialyzed IgG, but not the IgM, neutralized the total unsaturated vitamin $B_{12}$-binding capacity of human gastric juice, saliva, and serum, they suggested that the IgG contained blocking antibodies (Type II) against IF, TC II, TC I, and other R-type binders. Binding antibodies (Type I) were also demonstrated. Where only immuno-complexes of Type II, which prevent vitamin $B_{12}$ from complexing with IF, occur, a high serum vitamin $B_{12}$ is expected. In this case, the presence of both types I and II antibodies resulted in lower serum vitamin $B_{12}$ levels.

Carmel and Johnson (1978) suggest the earlier age at onset of pernicious anemia in black and possibly in Latin-American women may have an autoimmune basis.

## FOLIC ACID DEFICIENCY

### Occurrence

Folic acid deficiency, next to iron deficiency, is probably the most common cause of anemia in man. As noted in the etiologic classification, Table I, causes of folic acid deficiency include dietary lack of folate-containing foods, as a complication of gastrointestinal disease, in association with pregnancy, or as a side effect of certain drugs.

Sauberlich (1977) concluded that the indidence throughout the world was uncertain, but estimates have probably been understated. Herbert (1970b) found that as many as 50% of pregnant women have megaloblastosis by the third trimester, largely in the low-income population groups who consume basic diets of beans and rice. Rothman (1970) reviewed 27 studies on pregnant women and noted the incidence of megaloblastic anemia ranged from 0.5 to 54%. Chanarin et al. (1965) found a higher incidence of folate depletion in pregnant women who had not received iron supplementation than in those who had been supplemented. It must be remembered that it was in pregnant women in Southern India, that Wills (1931, 1932, 1933) described the differences in response of megaloblastic anemia to purified liver extract and to autolyzed yeast which led to the discovery of folic acid.

In Southern India, Baker and Mathan (1971) found 64% of patients with tropical sprue had megaloblastic anemia, more than half of whom had pure folate deficiency. Sprue is known to be endemic in tropical countries (Klipstein et al., 1966). Herbert (1968) attributed much of the folate deficiency in Carribean population groups to the common practice of cooking foods in boiling water for a time in excess of one-half hour.

In 95 children with kwashiorkor, studied by Pereira and Baker (1966), serum folic acid levels were significantly lower ($p=0.001$) than in normal children and ranged from 1.4 - 2.4 ng/ml. High serum vitamin $B_{12}$ levels were also found. Baker and Mathan (1971) have associated high serum vitamin $B_{12}$ levels in sprue with liver damage.

In a study of randomly selected municipal hospital adult patients, Leevy et al. (1965) found 45% had subnormal folate levels. Nineteen, of the 117 patients studied, chronically consumed alcohol to excess; 80 drank socially. Herbert et al. (1963) found up to 80% of alcoholic patients had folic acid deficiency. Further references may be found in a review of folic acid deficiency (Streiff, 1970a).

## Biochemical Considerations

Several reviews of the biochemistry of folic acid are available (Chanarin, 1969; Beck, 1977c; Rosenberg, 1976; Hoffbrand, 1975; Blakley, 1977).

Pteroylglutamic acid (folic acid) is the generic substance of a large group of compounds known collectively as folates. The molecule (peteroylmonoglutamic acid) is composed of a 2,4,6 substituted pterin ring, paraminobenzoic acid, and an L-glutamic acid residue. In nature, folic acid occurs as conjugates in which varying numbers of glutamic acid residues are attached by peptide linkages to the $\gamma$-carboxyl group of the adjacent glutamic acid moiety. In liver, the major folate is pentaglutamyl conjugate (Shin et al., 1972); in plants, heptaglutamyl conjugate (Rosenberg and Godwin, 1971).

Folic acid is not active as the monoglutamate in mammals, but must be reduced by conjugase to release tetrahydrofolic acid, which serves as the coenzyme in one-carbon transfer acceptance. The reactions in production of tetrahydrofolic acid are catalyzed by dihydrofolate reductase, an NADPH-linked enzyme.

Reduced folic acid derivatives (coenzymes) differ in the one-carbon unit and its site of attachment. With one exception, $N^5$-formyltetrahydrofolic acid, these are unstable to oxidation and to heat. Alternative names for $N^5$-formyltetrahydrofolate are citrovorum factor, leucovorin, and folinic acid.

Folic acid coenzymes are required in metabolic reactions; probably the most important of which is the methylation of deoxyuridylate to thymidylate in pyrimidine nucleotide biosynthesis, essential to formation of DNA. If this process is impaired, megaloblastosis results. Other reactions for which folate coenzymes are necessary are the interconversion of serine and glycine (also requires vitamin $B_6$), the conversion of histidine to glutamic acid, the conversion of homocysteine to methionine (also requires vitamin $B_{12}$), the formation of glycinamide ribonucleotide, the reaction in which 4-imidazole carboxamide ribonucleotide reacts with the coenzyme to form 5-formamido-4-imidazole carboxamide ribonucleotide, reduction of dihydrofolate, reduction of 5, 10-methylene tetrahydrofolate, formiminoglutamic acid conversion to glutamic acid, generation of and utilization of the formate pool, methylation of uracil residues in transfer RNA, and folate-dependent methylation not involving S-adenosyl methionine.

Evidence has been found for feedback inhibition and end-product repression of folate coenzymes (Blakley, 1977). Krumdieck et al. (1977) consider the peptide chain of naturally occurring folates important in the regulation of one-carbon metabolism. There is evidence that these folates affect passage of coenzymes across biological membranes, the recognition of proper coenzyme-carbon-fragment complex, and the activation or inhibition of enzymes involved in one-carbon metabolism.

## Dietary Considerations

Food folates are primarily found in green and leafy vegetables, fruits, organ meats, and yeast (Toepfer et al., 1951; Chanarin, 1969). The analyses of folic acid in foods are dependent upon extraction procedures which do not destroy the labile reduced forms; the assay method used, since microorganisms vary in requirements and utilization of one-carbon substituents of tetrahydrofolate and the short polyglutamates; and pretreatment with conjugase, as none of the microorganisms are able to utilize polyglutamates. Cooking procedures need also to be considered, as excessive boiling and use of large quantities of water result in folate destruction (Herbert, 1963).

Need for additional studies on food folate content is recognized as analyses done prior to the use of ascorbate or other substance to prevent heat lability (Toepfer et al., 1951) may be low. Evidence also has been found that folates may be non-specifically bound to cellular fibers (Santini et al., 1962), since, after removal from fibrous tissue, conjugase treatment of diets results in an increase of folate assayed. This has not been evaluated for availability of folate for man.

Tamura and Stokstad (1973) and Stokstad et al. (1977) reported studies on availability of food folate. Low availability was found in orange juice, 31%, romaine lettuce, 25%, and egg yolk, 39%. Higher availability was found in bananas, 82%, dried lima beans, 70%, and frozen lima beans, 96%. High acidity was attributed to be the cause of low folate availability from orange juice. *In vitro* studies by Herbert and Jacob (1974) indicated that pharmacologic amounts of ascorbic acid added to diets assayed for vitamin $B_{12}$ destroyed 43% of that vitamin.

Rosenberg (1975) studied absorption by man of tritiated mono- and heptaglutamic derivatives of folate. After a flushing dose of non-tritiated folic acid, 70% of the monoglutamate and 56% of the polyglutamate were recovered in the urine in 48 hours. He concluded that dietary folate, 80% of which is in polyglutamate form, was substantially utilized although various foods may differ in efficiency.

Waxman et al. (1970a) evaluated the evidence that amino acids could effect megaloblastosis. Methionine, homocysteine, serine, and glycine are required in the production of DNA thymine. In *in vitro* human bone marrow culture studies, methionine and glycine provoked megaloblastosis; whereas, homocysteine and serine appeared to reduce the DNA defect.

Hoppner et al. (1977) called for evaluation and standardization of methodology in dietary folate analysis, and additional data for folate content of foods. They concluded from data presently available that from the "average composite" diet, adolescents, pregnant, and nonlactating adults may not obtain adequate amounts of folic acid.

## Dietary Folic Acid Requirements

Herbert (1962b) showed that 50 μg folic acid per day was the minimal daily amount needed to maintain serum folate levels in man. The classic depletion study produced megaloblastic changes in 133 days in a normal male on a diet containing 5 μg folic acid per day (Herbert, 1962a) indicating body stores are depleted in about four to five months.

Official NRC-NAS (1974) requirements are set at 400 μg per day to meet needs for good nutrition of adults and double that amount in pregnancy. Infant requirements are 40 to 50 μg per day. Waslein (1977) concluded the FAO-WHO recommendation of 5.0 μg/kg body weight is adequate for the normal child, but for maximal repletion in protein-calorie malnutrition two to three times that amount are needed.

Considerable attention has been given to the study of folic acid requirement in pregnancy. Up to 50% or more demonstrate megaloblastic changes by the third trimester (Herbert, 1970b; Beck, 1977c). In one study, seventy per cent of the pregnant women with folic acid deficiency were multiparas (Stone et al., 1967). The subject has been reviewed extensively (Whiteside et al., 1968; Chanarin, 1969; Davis et al., 1969; Kitay, 1969; Rothman, 1970).

Contributing to the high frequency of folic acid deficiency in pregnancy are physiologically increased requirements, adolescent pregnancies, poor diet, anorexia, infections, coexisting hemolytic conditions, and drugs such as prior use of oral contraceptive agents or anticonvulsants.

Complications of hemorrhage, abortion and congential malformations were noted by Hibbard and Smithells (1965). Stone (1968) found 60% with abruptio placenta had positive formiminoglutamic acid excretion. Pritchard et al. (1969) were unable to confirm these findings. The latter compared the baby's hemoglobin at birth with that of the mother. For mothers with either more or less than 6.9 g/dl, the infants' hemoglobins were 18.9 g/dl. In a patient seen by Beck (1977c) the mother had a megaloblastic anemia at term with only 6 g/dl hemoglobin level, while the baby, delivered before folic acid could be given, had a hemoglobin level of 28. He attributed the high hemoglobin level to anoxia. Another explanation may be for folic acid to go preferentially to the fetus. No firm explanation is available. However, Cooper et al. (1970) observed an abrupt rise in serum folic acid seven days postpartum and postulated a shunt across the placenta and also a low rate of transfer of liver folate to the plasma of the mother.

Routine folic acid supplementation in pregnancy is generally accepted with amounts recommended varying from 0.5 mg to 1.0 mg per day.

Another possible complication observed in megaloblastic anemia in pregnancy is the histological change of the uterine cervix smears (Kitay and Marshall, 1968; Kitay, 1969) which could possibly be mistaken for malignant change (O'Brien, 1962; van Niekirk, 1966). The changes may be a result of localization of conjugase to target tissues or to other systems whenever there is a cycle of cell multiplication. Krumdieck et al. (1977) have found the conjugase activity of the rat uterus is increased three fold at the time of maximum estrogen secretion. That the uterus is a target tissue for folic acid is probably borne out in the use of folic acid antagonists to induce abortions (Thiersch, 1952) which was renounced by Goetsch (1962) as hazardous and possibly teratogenic if abortion was not induced. While folic acid deficiency is definitely teratogenic in animals, it is not yet confirmed in the human. The most convincing evidence to date seems to be that of Hibbard and Smithells (1965) in which five times as many mothers with abnormal formiminoglutamic acid excretion gave birth to infants with severe malformations, primarily of the CNS, as did mothers with normal excretion levels.

While the infant may develop folate deficiency in the second six months of life as a result of dietary deficiency, the low-birth-weight, preterm baby may develop overt megaloblastic anemia in the first six months of life. In the latter, no associated causative factors, such as infection, general malnutrition or abnormally low folate diet, may be necessary (Hoffbrand, 1970;

Dallman, 1974).

During lactation, folate is excreted at a level of 50 μg/ liter breast milk (WHO, 1972). Metz (1970) has noted that 100 μg folic acid per day supplementation to lactating women resulted in a rise in folic acid in breast milk, but the serum folic acid levels and reticulocyte counts remained the same. It was concluded that folic acid preferentially goes to the milk secretion.

## Impaired Absorption

Absorption of folic acid occurs in the proximal jejunum. Butterworth et al. (1969) demonstrated that the mucosal cell accepts only monoglutamate across the cell membrane. That the brush border converts polyglutamates to monoglutamates by means of conjugase was suggested by Rosenberg et al. (1969). Beck (1977c) has pointed out, however, that the assay microorganism used, *L. casei*, also utilizes diglutamates and triglutamates.

In studies on isolated intestinal segments of the dog, Baugh et al. (1971) found pteroylmono- and di-glutamates crossed the intestinal mucosa intact. Reduction and methylation were not required. Rosenberg (1975, 1977) has expanded on the hypothesis proposed by Rosenberg and Godwin (1971). In this scheme, intestinal enzymes hydrolyze heptaglutamate to monoglutamate. The overall rate of transport into the mesenteric circulation is controlled by the movement of monoglutamate. Under appropriate conditions, the monoglutamate is reduced and methylated in the intestinal cell and appears in the circulation as methyltetrahydrofolate.

These conflicting results are possibly related to variable doses of folate, possible presence of folate inhibitors, and the form of the glutamates (Baker and DeMaeyer, 1979). It is likely that absorption and transport systems may be resolved by study of specific genetic defects (Rosenberg, 1975).

Three reports have presented a specific congenital defect in gastrointestinal absorption of folates (Luhby et al., 1961; Lanzkowsky et al., 1969; Santiago-Borrero et al., 1973). The case of Luhby et al. responded to high doses of folic acid orally. That of Lanzkowsky et al., a product of consanguineous parents, had associated mental retardation and calcification of the basal ganglia with absent folate in the CSF. Santini et al. (1973) have presented a follow-up of the case documented by Santiago-Borrero et al. (1973).

Active transport mechanism across a gradient has been suggested by the work of Hepner et al. (1968) and reviewed by Cooper (1977). Hoffbrand et al. (1977) pointed out that folate conjugase is localized in the cell lysozomes and may exist in many forms (possibly isozymes) in human tissue. A recent report of 2332 patients routinely admitted to one medical center (Magnus, 1977) noted elevated plasma conjugase activities in patients with metastatic cancer, with active liver disease and in those receiving phenytoin. Normal activities were found in pernicious anemia patients. Low activities accompanied diagnoses of wide-spread malignancies, hepatic coma, aplastic anemia, diabetes mellitus, and severe malnutrition. It was suggested that plasma conjugase may be of value in the diagnosis or monitoring of certain diseases.

Development of the triple lumen tube to examine intestinal perfusion offers possibilities for study of transport mechanisms. In one study, using this technique, glucose was found to enhance folic acid absorption (Gerson et al., 1971). In other studies, oral folic acid, but not intramuscular folic acid or oral vitamin $B_{12}$, increased jejunal glycolytic enzyme activity. There was no effect on disaccharidase activity.

Folic acid metabolites are primarily excreted by the urine and in the bile. A small amount is excreted in the feces. In malabsorption syndromes folic acid depletion may be hastened by means of biliary loss with little reabsorption of folic acid.

Recognition of folate binders in milk, liver, kidney, intestinal epithelial cells, leukemic granulocytes, and folate binders from pregnant women or women taking oral contraceptive agents may be of importance in conservation of folic acid (Rothenberg et al., 1977).

## Malabsorption Syndromes

Malabsorption of folic acid may occur with any lesion which interferes with the absorptive mucosal surface of the jejunum by means of reducing the available number and surface area of the microvilli, by impairment of the action of conjugase, by presence of folic acid antagonists, by presence of bacteria or infection, by competition by intestinal bacteria with reduction of available folate, by folate binders, or by possible imbalance of nutrients presented to the jejunum, or to unknown factors.

Where malabsorption is known to exist by clinical evidence, Beck (1977c) suggests the serum folate be assayed to determine folic acid deficiency. If the serum folate is low, additional tests are unnecessary to establish folic acid deficiency. In certain cases where absorption of D-xylose is normal (the test which is most frequently abnormal in intestinal malabsorption) further studies as suggested by Beck (1977c) may be indicated: microbiologic or isotopic procedures to measure the serum, urine, or fecal content of folic acid following a test dose of the vitamin.

Interference with the absorptive surface of the jejunal microvilli is seen in tropical sprue, non-tropical sprue, jejunal atrophy, regional enteritis, lymphoma of the small bowel, Whipple's disease, diabetes mellitus, scleroderma, amyloidosis, and after resection or bypass of small bowel as utilized in treating intractable obesity. (Refer to Table I).

Manson in the 1880's first described and named the condition, sprue. It manifests as a wasting disease accompanied by the symptomatology of megalobastic anemia and steatorrhea in persons living in tropical countries. Sometimes there is delayed onset months after migration from the tropics. The geographic predilection for certain endemic areas, eg., Puerto Rico, Southern India, and other areas, and the less common occurrence in other tropical areas is not understood. Krumdieck et al. (1977) noted that diets high in beans and pulses, such as used in Puerto Rico, contain naturally occurring conjugase inhibitors. Plasma conjugase activity decreased by four fold within 20 minutes after ingestion of red kidney beans in the subjects studied. Further investigation to determine the significance of this finding is needed. Extensive reviews of the literature (Gardner, 1958; Klipstein, 1968; Chanarin, 1969; Baker and Mathan, 1968; Gorback et al., 1970; Baker, 1972; Klipstein, 1972; Butterworth et al., 1974; Corcino et al., 1975) have examined etiologic possibilities.

Corcino et al. (1976) studied six Puerto Ricans with tropical sprue. Manifestations included steatorrhea, vitamin $B_{12}$ and folic acid deficiencies, impaired D-xylose absorption, and jejunal histological abnormalities. From studies done before and after treatment with vitamin $B_{12}$ and folic acid, they concluded that folate conjugase deficiency was not the cause, as the activity of conjugase was greater prior to therapy. In addition, there was no impairment of ability to convert polyglutamates to lower glutamic acid residues. Disaccharidases were elevated after therapy. They concluded the defect was a failure to transport folate across the mucosal cell.

Klipstein et al. (1973) have explained the lesion in part as a result of depressed gastric juice secretion with resultant inability to maintain sterility of the normal small bowel. Bhat et al. (1972) found, on peroral intubation and examination of stool bacteria, a correlation of severity of histological lesions and the number of bacteria. No correlation with response to treatment was noted. Klipstein (1972) emphasized the perpetuation of the gastrointestinal lesion by increased mucosal cell turnover resulting in increased folate requirements and compromised absorptive ability.

Butterworth et al. (1974) proposed the theory that an imbalance between "conjugated" or polyglutamate forms and the "free" or monoglutamate forms of folate occurs in the small bowel epithelium and other tissues with high cell proliferation rates.

Most workers agree that morphological lesions continue for a long period of time after treatment is begun with folate. Swanson et al. (1966) found evidence of changes up to a year. Others (Rickles et al, 1972) noted the relapsing nature of the disease or persistence of the lesions, even with the addition of antibiotic therapy which is thought to be curative in some cases.

Manifesting a similar symptomatology, non-tropical sprue (idiopathic steatorrhea, gluten-induced enteropathy, or coeliac disease) also is of unknown etiology. Serum folate concentrations are low; steatorrhea is present with fat droplets in the stool; megaloblastic anemia may or may not be present; atrophy of jejunal villi is present; symptomatic and laboratory response occurs with a diet free of gluten.

While the etiology is unknown, gluten-induced enteropathy has been demonstrated to be associated with abnormal immunological findings. Mann et al. (1970) found immunoglobulin M significantly diminished in 12 patients. An additional patient had a selected serum and exocrine immunoglobulin A deficiency. Cooper et al. (1978) found 63 of 314 patients with coeliac disease had a total of 75 identified immunological disorders. Of interest is the implied association of zinc deficiency with small bowel malabsorption, protein-calorie malnutrition, and decreased cellular mediated immunity (Nutrition Reviews, 1979).

Impaired intestinal folate absorption occurs also in alcoholics. In addition, hematopoiesis was suppressed in alcoholic patients by a possible toxic action even when folic acid or folinic acid was given in large doses (Sullivan and Herbert, 1964). The effect of alcohol on jejunal uptake has been studied extensively (Halsted et al., 1967; Halsted et al., 1971; Halsted et al., 1973). Hermos et al. (1972) found abnormal peroral biopsy specimens of duodenal-jejunal junction in two of three folate deficient alcoholics with demonstration of shortened villi, enlarged crypts, and villous epithelial cells.

Systemic bacterial infection appeared to be a factor in impaired folate absorption (Cook et al., 1974) in Zambian patients, contributing to megaloblastic anemia. The authors advise further study using mono- rather than a non-reduced form of folate which they used to determine absorption.

Anticonvulsant therapy, another cause for impaired folate absorption, may act by interference with absorption but the exact mechanism is unclear. Druskin et al. (1962) considered the drugs interfere with deconjugation of dietary folate, and were able to overcome folate deficiency with 25 μg folic acid daily. Many workers have demonstrated folate deficiency with anticonvulsant therapy and its treatment with folic acid. Stebbins and Bertino (1976) reviewed the several theories of action and have concluded that the triple lumen perfusion studies of absorption have demonstrated definite folate inhibition in the presence of phenytoin. Reynolds (1976), in the same monograph, reviewed the neurological aspects of folate and cautioned that seizures may be precipitated in convulsive disorders by folate. One death from severe anemia has been reported in a patient in whom anticonvulsant-induced anemia was superimposed upon Hemoglobin H thalassemia disease (a congenital abnormal hemoglobin with associated hemolysis) and resulted in terminal heart failure (Pinkhas et al., 1973).

Numerous but conflicting reports have implicated oral contraceptives, especially those containing mestranol, with folic acid deficiency, low serum folate concentrations, megaloblastic anemia, or with local cytological changes of the cervix. The latter is thought to be an end-organ manifestation of conjugase inhibition (O'Malley and Means, 1974) which has been shown to be reversible within three weeks by pharmacological doses of folic acid (Whitehead et al., 1973). While the mechanism is unclear, Streiff (1970b) demonstrated in patients taking oral contraceptives poor absorption of food polyglutamates but normal absorption of monoglutamates. Streiff and Greene (1970) reported inhibition of conjugase activity _in vitro_ by mestranol. Stephens et al. (1972) were unable to confirm _in vitro_ inhibition. Further, no difference in absorption between control subjects and women taking oral contraceptive agents was found provided they were saturated with folic acid. Without presaturation, a significant absorptive reduction occurred in the patients taking oral contraceptive agents. Recent studies by daCosta and Rothenberg (1974) have demonstrated a folate binder in sera of 24 of 51 pregnant women, 8 of 10 on oral contraceptives, and none of 15 women who were not pregnant nor taking oral contraceptive agents. They attributed these changes to a potentiation by oral contraceptive agents of inadequate folate intake or folate need by removal of folate from the body by the folate binder. Other workers have indicated the effect of oral contraceptives may be related to borderline diet adequacy of folate (Castrén and Rossi, 1970; Necheles and Snyder, 1970; Pritchard et al., 1971).

Other drugs have been associated with folate deficiency: antimalarials (Strickland and Kostinas, 1970; Sheehy and Dempsey, 1970), azulfidine (Franklin and Rosenberg, 1973), triamterene (Corcino et al., 1970), tetracycline (Jones, 1973), methyl-dopa (Schneerson and Gazzard, 1977), and neomycin (Jacobson et al., 1960).

## Folic Acid Requirements in Hemolytic and Other Diseases

Folic Acid requirements are significantly increased in any disease in which a hemolytic process is associated. These include sickle cell anemia (Huntsman, 1974), thalassemia (Jandl and Greenberg, 1959), myeloproliferative disorders (Chanarin, 1970), sideroblastic anemias (Mollin and MacGibbon, 1972), and aplastic anemia (Branda et al., 1978). In the aplastic anemia patient studied by Branda et al. (1978), a high dose of folic acid, 20 mg per day, induced a slow erythropoietic response with no further need for transfusions. The patient continued to have leucopenia and thrombocytopenia. This work is important in that family studies demonstrated, among four generations of the patient's family, 20 members with histories suggestive of severe hematologic disorders. The patient's PHA stimulated leucocytes demonstrated marked suppression of $^{14}C$-methyltetrahydrofolate uptake but normal $^{14}C$-thymidine uptake. Uptake of bone marrow cells reacted similarly. Five other family members tested, including the patient's son, demonstrated decreased $^{14}C$-methyltetrahydrofolate uptake by stimulated lymphocytes.

An unexplained lack of responsiveness to folate therapy in thalassemic patients among Chinese in Hong Kong, even though subnormal serum folate levels were noted, has been reported by Tso (1976). That megaloblastosis due to folic acid deficiency is uncommon among Chinese has been related to the high level of green and leafy vegetables in the diet as well as to their cooking practices.

Folic acid metabolism in renal disease patients has been studied by Retief et al. (1977) and Colman and Herbert (1974), Folate binding protein was elevated in the serum of uremic patients, probably as a result of impaired reabsorption by the tubules. Serum binders may withhold folic acid from the tissues.

Exfoliative dermatitis is another etiologic cause of increased requirement of folic acid as Hild (1969) has found losses of folic acid of 5 μg to 20 μg per day. The folic acid content of the skin was greater than that in normal skin. Changes of the epidermis have been noted with vitamin $B_{12}$ deficiency (Gilliam and Cox, 1973).

Folic acid antagonists have been used effectively in treating neoplastic diseases, acute leukemia in children, choriocarcinoma, head and neck neoplasias (usually for squamous cell carcinoma), and in adjuvant chemotherapy in conjunction with radiation therapy, and with other antimetabolic drugs. Concentration of folic acid in leukemic cells was demonstrated to be elevated in both acute and, less so, in chronic leukemia by Swenseid et al. (1951). Folic acid antagonists inhibit the reduction of dihydrofolate to tetrahydrofolate. As a consequence, DNA, RNA, and protein synthesis are inhibited.

The chemistry, enzymology, mechanisms of action and use of anti-folates have been reviewed in a recent symposium (Bertino, 1971).

More recently, Bertino et al. (1977) have reviewed augmentation of anti-folate action by metabolic transport inhibitors which prevent efflux of the antimetabolite from the cells. Another new and potentially useful possibility for cancer chemotherapy has been suggested by Huennekens et al. (1976) by alteration of the methionine synthetase content of cells and thus interfering with the vitamin $B_{12}$-folate acid interrelationship (see VITAMIN $B_{12}$ DEFICIENCY, Biochemical Considerations).

Degradation of folic acid to form pterin-6-aldehyde in cancer patients has been reported by Halpern et al. (1977). In no hospital patient tested was the test positive in the absence of cancer. The significance of this finding remains to be shown.

A major breakthrough in the use of high doses of anti-folate drugs in cancer chemotherapy was the use of citrovorum factor to "rescue" the normal cells by selective action (Frei et al., 1975).

## Neurological Considerations of Folic Acid Deficiency

In the last decade, a growing body of evidence indicates that folic acid deficiency results in neurological symptomatology. The subject has been reviewed by Reynolds (1976).

That anticonvulsants interfere with folic acid metabolism and absorption is well documented. There is a high correlation of anticonvulsant medication with low serum folic acid concentrations and with CSF folate (Reynolds, 1971). A study on marrow cells from epileptic patients taking anticonvulsants reported no correlation between the degree of megaloblastosis and the deoxyuridine suppression test (Taguchi et al., 1977). The authors suggested that anticonvulsants may exert their effect at a stage of DNA synthesis beyond that at which folate coenzymes function. Treatment of epileptics with folic acid has resulted in increase in seizure attacks (Reynolds, 1973) and a return to a normoblastic marrow in those with megaloblastic marrows (Reynolds, 1968). The mechanism of action of folate to increase seizure frequency is not known. Mayersdorf et al. (1971) have found, however, that folate concentration in the region of epileptic foci induced by metallic cobalt powder is higher than in other regions of the brain cortex.

The mechanism of transport of reduced folate to the CSF is thought to be by way of the choroid plexus (Spector and Lorenzo, 1975; Bertino et al., 1977; Spector, 1977).

The depression, irritability, and other mental disturbances reported in experimentally induced folate deficiency by Herbert (1962a) and noted in diseases accompanied by folate deficiency have been thought to be of significance in psychiatric illness (Reynolds, 1976). Shulman (1972), however, considers folate disorders in psychiatric illness to be related to malnutrition, chronic diseases, or to alcohol and other drug use.

It is of significance that one case of well-documented folate deficiency presented with subacute combined system disease, which did not respond either hematologically or neurologically to vitamin $B_{12}$, whereas folic acid administration produced a marked recovery (Pincus et al., 1972).

## INBORN ERRORS OF METABOLISM RESULTING IN MEGALOBLASTIC ANEMIA AND MEGALOBLASTIC ANEMIA IN CHILDREN

In 1960, the first case of an absorptive defect of vitamin $B_{12}$ in children was reported by Imerslund (1960). Ten patients from six families were found. Onset ranged from age five to six months up to four years. The findings differed from those of Addisonian pernicious anemia in that the patients had gastric hydrochloric acid secretion after histamine, normal IF activity, and an associated proteinuria. Gräsbeck et al. (1960) proposed a recessive gene inheritance for this condition. It responds promptly to vitamin $B_{12}$ therapy.

Congenital IF deficiency is also a rare disease in which the impaired vitamin $B_{12}$ absorption is correctable by administration of IF. Patients have normal gastric acid secretion, normal gastric mucosa, and no associated endocrinopathy (McIntyre et al., 1965; Miller et al., 1966). Katz et al. (1972) reported malabsorption due to a biologically inert IF.

A second group of pernicious anemia in childhood is that in which gastric acidity is absent; gastric mucosal atrophy is present; and parietal cell and IF antibodies are occasionally present (Farrell et al., 1979). Earlier, Herbert et al. (1964) considered this anemia not to have an autoimmune mechanism. Oliver and Baker (1969) reported a 21 year old man with sudden onset of pernicious anemia in whom hypothyroidism had been diagnosed at age six years. The patient's father and aunt had Addisonian pernicious anemia. Gastric and thyroid antibodies were demonstrated in all three and also in the patient's mother.

Methylmalonic aciduria responsive to vitamin $B_{12}$ has been described prenatally at midterm by Ampola et al. (1975). Treatment of the mother with large doses of vitamin $B_{12}$ resulted in successful termination of the pregnancy nine weeks after start of therapy. The enzymatic block was found to be a deficiency of 5'deoxyadenosylcobalamin. Prenatal diagnosis was suspected as the mother had had a first child with methylmalonic aciduria who died at three months of age. The child reported was maintained subsequently with a low protein diet.

By studies of the synthesis of 5'-deoxyadenosylcobalamin in fibroblast extracts from patients with inherited methylmalonic acidemia, Mahoney et al. (1975) showed biochemical heterogeneity in the phenotype with deficient accumulation of adenosylcobalamin. Thus, genetic heterogeneity is implied by their finding of at least two mutant classes in this phenotype.

Linnell et al. (1976) used two dimensional chromatobioautography to study cobalamin metabolism in cultured fibroblasts and skin biopsies taken from normal control subjects and from patients with methylmalonic aciduria. The latter fell into three classifications of abnormality: Group A, defective methylmalonyl-CoA mutase apoenzyme, but no abnormality of cobalamin metabolism; Group B, failure to accumulate adenosylcobalamin; and Group C, failure to accumulate either adenosylcobalamin or methylcobalamin.

Congenital disorders of vitamin $B_{12}$ transport and storage have been reported and reviewed by Cooper (1976). TC II deficiences lead to severe megaloblastic anemia, neurological deterioration and infections (Hakami et al., 1971; Gimpert et al., 1975; Burman et al., 1979). An autosomal recessive disorder, TC II deficiency is responsive to massive doses of vitamin $B_{12}$. TC R-type binder deficiency has been reported but has no known clinical manifestaions (Carmel and Herbert, 1969).

Inborn errors of folate metabolism have been classified according to whether the defect is of uptake, interconversion, or utilization of folates (Erbe, 1975). These are: congenital malabsorption of folate; interconversion defects: dihydrofolate reductase deficiency, methenyl-tetrahydrofolate cyclohydrolase deficiency, methylene-tetrahydrofolate reductase deficiency; and defective folate utilization: 5-methyltetrahydrofolate homocysteine methyltransferase deficiency, and glutamate formiminotransferase deficiency.

Erbe (1975) also reviewed related inborn errors affecting folate metabolism: inborn errors of vitamin $B_{12}$ metabolism, methionine adenosyl transferase deficiency, dihydropteridine reductase deficiency, non-ketotic hyperglycinemia, Lesch-Nyhan syndrome, orotic aciduria, and cystathionine synthetase deficiency.

Another unusual case presentation has recently reported the prenatal diagnosis of a deficiency of propionyl-CoA carboxylase (Sweetman et al., 1979). This enzyme mediates propionic acid metabolism, the immediate precursor of methylmalonic acid. Rosenberg (1969) reviewed vitamin-dependent aminoacidopathies.

Additional errors of metabolism have been mentioned in preceding sections.

While these defects are rare, more may be diagnosed and effectively treated as techniques for in utero diagnosis become available. At the present time, Milunsky (1976) includes the following in which this is possible and which are associated with defective vitamin B12 or folate metabolism: methylmalonic aciduria responsive to vitamin $B_{12}$, vitamin $B_{12}$ metabolic defect, and methylene tetrahydrofolate reductase deficiency. Another important aspect of the inborn errors of metabolism may be the opportunity to define more clearly the basic cellular and subcellular mechanisms which are not clearly understood at present.

SUMMARY

This chapter has reviewed some of the remarkable history in the advances in understanding of vitamin $B_{12}$ and folic acid. Newer knowledge in the mechanisms of action and their interrelationships through the disciplines of nutrition, hematology, medicine, biochemistry, physics, bacteriology and physiology continues to progress.

In the past decade, development of new techniques has led to the identification of the vitamin $B_{12}$ transport proteins, to a better knowledge of how the vitamin enters the cell, the enzymatic mechanisms through which the two vitamins interrelate, and the genetic basis by which certain of the deficiencies develop.

The study of the manifestations of the impairment of DNA synthesis, which lack of vitamin $B_{12}$ and folic acid causes, continues to lead to solutions which have implication for wider application such as in the study and treatment of cancer, genetic defects, and disorders of immune mechanisms.

ACKNOWLEDGEMENTS

Dr. Ralph Carmel kindly supplied references for work on the megaloblastic effect of anesthetic gases.

REFERENCES

Allen, R. H., 1975, Human vitamin $B_{12}$ transport proteins, in "Progress in Hematology" (E. B. Brown, ed.), pp. 57-84, Grune and Stratton, New York.

Allen, R. H., and Majerus, P. W., 1972a, Isolation of vitamin $B_{12}$-binding proteins using affinity chromatography. I. Preparation and properties of vitamin $B_{12}$ Sepharose, J. Biol. Chem., 247:7695.

Allen, R. H., and Majerus, P. W., 1972b, Isolation of vitamin $B_{12}$-binding proteins using affinity chromatography. II. Purification and properties of a human granulocyte vitamin $B_{12}$-binding protein, J. Biol. Chem., 247:7702.

Allen, R. H., and Majerus, P. W., 1972c, Isolation of vitamin $B_{12}$-binding proteins using affinity chromatography. III. Purification and properties of human plasma transcobalamin II, J. Biol. Chem., 247:7709.

Allen, R. H., and Mehlman, C. S., 1973, Isolation of gastric vitamin $B_{12}$ binding proteins using affinity chromatography. I. Purification and properties of human intrinsic factor, J. Biol. Chem., 248:3660.

Alperin, J. B., Haggard, M. E., and Haynie, T. P., 1970, A study of vitamin $B_{12}$ requirements in a patient with pernicious anemia and thyrotoxicosis; evidence of an increased need for vitamin $B_{12}$ in the presence of hyperthyroidism, Blood, 36:632.

Amess, J. A. L., Rees, G. M., Burman, J. F., Nancekieviel, D. G., and Mollin, D. L., 1978, Megalobastic haemopoiesis in patients receiving nitrous oxide, Lancet, 2:339.

Ampola, M. G., Mahoney, M. J., Nakamura, E., and Tanaka, K., 1975, Prenatal therapy of a patient with vitamin-$B_{12}$ responsive methylmalonic acidemia, New Engl. J. Med., 293:313.

Angier, R. B., Boothe, J. H., Hutchings, B. L., Mowat, J. H., Semb, J., Stokstad, E. L. R., Subbarow, Y., Waller, C. W., Cosulich, D. B., Fahrenbach, M. J., Hultquist, M. E., Kuh, E., Northey, E. H., Seeger, D. R., Sickels, J. P., and Smith, J. M., Jr., 1945, Synthesis of a compound identical with the L. casei factor isolated from liver, Science, 102: 227.

Ardeman, S., and Chanarin, I., 1965, Steoids and Addisonian pernicious anemia, New Engl. J. Med., 273:1352.

Ardeman, S., and Chanarin, I., 1966, Gastric intrinsic factor after partial gastrectomy, Gut, 7:217.

Ardeman, S., Chanarin, I., Krafchik, B., and Singer, W., 1966, Addisonian pernicious anemia and intrinsic factor antibodies in thyroid disorders, Quart. J. Med., 35:421.

Arvanitakis, C., Holmes, F. F., and Hearne, E., III, 1979, A possible association of pernicious anemia with neoplasia, Oncology, 36:127.

Astaldi, G., 1960, Differentiation, proliferation and maturation of haematopoietic cells studied in tissue culture, in "Ciba Foundation Symposium on Haemopoiesis: Cell Production and its Regulation" (G. E. W. Wolstenholme and M. O'Connor, eds.), p. 60, Little, Brown and Co., Boston.

Badenoch, J., Bedford, P. D., and Evans, J. R., 1955, Massive diverticulosis of the small intestine with steatorrhoea and megaloblastic anemia, Quart. J. Med., 24:321.

Baker, S. J., 1972, Vitamin $B_{12}$ and tropical sprue, Br. J. Haematol., 23 (Supplement):135.

Baker, S. J., and DeMaeyer, E. M., 1979, Nutritional anemia: its understanding and control with special reference to the work of the World Health Organization, Amer. J. Clin. Nutr., 32:368.

Baker, S. J., and Mathan, V. I., 1968, Syndrome of tropical sprue in South India, Amer. J. Clin. Nutr., 21:984.

Baker, S. J., and Mathan, V. I., 1971, "Tropical Sprue and Megaloblastic Anaemia", p. 189, Wellcome Trust, Churchill, London.

Barker, H. A., 1967, Biochemical functions of corrinoid compounds, Biochem. J., 105:1.

Barker, H. A., Weissbach, H., and Smyth, R. D., 1958, A coenzyme containing pseudovitamin $B_{12}$, Proc. Natl. Acad. Science, U. S. A., 44:1093.

Bar-Shany, S., and Herbert, V., 1967, Transplacentally acquired antibody to intrinsic factor with vitamin $B_{12}$ deficiency, Blood, 30:777.

Bash, R. I., and Rosner, F., 1979, Pernicious anemia in young black women, Arch. Intern. Med., 139:829.

Baugh, C. M., Krumdieck, C. L., Baker, H. J., and Butterworth, C. E., Jr., 1971, Studies on the absorption and metabolism of folic acid. I. Folate absorption in the dog after exposure of isolated intestinal segments to synthetic pterolypolyglutamates of various chain lengths, J. Clin. Invest., 50:2009.

Beck, W. S., 1977a, General considerations of megaloblastic anemias, in "Hematology" (W. J. Williams, E. Beutler, A. J. Erslev, and R. W. Rundles, eds.), Second Edition, pp. 300-307, McGraw-Hill, New York.

Beck, W. S., 1977b, Vitamin $B_{12}$ deficiency, in "Hematology" (W. J. Williams, E. Beutler, A. J. Erslev, and R. W. Rundles, eds.), Second Edition, pp. 307-334, McGraw-Hill, New York.

Beck, W. S., 1977c, Folic acid deficiency, in "Hematology" (W. J. Williams, E. Beutler, A. J. Erslev, and R. W. Rundles, eds.), Second Edition, pp. 334-355, McGraw-Hill, New York.

Begley, J. A., and Trachtenberg, A., 1979, An assay for intrinsic factor based on blocking of the R binder of gastric juice by cobamide, Blood, 53:788.

Berk, L., Castle, W. B., Welch, A. D., Heinle, R. W., Anker, R., and Epstein, M., 1948, Observations on the etiologic relationship of achylia gastrica to pernicious anemia. X. Activity of vitamin $B_{12}$ as food (extrinsic) factor, New Engl. J. Med., 239:911.

Bernier, G., and Hines, J. D., 1967, Immunologic heterogeneity of autoantibodies in patients with pernicious anemia, New Engl. J. Med., 277:1386.

Bertino, J. R., ed., 1971, Folate Antagonists as Chemotherapeutic Agents, Annals New York Acad. Science, 186:1.

Bertino, J. R., Nixon, P. F., and Nahas, A., 1977, Mechanism of uptake of folate monoglutamates and their metabolism, in "Folic Acid Biochemistry and Physiology in Relation to the Human Nutrition Requirement, Proceedings of a Workshop on Human Folate Requirements, Washington, D. C., June 2-3, 1975", pp. 178-187, Food and Nutrition Board, National Research Council, National Academy of Sciences, Washington, D. C.

Bhat, P., Shantakumari, S., Rajan, D., Mathan, V. I., Kapadia, C. R., Swarnabai, C., and Baker, S. J., 1972, Bacterial flora of the gastrointestinal tract in southern Indian control subjects and patients with tropical sprue, Gastroenterology, 62:11.

Bianchi, A., Chipman, D. W., Dreskin, A., and Rosenzweig, N. S., 1970, Nutritional folic acid deficiency with megaloblastic changes in the small bowel epithelium, New Engl. J. Med., 282:859.

Blackburn, E. K., Callender, S. T., Dacie, J. V., Doll, R., Girdwood, R. H., Mollin, D. L., Saracci, R., Stafford, J. L., Thompson, R. B., Varadi, S., and Wetherley-Mein, G., 1968, Possible association between pernicious anaemia and leukemia: a prospective study of 1,625 patients with a note on the very high incidence of stomach cancer, Internat. J. Cancer, 3:163.

Blakley, R. L., 1977, Folic acid biochemistry: present status and future direction, in "Folic Acid Biochemistry and Physiology in Relation to the Human Nutrition Requirement, Proceedings of a Workshop on Human Folate Requirements, Washington, D. C., June 2-3, 1975", pp. 3-24, Food and Nutrition Board, National Research Council, National Academy of Sciences, Washington, D. C.

Boger, W. P., Bayne, G. M., Wright, L. D., and Beck, G. D., 1957, Differential serum vitamin $B_{12}$ concentrations in mothers and infants, New Engl. J. Med., 256:1085.

Booth, C. C., and Mollin, D. L., 1959, The site of absorption of vitamin $B_{12}$ in man, Lancet, 1:18.

Booth, C. C., Tabaqchali, S., and Mollin, D. L., 1968, Comparison of stagnant-loop syndrome with chronic tropical sprue, Amer. J. Clin. Nutr., 21:1097.

Branda, R. F., Moldow, C. F., MacArthur, J. R., Wintrobe, M. M., Anthony, B. K., and Jacob, H. S., 1978, Folate-induced remission in aplastic anemia with familial defect of cellular folate uptake, New Engl. J. Med., 298:469.

Brandt, L. J., Bernstein, L. H., and Wagle, A., 1977, Production of vitamin $B_{12}$ analogues in patients with small-bowel bacterial overgrowth, Ann. Intern. Med., 87:546.

Brody, E. A., Estrén, S., and Herbert, V., 1966, Coexistent pernicious anemia and malabsorption in four patients: including one whose malabsorption disappeared with vitamin $B_{12}$ therapy, Ann. Intern. Med., 64:1246.

Buchanan, J. M., 1964, The function of vitamin $B_{12}$ and folic acid coenzymes in mammalian cells, Medicine, 43:697.

Buchanan, J. M., Elford, H. L., Loughlin, R. E., McDougall, B. M., and Rosenthal, S., 1964, The role of vitamin $B_{12}$ in methyl transfer to homocysteine, Ann. New York Acad. Science, 112:756.

Burger, R. L., Mehlman, C. S., and Allen, R. H., 1975, Human plasma R-type vitamin $B_{12}$-binding proteins. I. Isolation and characterization of transcobalamin I, transcobalamin III and the normal granulocyte vitamin $B_{12}$-binding protein, J. Biol. Chem., 250:7700.

Burman, J. F., Mollin, D. L., Sourial, N. A., and Sladden, R. A., 1979, Inherited lack of transcobalamin II in serum and megaloblastic anemia: a further patient, Br. J. Haematol., 43:27.
Burnier, E., Zwahlen, A., and Cruchard, A., 1976, Nonmalignant monoclonal immunoglobulinemia, pernicious anemia and gastric carcinoma. A model of immunologic dysfunction. Report of two cases and review of the literature, Amer. J. Med., 60:1019.
Butterworth C. E., Baugh, C. M., and Krumdieck, C., 1969, A study of folate absorption and metabolism in man utilizing carbon-14-labeled polyglutamates synthesized by the solid phase method, J. Clin. Invest., 48:1131.
Butterworth, C. E., Jr., Newman, A. J., and Krumdieck, C. L., 1974, Tropical sprue: a consideration of possible etiologic mechanisms with emphasis on pteroylpolyglutamate metabolism, Trans. Amer. Clin. Climatol. Assoc., 86:11.
Carmel, R., 1975, Extreme elevation of serum transcobalamin I in patients with metastatic cancer, New Engl. J. Med., 292:282.
Carmel, R., 1979, Macrocytosis, mild anemia and delay in the diagnosis of pernicious anemia, Arch. Intern. Med., 139:47.
Carmel, R., and Herbert, V., 1967, Correctable intestinal defect of vitamin $B_{12}$ absorption in pernicious anemia, Ann. Intern. Med., 67:1201.
Carmel, R., and Herbert, V., 1969, Deficiency of vitamin $B_{12}$-binding alpha globulin in two brothers, Blood, 33:1.
Carmel, R., and Johnson, C. S., 1978, Racial patterns in pernicious anemia. Early age at onset and increased frequency of intrinsic factor antibody in black women, New Engl. J. Med., 298:647.
Carmel, R., and Schurafa, M., 1977, Circulating immunoglobulin-transcobalamin I complex in patients with elevated serum vitamin $B_{12}$ levels, Fed. Proc., 36:1121 (Abstract).
Castle, W. B., 1929, Observations on etiologic relationship of achylia gastrica to pernicious anemia. I. Effect of administration to patients with pernicious anemia of contents of normal human stomach recovered after ingestion of beef muscle, Amer. J. Med. Science, 178:748.
Castle, W. B., 1961, The Gordon Wilson Lecture, a century of curiosity about pernicious anemia, Trans. Amer. Clin. Climatol. Assoc., 73:53.
Castle, W. B., and Townsend, W. C., 1929, Observations on etiologic relationship of achylia gastrica to pernicious anemia. II. Effect of administration to patients with pernicious anemia of beef muscle after incubation with normal human gastric juice, Amer. J. Med. Science, 178:764.
Castle, W. B., Townsend, W. C., and Heath, C. W., 1930, Observations on etiologic relationship of achylia gastrica to pernicious anemia. III. Nature of reaction between normal human gastric juice and beef muscle leading to clinical improvement and increased blood formation similar to effect of liver feeding, Amer. J. Med. Science, 180:305.

Castle, W. B., Heath, C. W., and Strauss, M., 1931, Observations on the etiologic relationship of achylia gastrica to pernicious anemia. IV. A biologic assay of the gastric secretion of patients with pernicious anemia having free hydrochloric acid and that of patients without anemia or with hypochromic anemia having no free hydrochloric acid, and of the role of intestinal impermeability to hematopoietic substances in pernicious anemia, Amer. J. Med. Science, 182:741.

Castrén, O. M., and Rossi, R. R., 1970, Effect of oral contraceptives on serum folic acid content, J. Obstet. Gynaecol. Br. Commonw., 77:548.

Chanarin, I., 1969, "The Megaloblastic Anaemias", F. A. Davis, Co., Philadelphia, Blackwell Scientific Publications, Great Britain.

Chanarin, I., 1970, Folate deficiency in the myeloproliferative disorders, Amer. J. Clin. Nutr., 23:855.

Chanarin, I., 1972, Pernicious anaemia as an autoimmune disease, Br. J. Haematol., 23 (Supplement):101.

Chanarin, I., and Perry, J., 1968, Metabolism of 5-methyltetrahydrofolate in pernicious anemia, Br. J. Haematol., 14:297.

Chanarin, I., and Perry, J., 1977, Mechanisms in the production of megaloblastic anemia, in "Folic Acid Biochemistry and Physiology in Relation to the Human Nutrition Requirement, Proceedings of a Workshop on Human Folate Requirements, Washington, D. C., June 2-3, 1975", pp. 156-168, Food and Nutrition Board, National Research Council, National Academy of Sciences, Washington, D. C.

Chanarin, I., Perry, J., and Lumb, M., 1974, The biochemical lesion in vitamin $B_{12}$ deficiency in man, Lancet, 1:1251.

Chanarin, I., Rothman, D., and Berry, V., 1965, Iron deficiency and its relation to folic acid status in pregnancy. Results of a clinical trial, Br. Med. J., 1:480.

Charache, P., Hodkinson, B. A., Lambiotte, B., and McIntyre, P. A., 1968, Genetic and auto-immune features of pernicious anemia. II. Effect of transplacental transfer of antibody to intrinsic factor, Johns Hopkins Med. J., 122:184.

Christensen, J. M., Hippe, E., Olesen, H., Rye, M., Haber, E., Lee, L., and Thomsen, J., 1973, Purification of human intrinsic factor by affinity chromatography, Biochim. Biophys. Acta, 303:319.

Colman, N., and Herbert, V., 1974, Evidence for granulocyte-related and liver-related folate binders in human serum, and renal glomerular filtration of folate binder, Clin. Res., 22:700 A (Abstract).

Cook, G. C., Morgan, J. O., and Hoffbrand, A. V., 1974, Impairment of folate absorption by systemic bacterial infections, Lancet, 2:1416.

Cooper, B. A., 1976, Megalobastic anaemia and disorders affecting utilization of vitamin $B_{12}$ and folate in childhood, in "Clinics in Haematology, Megaloblastic Anaemia" (A. V. Hoffbrand, ed.), pp. 631-659, W. B. Saunders Co. Ltd., London.

Cooper, B. A., 1977, Physiology of absorption of monoglutamyl folates from the gastrointestinal tract, in "Folic Acid Biochemistry and Physiology in Relation to the Human Nutrition Requirement, Proceedings of a Workshop on Human Folate Requirements, Washington, D. C., June 2-3, 1975", pp. 188-197, Food and Nutrition Board, National Research Council, National Academy of Sciences, Washington, D. C.

Cooper, B. A., and Castle, W. B., 1960, Sequential mechanisms in the enhanced absorption of vitamin $B_{12}$ by intrinsic factor in the rat, J. Clin. Invest., 39:199.

Cooper, B. A., and White, J. J., 1968, Absence of intrinsic factor from human portal plasma during $^{57}Co$ $B_{12}$ absorption in man, Br. J. Haematol., 14:73.

Cooper, B. A., and Whitehead, V. M., 1978, Evidence that some patients with pernicious anemia are not recognized by radiodilution assay for cobalamin in serum, New Engl. J. Med., 299:816.

Cooper, B. A., Cantlie, G. S. D., and Brunton, L., 1970, The case for folic acid supplements during pregnancy, Amer. J. Clin. Nutr., 23:848.

Cooper, B. T., Holmes, G. K. T., and Cooke, W. T., 1978, Coeliac disease and immunological disorders, Br. Med. J., 1:537.

Corcino, J., Waxman, S., and Herbert, V., 1970, Mechanism of triamterene-induced megaloblastosis, Ann. Intern. Med., 73:419.

Corcino, J. J., Zalusky, R., Greenberg, M., and Herbert, V., 1971, Coexistence of pernicious anemia with chronic myeloid leukemia; an experiment of nature involving $B_{12}$ metabolism, Br. J. Haematol., 20:511.

Corcino, J. J., Coll, G., and Klipstein, F. A., 1975, Pteroylglutamic acid malabsorption in tropical sprue, Blood, 45:577.

Corcino, J. J., Reisenauer, A. M., and Halsted, C. H., 1976, Jejunal perfusion of simple and conjugated folates in tropical sprue, J. Clin. Invest., 58:298.

Cox, E. V., and White, A. M., 1962, Methylmalonic acid excretion: an index of vitamin $B_{12}$ deficiency, Lancet, 2:853.

Cox, E. V., Robertson-Smith, D., Small, M., and White, A. M., 1968, The excretion of propionate and acetate in vitamin $B_{12}$ deficiency, Clin. Science, 35:123.

da Costa, M., and Rothenberg, S. P., 1974, Appearance of a folate binder in leucocytes and serum of women who are pregnant or taking oral contraceptives, J. Lab. Clin. Med., 83:207.

Dallman, P. R., 1974, Iron, vitamin E, and folate in the preterm infant, J. Pediatr., 95:742.

Darby, W. J., Bridgforth, W. B., LeBrocquy, J., Clark, S. L. Jr., DeOlivera, J. D., Kevany, J., McGanity, W. J., and Perez, C., 1958, Vitamin $B_{12}$ requirement of man, Amer. J. Med., 25:726.

Das, K., and Herbert, V., 1976, Vitamin $B_{12}$-folate interrelations, in "Clinics in Haematology, Megaloblastic Anaemia" (A. V. Hoffbrand, ed.), pp. 697-725, W. B. Saunders Co., Ltd., London.

Davidson, G. P., and Townley, R. R. W., 1977, Structural and functional abnormalities of the small intestine due to nutritional folic acid deficiency in infancy, J. Pediat., 90:590.
Davis, R. E., Stenhouse, N. S., and Woodliff, H. J., 1969, Serum folate levels in pregnancy, Med. J. Austral., 1:52.
Day, P. L., Mims, V., Totter, J. R., Stokstad, E. L. R., Hutchings, B. L., and Sloane, N. H., 1945, Successful treatment of vitamin M deficiency in monkey with highly purified Lactobacillus casei factor, J. Biol. Chem., 157:423.
Deacon, R., Perry, J., Lumb, M., Chanarin, I., Mintz, B., Halsey, M. J., and Nunn, J. F., 1978, Selective inactivation of vitamin $B_{12}$ in rats by nitrous oxide, Lancet, 2:1023.
Deller, D. J., and Witts, L. J., 1962, Changes in the blood after partial gastrectomy with special reference to vitamin $B_{12}$. I. Serum vitamin $B_{12}$, haemoglobin, serum iron and bone marrow, Quart. J. Med., 31:71.
Deller, D. J., Richards, W. C. D., and Witts, L. J., 1962, Changes in the blood after partial gastrectomy with special reference to vitamin $B_{12}$. II. The cause of the fall in serum vitamin $B_{12}$. Quart. J. Med., 31:89.
Dellipiani, A. W., Samson, R. R., and Girdwood, R. H., 1968, The uptake of vitamin $B_{12}$ by E. coli. Possible significance in relation to the blind loop syndrome, Amer. J. Digest. Dis., 13:718.
De Oliviera, J. E. D., 1979, The stone that the builders rejected, Am. J. Clin. Nutr., 32:1566.
Deren, J. J., Arora, B., Toskes, P. P., Hansell, J., and Sibinga, M. S., 1973, Malabsorption of crystalline vitamin $B_{12}$ in cystic fibrosis, New Engl. J. Med., 288:949.
Doig, A., Girdwood, R. H., Duthie, J. J. R., and Knox, J. D. E., 1957, Response of megaloblastic anemia to prednisolone, Lancet, 2:966.
Donaldson, R. M., Jr., 1968, Significance of small bowel bacteria, Am. J. Clin. Nutr., 21:1088.
Donaldson, R. M., Jr., 1978, "Serum $B_{12}$" and the diagnosis of cobalamin deficiency, New Engl. J. Med., 299:827.
Doscherholmen, A., and Swaim, W. R., 1973, Impaired assimilation of egg $Co^{57}$ vitamin $B_{12}$ in patients with hypochlorhydria and achlorhydria and after gastric resection, Gastroenterology, 64:913.
Doscherholmen, A., Hagen, P. S., and Olin, L., 1959, Delay of absorption of radiolabeled cyanocobalamin in the intestinal wall in the presence of intrinsic factor, J. Lab. Clin. Med., 54:434.
Druskin, N. S., Waller, M. H., and Bonagura, L., 1962, Anticonvulsant associated megaloblastic anemia. Response to 25 microgm. of folic acid administered by mouth daily, New Engl. J. Med., 267:483.
Eastwood, D. W., Green, C. D., Lamdin, M. A., and Gardner, R., 1963, Effect of nitrous oxide on the white cell count in leukemia, New Engl. J. Med., 268:297.

Editorial, 1978, Nitrous Oxide and the bone-marrow, Lancet, 2:613.
Engel, A. G., and Stickney, J. M., 1972, Pernicious anemia and polycythemia vera in one patient, Arch. Intern. Med., 109:168.
Erbe, R. W., 1975, Inborn errors of folate metabolism, second of two parts, New Engl. J. Med., 293:807.
Farrell, M., Farrell, A., Murphy, C., and Dundon, S., 1979, Pernicious anemia in an eleven-year-old male, Acta Haematol., 61:175.
Finch, C. A., Coleman, D. H., Motulsky, A. G., Donohue, D. M., and Reiff, R. H., 1956, Erythrokinetics in pernicious anemia, Blood, 11:807.
Fink, B. R., Shepard, T. H., and Blandau, R. J., 1967, Teratogenic activity of nitrous oxide, Nature, 214:146.
Finkler, A. E., and Hall, C. A., 1967, Nature of the relationship between vitamin $B_{12}$ binding and cell uptake, Arch. Biochem. and Biophys., 120:79.
Fisher, J. M., and Taylor, K. B., 1965, A comparison of autoimmune phenomena in pernicious anemia and chronic atrophic gastritis, New Engl. J. Med., 272:499.
Franklin, J. L., and Rosenberg, I. H., 1973, Impaired folic acid absorption in inflammatory bowel disease; effects of salicylazosulfapyridine (azulfidine), Gastroenterology, 64:517.
Fraser, K. J., 1969, Multiple myeloma and pernicious anemia, Med. J. Austral., 1:298.
Frater-Schröder, M., Hitzig, W. H., and Bütler, R., 1979, Studies on transcobalamin (TC). 1. Detection of TC II isoproteins in normal serum, Blood, 53:193.
Frei, E., III, Jaffe, N., Tattersall, M. H. N., Pitman, S., and Parker, L., 1975, New approaches to cancer chemotherapy with methotrexate, New Engl. J. Med., 292:846.
Frenkel, E. P., 1973, Abnormal fatty acid metabolism in peripheral nerves of patients with pernicious anemia, J. Clin. Invest., 52:1237.
Frenkel, E. P., and Kitchens, R. L., 1977, Applicability of an enzymatic quantitation of methylmalonic, propionic and acetic acids in normal and megaloblastic states, Blood, 49:125.
Friedman, P. A., Shia, M. A., and Wallace, J. K., 1977, A saturable high affinity binding site for transcobalamin II-vitamin $B_{12}$ complexes in human placental membrane preparations, J. Clin. Invest., 59:51.
Gardner, F. H., 1956, Observations on the cytology of gastric epithelium in tropical sprue, J. Lab. Clin. Med., 47:529.
Gardner, F. H., 1958, Tropical sprue, New Engl. J. Med., 258:791, 835.
Garrido-Pinson, G. C., Turner, M. D., Crookston, J. H., Samloff, I. M., Miller, L. L., and Segal, H. L., 1966, Studies of human intrinsic factor auto-antibodies, J. Immunol., 97:897.
Gerson, C. D., Cohen, N., Hepner, G. W., Brown, N., Herbert, V., and Janowitz, H. D., 1971, Folic acid absorption in man: enhancing effect of glucose, Gastroenterology, 61:224.

Glass, G. B. J., 1974, "Gastric Intrinsic Factor and Other Vitamin $B_{12}$ Binders, Biochemistry, Physiology and Relation to Vitamin $B_{12}$ Metabolism", George Thieme, Stuttgart.
Gilliam, J. N., and Cox, A. J., 1973, Epidermal changes in vitamin $B_{12}$ deficiency, Arch. Dermatol., 107:231.
Gimpert, E., Jakob, M., and Hitzig, W. H., 1975, Vitamin $B_{12}$ transport in blood. Congenital deficiency of transcobalamin II, Blood, 45:71.
Goetsch, C., 1962, An evaluation of aminopterin as an abortifacient, Am. J. Obstet. Gynecol., 83:1474.
Goldberg, L. S., Barnett, E. V., and Desai, R., 1967, Effect of transplacental transfer of antibody to intrinsic factor, Pediatrics, 40:851.
Gompertz, D., 1971, The metabolic effects of an impaired methylmalonyl CoA mutase, in "The cobalamins" (H. R. V. Arnstein, and R. J. Wrighton, eds.), p. 101, Churchill Livingstone, Edinburgh.
Gorbach, S. L., Banwell, J. G., Jacobs, B., Chatterjee, B. D., Mitra, R., Sen, N. N., and Mazumber, D. N. G., 1970, Tropical sprue and malnutrition in West Bengal. I. Intestinal microflora and absorption, Am J. Clin. Nutr., 23:1545.
Gottlieb, C., Lau, K.-S., Wasserman, L. R., and Herbert, V., 1965, Rapid charcoal assay for intrinsic factor (IF), gastric juice, unsaturated $B_{12}$ binding capacity, antibody to IF, and serum unsaturated $B_{12}$ binding capacity, Blood, 25:875.
Gräsbeck, R., 1959, Influence of some specific group inhibitors on rat intrinsic factor, Acta Physiol. Scand., 45:116.
Gräsbeck, R., 1960, Physiology and pathology of vitamin $B_{12}$ absorption, distribution and excretion, Advan. Clin. Chem., 3:299.
Gräsbeck, R., 1969, Intrinsic factor and the other vitamin $B_{12}$ transport proteins, in "Progress in Hematology", Vol. VI (E. B. Brown and C. V. Moore, eds.), pp. 233-260, Grune and Stratton, New York.
Gräsbeck, R., 1975, Absorption and transport of vitamin $B_{12}$, Br. J. Haematol., 31 (Supplement):103.
Gräsbeck, R., Gordin, R., Kantero, I., and Kuhlbäck, B., 1960, Selective vitamin $B_{12}$ malabsorption and proteinuria in young people, Acta Med. Scand., 167:289.
Gräsbeck, R., Simons, K., and Sinkkonen, I., 1966, Isolation of intrinsic factor and its probable degradation product, as their vitamin $B_{12}$ complexes, from human gastric juice, Biochem. Biophys. Acta, 127:47.
Greipp, P. R., 1978, Hyperpigmentation syndromes (Diffuse hypermelanosis), Arch. Intern. Med., 138:356.
Guest, J. R., Friedman, S., Woods, D. D., and Smith, E. L., 1962, A methyl analogue of cobamide coenzyme in relation to methionine synthesis by bacteria, Nature, 195:340.
Guest, J. R., Foster, M. A., and Woods, D. D., 1964, Methyl derivatives of folic acid as intermediates in the methylation of homocysteine by Escherichia coli, Biochem. J., 92:488.

Gutteridge, J. M. C., and Wright, E. B., 1970, A simple and rapid thin layer technique for the detection of methylmalonic acid in urine, Clin. Chim. Acta, 27:289.

Hakami, N., Neiman, P. E., Canellos, G. P., and Lazerson, J., 1971, Neonatal megaloblastic anemia due to inherited transcobalamin II deficiency in two siblings, New Engl. J. Med., 285:1163.

Hall, C. A., 1971, Vitamin $B_{12}$-binding proteins of man, Ann. Intern. Med., 75:297.

Hall, C. A., 1973, Congenital disorders of vitamin $B_{12}$ transport and their contribution to concepts, Gastroenterology, 65:684.

Hall, C. A., and Finkler, A. E., 1969, Vitamin $B_{12}$-binding protein in polycythemia vera plasma, J. Lab. Clin. Med., 73:60.

Hall, C. A., Hitzig, W. H., Green, P. D., and Begley, J. A., 1979, Transport of therapeutic cyanocobalamin in the congenital deficiency of transcobalamin (TC II), Blood, 53:251.

Halpern, R., Halpern, B. C., Stea, B., Dunlap, A., Conklin, K., Clark, B., Ashe, H., Sperling, L., Halpern, L. A., Hardy, D., and Smith, R. A., 1977, Pterin-6-aldehyde. A cancer cell catabolite: identification and application in diagnosis and treatment of human cancer, Proc. Soc. Nat. Acad. Sci. U. S. A., 74:587.

Halsted, C. H., Griggs, R. C., and Harris, J. W., 1967, The effect of alcoholism on the absorption of folic acid ($^3$H-PGA) evaluated by plasma levels and urine excretion, J. Lab. Clin. Med., 69:116.

Halsted, C. H., Robles, E. A., and Mezey, E., 1971, Decreased jejunal uptake of labeled folic acid ($^3$H-PGA) in alcoholic patients: roles of alcohol and nutrition, New Engl. J. Med., 285:701.

Halsted, C. H., Robles, E. A., and Mezey, E., 1973, Intestinal malabsorption in folate-deficient alcoholics, Gastroenterology, 64:526.

Harker, L. A., and Finch, C. A., 1969, Thrombokinetics in man, J. Clin. Invest., 48:963.

Hart, R. J., Jr., and McCurdy, P. R., 1971, Pernicious anemia in Negroes. Ann. Intern. Med., 74:448.

Haurani, F. I., 1973, Vitamin $B_{12}$ and the megaloblastic development, Science, 182:78.

Heinrich, H. C., 1964, Metabolic basis of the diagnosis and therapy of vitamin $B_{12}$ deficiency, Seminar. Hematol., 1:199.

Helman, N., 1973, Macrocytosis and cigarette smoking, Ann. Intern. Med., 79:287.

Hepner, G. W., Booth, C. C., Cowan, J., Hoffbrand, A. V., and Mollin, D. L., 1968, Absorption of crystalline folic acid in man, Lancet, 2:302.

Herbert, V., 1959, Mechanism of intrinsic factor action in everted sacs of rat small intestine, J. Clin. Invest., 38:102.

Herbert, V., 1962a, Experimental nutritional folate deficiency in man, Trans. Assoc. Am. Physicians, 75:307.

Herbert, V., 1962b, Minimal daily adult folate requirement, Arch. Intern. Med., 110:649.
Herbert, V., 1963, A palatable diet for producing experimental folate deficiency in man, Am. J. Clin. Nutr., 12:17.
Herbert, V., 1968, Megaloblastic anemia as a problem in world health, Am. J. Clin. Nutr., 21:1115.
Herbert, V., 1970a, Introduction to the "nutritional anemias", Seminar. Hematol., 2:2.
Herbert, V., 1970b, Symposium, Folic acid deficiency. I. Introduction, Am. J. Clin. Nutr., 23:841.
Herbert, V., 1972a, Metformin and $B_{12}$ malabsorption, Ann. Intern. Med., 76:140.
Herbert, V., 1972b, Malabsorption syndrome secondary to $B_{12}$ deficiency, in "Hematopoietic and Gastrointestinal Investigations with Radionuclides" (J. A. Gilson, W. M. Smoak, and M. B. Weinstein, eds.), pp. 287-293, C. C. Thomas, Springfield, Ill.
Herbert, V., 1972c, Detection of malabsorption of vitamin $B_{12}$ due to gastric or intestinal dysfunction, Seminar. Nucl. Med., 2:220.
Herbert, V., 1973, The five possible causes of all nutrient deficiency; illustrated by deficiencies of vitamin B12 and folic acid, Am. J. Clin. Nutr., 26:77.
Herbert, V., 1975, Drugs effective in megaloblastic anemias. Vitamin $B_{12}$ and folic acid, in "The Pharmacological Basis of Therapeutics" Fifth Edition (L. S. Goodman and A. Gilman, eds.), pp. 1324-1349, MacMillan Co., New York.
Herbert, V., 1979, Megaloblastic anemias, in "Cecil, Textbook of Medicine" (P. B. Beeson, W. McDermott, Jr., and J. B. Wyngaarden, eds.), pp. 1719-1729, W. B. Saunders, Co., Philadelphia.
Herbert, V., and Castle, W. B., 1961, Divalent cation and pH dependence of rat intrinsic factor action in everted sacs and mucosal homogenates of rat small intestine, J. Clin. Invest., 40:1978.
Herbert, V., and Das, K. C., 1976, The role of vitamin $B_{12}$ and folic acid in hemato- and other cell-poiesis, in "Vitamins and Hormones, Advances in Research and Applications", Vol. 34, (P. L. Munson, E. Diczfalusy, J. Glover, and R. E. Olson, eds.), pp. 1-30, Academic Press, New York.
Herbert, V., and Jacob, E., 1974, Destruction of vitamin $B_{12}$ by ascorbic acid, J. Am. Med. Assoc., 230:241.
Herbert, V., and Zalusky, R., 1962, Interrelation of vitamin $B_{12}$ and folic acid metabolism: folic acid clearance studies, J. Clin. Invest., 41:1263.
Herbert, V., Zalusky, R., and Davidson, C. S., 1963, Correlation of folate deficiency with alcoholism and associated macrocytosis, anemia, and liver disease, Ann. Intern. Med., 58:977.

Herbert, V., Streiff, R. R., and Sullivan, L. W., 1964, Notes on vitamin $B_{12}$ absorption: autoimmunity and childhood pernicious anemia; relation of intrinsic factor to blood group substance, Medicine, 43:679.

Herbert, V., Tisman, G., Go, L. T., and Brenner, L., 1973, The dU suppression test using $^{125}$I-UdR to define biochemical megaloblastosis, Br. J. Haematol., 24:713.

Hermos, J. A., Adams, W. H., Lui, Y. K., Sullivan, L. W., and Trier, J. S., 1972, Mucosa of the small intestine in folate-deficient alcoholics, Ann. Intern. Med., 76:957.

Heyssel, R. M., Bozian, R. C., Darby, W. J., and Bell, M. C., 1966, Vitamin $B_{12}$ turnover in man: the assimilation of vitamin $B_{12}$ from natural foodstuff by man and estimates of minimal daily dietary requirements, Am. J. Clin. Nutr., 18:176.

Hibbard, E. D., and Smithells, R. W., 1965, Folic acid metabolism and human embryopathy, Lancet, 1:1254.

Higginbotham, M. C., Sweetman, L., and Nyhan, W. L., 1978, A syndrome of methylmalonic aciduria, homocystinuria, megaloblastic anemia and neurologic abnormalities of a vitamin $B_{12}$-deficient breast-fed infant of a strict vegetarian, New Engl. J. Med., 299:317.

Hild, D., 1969, Folate losses from the skin in exfoliative dermatitis, Arch. Intern. Med., 123:51.

Hines, J. D., 1966, Megaloblastic anemia in adult vegan, Am. J. Clin. Nutr., 19:260.

Hines, J. D., Hoffbrand, A. V., and Mollin, D. L., 1967, The hematologic complications following partial gastrectomy. A study of 292 patients, Am. J. Med., 43:555.

Hoedemaker, P. J., Abels, J., Wachters, J. J., Arends, A., and Nieweg, H. O., 1964, Investigations about the site of production of Castle's gastric intrinsic factor, Lab. Invest., 13:1394.

Hoffbrand, A. V., 1970, Folate deficiency in premature infants, Arch. Dis. Child., 45:441.

Hoffbrand, A. V., 1975, Synthesis and breakdown of natural folates (folate polyglutamates), in "Progress in Hematology", Vol. IX, (E. B. Brown, ed.), pp. 85-105, Grune and Stratton, New York.

Hoffbrand, A. V., Chanarin, I., Kremenchuzky, S., Szur, L., Waters, A. H., and Mollin, D. L., 1968, Megaloblastic anaemia in myelosclerosis, Quart. J. Med., 37:493.

Hoffbrand, A. V., Tripp, E., Lavoié, A., 1977, Folate polyglutamate synthesis and breakdown in human cells, in "Folic Acid Biochemistry and Physiology in Relation to the Human Nutrition Requirement, Proceedings of a Workshop on Human Folate Requirements, Washington, D. C., June 2-3, 1975", pp. 110-121, Food and Nutrition Board, National Research Council, National Academy of Sciences, Washington, D. C.

Hooper, D. C., Alpers, D. H., Burger, R. L., Mehlman, C. S., and Allen, R. H., 1973, Characterization of ileal vitamin $B_{12}$ binding using homogeneous human and hog intrinsic factors, J. Clin.

Invest., 52:3074.
Hoppner, K., Lampi, B., and Smith, D. C., 1977, Data on folacin activity in foods: availability, applications, and limitations, in "Folic Acid Biochemistry and Physiology in Relation to the Human Nutrition Requirement, Proceedings of a Workshop on Human Folate Requirements, Washington, D. C., June 2-3, 1975", pp. 69-81, Food and Nutrition Board, National Research Council, Washington, D. C.
Horton, M. A., and Oliver, R. T. D., 1976, HLA and pernicious anemia, New Engl. J. Med., 294:396.
Huennekens, F. M., 1966, Biochemical functions and interrelationships of folic acid and vitamin $B_{12}$, in "Progress in Hematology", Vol. 5, (E. B. Brown, and C. V. Moore, eds.), pp. 83-104, Grune and Stratton, New York.
Huennekens, F. M., DiGirolamo, P. M., Fujii, K., Jacobsen, D. W., and Vitols, K. S., 1976, $B_{12}$-methionine synthetase as a potential target for cancer chemotherapy, Adv. Enzyme Regul., 14:187.
Huntsman, R. G., 1974, Treatment of sickle-cell disease, Trans. Roy. Soc. Trop., Med. Hygiene, 68:80.
Imerslund, O., 1960, Idiopathic chronic megaloblastic anemia in children, Acta Pediat., 49 (Supplement 119):1.
Ingeberg, S., and Stoffersen, E., 1979, Platelet dysfunction in patients with $B_{12}$ deficiency, Acta Haematol., 61:75.
Irvine, W., 1963, Gastric antibodies studied by fluorescence microscopy, Quart. J. Med., 48:427.
Irvine, W., 1965, Immunologic aspects of pernicious anemia, New Engl. J. Med., 273:432.
IUPAC-IUB, 1966, Tentative Rules, nomenclature and symbols for folic acid and related compounds. Nomenclature of corrinoids, J. Biol. Chem., 241:2991.
IUPAC-IUB, 1974, The nomenclature of corrinoids (1973 Recommendations), Biochemistry, 13:1555.
Jacob, E., and Glass, G. B. J., 1969, The participation of complement in the parietal cell antigen-antibody reaction in pernicious anaemia and atrophic gastritis, Clin. Exper. Immunol., 5:141.
Jacob, E., and Glass, G. B. J., 1971, Localization of intrinsic factor and complement-fixing intrinsic factor-intrinsic factor antibody complex in parietal cell of man, Clin. Exper. Immunol., 8:517.
Jacob, E., Herbert, V., Burger, R. L., and Allen, R. H., 1977, Atypical plasma factor associated with bronchogenic carcinoma and complexing with R-type vitamin $B_{12}$-binding proteins, New Engl. J. Med., 296:915.
Jacobson, E. D., Chodos, R. B., and Faloon, W. W., 1960, An experimental malabsorption syndrome induced by neomycin, Am. J. Med., 28:524.

Jandl, J. H., and Greenberg, M. S., 1959, Bone marrow failure due to relative nutritional deficiency in Cooley's hemolytic anemia, New Engl. J. Med., 260:461.

Jeffries, G. H., 1965, Recovery of gastric mucosal structure and function in pernicious anemia during prednisone therapy. Gastroenterology, 48:371.

Jeffries, G. H., and Sleisenger, M. H., 1965, Studies of parietal cell antibody in pernicious anemia, J. Clin. Invest., 44:2021.

Johns, D. G., and Bertino, J. R., 1965, Folates and megaloblastic anemia: a review, Clin. Pharmacol. Ther., 6:372.

Jones, C. C., 1973, Megaloblastic anemia associated with long-term tetracycline therapy, report of a case, Ann. Intern. Med., 78:910.

Kaplan, H. S., and Rigler, L. G., 1945, Pernicious anemia and carcinoma of the stomach. Autopsy studies concerning their interrelationship, Am. J. Med. Sci., 209:339.

Kass, L., 1968, A clockface chromatin pattern in the intermediate megaloblast of vitamin $B_{12}$ or folate deficiency, Blood, 32:711.

Kass, L., 1976a, Unusual morphologic abnormalities of megaloblasts in pernicious anemia and folate deficiency, Am. J. Clin. Pathol., 65:195.

Kass, L., 1976b, Pernicious Anemia, Vol. VII in "Major Problems in Internal Medicine" (L. H. Smith, Jr., ed.), W. B. Saunders, Co., Philadelphia.

Kass, L., 1978, William B. Castle and intrinsic factor, Ann. Intern. Med., 89:983.

Katz, M., and Cooper, B. A., 1974, Solubilized receptor for vitamin $B_{12}$-intrinsic factor complex from human intestine, Br. J. Haematol., 26:569.

Katz, M., Lee, S. K., and Cooper, B. A., 1972, Vitamin $B_{12}$ malabsorption due to a biologically inert intrinsic factor, New Engl. J. Med., 287:425.

Katz, M., Mehlman, C. S., and Allen, R. H., 1974, Isolation and characterization of an abnormal human intrinsic factor, J. Clin. Invest., 53:1274.

Killmann, S. A., 1964, Effect of deoxyuridine on incorporation of tritiated thymidine; difference between normoblasts and megaloblasts, Acta Med. Scand., 175:483.

Kiossoglou, K. A., Mitus, W. J., and Dameshek, W., 1965, Chromosomal aberrations in pernicious anemia. Study of three cases before and after therapy, Blood, 25:662.

Kitay, D. Z., 1969, Folic acid deficiency in pregnancy. On the recognition, pathogenesis, consequences, and therapy of the deficiency state in human reproduction, Am. J. Obstet. Gynecol., 104:1067.

Kitay, D. Z., and Marshall, J. S., 1968, Remission of folic acid deficiency in pregnancy, Am. J. Obstet. Gynecol., 102:297.

Klipstein, F. A., 1968, Progress in gastroenterology: tropical sprue, Gastroenterology, 54:275.

Klipstein, F. A., 1972, Folate in tropical sprue, Br. J. Haematol., 23 (Supplement):119.

Klipstein, F. A., Samloff, I. M., and Schenk, E. A., 1966, Tropical sprue in Haiti, Ann. Intern. Med., 64:575.

Klipstein, F. A., Holdeman, L. V., Corcino, J. J., and Moore, W. E. C., 1973, Enterotoxigenic intestinal bacteria in tropical sprue, Ann. Intern. Med., 79:632.

Kolhouse, J. F., Kondo, H., Allen, N. C., Podell, E., and Allen, R. H., 1978, Cobalamin analogues are present in human plasma and can mask cobalamin deficiency because current radioisotope dilution assays are not specific for true cobalamin, New Engl. J. Med., 299:785.

Krumdieck, C. L., Cornwell, P. E., Thompson, R. W., and White, W. E., Jr., 1977, Studies on the biological role of folic acid polyglutamates, in "Folic Acid Biochemistry and Physiology in Relation to the Human Nutrition Requirement, Proceedings of a Workshop on Human Folate Requirements, Washington, D. C., June 2-3, 1975", pp. 25-42, Food and Nutrition Board, National Research Council, National Academy of Sciences, Washington, D. C.

Lane, G. A., Nahrwold, M. L., Tait, A. R., Taylor, M. D., Beaudoin, A. R., Cohen, P. J., 1979, Nitrous oxide is teratogenic: xenon is not!, Anesthesiol., 51 (Supplement No. 3S):S260.

Lanzkowsky, P., Erlandson, M. E., and Bezan, A. I., 1969, Isolated defect of folic acid absorption associated with mental retardation and cerebral calcification, Blood, 34:452.

Larrabee, A. R., Rosenthal, S., Cathou, R. E., and Buchanan, J. M., 1961, A methylated derivative of tetrahydrofolate as an intermediate of methionine synthesis, J. Am. Chem. Soc., 83:4094.

Larrabee, A. R., Rosenthal, S., Cathou, R. E., and Buchanan, J. M., 1963, Enzymatic synthesis of the methyl group of methionine. IV. Isolation, charaterization and role of 5-methyltetrahydrofolate, J. Biol. Chem., 238:1025.

Larsson, S. O., 1962, Myeloma and pernicious anemia, Acta Med. Scand., 172:195.

Lassen, H. C. A., Henriksen, E., Neukirch, F., and Kristensen, H. S., 1956, Treatment of tetanus: severe bone-marrow depression after prolonged nitrous oxide anaesthesia, Lancet, 1:527.

Lawson, D. H., Murray, R. M., and Parker, J. L. W., 1972, Early mortality in the megaloblastic anaemias, Quart. J. Med., 41:1.

Layzer, R. B., 1978, Myeloneuropathy after prolonged exposure to nitrous oxide, Lancet, 2:1227.

Leevy, C. M., Cardi, L., Frank, O., Gellene, R., and Baker, H., 1965, Incidence and significance of hypovitaminemia in a randomly selected municipal hospital population, Am. J. Clin. Nutr., 17:259.

Levine, P. H., 1973, A qualitative platelet defect in severe vitamin $B_{12}$ deficiency, Ann. Intern. Med., 78:533.

Lewis, S. M., and Verwilghen, R. L., 1973, Dyserythropoiesis and dyserythropoietic anemias, in "Progress in Hematology", Vol. VIII (E. B. Brown, ed.), Grune and Stratton, Inc., New York.

Lillie, R. S., 1914, The action of various anaesthetics in suppressing cell-division in sea-urchin eggs, J. Biol. Chem., 17:121.

Lindebaum, J., Pezzimenti, J. F., and Shea, N., 1974, Small-intestinal function in vitamin $B_{12}$ deficiency, Ann. Intern. Med., 80:326.

Linnell, J. C., Matthews, D. M., Mudd, S. H., Uhlendorf, B. W., and Wise, I. J., 1976, Cobalamins in fibroblasts cultured from normal control subjects and patients with methylmalonic aciduria, Ped. Res., 10:179.

London, I. M., and West, R., 1950, The formation of bile pigment in pernicious anemia, J. Biol. Chem., 184:359.

Loughlin, R. E., Elford, H. L., and Buchanan, J. M., 1964, Enzymatic synthesis of the methyl group of methionine. VII. Isolation of a cobalamin-containing transmethylase (5-methyltetrahydrofolate-homocysteine) from mammalian liver, J. Biol. Chem., 239:2889.

Lowe, W. C., 1976, Hypogammaglobulinemia, pernicious anemia. With carcinomas of urinary bladder and lung, New York State J. Med., 76:926.

Luhby, A. L., Eagle, F. J., Roth, E., and Cooperman, J. M., 1961, Relapsing megaloblastic anaemia in an infant due to a specific defect in gastrointestinal absorption of folic acid, Am. J. Dis. Child., 102:482 (Abstract).

Mackenzie, I. L., and Donaldson, R. M., Jr., 1972, Effect of divalent cations and pH on intrinsic factor-mediated attachment of vitamin $B_{12}$ to intestinal microvillous membranes, J. Clin. Invest., 51:2465.

Magnus, E. M., 1977, Plasma conjugase activity in health and disease, in "Folic Acid Biochemistry and Physiology in Relation to the Human Nutrition Requirement, Proceedings of a Workshop on Human Folate Requirements, Washington, D. C., June 2-3, 1975", pp. 147-151, Food and Nutrition Board, National Research Council, National Academy of Sciences, Washington, D. C.

Mahmud, K., Ripley, D., and Doscherholmen, A., 1971, Vitamin $B_{12}$ absorption tests. Their unreliability in postgastrectomy states, J. Am. Med. Assoc., 216:1167.

Mahoney, M. J., Hart, A. C., Steen, V. D., and Rosenberg, L. E., 1975, Methylmalonic acidemia: biochemical heterogeneity in defects of 5'-deoxyadenosylcobalamin synthesis, Proc. Natl. Acad. Science, U. S. A., 72:2799.

Mann, J. G., Brown, W. R., and Kern, F., Jr., 1970, The subtle and variable clinical expressions of gluten-induced enteropathy (adult celiac disease, nontropical sprue) an analysis of twenty-one cases, Am. J. Med., 48:357.

Marcoullis, G., Parmentier, Y., and Nicolas, J.-P., 1979, Blocking and binding type antibodies against all major vitamin $B_{12}$-binders in a pernicious anaemia serum, Br. J. Haematol., 43:15.

Mathan, V. I., Babior, B. M., and Donaldson, R. M. J., 1974, Kinetics of the attachment of intrinsic factor-bound cobamides to ileal receptors, J. Clin. Invest., 54:598.

Mayersdorf, A., Streiff, R. R., Wilder, B. J., and Hammer, R. H., 1971, Folic acid and vitamin $B_{12}$ alterations in primary and secondary epileptic foci induced by metallic cobalt powder, Neurology, 21:418 (Abstract).

McIntyre, R. O., Sullivan, L. W., Jeffries, G. H., and Silver, R. H., 1965, Pernicious anamia in childhood, New Engl. J. Med., 272:981.

McPhedran, P., Barnes, M. G., Weinstein, J. S., and Robertson, J. S., 1973, Interpretation of electronically determined macrocytosis, Ann. Intern. Med., 78:677.

Menzies, R. C., Crossen, P. E., Fitzgerald, P. H., and Gunz, F. W., 1966, Cytogenetic and cytochemical studies on marrow cells in $B_{12}$ and folate deficiency, Blood, 28:581.

Metz, J., 1970, Folate deficiency conditioned by lactation, Am. J. Clin. Nutr., 23:843.

Metz, J., Randall, T. W., and Kniep, C. H., 1961, Addisonian pernicious anemia in young Bantu females, Br. Med. J., 1:178.

Miller, D. R., Bloom, G. E., Streiff, R. R., LoBuglio, A. F., and Diamond, L. K., 1966, Juvenile "congenital" pernicious anemia: clinical and immunological studies, New Engl. J. Med., 275:978.

Milunsky, A., 1976, Prenatal diagnosis of genetic disorders, New Engl. J. Med., 295:377.

Minot, G. R., and Castle, W. B., 1935, The interpretation of reticulocyte reactions: their value in determining the potency of therapeutic materials, especially in pernicious anemia, Lancet, 2:319.

Minot, G., and Murphy, W. P., 1926, Treatment of pernicious anemia by a special diet, J. Am. Med. Assoc., 87:470.

Mitchell, H. K., Snell, E. E., and Williams, R. J., 1941, The concentration of "folic acid", J. Am. Chem. Soc., 63:228.

Mollin, D. L., and MacGibbon, B. H., 1972, Sideroblastic and megaloblastic anaemias, Br. J. Haematol., 23 (Supplement):147.

Myhre, E., 1964a, Studies on megaloblasts in vitro. I. Proliferation and destruction of nucleated red cells in pernicious anemia before and during treatment with vitamin $B_{12}$, Scand. J. Clin. Lab. Invest., 16:307.

Myhre, E., 1964b, Studies on megaloblasts in vitro. II. Maturation of nucleated red cells in pernicious anemia before and during treatment with vitamin $B_{12}$, Scand. J. Clin. Lab. Invest., 16:320.

Nathan, D. G., and Gardner, F. H., 1962, Erythroid cell maturation and hemoglobin synthesis in megaloblastic anemia, J. Clin. Invest., 41:1086.

National Nutrition Survey in California, Termination of Contract Report, 1971, California State Department of Public Health, Berkeley, California, November.

NRC/NAS, 1974, "Recommended Dietary Allowances", Eighth Revised Edition, National Research Council, National Academy of Sciences, Washington, D. C.

Necheles, T. F., and Snyder, L. M., 1970, Malabsorption of folate polyglutamates associated with oral contraceptive therapy, New Engl. J. Med., 282:858.

Nixon, P. F., and Bertino, J. R., 1970, Interrelationships of vitamin $B_{12}$ and folate in man, Am. J. Med., 48:555.

Nixon, P. F. and Bertino, J. R., 1972, Impaired utilization of serum folate in pernicious anemia. A study with radiolabeled 5-methyltetrahydrofolate, J. Clin. Invest., 51:1431.

Nunn, J. F., Sturrock, J. E., and Howell, A., 1976, The effect of inhalation anesthetics on division of bone marrow cells in vitro, Br. J. Anesth., 48:75.

Nutrition Reviews, 1979, Altered immune responses in zinc deficiency in rodents, Nutrition Reviews, 37:234.

Nybert, W., 1960, The influence of Diphyllobothrium latum on the vitamin $B_{12}$-intrinsic factor complex. I. In vivo studies with Schilling test technique, Acta Med. Scand., 167:185.

O'Brien, J. S., 1962, The role of the folate coenzymes in cellular division: a review, Cancer Res., 22:267.

Ogbuawa, O., Trowell, J., Williams, J. T., Bradley, C., Archer, J., and Henry, W. L., 1978, Hyperpigmentation of pernicious anemia in blacks, Arch. Intern. Med., 138:388.

Oliver, R. A. M., and Baker, G. P., 1969, Juvenile pernicious anemia and hypothyroidism, a family study, Br. Med. J., 2:27.

O'Malley, B. W., and Means, A. R., 1974, Female steroid hormones and target cell nuclei, Science, 183:610.

Osler, W., 1919, Observations on the severe anaemias of pregnancy and the post-partum state, Br. Med. J., 1:1.

Pereira, S. M., and Baker, S. J., 1966, Haematologic studies in kwashiorkor, Am. J. Clin. Nutr., 18:413.

Pfiffner, J. J., Binkley, S. B., Bloom, E. S., Brown, R. A., Bird, O. D., and Emmett, A. D., 1943, Isolation of antianemia factor-(vitamin $B_c$) in crystalline form from liver, Science, 97:404.

Pincus, J. H., Reynolds, E. H., and Glaser, G. H., 1972, Subacute combined system degeneration with folate deficiency, J. Am. Med. Assoc., 221:496.

Pinkhas, J., Ben-Bassat, M., and De Vries, A., 1973, Death in anticonvulsant-induced megaloblastic anemia, J. Am. Med. Assoc., 224:246.

Powsner, E. R., and Berman, L., 1965, Human bone marrow chromosomes in megaloblastic anemia, Blood, 26:784.

Prasad, A. S., Miale, A., Farid, Z., Sandstead, H. H., Schulert, A. R., and Darby, W. J., 1963, Biochemical studies on dwarfism, hypogonadism and anemia, Arch. Intern. Med., 111:407.

Pritchard, J. A., Whalley, P. J., and Scott, D. E., 1969, The influence of maternal folate and iron deficiencies on intrauterine life, Am. J. Obstet. Gynecol., 104:388.

Pritchard, J. A., Scott, D. E., and Whalley, P. J., 1971, Maternal folate deficiency and pregnancy wastage. IV. Effects of folic acid supplements, anticonvulsants and oral contraceptives, Am. J. Obstet. Gynecol., 109:341.

Reizenstein, P. G., 1959, Excretion of non-labeled vitamin $B_{12}$ in man, Acta Med. Scand., 165:313.

Reizenstein, P., Ek, G., and Matthews, C. M. E., 1966, Vitamin $B_{12}$ kinetics in man. Implications on total-body-$B_{12}$-determinations, human requirements and normal and pathological cellular $B_{12}$ uptake, Phys. Med. Biol., 11:295.

Retief, F. P., Heyns, A. du P., Oosthuizen, M., and van Reenen, O. R., 1977, Aspects of folate metabolism in renal failure, Br. J. Haematol., 36:405.

Reynolds, E. H., 1968, Mental effects of anticonvulsants and folate metabolism, Brain, 91:197.

Reynolds, E. H., 1971, Anticonvulsant drugs, folic acid metabolism, frequency and psychiatric illness, Psychiatr. Neurol. Neurochir., 74:167.

Reynolds, E. H., 1973, Anticonvulsants, folic acid, and epilepsy, Lancet, 1:1376.

Reynolds, E. H., 1976, Neurological aspects of folate and vitamin $B_{12}$ metabolism, in "Clinics in Haematology, Megaloblastic Anemia", Vol. 5, No. 3, (A. V. Hoffbrand, ed.), pp. 661-696, W. B. Saunders, Co., Ltd., London.

Rickes, E. L., Brink, N. G., Koniuszy, F. R., Wood, T. R., and Folkers, K., 1948, Crystalline vitamin $B_{12}$, Science, 107:396.

Rickles, F. R., Klipstein, F. A., Tomasini, J., Corcino, J. J., and Maldonado, N., 1972, Long-term follow-up of antibiotic-treated tropical sprue, Ann. Intern. Med., 76:203.

Riddell, E. M., and Davidson, R. J. L., 1968, Coexistence of pernicious anaemia and acute erythraemic myelosis, J. Clin. Path., 21:590.

Rose, M. S., Chanarin, I., Doniach, D., Brostoff, J., and Ardeman, S., 1970, Intrinsic-factor antibodies in absence of pernicious anaemia: 3-7 year follow-up, Lancet, 2:9.

Rosenberg, L. E., 1969, Inherited aminoacidopathies demonstrating vitamin dependency, New Engl. J. Med., 281:145.

Rosenberg, I., 1975, Folate absorption and malabsorption, New Engl. J. Med., 293:1303.

Rosenberg, I. H., 1976, Absorption and malabsorption of folates, in "Clinics in Haematology, Megaloblastic Anaemia", Vol. 5, No. 3, (A. V. Hoffbrand, ed.), pp. 589-618, W. B. Saunders, Co., Ltd., London.

Rosenberg, I. H., 1977, Role of intestinal conjugase in the control of the absorption of polyglutamyl folates, in "Folic Acid Biochemistry and Physiology in Relation to the Human Nutrition Requirement, Proceedings of a Workshop on Human Folate

Requirements, Washington, D. C., June 2-3, 1975", pp. 136-146, Food and Nutrition Board, National Research Council, National Academy of Sciences, Washington, D. C.

Rosenberg, I. H., and Godwin, H. A., 1971, The digestion and absorption of dietary folate, Gastroenterology, 60:445.

Rosenberg, I. H., Streiff, R. R., Godwin, H. A., and Castle, W. B., 1969, Absorption of polyglutamic folate: participation of deconjugating enzymes of the intestinal mucosa, New Engl. J. Med., 280:985.

Rosenberg, H., Orkin, F. K., and Springstead, J., 1979, Abuse of nitrous oxide, Anesth. Analg., 58:104.

Rothenberg, S. P., Da Costa, M., and Fischer, G., 1977, Use and significance of folate binders, in "Folic Acid Biochemistry and Physiology in Relation to the Human Nutrition Requirement, Proceedings of a Workshop on Human Folate Requirements, Washington, D. C., June 2-3, 1975", pp. 82-97, Food and Nutrition Board, National Research Council, National Academy of Sciences, Washington, D. C.

Rothman, D., 1970, Folic acid in pregnancy, Am. J. Obstet. Gynecol., 108:149.

Rucker, R. W., and Harrison, G. M., 1973, Vitamin $B_{12}$ deficiency in cystic fibrosis, New Engl. J. Med., 289:329.

Samson, D. C., Swisher, S. N., Christian, R. M., and Engel, G. L., 1952, Cerebral metabolic disturbance and delirium in pernicious anemia, Arch. Intern. Med., 90:4.

Sando, M. J. W., and Lawrence, J. R., 1958, Bone marrow depression following treatment with protracted nitrous-oxide anaesthesia, Lancet, 1:588.

Santiago-Borrero, P. J., Santini, R., Jr., Pérez-Santiago, E., Maldonado, N., Millan, S., and Coll-Camález, G., 1973, Congenital isolated defect of folic acid absorption, J. Pediat., 92:450.

Santini, R., Jr., Berger, F. M., Berdasco, G., Sheehy, T. W., Aviles, J., and Davila, I., 1962, Folic acid activity in Puerto Rican Foods, J. Am. Dietet. Assoc., 41:562.

Santini, R., Millan, S., and Santiago-Borrero, P. J., 1973, Additional information on congenital defect in absorption of folic acid, J. Pediat., 83:345.

Sauberlich, H. E., 1977, Detection of folic acid deficiency in populations, in "Folic Acid Biochemistry and Physiology in Relation to the Human Nutrition Requirement, Proceedings of a Workshop on Human Folate Requirements, Washington, D. C., June 2-3, 1975", pp. 213-231, Food and Nutrition Board, National Research Council, National Academy of Sciences, Washington, D. C.

Sauer, H., and Wilmanns, W., 1977, Cobalamin dependent methionine synthesis and methyl-folate-trap in human vitamin $B_{12}$ deficiency, Br. J. Haematol., 36:189.

Schade, S. G., Feick, P., Muckerheide, M., and Schilling, R. F., 1966, Occurrence in gastric juice of antibody to a complex of intrinsic factor and vitamin $B_{12}$, New Engl. J. Med., 275:528.

Schade, S., Abels, J., Schilling, R. F., Feick, P., and Muckerheide, M., 1967, Studies on antibody to intrinsic factor, J. Clin. Invest., 46:615.

Schilling, R. F., 1953, Intrinsic factor studies. II. The effect of gastric juice on the urinary excretion of radioactivity after oral administration of radioactive vitamin $B_{12}$, J. Lab. Clin. Med., 42:860.

Schneerson, J. M., and Gazzard, B. G., 1977, Reversible malabsorption caused by methyldopa, Br. Med. J., 2:1456.

Schwartz, M., 1958, Intrinsic factor-inhibiting substance in serum of orally treated patients with pernicious anaemia, Lancet, 2:61.

Sheehy, T., and Dempsey, H., 1970, Methotrexate therapy for Plasmodium vivax malaria, J. Am. Med. Assoc., 214:109.

Shin, Y. S., Williams, M. A., Stokstad, E. L. R., 1972, Identification of folic acid compounds in rat liver, Biochim. Biophys. Res. Commun., 47:35.

Shulman, R., 1972, The present status of vitamin $B_{12}$ and folic acid deficiency in psychiatric illness, Canad. Psychiatr. Assoc. J., 17:205.

Silber, R., and Moldow, C. F., 1970, The biochemistry of $B_{12}$-mediated reactions in man, Am. J. Med., 48:549.

Silink, M., Reddel, R., Bethal, M., and Rowe, P. B., 1975, γ-Glutamyl hydroxylase (conjugase), J. Biol. Chem., 250:5982.

Smith, E. L., 1965, "Vitamin $B_{12}$", 3rd Edition, Wiley, New York.

Smith, E. L., and Parker, L. F. J., 1948, Purification of anti-pernicious anaemia factor, Biochem. J., 43:viii (Abstract).

Snell, E. E., and Peterson, W. H., 1940, Growth factors for bacteria. X. Additional factors required by certain lactic acid bacteria, J. Bacteriol., 39:273.

Solanki, D. L., Jacobson, R. J., McKibbon, J., and Green, R., 1978, Racial patterns in pernicious anemia, New Engl. J. Med., 298:1365.

Spector, R., 1977, Vitamin homeostasis in the central nervous system, New Engl. J. Med., 296:1393.

Spector, R., and Lorenzo, A. V., 1975, Folate transport by the choroid plexus in vitro, Science, 187:540.

Spencer, J. D., Raasch, F. O., and Trefny, F. A., 1976, Halothane abuse in hospital personnel, J. Am. Med. Assoc., 235:1034.

Stadtman, T. C., 1971, Vitamin $B_{12}$, biochemical studies elucidate the role of this complex molecule in diverse metabolic processes, Science, 171:859.

Stebbins, R., and Bertino, J. R., 1976, Megaloblastic anaemia produced by drugs, in "Clinics in Haematology, Megaloblastic Anaemia", Vol. 5, No. 3, (A. V. Hoffbrand, ed.), pp. 619-630, W. B. Saunders Co., Ltd., London.

Stenman, U.-H., 1971, Isoelectric focusing of R-type vitamin $B_{12}$-binding proteins, Scand. J. Clin. Lab. Invest., 27 (Supplement 116):26.
Stephens, M. E. M., Craft, I., Peters, T. J., and Hoffbrand, A. V., 1972, Oral contraceptives and folate metabolism, Clin. Science, 42:405.
Stokstad, E. L. R., 1943, Some properties of a growth factor for Lactobacillus casei, J. Biol. Chem., 149:573.
Stokstad, E. L. R., Shin, Y. S., and Tamura, T., 1977, Distribution of folate forms in food and folate availability, in "Folic Acid Biochemistry and Physiology in Relation to the Human Nutrition Requirement, Proceedings of a Workshop on Human Folate Requirements, Washington, D. C., June 2-3, 1975", pp. 56-68, Food and Nutrition Board, National Research Council, National Academy of Sciences, Washington, D. C.
Stone, M. L., 1968, Effects on the fetus of folic acid deficiency in pregnancy, Clin. Obstet. Gynecol., 11:1143.
Stone, M. L., Luhby, A. L., Feldman, R., Gordon, M., and Cooperman, J. M., 1967, Folic acid metabolism in pregnancy, Am. J. Obstet. Gynecol., 99:638.
Streiff, R. R., 1970a, Folic acid deficiency anemia, Seminar. Hematol., 7:23.
Streiff, R. R., 1970b, Folate deficiency and oral contraceptives, J. Am. Med. Assoc., 214:105.
Streiff, R. R., and Greene, B., 1970, Drug inhibition of folate conjugase, Clin. Res., 18:418 (Abstract).
Strickland, G. T., and Kostinas, J. E., 1970, Folic acid deficiency in malaria, Clin. Res., 18:418 (Abstract).
Sturrock, J. E., and Nunn, J. F., 1974, Effect of anaesthesia on DNA synthesis in Chinese hamster fibroblasts, Br. J. Anaesth., 46:316.
Sturrock, J. E., and Nunn, J. F., 1975, Mitosis in mammalian cells during exposure to anesthetics, Anesth., 43:21.
Sullivan, L. W., 1970a, Vitamin $B_{12}$ metabolism and megaloblastic anemia, Seminar. Hematol., 7:6.
Sullivan, L. W., 1970b, Differential diagnosis and management of the patient with megaloblastic anemia, Am. J. Med., 48:609.
Sullivan, L. W., and Herbert, V., 1964, Suppression of hematopoiesis by ethanol, J. Clin. Invest., 43:2048.
Sullivan, L. W., and Herbert, V., 1965, Studies on the minimun daily requirement for vitamin $B_{12}$: hematopoietic response for 0.1 microgram of coenzyme $B_{12}$ and comparison of their relative potency, New Engl. J. Med., 272:340.
Sullivan, L. W., Herbert, V., and Castle, W. B., 1963, In vitro assay for human intrinsic factor, J. Clin. Invest., 42:1443.
Swanson, V. L., Wheby, M. S., and Bayless, T. M., 1966, Morphologic effects of folic acid and vitamin $B_{12}$ on jejunal lesion of tropical sprue, Am. J. Path. 49:167.
Sweetman, L., Weyler, W., Shafai, T., Young, P. E., and Nyhan, W. L., 1979, Prenatal diagnosis of propionic acidemia, J. Am.

Med. Assoc., 242:1048.

Swenseid, M. E., Bethell, F. H., and Bird, O. D., 1951, The concentration of folic acid in leukocytes: observations on normal subjects and persons with leukemia, Cancer Res., 11:864.

Taguchi, H., Laundy, M., Reid, C., Reynolds, E. H., and Chanarin, I., 1977, The effect of anticonvulsant drugs on thymidine and deoxyribosenucleic acid synthesis by human marrow cells, Br. J. Haematol., 36:81.

Tamura, T., and Stokstad, E. L. R., 1973, The availability of food folate in man, Br. J. Haematol., 25:513.

Tamura, T., Shane, B., Baer, M. T., King, J. C., Margen, S., and Stokstad, E. L. R., 1978, Absorption of mono- and polyglutamyl folates in zinc-depleted man, Am. J. Clin. Nutr., 31:1984.

Taylor, K. B., 1965, The localization of gastric intrinsic factor, Gastroenterology, 48:853 (Abstract).

Taylor, K. B., 1976, Immune aspects of pernicious anaemia and atrophic gastritis, in "Clinics in Haematology, Megaloblastic Anaemia", Vol. 5, No. 3, (A. V. Hoffbrand, ed.), pp. 497-519, W. B. Saunders, Co., Ltd., London.

Taylor, K. B., and Morton, J. A., 1958, An antibody to Castle's intrinsic factor, Lancet, 1:29.

Ternberg, J. L., and Eakin, R. E., 1949, Erythein and apoerythein and their relation to the antipernicious anemia principle, J. Am. Chem. Soc., 71:3858.

Thiersch, J. B., 1952, Therapeutic abortions with a folic acid antagonist, 4, aminopteroylglutamic acid (4-amino PGA) administered by the oral route, Am. J. Obstet. Gynecol., 63:1298.

Thom, S. R., 1979, Effects of anesthetic gases and helium on microbial growth, Clin. Res., 27:607A (Abstract).

Thorn, G. W., Forsham, P. H., Frawley, R. F., Hill, S. R., Jr., Toche, M., Staehelin, D., and Wilson, D. L., 1950, The clinical usefulness of ACTH and cortisone, New Engl. J. Med., 242:865.

Toepfer, E. W., Zook, E. G., Orr, M. L., and Richardson, L. R., 1951, "Folic Acid Content of Foods", Agric. Handbook No. 29, U.S. Dept. Agric., Washington, D. C.

Toskes, P. P., and Deren, J. J., 1972, The role of the pancreas in vitamin $B_{12}$ absorption: studies of vitamin $B_{12}$ absorption in partially pancreatectomized rats, J. Clin. Invest., 51:216.

Toskes, P. P., and Deren, J. J., 1973, Vitamin $B_{12}$ absorption and malabsorption, Gastroenterology, 65:662.

Toskes, P. P., Hansell, J., Cerda, J., and Deren, J. J., 1971, Vitamin $B_{12}$ malabsorption in chronic pancreatic insufficiency, New Engl. J. Med., 284:627.

Toskes, P. P., Deren, J. J. and Conrad, M. E., 1973, Trypsin-like nature of the pancreatic factor that corrects vitamin $B_{12}$ malabsorption associated with pancreatic dysfunction, J. Clin. Invest., 52:1660.

Trudell, J. R., Hong, K., O'Neil, J. R., and Cohen, E. N., 1979, Metabolism of nitrous oxide by human and rat intestinal contents, Anesthesiology, 51 (Supplement 38):S258.

Tso, S. C., 1976, Significance of subnormal red-cell folate in thalassemia, J. Clin. Path., 29:140.

van Dommelen, C. K. V., Slagboom, G., Meester, G. T., and Wadman, S. K., 1963, Reversible hypogammaglobulinemia in cyanocobalamin (vitamin $B_{12}$) deficiency, Acta Med. Scand., 174:193.

van Niekirk, W. A., 1966, Cervical cytological abnormalities caused by folic acid deficiency, Acta Cytol., 10:67.

Vilter, R. W., Horrigan, D., Mueller, J. F., Jarrold, T., Vilter, C. F., Hawkins, V., and Seaman, A., 1950, Studies on the relationships of vitamin $B_{12}$, folic acid, thymine, uracil and methyl group donors in persons with pernicious anemia and related megaloblastic anemias, Blood, 5:695.

Visuri, K., and Gräsbeck, R., 1973, Human intrinsic factor, isolation by improved conventional methods and properties of the preparation, Biochim. Biophys. Acta, 310:508.

Von der Lippe, G., 1977, Absorption of vitamin $B_{12}$ in chronic pancreatic insufficiency, Scand. J. Gastroenterology, 12:257.

Waslien, C. I., 1977, Folacin requirement of infants, in "Folic Acid Biochemistry and Physiology in Relation to the Human Nutrition Requirement, Proceedings of a Workshop on Human Folate Requirements, Washington, D. C., June 2-3, 1975", pp. 232-246, Food and Nutrition Board, National Research Council, National Academy of Sciences, Washington, D. C.

Waxman, S., and Gilbert, H. S., 1973, A tumor-related vitamin $B_{12}$ binding protein in adolescent hepatoma, New Engl. J. Med., 289:1053.

Waxman, S., Corcino, J., and Herbert, V., 1970a, Aggravation or initiation of megaloblastosis by amino acids in the diet, J. Am. Med. Assoc., 214:101.

Waxman, S., Corcino, J. J., and Herbert, V., 1970b, Drugs, toxins, and dietary amino acids affecting vitamin $B_{12}$ or folic acid absorption or utilization, Am. J. Med., 48:599.

West, R., and Reisner, E. H., Jr., 1949, Treatment of pernicious anemia with crystalline vitamin $B_{12}$, Am. J. Med., 6:643.

Whitehead, V. M., and Cooper, B. A., 1977, Failure of radiodilution assay for vitamin $B_{12}$ to detect deficiency in some patients, Blood, 50 (Supplement 1):99.

Whitehead, N., Reyner, F., and Lindenbaum, J., 1973, Megaloblastic changes in the cervical epithelium, association with oral contraceptive therapy and reversal with folic acid, J. Am. Med. Assoc., 226:1421.

Whiteside, M. G., Ungar, B., and Cowling, D. C., 1968, Iron, folic acid, and vitamin $B_{12}$ levels in normal pregnancy and their influence on birth weight and duration of pregnancy, Med. J. Austral., 1:338.

Wickramasinghe, S. N., and Longland, J. E., 1974, Assessment of deoxyuridine suppression test in diagnosis of vitamin $B_{12}$ or folate deficiency, Br. Med. J., 3:148.

Wickramasinghe, S. N., and Pratt, J. R., 1970, Myelocyte proliferation in pernicious anaemia, Acta Haematol., 44:37.

Wickramasinghe, S. N., Chalmers, D. G., and Cooper, E. H., 1967, Disturbed proliferation of erythropoietic cells in pernicious anaemia, Nature, 215:189.

Wickramasinghe, S. N., Cooper, E. H., and Chalmers, D. G., 1968, A study of erythropoiesis by combined morphologic, quantitative, cytochemical and autoradiographic methods, Blood, 31:304.

Williams, H. H., Erickson, B. N., Bernstein, S., Hummel, F. C., and Macy, I. G., 1937, The lipid and mineral distribution of the serum and erythrocytes in pernicious anemia, J. Biol. Chem., 118:599.

Williams, R. B., Mills, C. F., and Davidson, R. J. L., 1973, Relationships between zinc deficiency and folic acid status of the rat, Proc. Nutr. Soc., 32:2A (Abstract).

Wills, L., 1931, Treatment of "pernicious anemia of pregnancy" and "tropical anaemia", Br. Med. J., 1:1059.

Wills, L., 1932, Tropical macrocytic anaemia, Proc. Roy. Soc. Med., 25:1720.

Wills, L., 1933, The nature of the haemopoietic factor in marmite, Lancet, 1:1283.

Wintrobe, M. M., Cartwright, G. E., Palmer, J. G., Kuhns, W. J., and Samuels, L. T., 1951, Effect of corticotrophin and cortisone on the blood in various disorders in man, Arch. Intern. Med., 88:310.

WHO, Nutritional Anaemias, Report of a WHO group of experts, 1972, WHO Technical Report Series No. 503.

Young, V. R., and Scrimshaw, N. S., 1979, Genetic and biological variability in human nutrient requirements, Am. J. Clin. Nutr., 32:486.

Zarafonetis, C. J. D., Overman, R. L., and Molthan, L., 1957, Unique sequence of pernicious anemia, polycythemia and acute leukemia, Blood, 12:1011.

Zittoun, R., Zittoun, J., Seignalet, J., and Dausset, J., 1975, HL-A and pernicious anemia, New Engl. J. Med., 293:1324.

# TRACE METALS IN HEALTH AND DISEASE

Raymond J. Shamberger, Ph.D.

Department of Biochemistry
The Cleveland Clinic Foundation
9500 Euclid Avenue
Cleveland, Ohio

## INTRODUCTION

Trace elements often have a bimodal or even a trimodal effect. Severe deficiency of certain trace metals can result in death or severe crippling of an animal or in birth defects of the newborn. The next level of intake is the nutritional level where chronic deficiency over a lifetime may cause major diseases such as cancer and heart disease. The next level of intake is the toxic level which may result in severe crippling or death of the animal. The primary area of concern in this chapter will be the nutritional intake level and the deficiency level and their relationship to animal or human disease. The trace elements described in this chapter are: chromium; cobalt; copper; fluorine; iodine; iron; manganese; molybdenum; nickel; silicon; selenium; tin; vanadium; and zinc.

## CHROMIUM

### Glucose Metabolism

The most prominent feature of chromium deficiency in rats (Schwarz and Mertz, 1961) and other animals (Davidson and Blackwell, 1968) is impairment of glucose tolerance. After a few weeks on a torula yeast diet, glucose removal rates declined almost 50%. The decline can be reversed quickly by one oral dose of 20 µg or an intravenous dose of 0.25 µg/100 g body weight of trivalent chromium (Mertz et al., 1961; Mertz et al., 1965). Chromium deficiency in a more severe degree leads to a syndrome like mild diabetes mellitus, including glycosuria and fasting hyperglycemia.

Since 1966, evidence for a human nutritional requirement for chromium has been looked at in patients with an impairment of glucose tolerance including "maturity onset" diabetes, middle-aged and elderly subjects with impaired glucose tolerance, and infants with protein-calorie malnutrition. In the first therapeutic trial (Glinsmann and Mertz, 1966) four of six maturity onset diabetics improved after administration of 180 to 1000 μg of chromium as $CrCl_3$ for periods of 7 to 13 weeks. In another study (Sherman et al., 1968) treated 10 adult diabetics with 150 μg daily for 16 weeks and observed no improvement of glucose tolerance. In another experiment 4 of 12 diabetics treated with 1 mg chromium each day for six months had improved glucose tolerance, but 16 treated for a shorter time period with 150 μg chromium daily did not improve (Schroeder, 1968). In 10 elderly patients (over 70) treated up to four months with 150 μg of chromium, glucose tolerance was restored to normal in four (Levin et al., 1968). Half of a group of middle-aged subjects treated with 150 μg daily for 6 months had a marked improvement (Hopkins and Price, 1968). Benjanuvatra and Bennien (1975) have measured hair chromium analyses for 28 subjects with adult-onset diabetes mellitus and 28 nondiabetic control subjects from Bangkok, Thailand. Hair chromium concentrations were significantly lower in the diabetic than in the control group.

Davidson and Burt (1973) have suggested that the "diabetogenic" effect of pregnancy is related to a significantly lower concentration of plasma chromium of pregnant women. The "diabetogenic" effect of pregnancy is characterized by impairment of peripheral glucose metabolism, decreased glucose tolerance, and an exaggerated insulin excretion in response to glucose challenge.

Morgan (1972) has measured liver chromium content from postmortem examinations in elderly patients with diabetes, with ischemic heart disease, with hypertensive cardiovascular disease, and without disease. The results in μg/g were controls 12.7; arteriosclerotic 9.96; hypertensive 10.2; and diabetics 8.59. The diabetic group was significantly decreased in regard to the control population ($p = 0.05$). Serum chromium has been related (Newman et al., 1978) to angiographically determined coronary artery disease. Human aortas sampled from 17 adults where there is little advanced atheromatous plaque formation contain higher concentrations of chromium than do aortas from 15 adults in which atheromatosis is prevalent.

In a study on a patient with long term total parenteral nutrition Jeejeebhoy et al., (1977) have reported that a chromium deficiency in man, causes: (1) glucose intolerance, (2) inability to utilize glucose for energy, (3) neuropathy with normal insulin concentrations, (4) high free fatty acid concentrations and low respiratory quotient and, (5) abnormalities of nitrogen metabolism.

Lipid Metabolism

Serum cholesterol concentrations in rats were lowered by addition of chromium to a low-chromium diet and chromium inhibited the tendency of these cholesterol concentrations to increase with age. Male rats fed 1 μg Cr/ml and females fed 5 μg/ml in the drinking water had significantly depressed serum cholesterol (Schroeder et al., 1962; Schroeder and Balassa, 1965; Schroeder, 1969). Brown sugar, which contains chromium, also lowered serum cholesterol concentrations, but cholesterol was elevated and increased with age in the rats receiving white sugar, which is low in chromium (Schroeder, 1969; Staub et al., 1969). Significant decrease of serum cholesterol was seen in institutionalized patients fed 2 mg Cr for 5 months, but other patients treated similarly showed no response (Schroeder, 1968).

Significantly lower (2%) incidence of spontaneous plaques were observed in the chromium-fed animals than chromium-deficient animals (19%) (Schroeder and Balassa, 1965). Decreased amounts of stainable lipids and of fluorescent material in the aorta were observed. Similar observations have been made in man in that diabetes is associated with an increased incidence of vascular lesions (Goldenberg and Blumenthal, 1964). Curran (1954) has observed that trivalent chromium enhances the incorporation of acetate into the formation of cholesterol and fatty acids. Mertz et al. (1965) have observed that chromium plus insulin significantly increases glucose uptake and incorporation of glucose into epidedymal fat in chromium-deficient rats.

Protein Synthesis

Diets deficient in chromium and protein impair the rats' capacity to incorporate α-amino isobutyric acid, glycine, serine and methionine into heart protein (Roginski and Mertz, 1967a; Roginski and Mertz, 1967b). No chromium effect was observed with lysine, phenylalanine or a mixture of 10 other amino acids. Insulin *in vivo* enhanced the cell transport of an amino acid analog to a greater degree in rats fed a low-protein, chromium-supplemented diet than it did in Cr-deficient controls (Roginski and Mertz, 1969).

COBALT

Cobalt deficiency is not know in man. However, cobalt is an essential component of vitamin B-12 which normally occurs in the serum in the range of 150-900 pg/ml. Concentrations of serum B-12 from 0-150 pg/ml are considered low. The following is a general discussion about B-12. Far greater details are contained in this monograph in the chapter by Steinkamp.

Vitamin B-12 deficiency secondary to inadequate intake is rare in the western world except among Seventh Day Adventists, food faddists and infants breast-fed by vegetarian mothers. B-12 deficiency occurs only after a prolonged period of inadequate intake since such subjects secrete adequate quantities of gastric intrinsic factor (GIF) and thus continue to reabsorb B-12 that is secreted into the small intestine via bile. Pernicious anemia (PA) patients develop a deficiency after a shorter depletion period because they are unable to reabsorb endogenous vitamin B-12 into the bile because of the lack of GIF. When normal gastric function exists, the release of vitamin B-12 from dietary protein does not restrict the rate of B-12 absorption. However, in gastric atrophy or after subtotal gastrectomy, acid-peptic digestion is depressed and the release of free vitamin B-12 from dietary sources may be impaired.

Vitamin B-12 may fail to attach to GIF. The most common cause is the lack of GIF associated with Addisonian pernicious anemia. GIF secretion is markedly depressed, as part of the generalized atrophy of the acid pepsin portion of the stomach. Patients with idiopathic gastric atrophy or after gastric surgery may have depressed secretion of GIF in the basal state. GIF antibodies may also occur, but are rarely found in diseases other than pernicious anemia. A few patients with adrenal insufficiency, diabetes mellitus, hypothyroidism, and thyrotoxicosis have serum GIF antibodies.

Another group of malabsorptive diseases are associated with the passage of the GIF-vitamin B-12 complex down the small intestine. The blind loop syndrome defect appears to be associated with bacterial overgrowth. The absorptive defect appears to result from the intraluminal competition of the bacteria with the host for dietary B-12. The mechanism of B-12 malabsorption after infestation with the fish tapeworm _Diphyllobothrium latum_ or patients with pancreatic exocrine insufficiency has not been explained. A damaged pancreas may not be able to buffer the gastric acid output of the stomach and would impair B-12 absorption.

Certain B-12 malabsorption diseases are associated with the passage of B-12 through the ileal epithelial cell. Children with familial selective vitamin B-12 malabsorption (Imerslund or Grasbeck syndrome) may lack the specific ileal receptors for the GIF-vitamin B-12 complex. Malabsorptive disease may be associated with the exit of B-12 from the epithelial cell. Transcobalamin-II deficiency has been described. A number of therapeutic agents have been shown to interfere with the absorption of B-12 including alcohol, p-aminosalicylic acid, colchicine, metformin and neomycin.

Serum B-12 concentrations well over 1000 pg/ml of serum suggest either liver disease or a myeloproliferative disorder (such as granulocytic leukemia, myeloid metaplasia or polycythemia vera; values above 4000 pg/ml are unusual in liver disease but common in myeloproliferative disorders.

## COPPER

### Deficiency Anemia

Malnourished infants with anemia and neutropenia caused by copper deficiency have been reported (Cordano et al., 1964; Cordano et al., 1966; Al-Rashid and Spangler, 1971; Karpel and Peden, 1972; Graham and Cordano, 1969). In general, copper is widely distributed in foods making it virtually impossible for severe copper deficiency to develop on a nutritional basis in the adult. Dunlop (1974) has reported a copper deficiency in an adult woman and an adolescent girl who had experienced extensive bowel surgery and received long-term parenteral hyperalimentation. Another case was reported in a 56-year old woman on intravenous hyperalimentation (Dunlop, 1974). Megaloblastoid changes in the bone marrow were also seen in some of the infants with copper deficiency (Cordano et al., 1964; Al-Rashid and Spangler, 1971). The megaloblastic changes were corrected by copper, which implies an effect on RNA/DNA synthesis in the _de nova_ pathway of deoxyuridine to thymine. The neutropenia, which is thought to be the most constant sign of copper deficiency, was reversed by copper therapy. Neutropenia has been reported in all of the copper-deficient infants (Cordano et al., 1964; Cordano et al., 1966; Al-Rashid and Spangler, 1974; Karpel and Peden, 1972; Graham and Cordano, 1969). Neutropenia has also been reported in copper-deficient swine (Wintrobe et al., 1953).

### Cardiovascular Disorders

Bennetts and coworkers (Bennets and Hall, 1939; Bennetts et al., 1941; Bennetts et al., 1942) first noticed cardiac lesions in western Australian cattle suffering from "falling disease". In these cattle there was an atrophy of the myocardium with replacement fibrosis with dense collagenous tissue. Death resulted due to cardiac failure after mild exercise or excitement. In other animals such as sheep, horses and pigs the myocardium is rarely affected. In chicks (O-Dell et al., 1961; Carlton and Henderson, 1963; Simpson and Harms, 1964; Hunt et al., 1970), pigs (Carnes et al., 1961; Coulson and Carnes, 1963; Shields et al., 1961) guinea pigs (Everson et al., 1967) and rats (Kelly et al., 1974) on copper-deficient diets, ruptures of the coronary and pulmonary arteries have been observed. The tensile strength of the aorta in these species was markedly reduced and the myocardium was abnormally friable. A derangement of the elastic membranes of the aorta was evident. Similar studies have not been done on human patients with aneurysms, but such a study would be of interest. Copper deficiency in rats resulted in a 129% increase in serum cholesterol (Allen and Klevay, 1978) and a copper deficiency in rats over 3 generations produced a 56% decrease in myelin (Zimmerman et al., 1976). Klevay and Forebush (1976) have postulated that the zinc/copper ratio of human dietary intakes is related to heart disease.

## Wilson's Disease

Wilson's disease or hepatolenticular degeneration is an inherited autosomal recessive disorder, described first by Wilson (1912). This disease affects mostly the liver, basal ganglia of the brain, cornea, and the kidneys. Chronic degeneration of the basal nuclei is associated with degeneration of the hepatic parenchyma, which leads to eventual cirrhosis. This disease is progressive and ultimately fatal. The most consistent feature has been severe dimunition or occasional absence of serum ceruloplasmin. Two major hypotheses have been advanced to account for the pathological findings of Wilson's disease. Hypothesis A: An abnormal gene suppresses the synthesis of normal ceruloplasmin (Bearn, 1953; Scheinberg and Gitlin, 1952; Cartwright et al., 1954; Earl et al., 1954). As a consequence of ceruloplasmin deficiency, copper absorption from the gut is accelerated leading to excessive copper being deposited in the tissues, resulting in structural damage and functional impairment in the organs involved. A second hypothesis that an abnormal protein, presumably determined by the mutant gene, has an enhanced affinity for copper. This deprives ceruloplasmin of its copper, blocking its formation, hence the deficiency (Iber et al., 1954; Uzman et al., 1956). Neither hypothesis has been proven. The biochemical defect could also be related to a decreased cytochrome oxidase activity. Individuals with Wilson's disease have extremely low activities of leucocyte cytochrome oxidase, comparable to the low levels of ceruloplasmin in their serum (Shokeir and Shreffler, 1969). Carriers of Wilson's disease display intermediate activities of cytochrome oxidase in the leucocytes and ceruloplasmin in the serum.

Torsion dystony is somewhat similar to Wilson's disease and is also a copper metabolic defect (Herishanu and Loewinger, 1972). This pathological and clinical condition is also called dystonia lenticularis.

## Menkes Kinky Hair Syndrome

Menkes syndrome (Trichopoliodystrophy) is a progressive brain disease of male infants, which usually causes death before 3 years of age and which is characterized by severely retarded growth and development, "kinky hair" (pili torti), hypothermia, scorbutic bone changes, arterial tortuosity and cerebral gliosis with cystic degeneration (Menkes, 1972). Alterations of free sulfhydryl groups of hair may result in its tendency to twist (Danks et al., 1972). Menkes syndrome is associated with profound hypocupremia, hypoceruloplasminemia, and diminished concentrations of copper in hair (Danks et al., 1972a; Danks et al., 1972b; Danks et al., 1973; Singh and Bresnan, 1973). Copper deficiency may be related to the changes in the elastic fibers of arterial walls and scorbutic bone deformities (Danks et al., 1972a; Danks et al., 1972b). There is also a defect in intestinal copper absorption which could be caused by either a disturbance of the intracellular transport of copper in the duodenal mucosa or an impairment of copper transport across the serosal cell membrane (Danks et al, 1973). Beneficial results may be derived from parenteral administration of copper (Danks et al., 1972a; Danks et al., 1972b; Danks et al., 1973). Screening of infants for Menkes syndrome as well as Wilson's disease by analyses of serum copper could become a major application of clinical pathology laboratories (Sunderman, 1974).

### Hodgkins Disease

Measurements of serum copper provide valuable laboratory information in monitoring relapse in patients with Hodgkin's lymphoma because there is a good correlation between the concentration of serum copper and the clinical activity of the lymphoma (Hrgovcic et al., 1968; Warren et al., 1969). Hyperceruloplasminemia occurs as a manifestation of the "acute phase reaction" during active Hodgkins disease. Serum concentrations of copper may help guide the physician in anticipating the clinical response to treatment in patients with Hodgkins disease. Increased serum copper concentration is not specific for Hodgkins disease activity, but can occur with increased estrogen and with various nonmalignant processes including heart disease and anemia. Wilimas et al. (1978) have found serum copper concentrations to be unreliable as a diagnostic aid in the individual child. A large number of patients had elevated serum copper concentrations and were free of the disease. In addition, copper was elevated in the serum of 19 patients with sarcomas (Fisher et al., 1976). Copper was also elevated in the serum of 56% of patients with schizophrenia (Chugh et al., 1973). Patients with psoriasis tend to have higher than normal serum copper concentrations (Lipkin et al., 1962; Kekki et al., 1966; Molokhia and Portnoy, 1970).

## FLUORINE

### Dental Caries

A relationship between dental caries and increasing fluoride concentration was first shown by Dean (1942). In a study of 7,257 children, 12-14 years old, caries decreased in permanent teeth as the drinking water fluoride concentration increased above 1.3 ppm. In many countries an inverse relationship has been observed between naturally present fluoride in the water and dental caries (Forrest et al., 1-51; McKay, 1948; Russell and Elvove, 1951).

The water supplies of hundreds of communities have been treated with sodium fluoride or fluorosilicate at 0.8 to 1.2 ppm. When these communities were compared to adjacent communities without fluoridation, the value of fluoridation was observed (Arnold, 1957; Naylor, 1969). Topical applications of fluorides have achieved some success (Harris, 1959; Knutson and Armstrong, 1943; 1945; 1946; 1947; Jordan et al., 1958).

## Osteoporosis and Aortic Calcification

Osteoporosis is a disease of the elderly characterized by a decrease in bone density often accompanied by collapsed vertebrae; large amounts of fluoride improve osteoporosis in these patients (Aeschlimann et al., 1966; Cass et al., 1966; Cohen and Gardner, 1966; Purves, 1962; Rich, et al., 1964). In two comparative studies Leone et al., (1960 have found less osteoporosis in a high fluoride area of Texas (8 ppm in water) than in a low-fluoride area of Massachusetts (0.09 ppm in the water) and Bernstein et al. (1966) have found similar results in a comparative study of high- and low- flouride areas of North Dakota.

Leone (1964) has compared the mortality of a high and low-fluoride area of Texas. The low fluoride area had a much greater incidence of heart attacks. Calcification of the aorta was much greater in the low-fluoride area.

## Cancer Controversy

There has been a good deal of interest in a purported relation between community water fluoridation and cancer mortality. Burk (1975) and Yiamouyiannis (1975; 1977) reported on cancer mortality trends in 20 United States cities, half of which fluoridated their water supplies during 1952-1956. In 1950 the crude mortality rates for cities with fluoridated water supplies were very nearly equal to the rates in the cities with non-fluoridated supplies. In 1970 there was a differential of some 20 deaths per 100,000 persons, the higher rates being in the cities with fluoridated water. Re-analysis of this data pertaining to these cities (Doll and Kinlen, 1977; Taves, 1977) and a similar study of cancer mortality in selected counties in the United States with fluoridated and non-fluoridated water systems (Hoover et al., 1976) did not support the data of Burk (1975) and Yiamouyiannis (1975).

# IODINE

## Goiter

The first large scale trial on iodine deficiency and goiter in man was carried out in the schools of Ohio between 1916 and 1920. Similar prophylactic measures were soon initiated in goitrous areas in other countries with such success that iodized salt soon became a widely recognized form of control. Nonetheless, the actual number of goitrous individuals in the world is estimated in 1960 at close to 200 million (Kelly and Snedden, 1960). Women and children are more affected than adult males (Malamos et al., 1966). In the United States, Switzerland and New Zealand, the intensity of the disease declined markedly as a result of the use of iodized salt. In many Latin American countries, despite the simplicity of needed control measures, endemic goiter remains a serious public health problem, associated often with cretinism, feeblmindedness and deaf-mutism (Scrimshaw, 1960).

The healthy human adult probably contains a total of 10 to 20 mg iodine, of which a high proportion, 70-80%, is concentrated in the thyroid gland which is only about 0.05% of the body mass. Iodine occurs as organically bound iodine and as inorganic iodide. The latter occurs in most non-thyroid tissues in very low concentrations, of the order of 1 to 2 μg/100 g.

## Thyroid Function

The normal range of serum protein-bound iodine in adult man has been placed at 4 to 8 or 3 to 7.5 μg/100 ml with a mean close to 5 to 6 μg (Grossmann and Grossmann, 1955; Hallman et al., 1951; Rapport and Curtis, 1950; Wayne et al., 1964). Increased serum PBI values are highly characteristic of hyperthyroidism and decreased values of hypothyroidism in man. However, under a variety of conditions a discrepancy between the serum PBI and the patient's clinical thyrometabolic state was observed. The most common cause of spurious elevation of the PBI is exposure to a variety of iodine-containing drugs and radiographic contrast media and estrogen therapy. Development of $T_4$ tests (3,5,3,5,-Tetraiodothyronine), $T_3$ uptake (3,5,3-Triiodothyronine) and free thyroxine index has resulted in better diagnostic accuracy and has widely supplemented the PBI procedure.

# IRON

## Deficiency Anemia

Human iron deficiency is clinically characterized by listlessness and fatigue, palpitation on exertion, sore tongue on occasion, koilonychia, and angular dysphagia (Darby, 1951). In children, anorexia, depressed growth, and decreased resistance to infection are additionally commonly observed. In iron-deficient anemia abnormalities of the gastro-intestinal tract, including superficial gastritis and achlorhydria have been observed (Hawksley et al., 1934; Lees and Rosenthal, 1958). Anemia due to iron-deficiency is much more common in women than in men because women of fertile age are subject to additional iron losses in menstruation, pregnancy and lactation. The incidence of iron deficiency in women has been reported as 20-25% in different English and Swedish studies (Laufberger, 1937; Rybo, 1966). Among pregnant women in the United States, iron-deficiency anemia frequencies ranging from 15% to 50% have been observed (Allaire and Campagna, 1961; Pritchard and Hung, 1958). In adult men and postmenopausal women the principal cause of anemia is chronic bleeding due to infections, malignancy, bleeding ulcers, and hookworm infestation. Economic disadvantage (Bothwell and Finch, 1962; Mukherjee and Mukherjee, 1953), infection, vegetarian-type diets (Hussain et al., 1967), and excessive sweating can also lead to iron-deficiency anemia.

During the suckling period, both young animals and humans can also become iron deficient (von Bunge, 1899; Ezekiel, 1967). In the United States, iron-deficiency anemia may range as low as 8% to as high as 64% (Beal et al., 1962; Sturgeon, 1956). Between 4 and 24 months of age is when iron-deficiency anemia occurs most frequently. There is a regular need for supplementation with medicinal iron or iron-fortified foods (Beal et al., 1962; Sturgeon, 1956; Farquhar, 1963).

## States with Increased Serum Iron

Increases in serum or plasma iron are brought about by several conditions: increased red-cell destruction (decreased survival of red cells, hemolytic anemia), in cases of decreased utilization (decreased blood formation as in pyridoxine deficiency or in lead poisoning), in situations in which release of iron occurs (release of ferritin in necrotic hepatitis), in states where iron storage is defective (as in pernicious anemia), and in conditions in which there is an increased rate of absorption (e.g., hemosiderosis and homochromatosis).

## States with Decreased Serum Iron

Decreases in serum or plasma iron are generally due to a deficiency in the total amount of iron present in the body which may be caused by a lack of sufficient intake or absorption of iron (nephrosis or chronic blood loss) or by an increased demand of the body stores (pregnancy). Diminished iron levels may also be caused by a decreased release of iron from body stores (reticuloendothelial system) as seen in infections or abscesses.

## States with Altered Iron-Binding Capacity

An increased production of transferrin may be related to increases in the total iron-binding capacity of serum. Increases in the total iron-binding capacity may also be caused by an increased release of ferritin, as in hepatocellular necrosis. Decreases in the total iron-binding capacity may be caused by a deficiency in ferritin, as found in cirrhosis and hemochromatosis, or as a result of an excessive loss of protein (transferrin) occurs in nephrosis.

## MANGANESE

### Deficiency

Human requirements for manganese are not known, and there was even a doubt that a deficiency could occur in man until Doisy (1973) recognized the first case in a volunteer undergoing a study of vitamin K deficiency in a metabolic ward. Failure to add Mn to a purified diet mixture resulted in weight loss, transient dermatitis, nausea, and slow growth of hair and beard with changes in hair color, and biochemically there was striking hypocholesterolemia.

### Lipid Metabolism

Liver and bone fat are reduced in manganese- and choline-supplemented rats, and back fat is reduced in pigs (Plumlee et al., 1954; 1956). Because manganous ion is a necessary cofactor for the conversion of mevalonic acid to squalene by mevalonic kinase (Amdur et al., 1957), manganese stimulates the hepatic synthesis of cholesterol and fatty acids in rats (Curran, 1954).

### Carbohydrate Metabolism

Manganese may be important in gluconeogenesis through its presence in the metalloenzyme pyruvate decarboxylase. Aplasia or marked hypoplasia of guinea pig pancreas occurred in Mn-deficient animals (Everson and Shrader, 1968; Shrader and Everson, 1968), and decreased the number of pancreatic islets. A Mn-deficient guinea pig shows a decreased capacity to utilize glucose and exhibits a diabetic-like curve in response to glucose loading. The reduced glucose utilization was reversed by manganese supplementation. Manganese administration to diabetic subjects has a hypoglycemic effect (Belyaev, 1938; Rubenstein, 1962) and both pancreatectomy and diabetes have been correlated with decreased concentrations in blood and tissues (Babenko and Karplyuk, 1963; Kosenko, 1964).

## MOLYBDENUM

### Deficiency

Although molybdenum deficiency has never been reported in man, molybdenum has been shown to be essential for the diet of lambs, chicks and turkey poults (Ellis et al., 1958; Higgins et al., 1959; Reid et al., 1956). In 1953 (deRenzo et al., 1953a; deRenzo et al, 1953b, Richert and Westerfield, 1953) it was shown that the flavoprotein enzyme xanthine oxidase is a molybdenum-containing metalloenzyme which requires molybdenum for its activity

Dental Caries

Claims have been made based on epidemiological studies carried out in Hungary (Adler and Straub, 1953) and New Zealand (Ludwig et al., 1960; Healy et al., 1961) that molybdenum decreases the incidence and severity of dental caries and that it might enhance the well-established effect of fluoride. In two of the earliest experiments with rats, a reduction in caries was reported when molybdenum was supplied during tooth formation. Other experiments failed to show an effect when rats fed a cariogenic diet were given supplemental molybdenum at levels up to 50 ppm (Westerfield and Richert, 1953; Buttner, 1961; Buttner, 1963; Malthus et al., 1964).

NICKEL

Concentrations of serum nickel become increased in patients following acute myocardial infarction, stroke, burn and septicemia (Sunderman et al., 1970; McNeely et al., 1971; Sunderman et al., 1972). Diminished average concentrations of serum nickel are found in patients with hepatic cirrhosis and with chronic renal insufficiency (Sunderman et al., 1970; McNeely et al., 1971). Significant concentrations of nickel are present in DNA (Eichhorn, 1962; Wacker et al., 1959) and RNA (Bertrand and Macheboeuf, 1926; Allison and Lancaster, 1964; Wacker et al., 1963) from phylogenetically diverse sources. Nickel and other metals which are present may help stabilize the structure of nucleic acids.

SILICON

Aging

Leslie et al. (1962) have found a decrease in the silicon content of rat skin with age. Similarly, in human skin the silicon content of the dermis diminishes with age (Brown, 1927; MacCardle et al., 1943). The silicon content of the aorta decreases with age (Jones and Handreck, 1967), and arterial wall silicon decreases with the development of atherosclerosis (Loeper et al., 1966). Schwarz (1974) has suggested that silicon may have a role in mucopolysaccharide metabolism. Silicon is a constituent of certain glycosaminoglycans and polyuronides where it is apparently bound to the polysaccharide matrix (Schwarz, 1974). Silicon may link portions of either the same polysaccharides to each other, acid mucopolysaccharides to each other, or acid mucopolysaccharides to proteins. Silicon may function as a biological cross-linking agent and may contribute to the structure and resilience of connective tissue.

## SELENIUM

### Cancer

Sodium selenide, as well as the antioxidants vitamin E and ascorbic acid, prevented skin tumor formation in mice (Shamberger, 1966; Shamberger, 1970). Sodium selenite-supplemented torula yeast diets also significantly decreased mouse skin tumor formation induced by 7,12-dimethylbenzanthracene-croton oil (Shamberger, 1970). In addition, there is an inverse relationship between the forage crop content of selenium and human cancer mortality in several states (Shamberger and Willis, 1971; Shamberger et al., 1974a; Shamberger et al., 1976). When the mortality from several types of cancer are compared in the high- and low-selenium areas, there are much lower death rates from the types of cancer in organs which might come into contact with selenium than in the organs which do not come into contact directly with selenium. Digestive organ cancer death rates are much lower in the high selenium areas (Shamberger and Willis, 1971; Shamberger et al., 1974a; Shamberger et al., 1976). Similar results have been reported by Schrauzer et al. (1977) when he found an inverse relationship between selenium in blood bank blood and human cancer mortality.

Selenium as well as other antioxidants may be stabilizing or breaking down malonaldehyde, a product of the peroxidative breakdown of unsaturated fatty acids. Malonaldehyde is a carcinogen which is present in large amounts in certain types of meats, but not in fruits or vegetables (Shamberger et al., 1974b; Shamberger et al., 1977a). This finding may have some relationship to the elevated colon cancer death rates in countries with high beef consumption. Malonaldehyde has also been found to be mutagenic in the Ames system (Mukai and Goldstein, 1976; Shamberger et al., 1979a). The mutagenicity of malonaldehyde has been reduced by the antioxidants selenium, vitamin E, BHT and ascorbic acid (Shamberger et al., 1979a).

Selenium in the form of sodium selenite, as well as other antioxidants, reduced carcinogen-induced chromosome breakage when compared to the controls (Shamberger et al., 1973a). Lower blood selenium levels have been observed in patients with gastrointestinal cancer or liver metastases (Shamberger et al., 1973b). McConnell et al. (1975) have reported lower blood selenium concentrations in almost all types of cancer patients.

Several reductions in tumor incidence induced by different carcinogens have been observed with selenium supplemented diets: Clayton and Baumann (1949) found that the inclusion of 5 ppm of selenium in a purified diet reduced the incidence of liver tumors in rats induced by $N^1$-methyl-p-dimethyl-aminoazobenzene. Marshall et al. (1978) have reported similar results. Dietary selenium has also reduced N-2-fluorenyl-acetamide-induced cancer in vitamin-E supplemented rats (Harr et al., 1972). Griffin and Jacobs (1977) have reported a decrease of liver tumors induced by $3^1$-methyl-4-dimethylaminoazobezene. Schrauzer and Ishmael (1974) observed a lower incidence in spontaneous mammary tumors in $C_3H$/St mice supplemented with 2 ppm of selenium in the drinking water for 15 months. Jacobs et al. (1977) have observed a marked reduction in the number of colon tumors per animal and in the total number of tumors induced by 1,2-dimethylhydrazine when 4 ppm of selenium as sodium selenite was added to the drinking water.

Heart Disease - Animal Studies

Rats (Godwin, 1965) and lambs (Godwin and Fraser, 1966) fed Se deficient diets develop abnormal electrocardiograms accompanied by blood pressure changes. In lambs, progressive development of a characteristic electrocardiogram abnormality is seen, after the animals had been on the deficient diet for a few weeks. Just before death the ECG pattern became grossly abnormal in some cases, a rise in the T-wave giving way to an elevated S-T segment, similar to that seen frequently in myocardial infarction in man.

In piglets born of Se and vitamin E deprived sows, progressive vascular changes were observed. Connective tissue lesions of the heart were noted which characterize mulberry heart disease of swine and may result in sudden death of the animal. The disease has been termed "mulberry heart disease" because the affected animals frequently have extensive cardiac hemorrhage that gives the organ a reddish-purple gross appearance comparable to that of a mulberry.

Cadmium has been observed to cause hypertension in rats (Perry, 1973). The Cd pressor effect caused by 2.5 and 10 ppm of cadmium in the drinking water can be reversed by 3 ppm of selenium in the drinking water (Perry, 1974).

Heart Disease - Human Epidemiology

Kubota et al. (1967), after determining the selenium content of various forage crops, constructed a map of the United States showing the generalized regional distribution pattern of Se in forage crops. The male and female 54-64 age specific death rates per 100,000 for cardiovascular-renal, cerebrovascular, coronary, and hypertensive heart disease were all significantly lower ($P<0.001$) in the high selenium area than the low Se area (Shamberger, 1975). For the most part there was an inverse gradation in mortality and Se bioavailability. States with occasionally large amounts of selenium in the drinking water such as Colorado, North and South Dakota and Utah had the lowest hypertensive death rate. Although selenium in the drinking water could be an important factor in these states with the lowest white male hypertensive death rate, certainly this observation may be unrelated to selenium in the drinking water and may be due to as yet unrecognized environmental factors.

Comparisons among 9 types of heart mortality in 17 cities showed significantly lower death rates due to arteriosclerotic and hypertensive heart disease for both males and females (Shamberger et al., 1975). Non-white mortality rates were markedly greater for cerebrovascular and hypertensive heart disease. The non-white are located in the inner city areas where there is a greater chance of airborne cadmium pollution, which has been previously correlated with arteriosclerotic and hypertensive heart disease.

In another epidemiological study (Shamberger et al., 1978) selenium, cadmium, zinc, copper, chromium, arsenic and manganese estimated intakes in 25 countries were related to the 1973 age-specific mortality from hypertensive heart disease, cerebrovascular heart disease and diseases of the arteries and capillaries in those 25 countries. In general, a negative significant correlation coefficient was seen with Se and ischemic heart disease and a positive correlation was observed with Cd and ischemic heart disease. Both of these elements compete for the same sulfydryl binding site and could be interrelated. Some other metals, chromium and manganese, are also related to heart disease mortality. Japan and Hong Kong, which were not included in the study with the 25 countries because of racial differences, had a very low Cd/Se ratio and a low incidence of ischemic heart disease. In autopsy specimens, Shamberger et al. (1977b) found no difference in the Cd/Se ratio of the kidneys from those who died from heart disease and those who died from other causes. Voors et al. (1978) have done a similar study on autopsy kidneys but related heart weight as an index of hypertension of the quotient of the tissue concentrations of Cd/Se Zn. A positive correlation was observed ($p<0.01$, by a one-tailed test).

In another epidemiological experiment Shamberger et al. (1979b) observed that male and female 55-64 age-specific death rates per 100,000 for cardiovascular-renal, coronary and hypertensive heart disease had significant inverse relationships to the mean blood bank selenium concentration in 19 states. In addition, a positive Chi square association was observed between blood bank selenium and forage crop selenium. Bjorksten (1979) has recently reported that the countries in Finland having more selenium in the drinking water had only about 1/7 of the heart disease mortality than the countries with low selenium water.

## Clinical Trials

Patients taking selenium-tocopherol capsules (Tolsem) have shown reduced incidence of anginal pain in 22 of 24 cases

(Villalon, 1974). Keshan disease was first discovered in China in 1935. This is a type of congestive heart failure with the most susceptible people being children below 15 years of age, women of child-bearing age in the northern privinces, and children of 2-7 years of age in the southern affected areas. Encouraging results were obtained in trials using sodium selenite to prevent the disease by researchers at Sian Medical College in 1965. This led their researchers to conduct a four-year clinical trial conducted by the Keshan Disease Research Group of the Chinese Academy of Medical Sciences (1979). Children received sodium selenite tablets or placebo once a week, the dosage being: 1-5 year olds, 0.5 mg and 6-9 year olds, 1.0 mg. Of the 36,603 Se treated children, 21 cases occurred during the four years of investigation. Of these only 3 of them died and 1 became chronic up to the end of 1977, whereas of the 9,642 children in the control group, 107 cases occurred of which 53 were fatal and 6 still had insufficient heart function. These results indicate that administration of sodium selenite not only decreases morbidity but also alleviates clinical symptoms and improves the prognosis.

## TIN

Although the biological function of tin is unknown, tin has a number of chemical properties that offer possibilities for a biological function. Tetravalent tin has a strong tendency to form coordination complexes with four, five, six and possibly eight ligands. Tin (Schwarz et al., 1970) may contribute to the tertiary structure of proteins or other components of biological importance. Tin could participate in oxidation-reduction reactions in biological systems because the $Sn^{2+} \rightleftarrows Sn^{4+}$ potential of 0.13 volts is within the physiological range near the oxidation-reduction potential of flavine enzymes.

## VANADIUM

### Lipid and Carbohydrate Metabolism

Vanadium reduced the *in vivo* cholesterol concentration in plasma from rabbits and from human plasma (Curran et al., 1959; Mountain et al., 1956; Lewis, 1959; Curran and Costello, 1956; Korkor, 1965). Decreased plasma phospholipids were also observed. Workmen exposed to an industrial source of vanadium had slightly but significantly lower (205 mg/100 ml) concentrations of phospholipids than that of controls (228 mg/100 ml). In older human patients and in patients with hypercholesterolemia or ischemic heart disease, vanadium does not lower serum cholesterol (Schroeder et al., 1963; Curran and Burch, 1967; Dimond et al., 1963; Somerville and Davies, 1962; Azaroff and Curran, 1967). In younger subjects vanadium interferes with cholesterol synthesis by inhibiting utilization of mevalonic acid (Curran et al., 1959) and also accelerates cholesterol catabolism. Vanadium also inhibits the synthesis of cholesterol from acetate but not from squalene (Azaroff and Curran, (1967), because vanadium inhibits squalene synthetase catalyzing the conversion of labeled farnesyl pyrophosphate to squalene (Azaroff et al., 1961). Plasma triglyceride concentrations are significantly increased in vanadium-deficient chicks (Hopkins and Mohr, 1974; Hopkins, 1974). Glucose is more rapidly turned over after vanadium administration (Meeks et al., 1971). As glucose is a precursor of triglyceride, this may explain why triglyceride concentrations are lower.

### Dental Caries

Microradioautography indicates that $^{48}V_2O_5$ injected subcutaneously into mice is highly concentrated in the areas of rapid mineralization in tooth dentin and bone (Soremark and Ullberg, 1962). After injection, vanadium is also incorporated into the tooth structure of rats and retained in the molars up to 90 days (Thomassen and Leicester, 1964). Rygh (1949a; 1949b; 1949c) has reported that the addition of vanadium and of strontium to specially purified diets promoted mineralization of the bones and teeth and reduced the number of carious teeth in rats and guinea pigs. Vanadium protected against caries in hamsters fed a cariogenic diet when vanadium was administered as $V_2O_5$, either orally or parenterally (Geyer, 1953). Intraperitoneal administration of vanadium to rats during tooth development was effective in reducing dental caries (Kruger, 1958). However, in three other rat studies, administration of varying levels of vanadium in the drinking water was not successful in decreasing caries incidence (Buttner, 1963; Muhlemann and Konig, 1964; Muhler, 1957; Tank and Storrick, 1960). Ten ppm of vanadium actually increased caries when young hamsters were given a cariogenic diet (Hein and Wisotzky, 1955).

## ZINC

### Acrodermatitis Enteropathica

Acrodermatitis enteropathica (AE) is a serious illness which starts in early infancy after weaning. The disease becomes gradually more insidious and is recognized by changes in skin and hair. There are also gastrointestinal symptoms, especially diarrhea. Skin lesions are localized on the hands and feet, as well as near the orifices, especially in the perioral and in the diaper regions. During inactive phases the skin in these areas is dry, erythematous, peeling, sometimes psoriasis-like, while during the active periods, it is vesicopustular, red and erosive. The first treatment of this disease was with chlorquinaldol, but side effects such as optic atrophy have been observed. Moynahan (1974) has reported excellent results with oral zinc with several AE patients. Michaelsson (1974) has reported similar results. Oral zinc in combination with vitamin A has proven to be more effective in treating acne than vitamin A alone (Michaelsson et al., 1974).

The AE lesion in humans may be related to some of the known zinc deficiency symptoms in animals. Gross epithelial lesions and alopecia develop in rats and mice with zinc deficiency. Histological studies revealed a condition of parakeratosis, i.e., a hyperkeratinization of the epithelial cells of the skin (Follis et al., 1941). Rats with more severe zinc deficiency have scaling and cracking of the paws with deep fissures, coarseness, and loss of hair, and dermatitis (Orten, 1966). Parakeratosis occurs around the eyes and mouth and on the scrotum and lower parts of the legs (Miller et al., 1965). Zinc-deficient ruminants show similar patterns (Underwood and Somers, 1967; Blackmon et al., 1967). Zinc deficiency is evident in sheep through changes in the wool growth, hoof structure and horns (Mills et al., 1967; Underwood and Somers, 1969). The wool fibers lose their crimp and become thin and loose and readily shed.

In the Zn-deficient chick (O-Dell and Savage, 1957; O'Dell et al., 1958), turkey poult (Kratzer et al., 1-58), and pheasant (Scott et al., 1959), feathering is poor and abnormal and dermatitis is present. The epidermal changes seen in zinc-deficient animals may be related to vitamin A deficiency. Zinc is necessary to maintain a normal concentration of rat plasma vitamin A (Smith et al., 1973). Zinc is known to be necessary for normal mobilization of vitamin A from the liver.

Reproduction

Hypogonadism, with suppression in the development of the secondary sexual characteristics, is conspicuous of zinc deficiency of young men in parts of Iran and Egypt (Prasad et al., 1961; Prasad et al., 1963). Hypogonadism also occurs in Zn-deficient bull calves (Miller et al., 1966; Pitts et al., 1966), kids (Miller et al., 1964) and ram lambs (Underwood and Somers, 1969). Human spermatozoa have high zinc levels. Zinc-deficient male rats (Follis et al., 1941) have been observed to have atrophic seminiferous tubules. Zinc deficiency also retarded development of the testes, epididymus, prostate and pituitary glands with atrophy of the testicular germinal epithelium. High concentrations of zinc are necessary for the final stages of sperm maturation and seem to be essential for the maintenance of spermatogenesis and for the survival of germinal epithelium (Underwood and Somers, 1969; Millar et al., 1958, and Wetterdal, 1958). In female rats, zinc deficiency caused a severe disruption of estrus (Hurley and Swenerton, 1966). Zinc deficiency during gestation of rats greatly reduces the number of young that are born alive (Apgar, 1968a; Apgar, 1968b).

Hypogeusia

Several adult patients with low serum zinc developed decreased taste acuity (hypogeusia). Oral zinc supplementation relieved this program (Schechter et al., 1972; Cohen et al., 1973).

Atherosclerosis

Twelve of thirteen patients with advanced vascular disease who were given zinc sulfate orally up to 29 months, showed clinical improvement; nine returned to normal activity. In a similar study by Henzel et al. (1968), about half of the patients improved. Plasma and hair zinc concentrations are usually significantly subnormal in patients with atherosclerosis and myocardial infarction (Pories et al., 1967; Volkov, 1963; Wacker et al., 1956). Atherosclerosis may be an expression of inadequate arterial repair (Pories et al., 1967), which like wound healing in skin, muscle, and bone is related to zinc concentrations.

Within a day of onset of acute myocardial infarction serum zinc concentrations fall sharply and then rise to normal values within 7 to 10 days (Handjani et al., 1974).

Wound Healing

Rats fed zinc accidently have an accelerated rate of wound healing in rats (Pories et al., 1968; Pories and Strain, 1966). Seriously burned patients have subnormal hair zinc concentrations (Pories and Strain, 1966). Wound healing is impaired in Zn-deficient calves (Miller et al., 1965) and rats (Sandstead and Shepard, 1968). Although the mode of action of zinc in tissue repair is unknown, zinc is preferentially concentrated in healing tissues in rats, both in skin and muscle wounds (Savlov et al., 1962), and in bone fractures (Calhoun and Smith, 1968). This concentration suggests a heightened zinc requirement for tissue synthesis during the healing process. Wound healing showed a decreased rate in zinc-deficient animals (Rahmat et al., 1974). Some investigators have reported no effect of supplemental zinc on wound healing in rats (Murray and Rosenthal, 1968; Quarantillo, 1971).

Cancer

In patients with malignant tumors subnormal plasma zinc concentrations have been reported (Addink, 1960; Addink and Frank, 1959; Vidbladh, 1950). Leukocyte zinc concentrations are markedly reduced in patients with chronic leukemia. Zinc concentrations are raised to normal by zinc injections and rise to normal during clinical remission (Gibon et al., 1954). The zinc content of the leukocytes also decreases in patients with a variety of neoplastic diseases (Szmigielski and Litwin, 1964).

The inhibition of chemical carcinogenesis in the hamster cheek pouch by dietary zinc deficiency has been reported (Poswillo and Cohen, 1971). DeWys et al. (1970) reported the inhibition of Walker 256 carcinosarcoma by dietary zinc deficiency. McQuilty et al. (1970)have reported similar results. Because zinc is necessary for growth of normal tissues, zinc may also be necessary for tumor growth. The growth of P 388 leukemia in weanling mice was depressed after the third post-transplantation day (Barr and Harris, 1973). Growth of a transplantable hepatoma induced by $3^{1}$-methyl-4-dimethylaminoazobenzene was significantly reduced in rats maintained on a low-zinc diet (Duncan et al., 1974).

## Growth

Retardation of growth in young animals was observed in the original demonstration of zinc deficiency in rats (Todd et al., 1934) and has been a conspicuous feature of this deficiency in all subsequent studies. The growth inhibition results partly from reduced food intake and partly from an impaired utilization of the food that is consumed and absorbed. Male dwarfs living in restricted areas in the Middle East are the only evidence of zinc deficiency in man of sufficient severity to produce deficiency symptoms (Prasad et al., 1961).

## Sickle Cell Anemia

Zinc plasma concentrations were compared in patients with sickle cell anemia in a steady state, in sickle cell crisis and in controls (Niell et al., 1979). Lowest zinc plasma levels were observed during sickle cell crisis. Low values of plasma zinc were also seen when patients had their sickle cell anemia in a steady state. The drop in plasma zinc seen in sickle cell patients could be related to stress.

## NEED FOR FURTHER RESEARCH

Although several trace elements have tentatively been related to health and disease effects in animals and humans, much additional research is needed to firmly establish these relationships. To accomplish this goal one of the greatest needs is to develop analytical methods which will detect trace elements at concentrations which occur in biological materials. When these methods are developed rapid progress in trace element research could be made.

BIBLIOGRAPHY

Adler P., and Straub, J., 1953, Water-borne caries - protective agent other than fluorine, Acta Med. Acad. Sci. Hung. 4:221.

Aeschlimann, M.I., Grant, J.A., and Grigler, Jr., J.F., 1966, Effects of sodium fluoride on the clinical course and metabolic balance of an infant with osteogenesis imperfecta congenita, Metab. Clin. Exp. 15:905.

Addink, N.W.H., and Frank, L.J.P., 1959, Remarks apropos of analysis of trace elements in human tissues, Cancer 12:544.

Addink, N.W.H., 1960, Subnormal levels of carbonic anhydrase in blood of carcinoma patients, Nature 186:253.

Allaire, B.I., and Campagna, A., 1961, Iron-deficiency anemia in pregnancy. Evaluation of diagnosis and therapy by bone marrow hemosiderin, Obstet Gynecol. 17:605.

Allen, K.E.D., and Klevay, L.M., 1978, Copper deficiency and cholesterol metabolism in the rat, Atherosclerosis 31:259.

Allison, Jr., F., and Lancaster, M.G., 1964, Studies on factors which influence the adhesiveness of leukocytes "in vitro", Ann. N.Y. Acad. Sci. 116:936.

Al-Rashid, R.A., and Spangler, J., 1971, Neonatal copper deficiency N. Eng. J. Med. 285:841.

Amdur, B.H., Billing, H., and Bloch, K., 1957, Enzymatic conversion of mevalonic acid to squalene, J. Amer. Chem. Soc. 79:2646.

Apgar, J., 1968a, Effect of zinc deficiency on parturition in the rat, Amer. J. Physiol. 215:160.

Apgar, J., 1968b, Comparison of the effect of copper, manganese, and zinc deficiencies on parturition in the rat, Amer. J. Physiol. 215:1478.

Arnold, Jr., F.A., 1957, Grand Rapids Fluoridation study; results pertaining to the eleventh year of fluoridation, Amer. J. Pub. Health 47:539.

Azaroff, D.L., Brick, F.E., and Curren, G.L., 1961, A specific site of vanadium inhibition of cholesterol biosynthesis, Biochem. Biophys. Acta 51:397.

Azaroff, D.L., and Curran, G.L., 1967, Site of vanadium inhibition of cholesterol biosynthesis, J. Amer. Chem. Soc. 79:2968.

Babenko, G.O., and Karplyuk, Z.V., 1963, Vplyv vydalennia pidshlunkovoi zalozyna vmist marhantsiu v tkanynakh ta orhanakh sobak, Ukr. Biokhim. Zh. 35:732.

Barr, D.H., and Harris, J.W., 1973, Growth of the P388 leukemia as an ascites tumor in zinc-deficient mice, Proc. Soc. Exp. Biol. Med. 144:284.

Beal, V.A., Myers, A.J., and McCammon, R.W., 1962, Iron intake, hemoglobin, and physical growth during the first two years of life, Pediatrics 30:518.

Bearn, A.G., 1953, Genetic and biochemical aspects of Wilson's disease, Amer. J. Med. 15:442.

Belyaev, P.M., 1938, Trace elements in the human organism. V. Influence of Mn ions on blood-sugar content, J. Physiol. (U.S.S.R.) 25:741.

Benjanuvatra, N.K., and Bennien, M., 1975, Hair chromium concentration of Thai subjects with and without diabetes mellitus, Nutr. Rep. Int. 12:325.

Bennetts, H.W., Beck, A.B., and Harley, R., 1939, Falling disease of cattle, Aust. Vet. J. 15:52.

Bennetts, H.W., Beck, A.B., and Harley, R., 1941, Falling disease of cattle, II. Copper deficiency, Aust. Vet. J. 24:237.

Bennetts, H.W., Harley, R., and Evans, S.T., 1942, Copper deficiency of cattle-fatal termination, Aust. Vet. J. 18:50.

Bernstein, D.S., Sadowsky, N., Hegsted, D.M., Guri, C.D., and Stare, F.J., 1966, Prevalence of osteoporosis in high and low-fluoride areas in North Dakota, J. Amer. Med. Assoc. 198:499.

Bertrand, G., and Macheboeuf, M., 1926, Influence du nickel et du cobalt sur l'action excercee par l'insulin chez le lapin, C.R. Acad. Sci. 182:1504.

Bjorksten, J.A., 1979, The primary cause of the high myocardic death rate in Eastern Finland defined as selenium deficiency, Rejuvenation 7:61.

Blackmon, D.M., Miller, W.J., and Morton, J.D., 1967, Zinc deficiency in ruminents Occurrence, effects, diagnosis, and treatments. Vet. Med. 62:265.

Bothwell, T.H., and Finch, C.A., 1962, "Iron Metabolism," Boston, Little, Brown.

Brown, H., 1927, The mineral content of human skin, J. Biol. Chem. 75:789.

von Bunge, G., 1889, Uber die anufnahme des eisens in den organismus des sauglings, Hoppe-Seylers Z. Physiol. Chem. 13:399.

Burk, D., and Yiamouyiannis, J., 1975, Fluoridation and cancer, 94th Congress, First Session, Congressional Record 121:7172.

Buttner, W., 1961, Effects of some trace elements on fluoride retension and dental caries, Arch. Oral Biol. Suppl. 6:40.

Buttner, W., 1963, Action of trace elements on the metabolism of fluoride, J. Dent. Res. 42:453.

Calhoun, N.R., and Smith, J.C., 1968, Uptake of $65_{Zn}$ in fractured bones, Lancet 2:682.

Carlton, W.W., and Henderson, W., 1963, Cardiovascular lesions in experimental copper deficiency, J. Nutr. 81:200.

Carnes, W.H., Shields, G.S., Cartwright, G.E., and Wintrobe, M.M., 1961, Vascular lesions in copper deficient swine, Fed. Proc. 20:118.

Cartwright, G.E., Hodges, R.D., Gubler, C.J., Mahoney, J.P., Daum, K., Wintrobe, M.M., and Bean, W.B., 1954, Studies on copper metabolism, XIII. Hepatolenticular degeneration, J. Clin. Invest. 33:1487.

Cass, R.M., Croft, J.D., Perkins, P., Nye, W., Waterhouse, C., and Terry, R., 1966, New bone formation in osteoporosis following treatment with sodium fluoride, Arch. intern. Med. 118:111.

Chugh, T.D., Dhingra, R.K., Gulati, R.C., and Bathla, J.C., 1973, Copper metabolism in schizophrenia, Indian J. Med. Res. 61:1147.

Clayton, C.C., and Bauman, C.A., 1949, Diet and azo dye tumors; effect of diet during a period when the dye is not fed, Cancer Res. 9:575.
Cohen, P., and Gardner, F.H., 1966, Induction of skeletal fluorosis in two common demineralizing disorders, J. Amer. Med. Assoc. 195:962.
Cohen, I.K., Schechter, P.J., and Henkin, R.I., 1973, Hypogeusia, anorexia, and altered zinc metabolism following thermal burn, J. Amer. Med. Assoc. 223:914.
Cordano, A., Baertl, J.M., and Graham, G.G., 1964, Copper deficiency in infancy, Pediatrics 34:324.
Cordano, A., Placko, R.P., and Graham, G.G., 1966, Hypocupremia and neutropenia in copper deficiency, Blood 28:280.
Coulson, W.F., and Carnes, W.H., 1963, Cardiovascular studies on copper-deficient swine, V. The histogenesis of the coronary artery lesions, Am. J. Pathol. 43:945.
Curran, G.L., 1954, Effect of certain transition group elements on hepatic synthesis of cholesterol in the rat, J. Biol. Chem. 210:765.
Curran, G.L., and Costello, R.L., 1956, Reduction of excess cholesterol in the rabbit aorta by inhibition by endogenous cholesterol synthesis, J. Exp. Med. 103:49.
Curran, G.L., Azaroff, D.L., and Bolinger, R.E., 1959, Effect of cholesterol synthesis inhibition in normocholesteremic young men, J. Clin. Invest. 38:1251.
Curran, G.L., and Burch, R.E., 1967, "Proc. First Ann. Conf. Trace Substances Environ. Health," (D.D. Hemphill, ed), Univ. of Missouri.
Danks, D.M., Stevens, B.J., Campbell, P.E., Gillespie, J.M., Walker-Smith, J., Bloomfield, J., and Turner, B., 1972a, Menkes kinky hair syndrome. Lancet 1:1100.
Danks, D.M., Campbell, P.E., Stevens, B.J., Mayne, V., and Cartwright, E., 1972b, Menkes kinky hair syndrome: An inherited defect in copper absorption with widespread effects, Pediatrics 50:188.
Danks, D.M., Cartwright, E., Stevens, B.J., and Townley, R.R.W., 1973, Menkes kinky hair disease: Further definition of the defect in copper transport, Science 179:1140.
Darby, W.J., 1951, "Handbook of Nutrition," 2nd Ed., McGraw-Hill (Blakiston), New York.
Davidson, I.W.F., and Blackwell, W.L., 1968, Changes in carbohydrate metabolism of squirrel monkeys with chromium dietary supplementation, Proc. Soc. Exper. Biol. Med. 127:66.
Davidson, I.W.F., and Burt, R.L., 1973, Physiologic changes in plasma chromium of normal and pregnant women, Am. J. Obstet. Gynaecol. 116:601.
Dean, H.T., 1942, "Fluorine and dental health," American Assoc. Advan. Sci., Washington, D.C.

DeWys, W., Pories, W.J., Richter, M.C., and Strain, W.H., 1970 Inhibition of Walker 256 carcinosarcoma growth by dietary zinc deficiency, Proc. Soc. Exp. Biol. and Med. 135:17.

Dimond, E.G., Caravaca, J., and Benchimol, A., 1963, Vanadium. Excretion, tocicity, lipid effect in man, Amer. J. Clin. Nutr. 12:49.

Doisy, E.A., 1973, "Proceeding of the 6th Annual Conference on Trace Substances in Environmental Health," (Hemphill, D.D., ed.) University of Missouri, Columbia, Mo.

Doll, R., and Kinlen, L., 1977, Fluoridation of water and cancer mortality in the U.S.A., Lancet 1:1300.

Duncan, J.R., Dreosti, I.E., and Albrecht, C.F., 1974, Zinc intake and growth of a transplanted hepatoma induced by 3-methyl-4-dimethyl-aminoazobenzene in rats, J. Nat. Cancer Inst. 53:277.

Dunlap, W.M., James, G.W., and Hume, D.M., 1974, Anemia and neutropenia caused by copper deficiency, Ann. Intern. Med. 80:470.

Earl, C.J., Mouton, M.J., and Selverstone, V., 1954, Metabolism of copper in Wilson's disease and in normal subjects: Studies with $Cu^{64}$, Am. J. Med. 17:205.

Eichhorn, G.L., 1962, Metal ions as stabilizers or destabilizers of the deoxyribonucleic acid structure, Nature (London) 194:474.

Ellis, W.C., Pfander, W.H., Muhner, M.E., and Pickett, E.E., 1958, Molybdenum as a dietary essential for lambs, J. Anim. Sci. 17: 180.

Everson, G.J., Tsai, M.C., and Wang, T., 1967, Copper deficiency in the guinea pig, J. Nutr. 93:533.

Everson, G.J., and Shrader, R.D., 1968, Abnormal glucose tolerance in manganese-deficient guinea pigs, J. Nutr. 94:89.

Ezekiel, E., 1967, Intestinal iron absorption by neonates and some factors affecting it, J. Lab. Clin. Med. 70:138.

Farquhar, J.D., 1963, Iron supplementation during first year of life, Amer. J. Dis. Child. 106:201.

Fisher, G.L., Byers, V.S., Shifrine, M., and Levin, A.S., 1976, Copper and zinc levels in serum from human patients with sarcomas, Cancer 37;356.

Follis, Jr., R.H., Day, H.G., and McCollum, E.V., 1941, Histological studies of the tissue of rats fed a diet extremely low in zinc, J. Nutr. 22:223.

Forrest, J.R., Parfitt, G.J., and Bransby, E.R., 1951, Incidence of dental caries among adults and young children in 3 high- and 3 low-fluorine areas in England, Mon. Bull. Min. Health Pub. Health Serv. 10:104.

Geyer, C.F., 1953, Vanadium, caries-inhibiting trace element in Syrian hamster, J. Dent. Res. 32:590.

Gibson, J.G., Vallee, B.L., Fluharty, R.G., and Nelson, J.E., 1950, Studies of the zinc content of the leukocytes in myelogenous leukemia, Acta Unio Intern. Contra Cancrum. 6:1102.

Glinsmann, W.H., and Mertz., W., 1966, Effects of trivalent chromium on glucose tolerance, Metabolism 15:510.

Godwin, K.O., 1965, Abnormal electrocardiograms in rats fed a low-selenium diet, Quart. J. Exp. Phys. 50:282.

Godwin, K.O., and Fraser, F.J., 1966 Abnormal electrocardiograms blood pressure changes and some aspects of the histopathology of selenium deficiency in lambs, Quart. J. Exp. Phys. 51:94.

Goldenberg, S. Blumenthal, H.T., 1964, "Diabetes Mellitus, Diagnosis and Treatment, Amer. Diabetes Assoc., New York.

Graham, G.G., and Cordano, A., 1969, Copper depletion and deficiency in malnourished infants, The Johns Hopkins Med. J. 124:139.

Griffin, A.C., and Jacobs, M.M., 1977, Effects of selenium on azo dye hepatocarcinogenesis, Cancer Lett. 3:177.

Grossmann, A., and Grossmann, G.F., 1955, Protein-bound iodine by alkaline incineration and a method for producing a stable cerate color, J. Clin. Endocrinol. Metab. 15:354.

Hallman, B.L., Bondy, P.K., and Hagewood, M.A., 1951, Determination of serum protein-bound iodine as a routine clinical procedure, Arch. Intern. Med. 87:817.

Handjani, A.M., Smith, Jr., J.C., Herrmann, J.B., and Halsted, J.A., 1974, Serum zinc concentration in acute myocardial infarction, Chest 65:185.

Harr, J.R., Exon, J.H., Whanger, P.D., and Weswig, P.H., 1972, Effect of dietary selenium on N-2 fluorenyl-acetamide (FAA)-induced cancer in vitamin E supplemented, selenium depleted rats, Clin. Toxicol. 5:187.

Harris, R., 1959, Observations on the effect of topical sodium fluoride on caries incidnece in children, Aust. Dent. J. 4:257.

Hawksley, J.C., Lightwood, R., and Bailey, U.M., 1934, Iron-deficiency anaemia in children: Its association with gastrointestinal disease, achlorhydria and hemorrhage, Arch. Dis. Child 9:359.

Healy, W.B., Ludwig, T.G., and Losee, F.L., 1961, Soils and dental caries in Hawke's Bay, New Zealand, Soil Sci. 92:359.

Hein, J.W., and Wilsotzky, J., 1955, Effect of ten ppm vanadium drinking solution on dental caries in male and female Syrian hamsters, J. Dent. Res. 34:756.

Henzel, J.H., Holtman, B., Keitzer, F.W., DeWeese, M.S., and Lichti, E., 1968, Trace elements in atherosclerosis, efficacy of zinc medication as a therapeutic modality, 2nd Annual Conf. on Trace Sub. in Environ. Health., (D.D. Hemphill, ed.), Univ. of Missouri.

Herishanu, Y., and Loewinger, E., 1972, Torsion dystony and abnormal copper metabolism, Europ. Neurol. 8:251.

Higgins, E.S., Richert, D.A., Westerfield, W.W., 1956, Molybdenum deficiency and tungstate tungstate inhibition studies, J. Nutr. 59:53.

Hoover, R.N., McKay, F.W., and Fraumen, Jr., J.F., 1976, Fluoridated drinking water and the occurrence of cancer, J. Natl. Cancer Inst. 57:757.

Hopkins, L.L., Jr., and Price, M.G., 1968, Effect of chromium supplementation on middle-age diabetics, Proc. Western Hemisphere Nutr. Congr., Puerto Rico, Vol. 11:40.

Hopkins, Jr., L.L., 1974, "Trace Element Metabolism in Animals," Vol. 2., (W.G. Hoekstra, J.W. Suttie, H.E. Ganther, and W. Mertz, eds.), University Park Press, Baltimore.

Hopkins, Jr., L.L., and Mohr, H.E., 1974, Vanadium as an essential nutrient, Fed. Proc. 33:1773.

Hrgovcic, M., Tessmer, C.F., Minckler, T.M., Mosier, B., and Taylor, G.H., 1968, Serum copper levels in lymphoma and leukemia, Cancer 21:743.

Hunt, C.E., Landesman, J., and Newberne, P.M., 1970, Copper deficiency in chicks: effects of ascorbic acid on iron, copper, cytochrome oxidase activity, and aortic mucopolysaccharides, Brit. J. Nutr. 24:607.

Hurley, L.S., and Swenerton, H., 1966, Congenital malformation resulting from zinc deficiency in rats, Proc. Soc. Exper. Biol. Med. 123:692.

Hussain, R., Walker, R.B., Layrisse, M., Clark, P., and Finch, C.A., 1965, Nutritive value of food iron, Amer. J. Clin. Nutr. 16:464.

Iber, F.L., Chalmers, T.C., and Uzman, L.L., 1957, Studies of protein metabolism in hepatolenticular degeneration, Metabolism 6:388.

Jeejeebhey, K.N., Chu, R.C., Marliss, E.B., Greenberg, G.R., and Brue-Robertson, A., 1977, Chromium deficiency, glucose intolerance, and neuropathy reversed by chromium supplementation, in a patient receiving long-term total parenteral nutrition, Amer. J. Clin. Nutr. 30:531.

Jacobs, M.M., Jansson, B., and Griffin, A.C., 1977, Inhibitory effects of selenium on 1,2-dimethylhydrazine and methylazoxymethanol acetate induction of colon tumors, Cancer Lett. 2:133.

Jones, L.P.H., and Handreck, K.A., 1967, Silica in soils, plants and animals, Advan. Agron. 19:107.

Jordan, W.A., Snyder, J.R., and Wilson, V., 1958, Stannous fluoride clinical study in Olmsted County, Minnesota, Pub. Health Rep. 73:1010.

Karpel, J.T., and Peden, V.H., 1972, Copper deficiency in long-term parenteral nutrition, J. Pediatr. 80:32.

Kelly, F.C., and Snedden, W.W., 1960, Prevalence and geographical distribution of endemic goiter, World Health Organ., Mon. Ser. 44:27.

Kelly, W.A., Kesterson, J.W., and Carlton, W.W., 1974, Myocardial lesions in the offspring of female rats fed a copper deficient diet, Exper. Molec. Path. 20:40.

Kekki, M., Koskelo, P., and Lassus, A., 1966, Serum ceruloplasmin-bound, copper-bound copper and non-ceruloplasmin copper in uncomplicated psoriasis, J. Invest. Derm. 47:159.

Keshan Disease Research Group of the Chinese Academy of Science, 1979, Chinese Medical J. In press.

Knutson, J.W., and Armstrong, D.W., 1943, Effect of topically applied sodium fluoride on dental caries experience, Pub. Health Resp. 58:1701.

Knutson, J.W., and Armstrong, D.W., 1945, Effect of topically applied sodium fluoride on dental caries experience; report of findings for second study year, Pub. Health Rep. 60:1085.

Knutson, J.W., and Armstrong, D.W., 1946, Effect of topically applied sodium fluoride on dental caries experience; report of findings for third study year, Pub. Health Rep. 61:1683.

Knutson, J.W., Armstrong, D.W., and Feldman, F.M., 1947, Effect of topically applied sodium fluoride on dental caries experience; report of findings with 2, 4 and 6 applications, Pub. Health Rep. 62:425.

Korkov, V.V., 1965, Vanadiia Vfrofilaktike i lechenii eksperimental'nogo ateroskleroza, Farmakol Toksikol 28:83.

Klevay, L.M., and Forebush, J., 1976, Copper metabolism and the epidemiology of coronary heart disease, Nutr. Rep. Int. 14:221.

Kosenko, L.G., 1964, Soderzhanie nekotorykh microelementol v krevi bol'nykh sakharnym diabetom, Klin. Med. (Moscow) 42:113.

Kratzer, F.H., Vohra, P., Albred, J.B., and Davies, P.N., 1958, Effect of zinc upon growth and incidence of perosis in turkey poults, Proc. Soc. Exp. Biol. Med. 98:205.

Kruger, B.J., 1958, Effect of trace elements on experimental dental caries, II. Al, B, Fl, I, V, J. Aust. Dent. Assoc. 3:236.

Kubota, J., Allaway, W.H., Carter, D.L., Cary, E.E., and Lazar, V.A., 1967, Selenium in crops in the United States in relation to selenium responsive diseases of animals, J. Agric. Food Chem. 15:448.

Laufberger, V., 1937, Sur la cristallisation de la ferritine, Bull. Soc. Chim. Biol. 19;1575.

Lees, F., and Rosenthal, F.D., 1958, Gastric mucosal lesions before and after treatment in iron deficiency anaemia, Quart. J. Med. 27:19.

Leone, N.C., Stevenson, C.A., Beese, B., Hawes, L.E., and Dawber, T.A., 1960, The effects of the absorption of fluoride, II. A radiological investigation of five hundred and forth-six human residents of an area in which the drinking water contained only a minute trace of fluoride, Amer. Med. Assoc. Arch. Ind. Health 21:326.

Leone, N.C., Shimkin, M.B., Arnold, E.A., Stevenson, C.A., Zimmerman, E.R., Geiser, B.G., and Lieberman, J.J., 1964, In fluoridation as a Public Health Measure, Amer. Assoc. Advan. Sci., Washington, D.C.

Leslie, J.G., Kung-Ying, K., and McGavack, T.H., 1962, Silicon in biological material, II. Variations in silicon contents in tissues of rat at different ages, Proc. Soc. Exp. Biol. Med. 110:218.

Levine, R.A., Streeten, D.H.P., and Doisy, R.J., 1968, Effects of oral chromium supplementation on the glucose tolerance of elderly human subjects, Metabolism 17:114.

Lewis, C.E., 1959, The biological effects of vanadium, II. The signs and symptoms of occupational vanadium exposure, Amer. Med. Assoc. Arch. Int. Health 19:419.

Lipkin, G., Herrmann, F., and Mandol, L., 1962, Studies on serum copper, I. The copper content of blood serum in patients in psoriasis, J. Invest. Derm. 39:593.

Loeper, J., Loeper, J., and Lemaire, A., 1966, Etude du silicum en biologie animals et au cours de l'atherome, Presse Med. 74:865.

Ludwig, T.L., Healy, W.B., and Losee, F.L., 1960, An association between dental caries and certain soil conditions in New Zealand, Nature (London) 186:695.

MacCardle, R.C., Engman, Jr., M.F., and Engman, Sr., M.F., 1943, Mineral changes in neurodermatitis revealed by microincineration, Arch. Dermatol. Syphilol. 47:335.

Malamos, B., Koutras, D.A., Kostamis, P., Kralios, A.C., Rigopoulos, G., and Zerefos, N., 1966, Endemic goiter in Greece: Epidemiologic and genetic studies, J. Clin. Endocrinol. Metab. 26:688.

Malthus, R.S., Ludwig, T.G., and Healy, W.B., 1964, Effects of trace elements on dental caries in rats, N.Z. Dent. J. 60:291.

Marshall, M.V., Jacobs, M.M. and Griffin, A.C., 1978, Reduction in acetylaminoflurene (AAF) hepatocarcinogenesis by selenium Proc. Amer. Cancer Res. 19:75.

McConnell, K.P, Broghamer, Jr., W.L., Blotcky, A.J., and Hurt, O.J., 1975, Selenium levels in human blood and tissues in health and in disease, J. Nutr. 105:1026.

McKay, F.S., 1948, Mass control of dental caries through use of domestic water supplies containing fluorine, Amer. J. Pub. Health 38:328.

McNeely, M.D., Sunderman, Jr., F.W., Nechay, M.W., and Levine, H., 1971, Abnormal concentrations of nickel in serum in cases of myocardial infarction, stroke, burns, hepatic cirrhosis and uremia, Clin. Chem. 17:1123.

McQuilty, Jr., J.T., DeWys, W.D., Monaco, L., Strain, W.H., Rob, C.G., Apgar, J., and Pories, W.J., 1970, Inhibition of tumor growth by dietary zinc deficiency, Cancer Res. 30:1387.

Meeks, M.J., Lundelt, R.R., Kessler, W.V., and Born, G.S., 1971, Effect of vanadium on metabolism of glucose in the rat, J. Pharm. Sci. 60:482.

Menkes, J.H., 1972, Kinky hair disease, Pediatrics 50:181.

Mertz, W., Roginski, E.E., and Schwarz, K., 1961, Effect of trivalent chromium complexes on glucose uptake by epididymal fat tissue of rats, J. Biol. Chem. 236:318.

Mertz, W., Roginski, E.E., and Reba, R., 1965, Biological activity and fate of trace quantities of intravenous chromium (III) in the rat, Am. J. Physiol. 209:489.

Michaelsson, G., 1974, Zinc therapy in acrodermatitis enteropathica, Acta Derm. Venerol. (Stockh) 54:377.

Michaelsson, G., Juhlin, L., Vahlquist, A., 1977, Effects of oral zinc and vitamin A in acne, Arch, of Dermatology 113:31.

Millar, M.J., Fi scher, M.I., Elcoate, P.V., and Mawson, C.A., 1958, The effects of dietary zinc deficiency on the reproductive system of male rats, Canad. J. Biochem. Physiol. 36:557.

Millar, M.J., Fischer, M.I., Elcoate, P.V., and Mawson, C.A., 1960, Effect of testosterone and gonadotropin injections on the sex organ development of zinc-deficient male rats, Can. J. Biochem. Physiol. 38:1457.

Miller, W.J., Pitts, W.J., Clifton, C.M., and Schmittle, S.C., 1964, Zinc deficiency in the goat, J. Dairy Sci. 47:556.

Miller, W.J., Morton, J.D., Pitts, W.J., and Clifton, C.M., 1965, Effect of zinc deficiency and restricted feeding on wound healing in the bovine, Proc. Soc. Exp. Biol. Med. 118:427.

Miller, W.J., Blackmon, D.M., Gentry, R.P., Powell, G.W., and Perkins H.E., 1966, Influence of zinc deficiency on zinc and dry malter content of ruminant tissues and on excretion of zinc, J. Dairy Sci. 49:1446.

Mills, C.F., Dalgarno, A.C., Williams, R.B., and Quarterman, J., 1967, Zinc deficiency and the zinc requirement of calves and lambs. Brit. J. Nutr. 21:751.

Molokhia, M.M., and Portnoy, B., 1970, Neutron activation analysis of trace elements in skin, V. Copper and zinc in psoriasis, Brit. J. Derm. 83:376.

Morgan, J.m., 1972, Hepatic chromium content in diabetic subjects, Metabolism 21:313.

Mountain, J.T., Stockwell, F.R., and Stokinger, H.E., 1956, Effect of ingested vanadium on cholesterol and phospholipid metabolism in the rabbit, Proc. Soc. Exp. Biol. 92:582.

Moynahan, E.J., 1974, Acrodermatitis enteropathica: a lethal inherited human zinc deficiency disorder, Lancet 2:399.

Muhlemann, H.R., and Konig, K.G., 1964, Die fressgewohneiten der ratten im zusammenhang mit der kariesforschung, Int. Z. Vitamin Forsch 34:426.

Muhler, J.C., 1957, The effect of vanadium pentoxide, fluorides, and tin compounds on the dental caries experience in rats, J. Dent. Res. 36:787.

Mukai, F.H., and Goldstein, B.D., 1976, Mutagenicity of malonaldehyde, a decomposition product of peroxidized polyunsaturated fatty acids, Science 191:868.

Mukherjee, C., and Mukherjee, S.K., 1953, Studies in iron metabolism in anaemias in pregnancy; serum iron, J. Indian Med. Assoc. 22: 345.

Murray, J., and Rosenthal, S., 1968, The effects of locally applied zinc and aluminum on healing incised wounds, Surg. Gynecol. Obstet. 126:1298.

Naylor, M.N., 1969, Recent advances in the prevention of caries, Ann. Roy. Coll. Surg. Eng 45:305.

Newman, Howard, A.I., Leighton, R.F., Lanese, R.R., and Freedland, N.A., 1978, Serum chromium and angiographically determined coronary artery disease, Clin. Chem. 24:541.

Niell, H.B., Leach, B.E., and Kraus, A.P., 1979, Zinc metabolism in sickle cell anemia, J. Amer. Med. Assoc. 242:2686.

O'Dell, B.L., and Savage, J.E., 1957, Symptoms of zinc deficiency in the chick, Fed. Proc. 16:394.

O'Dell, B.L., Newberne, P.M., and Savage, J.E., 1958, Significance of dietary zinc for the growing chicken, J. Nutr. 65:503.

O'Dell, B.L., Hardwick, B.C., Reynolds, G., and Savage, J.E., 1961, Connective tissue defect in the chick resulting from copper deficiency, Proc. Soc. Exp. Biol. Med. 108:402.

Orten, J.M., 1966, :Zinc Metabolism," (Prasad, A.S., ed.), Thomas, Springfield, Illinois.

Perry, J. Jr., H.H., 1973, Minerals in cardiovascular disease, J. Am. Diet. Assoc. 62:631.

Perry, Jr., H.M., Perry, E.F., and Erlanger, M.W., 1974, Reversal of cadmium induced hypertension by selenium or hard water, in "Proc. Eighth Conf. Trace Sub. Environ. Health," (D.D. Hemphill, ed.), University of Missouri.

Pitts, W.J., Miller, W.J., Fosgate, O.T., Morton, J.D., and Clifton C.M., 1966, Effect of zinc deficiency and restricted feeding from two to five months of age on reproduction in Holstein bulls, J. Dairy Sci. 49:995.

Plumlee, M.P., Thrasher, D.M., Beeson, W.N., Andrews, F.N., and Parker, H.E., 1954, Effects of manganese deficiency on growth development and reproduction of swine, J. Anim. Sci. 13:996.

Plumlee, M.P., Thrasher, D.M., Beeson, W.N., Andrews, F.N., and Parker, H.E., 1956, The effects of a manganese deficiency upon the growth, development and reproduction of swine, J. Anim. Sci. 15:352.

Pories, W.J., and Strain, W.H., 1966, :Zinc Metabolism," (A.S. Prasad, ed.), Thomas, Springfield, Illinois.

Pories, W.J., Henzel, J.H., Rob, C.G., and Strain, W.H., 1967, Acceleration of healing with zinc sulfate, Ann. Surg. 165:432.

Pories, W.J., Henzel, J.H., and Hennessen, J.A., 1968, In Proc. First Ann. Conf. Trace Substances in Environ. Health, (D.D. Hemphill, ed.), University of Missouri.

Poswillo, D.E., and Cohen, B., 1971, Inhibition of carcinogenesis by dietary zinc, Nature 231:447.

Prasad, A.S., Halsted, J.A., and Nadimi, M., 1961, Syndrome of iron deficiency anemia, hepatosplenomegaly, hypogonadism, dwarfism, and geophagia, Am. J. Med. 31:532.

Prasad, A.S., Miale, A., Farid, Z., Sandstead, H.H., Schulert, A. and Darby, W.J., 1963, Biochemical studies on dwarfism, hypogonadism, and anemia, Arch. Intern. Med. 111:407.

Pritchard, J.A., and Hunt, C.E., 1958, A comparison of the hematologic responses following the routine prenatal administration of intramuscular and oral iron, Surg. Gynecol. Obstet. 106:516.

Purves, M.J, 1962, Some effects of administering sodium fluoride to patients with Paget's disease, Lancet 2:1188.

Quarantillo, Jr., E.P., 1971, Effect of supplemental zinc on wound healing in rats, Amer. J. Surg. 121:661.

Rahmet, A., Norman, J.N., and Smith, G., 1974, The effect of zinc deficiency on wound healing, Brit. J. Surg. 61:271.

Rapport, R.L., and Curtis, G.M., 1950, The clinical significance of the blood iodine, Endocrinol. 10:735.

Reid, B.L., Kurnich, A.A., Svacha, R.L., and Crouch, J.R., 1956, The effect of molybdenum on chick and poult growth, Proc. Soc. Exp. Biol. Med. 93:245.

deRenzo, E.C., Kaleita, E., Heytler, P., Oleson, J.J., Hutching, B.L., and Williams, J.H., 1953a, The nature of the zanthine oxidase factor, J. Amer. Chem. Soc 75:753.

deRenzo, E.C., Kaleita, E., Heytler, P., Oleson, J.J., Hutching, B.L., and Williams, J.H., 1953b, Indentification of xanthine oxidase factor as molybdenum, Arch. biochem. Biophys. 45:247.

Rich, C., Ensinck, J., and Ivanocitch, P., 1964, The effects of sodium fluoride on clacium metabolism of subject with metabolic bone disease, J. Clin. Invest. 43:545.

Richert, D.A., and Westerfield, W.W., 1953, The relationship of iron to xanthine oxidase, J. Biol. Chem. 203:915.

Roginski, E.E., and Mertz, W., 1967a, An eye lesion in rats fed low-chromium diets, J. Nutr. 93:249.

Roginski, E.E., and Mertz, W., 1967b, Dietary chromium and amino acid incorporation in rats on a low-protein ration, Fed. Proc. 26:255.

Roginski, E.E., and Mertz, W., 1969, Effects of chromium (III) supplementation on glucose and amino acid metabolism in rats fed a low protein diet, J. Nutr. 97:525.

Rubenstein, A.H., 1962, Hypoglycaemia induced by manganese, Nature 194:188.

Russell, A.L., and Elvore E., 1951, Domestic water and dental caries; study of fluoride-dental caries relationship in adult population, Pub. Health Rep. 66:1389.

Rybo, G., 1966, Menstrual blood loss in relation to parity and menstrual pattern, Acta Obstet. Gynecol. Scand. 45 (suppl.), 7:25.

Rygh, O., 1949a, Recherches sur les oligo-elements; de l'importance du strontium, du baryum et du zinc, Bull, Soc. Chim. Biol. 31: 1052.

Rygh, O., 1949b, Recherches sur les oligo-elements; de l'importance CRL du thallium et du vanadium, du silicon et du fluor, Bull. Soc. Chim. Biol. 31:1403.

Rygh, O., 1949c, Recherches sur les oligo-elements; sur l'importance du strontium, du vanadium, du baryum, du thallium et du zinc dans le scorbut, Bull. Soc. Chim. Biol. 31:1408.

Rygh, O., 1951, Sur les reactions des acides amines soufres, specialement en correlation avec le scorbut,et l'influence des deux facteurs antiscorbutiques naturels, Bull. Soc. Chim. Biol. 33: 133.

Sandstead, H.H., and Shepard, G.H., 1968, The effect of zinc deficiency on the tensile strength of healing surgical incisions in the integument of the rat, Proc. Soc. Exper. Biol. Med. 128: 687.

Savlov, E.D., Strain, W.H., and Huegin, F., 1962, Radiozinc studies in experimental wound healing, Jour. Surg. Research 2:209.

Schechter, P.J., Friedewald, W.T., Bronzert, D.A., Raff, M.S., and Henkin, R.I., 1972, Ideopathic hypogeusia: A description of the syndrome and a single blind study with zinc sulfate, International Review of Neurobiology, Suppl. 1:125 (Ed. by Pfeiffer, C.C.) New York, Academic Press.

Scheinberg, I.H., and Gittlin, D., 1952, Deficiency of ceruloplasmin in patients with hepatolenticular degeneration (Wilson's disease), Science 116:484.

Shrader, K.E., and Everson, G.J., 1968, Pancreatic pathology in manganese-deficient guinea pigs, J. Nutr. 94:269.

Schrauzer, G.N., and Ishmael, D., 1974, Effects of selenium ane arsenic on the genesis of spontaneous mammary tumors in in-bred C3H mice, Ann. Clin. Lab. Sci. 4:441.

Schrauzer, G.N., White, D.A., and Schneider, C.J., 1977, Cancer mortality correlation studies - III, Statistical association wtih dietary selenium intakes, Bioinorganic Chem. 7:23.

Schroeder, H.A., Vinton, W.H., Jr., and Balassa, J.J., 1962, Abnormal trace metals in man, J. Chronic Dis. 15:941.

Schroeder, H.A., Balassa, J.J., and Tipton, I.H., 1963, Abnormal trace metals in man-vanadium, J. Chronic Dis. 16:407.

Schroeder, H.A., and Balassa, J.J., 1965, Influence of chromium cadmium, and lead on rat aortic lipids and circulating cholesterol, Am. J. Physiol. 209:433.

Schroeder, H.A., 1968, The role of chromium in mammalian nutrition, Am. J. Clin. Nutr. 21:230.

Schroeder, H., 1969, Serum cholesterol and glucose levels in rats fed refined and less refined sugars and chromium, J. Nutr. 97: 237.

Schwarz, K., and Mertz, W., 1961, A physiological role of chromium (III) in glucose utilization (glucose tolerance factor), Fedetation Proc. 20 (Suppl. 2):III.

Schwarz, K., Milne, D.B., and Vineyard, E., 1970, Growth effects of tin compounds in rats maintained in a trace element controlled environment, Biochem. Biophys. Res. Commun. 40:22.

Schwarz, K., 1974, "Trace Element Metabolism in Animals," Vol. 2., W.G. Hoekstra, J.W. Suttie, H.E. Ganther, and W. Mertz, eds.), University Park Press, Baltimore.

Scott, M.L., Holm, E.R., and Reynolds, R.E., 1959, Studies on the niacin, riboflavine, choline, manganese and zinc requirements of young ringnecked pheasants for growth, feathering, and prevention of leg disorders, Poultry Sci. 38:1344.

Scrimshaw, N.S., 1960, Endemic goiter in Latin American, Pub. Health Rep. 75:731.

Shamberger, R.J., and Rudolph, G., 1966, Protection against cocarcinogenesis by sntioxidants, Experientia 22:116.

Shamberger, R.J., 1970, Relationship of selenium to cancer, I. Inhibitory effect of selenium on carcinogenesis, J. Nat. Cancer Inst. 44:931.

Shamberger, R.J., and Willis, C.E., 1971, Selenium distribution and human cancer mortality, C.R.C. Crit. Rev. Clin. Lab. Sci. 2:211.

Shamberger, R.J., Baughman, F.F., Kalchert, S.L. Willis, C.E., and Hoffman, G.C., 1973a, Carcinogen-induced chromosomal breakage decreased by antioxidants, Proc. Nat. Acad. Sci. (U.S.A.) 70:1461.

Shamberger, R.J., Rukovena, E., Longfield, A.K., Tytko, S.A., Deodhar, S., and Willis, C.E., 1973b, Antioxidants and Cancer. I. Selenium in the blood of normals and cancer patients, J. Nat. Cancer Inst. 50:863.

Shamberger, R.J., Tytko, S., and Willis, C.E., 1974a, Antioxidants and cancer, II. Selenium distribution and human cancer mortality in the United States, Canada and New Zealand, Proc. Seventh Ann. Conf. Trace Sub. in Environ. Health, (D.D. Hemphill, ed.), University of Missouri.

Shamberger, R.J., Andreone, T.L., and Willis, C.E., 1974b, Antioxidants and cancer, IV. Initiating activity of malonaldehyde as a carcinogen, J. Nat. Cancer Inst. 53:1771.

Shamberger, R.J., Tytko, S.A., and Willis, C.E., 1975, Selenium and heart disease, "The Proc. Ninth Trace Sub. in Environ. Health.," (D.D. Hemphill, ed.), University of Missouri.

Shamberger, R.J., Tytko, S.A., and Willis, C.E., 1976, Antioxidants and cancer, VI. Selenium and age-adjusted human cancer mortality, Arch. Environ. Health 31:235.

Shamberger, R.J., Shamberger, B.A., and Willis, C.E., 1977a, Malonaldehyde content of food, J. Nutr. 107:1404.

Shamberger, R.J., 1977b, Selenium and cadmium ratios of human kidneys, 3rd Int. Symp. of Trace Element metabolism in Man and Animals, (Kirchgessner, M., ed.), Freising-Weihenstephan, West Germany.

Shamberger, R.J., Gunsch, M.S. Willis, C.E., and McCormack, L.J., 1978, Selenium and heart disease, II. In the Proc. 12th Trace Sub. in Environ. Health, (D.D. Hemphill, ed.), University of Missouri.

Shamberger, R.J., Corlett, C.L., Beaman, K.D., and Kasten, B.L., 1979a, Antioxidants reduce the mutagenic effect of malonaldehyde and B-propiolactone, Mutation Res. 66:349.

Shamberger, R.J., Willis, C.E., and McCormack, L.J., 1979b, Selenium and heart disease, III. Blood selenium and heart mortality in 19 states, "Proc. 13th Conf. Trace Sub. in Environ. Health (D.D. Hemphill, ed.), University of Missouri, In press.

Sherman, L., Glennon, J. A. Brech, W.J., Klomberg, G.H., and Gordon E.S., 1968, Failure of trivalent chromium to improve hyperglycemia in diabetes mellitus, Metab. Clin. Exptl. 17:439.

Shields, G.S., Coulson, W.F., Kimball, D.A., Carnes, W.H., Cartwright, G.E., and Wintrobe, M.M., 1962, Studies on copper metabolism, 32, Cardiovascular lesions in copper-deficient swine, Amer. J. Path. 41:603.

Shokeir, M.H.F., and Shreffler, D.C., 1969, Cytochrome oxidase deficiency in Wilson's disease: a suggested ceruloplasmin function, Proc. Nat. Acad. Sci. (USA) 62:867.

Simpson, C.F., and Harms, R.H., 1964, Pathology of the aorta of chicks fed a copper-deficient diet, Exp. Mol. Pathol. 3:390.

Singh, S., and Bresnan, M.J. 1973, Menkes kinky-hair syndrome (trichopoliodystrophy), Low copper levels in the blood, hair and urine, Amer. J. Dis. Child 125:572.

Smith, Jr., J.C., McDaniel, E.G., and Fan, F.F., 1973, A trace element essential in vitamin A metabolism, Science 181:954.

Somerville, T., and Davies, B., 1962, Effect of vanadium on serum cholesterol, Amer. Heart J. 64:54.

Soremark, R., and Ullberg, S., 1962, In Use of Radioisotopes in Animal Biology and the Medical Sciences, (Fried, N. ed.), Academic, New York.

Staub, H.W., Reussner, G., and Thiessen, Jr., R., 1969, Serum cholesterol reduction by chromium in hypercholesterolemic rats, Science 166:746.

Sturgeon, P., 1956, Studies of iron requirements in infants and children, II. The influence on normal infants of oral iron in therapeutic doses, Pediatrics 17:341.

Sunderman, Jr., F.W., Nomoto, S., Pradhan, A.M., Levine, H., Bernstein, S.H., and Hirsch, R., 1970, Increased concentrations of serum nickel after acute myocardial infarction, N. Eng. J. Med. 283-896.

Sunderman, Jr., F.W., Decsy, M.I., and McNeely, M.D., 1972, Nickel metabilism in health and disease, Ann. N.Y. Acad. Sci. 199:300.

Sunderman, Jr., F.W., 1973, Atomic absorption spectrometry of trace metals in clinical pathology, Human Pathol. 4:549.

Szmigielski, S., and Litwin, J., 1964, The histochemical demonstration of zinc in blood granulocytes, The new test in diagnosis of neoplastic diseases, Cancer 17:1381.

Tank, G., and Storrick, C.A., 1960, Dental caries experiences of school children in Corvallis, Oregon, after 7 years of fluoridation, J. Pediat. 58:528.

Taves, D.R., 1977, Fluoridation and Cancer Mortality, Origins of Human Cancer (Cold Spring Harbor Conferences on Cell Proliferation, Vol. 4), (H.H. Hiatt, J.D. Watson, J.A. Winsten, eds.), New York, Cold Spring Harbor Laboratory.

Thomassen, P.R., and Leicester, H.M., 1964, Uptake of radioactive beryllium, vanadium, selenium, cerium and yttrium in the tissues and teeth of rats, J. Dent. Res. 43:346.

Todd, W.R., Elvehjem, C.A., and Hart, E.B., 1934, Zinc in the nutrition of the rat, Amer. J. Physiol. 107:146.

Underwood, E.J., and Somers, M., 1967, Studies of zinc nutrition in sheep, I. The relation of zinc to growth, testicular development and spermattogenesis in young rams, Aust. J. Agr. Res. 20:889.

Uzman, L.L., Iber, F.L., and Chalmers, T.C., and Knowlton, M., 1956, The mechanism of copper deposition in the liver hepatolenticular degeneration (Wilson's disease), Am. J. Med. Sci. 231:511.

Vikbladh, I., 1950, Studies on zinc in blood, Scand. J. Clin. Invest. 2:143.

Villalon, J.A.M., 1974, M.D. Thesis, Universidad Michocana de San Nicolas de Hidalgo, Morelia, Mexico, 23 pp., Cited by Frost, D.V. and Lish, P.M., 1975, Selenium in biology, Ann. Rev. Pharm. 15:259.

Volkov, N.F., 1963, Cobalt, manganese, and zinc content in the blood of atherosclerosis patient, Fed. Proc. (Trans Suppl.) 22:T897.

Voors, A.W., Johnson, W.D., Shuman, M.S., Blotcky, A.J., 1978, Adjusted cadmium levels in the renal cortex and human heart at autopsy, in "Proc. 13th Conf. Trace Sub. Environ. Health," (D.D. Hemphill, ed.), University of Missouri.

Wacker, W.E., Ulmer, D.D., and Vallee, B.L., 1956, Metalloenzymes and metalloenzymes and myocardial infarction, II. Malic and lactic activities and zinc concentration in serum dehydrogenase, New Eng. J. Med. 255:449.

Wacker, W.E.C., and Vallee, B.L., 1959, Nucleic acids and metals I. Chromium, manganese, nickel iron, and toher metals in ribonucleic acid from diverse biological sources, J. Biol. Chem. 234:3257.

Wacker, W.E.C., Gordon, M.P., and Huff, J.W., 1963, Metal content of tobacco mosaic virus and tobacco mosacic virus RNA, Biochemistry 2:716.

Warren, R.L., Jelliffe, A.M., Watson, J.V., and Hobbs, C.B., 1969, Prolonged observations on variations in the serum copper in Hodgkins disease, Clin. Radiol. 20:247.

Wayne, E.J., Koutras, D.A., and Alexander, W.D., 1964, "Clinical Aspects of Iodine Metabolism," Blackwell, Oxford.

Westerfield, W.W., and Richert, D.A., 1953, Distribution of the xanthine oxidase factor (molybdenum) in foods, J. Nutr. 51:85.

Wetterdal, B., 1958, Experimental studies on radioactive zinc in the male reproductive organs of the rat, Acta Radiol., Suppl., 156:5.

Willimas, J., Thompson, E., and Smith, K.L., 1978, Value of serum copper levels and erythrocyte sedimentation rates as indicators of disease activity in children with Hodgkins disease, Cancer 42:1929.

Wilson, S.A.K., 1912, Progressive lenticular degeneration: a familial nervous disease associated with cirrhosis of the liver Brain 34:295.

Wintrobe, M.M., Cartwright, G.E., and Gubler, C.J., 1953, Studies on the function and metabolism of copper, J. Nutr. 50:395.

Yiamouyiannis, J., and Burk, D., 1975, Cancer from our drinking water?, 94th Congress, First session, Congressional Record 121:12731.

Yiamouyiannis, J., and Burk, D., 1977, Flouridation and cancer: Age-dependence of cancer mortality related to artificial fluoridation, Fluoride 10:102.

Zimmerman, A.W., Matthieu, J.M., Quarles, R.H., Brady, R.O., and Hsu, J.M., 1976, Hypomyelination in copper-deficient rats, Archives Neur. 33:111.

# ROLE OF SPECIFIC NUTRITIONAL COMPONENTS ON PLASMA LIPIDS, LIPOPROTEINS AND CORONARY HEART DISEASE

Herbert K. Naito, Ph.D.

Head, Lipid-Lipoprotein Laboratories, Division of Laboratory Medicine and Division of Research, The Cleveland Clinic Foundation and Clinical Professor Department of Chemistry, Cleveland State University

## INTRODUCTION

Heart and blood vessel diseases are the major causes of death in this country. They accounted for over 50 percent of all deaths annually; nearly three times the death rate from cancer, the next highest cause. Cardiovascular diseases cause over 60% of all deaths among people over 65 years of age and over 150,000 deaths per year in individuals below age 65. An estimated 30 million persons in the United States have diseases of the heart and blood vessels; this results in a huge burden of acute and chronic illness and disability. Some 27 million of these victims of cardiovascular disorders suffer from hypertension, 4 million from coronary heart disease, and 1.8 million from rheumatic heart disease. Approximately eight out of every one thousand children are born with congenital heart disease and half of these do not survive past one year of age.

However, during the last 25 years there has been over a 30% decrease (Figure 1) in cardiovascular death rate (age adjusted).

The number of deaths per year from major cardiovascular disease dropped below one million for the first time since 1967 in 1977, although the population in the United States has increased and is comprised of a larger proportion of senior citizens. The reasons for this are not clear, but are more than likely due to a combination of factors: (1) improved and earlier detection, treatment and rehabilitation of patients with cardiovascular disease and its related risk factors, (2) better emergency care of victims of myocardial infarction attacks, (3) a larger segment of the mass population being more aware and concerned about

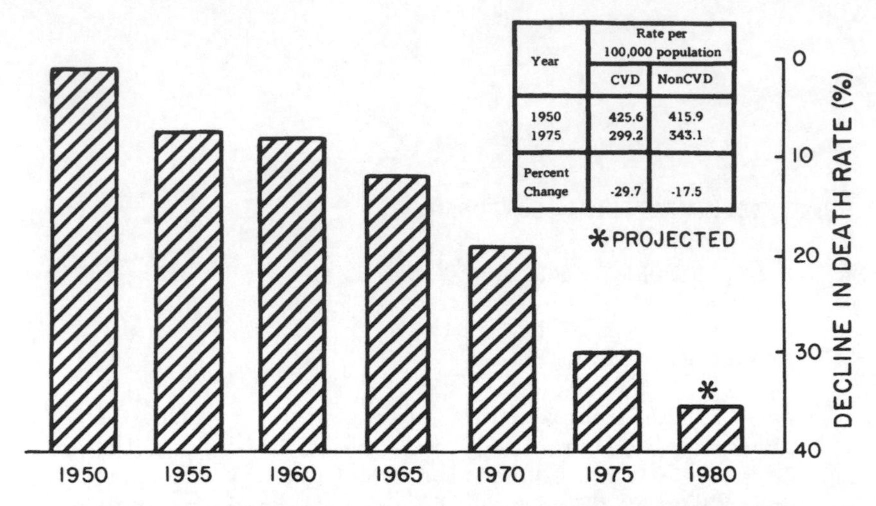


Fig. 1. Cardiovascular death rate (age adjusted over a 20 year period (adapted from the Fourth Report of the Director, NHLBI, 1977).

physical fitness, health, nutrition and diets and (4) more effective education of the mass population about controlling primary and secondary risk factors associated with coronary heart disease (CHD). Despite such progress, major cardiovascular disease still leads all other causes of death in the United States, with more than a half a million persons having been predicted to die of coronary heart disease in 1980.

This chapter will discuss the role of specific dietary components on plasma lipid and lipoprotein concentrations. The major plasma lipids of interest are total cholesterol (free cholesterol + ester cholesterol) and the triglycerides. Rarely is measurement of plasma phospholipids helpful, except for instances of obstructive liver disease. When one or more of these major classes of plasma lipids is elevated, a condition exists that is referred to as hyperlipidemia. A problem arises in defining what levels of plasma lipids separate persons with elevated blood fats from the rest of the "normal" population. Cut-off valves for plasma lipids are assured to represent the $95^{TH}$ percentile of their distribution in a given population. Thus, before one can address the question of what is adequate dietary or drug therapy for the control of serum lipid and lipoprotein levels, one needs to first consider the degree to which blood lipids should be lowered. In other words, at what level should the clinician consider a sample "hyperlipidemic?" Many clinical laboratories and practicing physicians still use normal ranges in evaluating whether or not a sample can be considered normal or abnormal. These normograms are based on sampling of "apparently normal" individuals and arbitrarily defining hyperlipidemia as being present when the plasma cholesterol and/or triglycerides are above the $95^{TH}$ percentile value for the population to which the persons belong. Unfortunately, because of the way we have defined "normal" in the past, many abnormal test results do not correlate very well with disease states or health-risk conditions on an individual basis. Thus, the "normal" range may not be a "healthy" level. While a cholesterol value of 250-280 mg/dl may be within the $95^{TH}$ percentile of the distribution of an apparently normal male population between the ages of 51-59 in the U. S., about 40-50% of these individuals eventually develop CHD. One of the most frequent and perhaps most frustrating questions confronting the clinician regarding laboratory tests is "What is normal?" In recent years the term "normal range" has been replaced by "reference range." "Upper limit of normal" is being replaced by "referent value," "critical value" or "cut-off point." There are at least five major considerations that the reference values depend upon: (1) the reference population and the way that it was chosen, (2) the environmental and physiological conditions under which the specimens were obtained, (3) the techniques and timing of specimen collection, transportation, preparation and storage, (4) the analytical method that was used, with data regarding its accuracy, precision, and quality control, and (5) the data set that was observed and the reference intervals that were derived (Galen, 1977). There also is the consideration for sub-selection based on ethnic background, geographical location, occupation, body weight, sex, age and other factors such as smoking, alcohol intake and dietary habits. The reference intervals, 2.5 to $97.5^{TH}$ percentiles, can be compared to the Gaussian estimation of 95% of the population, which is included within ± 2 S.D. from the mean. The 2.5 and $97.5^{TH}$ and 0.5 and $99.5^{TH}$ percentiles are frequently used to define the laboratory reference intervals for a chemistry test. The clinician dealing with a single patient, however, does not really care about standard deviations or percentiles. He is most interested in knowing whether the test result predicts a particular disease state or implies risk of morbidity or mortality. It is here that the predictive value approach may prove to be of utmost significance. Referent values for serum lipids and lipoproteins should be established that are highly predictive of disease or disease risk, irrespective of the "normal distribution."

There are four components that are important in determining the accuracy of a laboratory diagnostic test: sensitivity, specificity, predictive value, and efficiency. Sensitivity indicates the frequency of positive test results in patients with a particular disease, whereas specificity indicates the frequency of negative test results in patients without that specific disease. The predictive value of a positive test result indicates the frequency of diseased patients in all patients with positive test results. The predictive value of a negative test result indicates the frequency of nondiseased patients in all patients with negative test results. The efficiency of a test indicates the percent of patients correctly classified by the test. In screening for disease, we are most interested in predictive value of the positive result. In CHD, this means establishing predictive values for total cholesterol, triglycerides, LDL, LDL-cholesterol, HDL, HDL-cholesterol, etc. Epidemiological studies (Kannel et al., 1971; Glueck et al., 1976) show that CHD risk is linear down to a cholesterol concentration of 180 mg/dl. While this philosophical approach to patient care is pragmatic, it is an idealistic approach to use blood lipids as "predictors" or diagnostic aide of CHD on an individual basis. This is because coronary atherosclerosis is a complex disease which is multifaceted in origin. Although intensive research efforts have begun to elucidate various facets of the etiology, development and prognosis of the pathogenesis of atherosclerosis, we are still not sure how the initiating and accelerating factors associated with this degenerative disease process interact and manifest themselves. One reason is that the disease process itself is complex, probably developing through a multitude of processes. Large scale epidemiological studies from this country and others have clearly shown that certain predisposing factors are associated with CHD. These factors may not be causes of CHD, yet because of their close association, demonstrated by statistical means with cardiovascular disease, are called "risk factors," which have been categorized as primary and secondary. (Table I)

Table 1. Coronary Heart Disease Risk Factors

| Primary | Secondary |
|---|---|
| 1. Genetic predisposition for CHD | 1. Overweight |
| 2. High blood lipids | 2. No exercise |
| 3. Hypertension | 3. Chronic stress |
| 4. Smoking | 4. Imbalanced diet |
| | 5. Diabetes mellitus |
| | 6. Hypothyroidism |
| | 7. Renal disease |
| | 8. Birth control pill |

One can further define the risk factors into high, above average and normal risk as shown in Table 2.

Table 2. Degree of CHD Risk Based on Risk Factors

| Risk Factors | Risk for CHD | | |
|---|---|---|---|
| | Normal | Above Average | High |
| Total chol. (mg/dl) | <220 | 220-240 | >240 |
| Blood Press. (mm Hg) | | | |
| Diastolic | 80 | 90 | 105 |
| Systolic | 120 | 140 | 160 |
| Smoking (pks/day) | 0 | 1 | 2 |
| Fam. History for CHD | Normal | Mod. high | High |
| Body weight | Normal | 10% above normal | 20% above normal |
| Exercise | Regularly | Infreqently | Never |
| Chronic stress | Minimal | Mod. exposure | High exposure |
| Diet | Low chol. | Mod. chol. | High chol. |
| | Low sat. Fat | Mod. sat. fat | High sat. fat |
| | High fiber | Mod. fiber | Low fiber |

While this concept for categorizing risk for CHD into high, above average and normal risk is convenient, it is a simplistic approach to assessing a patient's risk for CHD. For example, it is well established that elevated serum cholesterol concentration is associated with increased risk of atherogenesis and CHD (Nichols et al., 1977, Glueck and Connor, 1977; Davis and Havlik, 1977 and Grundy, 1977). Persons with disorders characterized by elevated levels of serum cholesterol or low density lipoprotein (LDL) have a strikingly high incidence of CHD, which often is evident at a young age (Stone and Levy, 1972). If the familial β-hyperlipoproteinemic condition is of a homozygote origin, fatal complications of premature atherogenesis occur before age 30

(Fredrickson et al., 1967; Fredrickson and Levy, 1972). From epidemiological studies and basic and clinical research, there appears to be little doubt that hypercholesterolemia exists as an independent risk factor for coronary artery disease (CAD) (Medalie et al., 1973; Wilhelmsen et al., 1973; Gordon et al., 1974; Johnson et al., 1968). Data from Mainland China show that the mean serum cholesterol level in normal people is 136 mg/dl, while in CHD patients it is 190 mg/dl (Van Die Redaskie, 1973). Thus, in a country with a low rate of CHD, there also appears to be a positive relationship between CHD and cholesterol levels. The relationship is continuous and, in fact, is exponential as the level of serum cholesterol increases. There are no cut-off points to speak of. Therefore, in assessing a person's risk for CHD using risk factors, they should be considered as a continuum.

Another concept that should be remembered when using CHD risk factors is that statistically valid data obtained on large populations and in prospective epidemiological studies are not necessarily directly applicable to an individual. As discussed by Blackburn (1974), there is a large variation among individuals in these predisposing factors and within a single individual during his life span. Factors acting throughout the entire period of the developing degenerative disease are difficult, if not impossible, to evaluate. At any one time, primary risk factors (hypertension, smoking, elevated blood cholesterol level, family history for coronary artery disease) and secondary risk factors are simultaneously involved in the pathogenesis.

Still another difficulty is that differences in food habits have not been adequately demonstrated within the United States between people with and without coronary disease, although habitual diet is related to serum cholesterol concentrations in different populations. The Framingham Study seems to suggest that individual genetic, personal and environmental differences may override the small differences in dietary intake within any one culture (Kannel et al., 1971).

In spite of these possible overriding factors, there can be little doubt that the initiation, development, progression, and regression of atherosclerosis are closely associated with nutritional factors. The evidence accumulated over the last 40 years in animal, clinical and epidemiological studies is impressive and persuasive. Nutrition has an important influence on several of the risk factors, particularly serum lipids, lipoproteins, overweight, hypertension, endocrine imbalance and other metabolic conditions closely associated with atherosclerosis. With appropriate nutritional adjustments, these conditions associated with degenerative vascular disease can be altered and controlled. Whether the modification of these risk factors by diet will ultimately lead to lower mortality and morbidity from CHD is yet to be proven. Four primary prevention studies (Los Angeles Veterans Administration Study, Finnish Diet Study, Oslo Diet-Heart Study and U. S. Diet-Heart Study) using dietary measures have shown a trend toward favorable results (Dayton et al., 1969; Miettinen et al., 1972; Laren, 1966; National Diet-Heart Study Final Report, 1968). Recently, Frantz et al., 1975 from the Minnesota Coronary Survey reported that in the cholesterol-lowering dietary trial, men 50 years of age showed less mortality and morbidity from CHD events than the control group.

It is now believed that dietary treatment in this country should be initiated when the serum cholesterol level is >220 mg/dl in adults and above 200 mg/dl in children (Food and Nutrition Board, 1972; Connor et al., 1973). The upper limit of serum triglyceride levels is less clear. Perhaps levels in excess of 130-150 mg/dl can be considered unhealthy (Rhoads et al., 1976). The rationale for initiating dietary therapy when the above limits are exceeded is to prevent or minimize the development of CHD, since atherosclerosis seldom occurs with total cholesterol concentration <200 mg/dl over the life span of an individual (Connor and Connor, 1972), unless other risk factors such as genetics, high blood pressure, smoking and obesity are playing a dominant role.

While serum lipids are not to be neglected, the influence of other risk factors associated with CHD should also be simultaneously considered. They include smoking, hypertension, obesity, stress, lack of physical activity, hormonal imbalances and other primary disease states directly or indirectly associated with abnormal lipid and lipoprotein metabolism.

From a physiological viewpoint, it is difficult to talk about lipids as a separate entity from lipoproteins, since all lipids are associated with protein components to form soluble complexes in aqueous milieu (see Table 3). While it is easy and convenient to consider only the serum lipid levels when monitoring the effect of a prescribed diet, attention should be focused on the serum lipoproteins for at least two reasons:

a) Lipids are transported as lipoproteins and should be considered as lipid abnormality that is classifiable as different types of dyslipoproteinemic conditions.

b) LDL-cholesterol concentration is positively related to the risk of CHD while HDL-cholesterol concentration is negatively related to the risk of CHD. Thus, the progress of an individual on the diet should be assessed by not only the serum total cholesterol concentrations but by the LDL or LDL-cholesterol and HDL

Table 3. Physical, Chemical and Physiologic Description of Plasma Lipoproteins[a]

| | Chylomicron | Very-Low-Density Lipoproteins | Intermediate Density Lipoproteins | Low-Density Lipoproteins | High Density Lipoproteins |
|---|---|---|---|---|---|
| Electrophoretic Nomenclature | Chylomicron | Pre-β-Lipoprotein | Midband Lipoprotein | β-Lipoprotein | α-Lipoprotein |
| Density (g/ml) | 0.95 | 0.95-1.006 | 1.006-1.019 | 1.019-1.063 | $HDL_2$=1.063-1.10<br>$HDL_3$=1.10-1.25 |
| $S_f$ Rate | >400 | 20-400 | 12-20 | 0-10 | --- |
| Electrophoretic Mobility (Paper) | Origin | β-globulin | β-globulin | β-globulin | $\alpha_1$-globulin |
| Molecular Diameter ($A^o$) | 300-5000 | ∿400 | ∿245 | 200 | 70-100 |
| Major Lipid Transported | Exogenous Triglycerides | Endogenous Triglycerides | Cholesterol | Cholesterol | Phospholipids |
| Major Apoproteins | Apo B, C-I, II, III | Apo B, C-I, II, III, Arginine-rich | Apo B | Apo B | Apo A-I, II |
| Minor Apoproteins | Apo A-I, II | Thin-Line Protein<br>Apo A-I, II | | | Apo C-I, II, III<br>Thin-Line Protein<br>Arginine-Rich |

[a] From Jackson et al. (1976), Osborne and Brewer (1977) and Minamisono et al. (1978)

> or HDL-cholesterol levels. When hyperlipidemia is defined in terms of the class or classes of plasma lipoproteins that are considered to be elevated, then the term hyperlipoproteinemia is applied. The occurrence of hyperlipoproteinemia may be primary (genetically determined) or may be secondary to various metabolic dysfunctions, such as hypothyroidism, diabetes mellitus, nephrosis, renal failure, obstructive liver disease, idiopathic hypercalcemia, acute pancreatitis, gout, alcoholism, or dysglobulinemia. If an underlying cause is identified, then the condition is said to be secondary; otherwise, it is called primary hyperlipoproteinemia. (Table 4)

Elevations of plasma cholesterol, triglyceride, and lipoproteins are amenable to dietary control. Diet is fundamental to the treatment of all forms of primary hyperlipoproteinemia. Even when the physician deems it necessary to use a drug, the diet must be continued since the effect of the diet and medication are approximately additive. In many instances, elevations of plasma cholesterol or triglyceride, or both, can be satisfactorily controlled by diet.

Generally, when lowering one serum lipid (i.e. either cholesterol or triglyceride), the other decreases also. Concomitantly, the concentration of lipoprotein fractions rich in these materials (chylomicron, pre-β or β-lipoprotein) drop also. Serum total phospholipid levels usually parallel the changes in serum total cholesterol concentration; the cholesterol/phospholipid ratios under normal conditions remain about unity. However, the α-lipoprotein may not be as predictable. For years the importance of α-lipoproteins in the evaluation of dyslipoproteinemias has usually been neglected, mainly because more emphasis was given to the lipoproteins that carried the greatest proportion of triglycerides and cholesterol, namely chylomicron, pre-β-lipoprotein and β-lipoprotein. Nikkila (1953) and other investigators (Brunner et al., 1962) recognized many years ago that MI subjects had lower α-lipoprotein and higher β-lipoprotein values than did the normal controls. Recently, numerous reports on population studies indicate that plasma HDL-cholesterol levels are a major and independent risk factor for CHD (Rhoads et al., 1976 ; Miller et al, 1977; Castelli et al., 1977). The point is, it is important to examine the entire lipoprotein spectrum in evaluating a person's risk to CHD (Corday and Corday, 1975). A person may have a normal total cholesterol or β-lipoprotein concentration, but his HDL-cholesterol level may be inordinately low, thus placing the individual in a high-risk category for CHD. Whether having a moderately elevated serum cholesterol concentration (i.e. 275-325 mg/dl) with a high HDL-cholesterol (i.e. >65 mg/dl) places him in a non-risk category cannot be answered at this time. Table 5 summarizes factors that

Table 4. Classification of the Types of Primary* Hyperlipoproteinemia[a]

| CLINICAL MANIFESTATIONS | TYPE I | TYPE IIa | TYPE IIb | TYPE III | TYPE IV | TYPE V |
|---|---|---|---|---|---|---|
| **Incidence** | **Rare** | **Common** | **Common** | **Uncommon** | **Common** | **Uncommon** |
| Lipoprotein Abnormalities by Ultracentrifuge † | Severe chylomicronemia | ↑ LDL | ↑ LDL, ↑ VLDL | VLDL + LDL of abnormal composition | ↑ VLDL | ↑ VLDL Chylomicrons |
| Lipoprotein Abnormalities by Electrophoresis †† | Chylomicron band at origin | ↑ B-Band | ↑ B Pre-B | Broad B (or floating B) | ↑ Pre-B | ↑ Chylomicrons Pre-B |
| Plasma Appearance | Creamy layer over clear plasma | Clear or slightly turbid | Clear or slightly turbid | Turbid | Turbid | Creamy layer over turbid plasma |
| Cholesterol | ↔ | ↑ | ↑ | ↑ | ↔ | ↔ |
| Triglyceride | ↑ ↑ | ↔ | ↑ | ↑ | ↑ | ↑ |
| Age of Onset | Infancy and childhood. | At birth, if genetic; early childhood. | At birth, if genetic; early childhood. | Third decade; often after menopause in women. | Third decade or later, can occur in children with obesity or diabetes. | Early adulthood; rare in children. |
| Familial Forms | Deficiency of lipoprotein lipase. | "Familial Hyperlipoproteinemia" - also called "Familial Hypercholesterolemia", most common hyperlipoproteinemia; expressed in 2 forms: heterozygote (more common disorder, cholesterol concentration elevated, but LDL not grossly abnormal), homozygote (less common but cholesterol and LDL levels markedly exaggerated). Many mild examples are not obviously familial. | | Genetic mode uncertain. Often called Broad B disease or remnant disease. B-lipoprotein electrophoretic mobility, but has Pre-B density characteristics. This fraction has a high amount of triglycerides and low amounts of cholesterol. | Second most common form. Often half of adult close relatives will also have Type IV. | When familial, more than half of close relatives have either Type IV or Type V. |
| Clinical Presentation | Episodes of abdominal pain associated with ingestion of dietary fat, pancreatitis, eruptive xanthomas, hepatosplenomegaly, lipemia retinalis, SLE, lymphoma. | Premature vascular disease, tendon and tuberous xanthomas, arcus cornea in 50% of patients. | Severe forms are like Type IIa, milder Type IIb patterns tend to be accompanied by glucose intolerance, overweight. | Glucose intolerance, tuberoeruptive or planar xanthomas, premature vascular disease; worsened by alcohol. | Abnormal insulin levels, glucose intolerance in about 50%; excess caloric intake common; occasionally eruptive xanthomas, hyperuricemia, hepatosplenomegaly; worsened by alcohol. | Bouts of abdominal pain and pancreatitis, eruptive xanthomas, premature atherosclerotic heart disease, hepatosplenomegaly; excess caloric intake common; hyperuricemia, most patients have glucose intolerance; worsened by alcohol excess. |

Table 4. Classification of the Types of Primary* Hyperlipoproteinemia[a] (Continued)

| | | | | | | |
|---|---|---|---|---|---|---|
| Rule Out* | Dysgammaglobulinemia, diabetes mellitus, pancreatitis. | Thyroid dysfunction, nephrotic syndrome, obstructive liver disease. | | Dysgammaglobulinemia, thyroid dysfunction. | Diabetes mellitus, glycogen storage disease, nephrotic syndrome, pregnancy, Werner's syndrome. | Multiple myeloma, macroglobulinemia, nephrotic syndrome, alcoholism, pancreatitis, gout. |
| Drug Therapy | No effective drug therapy at present. | May require a combination of diet and drug therapy. | | | | |

*PRIMARY (FAMILIAL) FORMS OF HYPERLIPOPROTEINEMIA ARE GENETICALLY DEFINED AS THOSE WHICH RESULT FROM THE GENETIC DEFECT AND ARE NOT THE RESULT OF SEPARATE ENDOCRINE DISORDERS, METABOLIC DISORDERS, IMPROPER NUTRITION, DRUG INTERFERENCE, AND THE LIKE, WHICH ARE SECONDARY (OR ACQUIRED) HYPERLIPOPROTEINEMIAS. ALL SECONDARY FORMS OF HYPERLIPOPROTEINEMIA MUST BE RULED OUT BEFORE CONSIDERING TREATMENT OF PRIMARY HYPERLIPOPROTEINEMIA.

†LDL=Low Density (Beta) Lipoproteins; VLDL=Very Low Density (Pre-Beta) Lipoproteins.

††↑ = increased concentration. ↓ = decreased concentration. ⟷ = no change.

a - From The Cleveland Clinic Foundation Diet Manual: A Key to Nutritional Care (1978) are known to influence HDL-cholesterol levels.

Table 5. Factors Associated With HDL-Cholesterol Concentration

| Increase HDL-C Levels | Decrease HDL-C Levels |
|---|---|
| Genetics | Genetics |
| Weight reduction | Smoking |
| Physical exercise | High carbohydrate intake |
| Estrogen-treatment | Overweight |
| Atromid S(R) | Hypertriglyceridemia |
| Alcohol | |

[a] HDL-C = high-density lipoprotein cholesterol

There are other reasons for lowering serum lipids besides the association of high lipid levels with increased incidence in CHD. Hypertriglyceridemia increases the chances for development of pancreatitis, which is a very serious complication of hyperchylomicronemia. Individuals with marked hyperchylomicronemia (Type I hyperlipoproteinemia) may develop eruptive xanthomatosis and lipemia retinalis. With hypercholesterolemia, xanthomas may develop, with the most common type being tendon xanthomas. Other forms that can develop are exanthoma tuberosum, planar xanthomas, tuberoeruptive xanthomas and xanthelasma. Corneal arcus due to lipid deposits in the corneal stroma may occur.

## THE EFFECTS OF SPECIFIC NUTRITIONAL SUBSTANCES ON SERUM LIPID CONCENTRATIONS

Before discussing the role of various dietary factors on serum lipids and atherosclerosis, it must be emphasized that under practical conditions one cannot examine each dietary factor as a separate entity; the effect of each factor on serum lipids is the result of the total composition and amount of each dietary component in a meal. Proteins, carbohydrates, fats, minerals and vitamins all have effects of their own and can have additional effects on other dietary components, especially on efficiency of intestinal absorption, transport and metabolism of lipids. Thus, it is important to understand the interdependence of nutrients that are manipulated in fat-controlled diets. In this paper, the simplistic approach will be taken in discussing each nutritional factor as a separate entity. On occasion, attention will be drawn to the need for viewing the various nutritional components in relationship to each other.

There is little question that nutritional factors can affect serum lipid and lipoprotein levels and patterns. However, it is difficult to predict accurately the size of these changes, since the results are influenced by individual genetic differences in metabolism. Even in animals, such genetic differences occur. In a recent study (Malinow et al., 1976b) we showed that monkeys on a basal chow diet supplemented with 0.5% cholesterol for 30 days showed a wide range of serum cholesterol values (110-1180 mg/dl). The monkeys were then arbitrarily separated into hypo-, normo- and hyper-responders, depending upon their cholesterol level; the dividing concentration being 290 and 990 mg/dl, respectively. It is clear that hypo- and hyper-responders differ greatly in their response to dietary cholesterol and that, when they ingest a diet almost free of cholesterol, their basal cholesterol level is also lower in the former than in the latter (Figure 2). Similar findings have been reported in dogs (Mahley et al., 1974), rabbits (Fillios and Mann, 1956), rats (Imai and Matsumara, 1973), rhesus (Cox et al., 1958) and squirrel monkeys (Lofland et al., 1972), and man is no exception.

Secondly, the effect of each individual nutritional component is influenced by other nutritional components in the diet. It is important to understand the interdependence of nutrients that are manipulated in fat-controlled diets. None of these factors (i.e. total calories, total fats, cholesterol, saturated fat, polyunsaturated fat, proteins, carbohydrates, fiber) can be considered alone. For example, feeding a high-cholesterol diet with low total fat will produce different degrees of hypercholesterolemia as compared to a diet high in cholesterol and high in total fat (Naito and Gerrity, 1979). Furthermore, the proportion of saturated fat versus mono- and polyunsaturated fats in the total fat component of the diet will have differing degrees of influence (Hegsted et al., 1965).

Hegsted et al. (1965) have shown that serum cholesterol concentrations change at various levels of dietary cholesterol with three different dietary fats (coconut, olive, and safflower oils). All diets had 38% of the total calories as fat. In each series, three levels of cholesterol (100, 300 and 700 mg) were given to the subjects, who were fed a diet containing 38%, 17% and 48% of total calories as fat, protein and carbohydrates, respectively. Aside

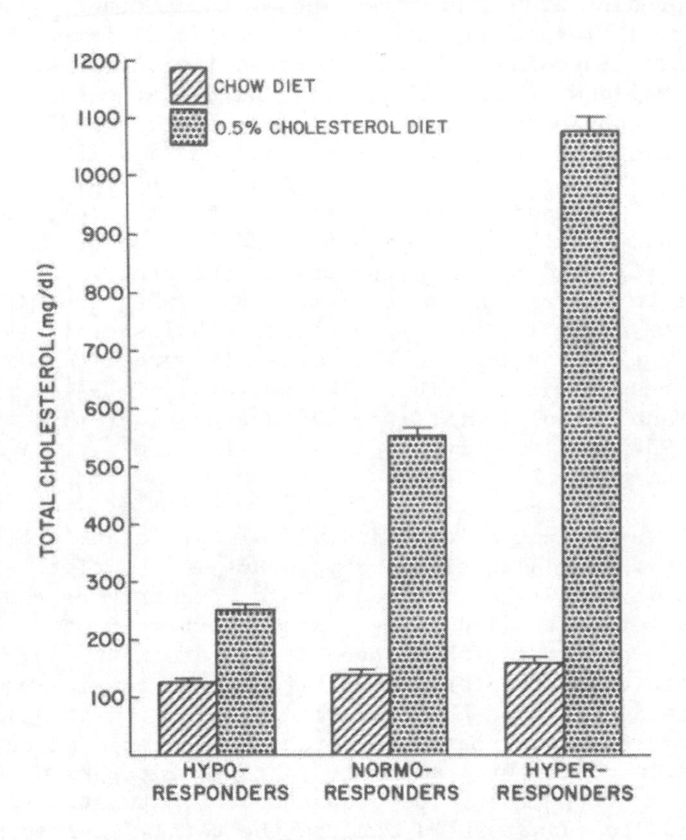


Fig. 2. Non-cholesterol and cholesterol-fed serum cholesterol values of hypo-, normo-, and hyper-responders (Cynomologus monkeys, Macaca fascicularis).

from the fat in lean meat, fish and poultry, all fat came from one of the three oils. The changes in serum cholesterol levels were compared to levels established when control subjects were fed a typical American diet. Regardless of the kind of oil in the diet, serum cholesterol concentration was reduced about 5 mg/dl for every 100 mg decrease in dietary cholesterol intake. With the olive oil (26% of calories as monounsaturated fat, with a polyunsaturated/saturated (P/S) ratio of 0.72%, the cholesterol level was similar to that of the control subjects. With the coconut oil (32% of calories as saturated fats, with a P/S ratio of 0.03), serum cholesterol increased by 40 mg/dl above control levels, while the safflower oil (25% of calories as polyunsaturated fats with a P/S ratio of 5.00) was 40 mg/dl below the control group at each of the respective levels of cholesterol intake. Thus, fat intake affects blood lipids depending on the total dietary fat and the quality of the fat components in the diet. For example:

1. Decreasing total dietary fat decreases total plasma cholesterol and triglyceride concentrations (Ahrens et al., 1957; Keys et al., 1957; Grundy, 1979; Connor and Connor, 1977). The degree of this hypolipaemic change is highly dependent upon the actual level and type of dietary fat consumed, the responsiveness of the individual, the genetic background, the individual's initial plasma lipid concentrations and the varying responsiveness of the different classes of plasma lipid and lipoproteins, i.e., serum triglycerides appear to be more responsive than total cholesterol and chylomicrons more responsive than LDL under most conditions;

2. Reducing dietary cholesterol mainly reduces the concentration of total plasma cholesterol and LDL in normals and in those individuals with elevated LDL (Grundy, 1979; Connor and Connor, 1977; Anderson et al., 1973). Some studies also indicate that increase in dietary cholesterol consumption has an effect on HDL levels, although slight (Connor and Connor, 1977);

3. Reducing dietary saturated fat reduces total plasma cholesterol, LDL and VLDL (Grundy, 1979; Connor and Connor, 1977; Brown, 1971). The hyper- or hypocholesterolemic effect due to the type of dietary fat is believed to be associated with the

particular type of fatty acid content. The saturated fatty acids of less than 12 carbon atoms and stearic acid ($C_{18:0}$) appear to have no influence, while lauric ($C_{12:0}$), myristic ($C_{14:0}$) and palmitic ($C_{16:0}$) appear to have a direct influence (Brown, 1971);

4. Ingesting large amounts of polyunsaturated fats in the diet (in some studies as high as 20 percent of the total calories) results in a fall in plasma total cholesterol and LDL levels, and, to a lesser extent, plasma triglycerides and VLDL, especially in persons with elevated triglyceride concentrations with a composition predominantly of saturated fat. The absolute fall in plasma cholesterol concentrations will vary depending on initial concentrations and a responsiveness factor, but, in general, persons with the highest initial plasma cholesterol show the greatest decrements (Grundy, 1979). Monounsaturated fats (i.e., oleic acid) have a neutral effect, i.e., little or no effect on plasma cholesterol concentrations. The cholesterol-reducing effect of polyunsaturated fat is one-half the cholesterol raising effect of saturated fats (Anderson et al., 1973, Hegsted et al., 1965; Keys et al., 1957a, 1957b);

5. "Carbohydrate-induced" hypertriglyceridemia, i.e., in studies where persons have been placed on a high carbohydrate diet (above 65 percent of total calories) from their usual mixed diet, there is a rise in the plasma triglyceride level (Ahrens et al., 1961). The degree of rise depends on whether the person already has an underlying hypertriglyceridemia. In most subjects, particularly those with initially normal plasma triglyceride levels, the "carbohydrate-induced" hyperlipidemia appears to be a short-term phenomenon (Connor and Connor, 1977).

It is obvious from the above discussion that both the quantity and quality of fat intake influences serum lipid and lipoprotein levels. Most people express the quality of fat as the amount of polyunsaturated to saturated fat in the diet, i.e., P/S ratio. At the Cleveland Clinic, we introduced the concept of "critical limit" which is the fatty acid composition of the diet at which the change from partial to full effectiveness in serum cholesterol reduction occurs (Brown, 1971). For example, critical limits for a diet having 38 percent of calories as fat and 200 mg of cholesterol intake, is 13 percent saturated fat and 14 percent polyunsaturated fatty acids, i.e. effective diets for reducing serum cholesterol levels contain up to 12 percent saturated fat and 14 to 22 percent polyunsaturates. With reduction in total dietary fat, these critical limits for the polyunsaturates are higher. The P/S ratio by itself may provide inadequate information. The critical limit concept accounts for the proportion of fatty acids in relation to total fat intake. Total calorie intake is important in its relation to body weight, i.e., hypertriglyceridemia may result from caloric excess when it is accompanied by gain in body weight. This is of particular significance if the individual already has an elevated plasma triglyceride concentration. Overweight is also associated with elevated total plasma cholesterol concentrations caused in part by elevated VLDL (Connor and Connor, 1977).

Studies with various animal models also have shown similar findings. For example, adult, male ground squirrels were fed a rodent chow diet (4.5% total fat) supplemented with 2% cholesterol (by weight) for a year and no elevation in serum cholesterol or triglycerides was observed (Naito and Gerrity, 1979). Like other previous studies, the following experiment emphasized the role of fats in influencing the coefficient of absorption of dietary cholesterol and/or decreased cholesterol catabolism and excretion to cause an elevation in the serum cholesterol level. Feeding a diet high in fat and cholesterol (egg-yolk diet), resembling a typical American diet, produced hypercholesterolemia in this animal model (the ground squirrel). Although the cholesterol content of this diet was 50% less than the low-fat rodent chow diet fortified with 2% cholesterol, the high fat content (30.4% vs. 4.5%) may have increased the coefficient of absorption of dietary cholesterol. It is well known (Grundy, 1979) that the intestinal absorption of dietary cholesterol is influenced by the amount and type of fat in the diet and the physical state of the dietary cholesterol (i.e., crystalline vs. noncrystalline). The lower proportion of polyunsaturated fats vs. saturated fats (P/S ratio) in egg-yolk (P/S = 0.44) vs. that of the rodent chow (P/S = 0.81) may have had an additional effect on the observed mild hypercholesterolemia.

While the effects of dietary cholesterol in inducing hypercholesterolemia in laboratory animals are convincing, the dietary cholesterol effects in man are less clear. Certain animals such as the rat and dog are resistant to hypercholesterolemia by cholesterol feeding alone. The coefficient of absorption of these two animal models is high (i.e. about 80-90%) but their ability to completely suppress cholesterol synthesis and increase cholesterol catabolism and excretion is remarkably high, thus enabling them to maintain serum cholesterol homeostasis (Dietchy and Wilson, 1970).

The rabbit, on the other hand, readily absorbs cholesterol, but seems to have difficulty in catabolizing and excreting exogenous cholesterol. We have found that the prairie dog may fit into this category since it, too, is extremely responsive to cholesterol feeding (Holzbach et al., 1976; Naito et al., 1977). In man, the coefficient of absorption is around 40%, depending upon the composition of the diet and the amount of dietary cholesterol. There is still some controversy as to how well the suppression of cholesterol synthesis works by the negative feedback system in man. It is believed that the 3-hydroxy-3-methyl glutaryl CoA reductase, the rate-limiting enzyme for cholesterol synthesis, can be suppressed by the end product negative feedback system, but only partially in man (Grundy et al., 1969). Mattson et al. (1972) have shown that by increasing the cholesterol content of a diet originally low in cholesterol, the serum cholesterol concentration increases by 5-12 mg/dl/100 mg dietary cholesterol/1,000 cal., providing all other dietary constituents remain unchanged. When the original diet contains substantial amounts of cholesterol, additional dietary cholesterol does not seem to greatly affect the serum concentrations (Kummerow et al., 1977). However, not all authors agree (Connor and Connor, 1977).

Plant sterols (i.e. campesterol, stigmasterol and β-sitosterol) are observed in the blood only to a slight extent, since most of them are excreted in the stool. Of these phytosterols, β-sitosterol usually predominates in the blood, constituting 50-60% of the total plant steroids. In a typical American diet, which may contain between 200-400 mg of β-sitosterol/day, its coefficient of absorption is about 20-30%. In pharmacologic amounts, β-sitosterol has been known to have a hypocholesterolemic effect because of interference with the absorption of cholesterol. Mattson et al. (1977) suggested that the extensive solubility of the phytosterol esters in fat, in contrast to the limited solubility of their unesterified form, provides a means for administering effective amounts of these hypocholesterolemic agents in dietary therapy.

Purified cholesterol is quite unstable when stored in air at room temperature. Recent data on the cytotoxic effects of products of cholesterol auto-oxidation (namely 25-hydroxycholesterol and cholestane-3β, 5α, 6β-triol) on aortic smooth muscle cells warrant mentioning (Imai et al., 1976; Peng et al., 1978), since it is conceivable that the effects might lead to initial lesion of atherosclerosis. Imai et al. (1976) showed that impurities in old USP-grade cholesterol stored at room temperature caused aortic lesions when a total of 1 gram of concentrated auto-oxidative products of cholesterol/kg body weight was administered to rabbits over a seven-week period. The possible implication of these findings is that foods containing cholesterol, particularly those that receive heat and oxygen treatment during processing (i.e. spray-dried eggs and whole milk) could prove to be hazardous to one's health. Their work further suggests that pure cholesterol (either endogenously synthesized by the animal or chemically isolated purified cholesterol) is not atherogenic. Only when cholesterol becomes a mixture of cholesterol plus spontaneously produced toxic derivatives does it become highly atherogenic. It appears that future studies should include an evaluation of commercially available foods that probably contain toxic derivatives of cholesterol.

Related to the cytotoxic effects of cholesterol auto-oxidation products is hypervitaminosis D. Vitamin $D_3$ feeding (intermittent or continuous) does not appear to influence serum cholesterol concentrations, but does result in development of atherosclerosis and calcified coronary arteries in animals with normal serum lipid concentrations (Kummerow, 1979). Kummerow et al. (1976) have noted the presence of cellular debris in the abdominal aortas of swine when given higher levels of vitamin $D_3$ than the NRC recommendations for swine. Additional findings suggest that high intake of vitamin $D_3$ will cause degeneration of smooth muscle cells in the aorta (Kummerow, 1979). This toxic substance is covered in greater detail in another chapter in this text.

## Carbohydrates

In discussing carbohydrates, it becomes important to attempt to separate out effects of the different types of carbohydrate, i.e., complex carbohydrates (starch, polysaccharides) versus simple sugars. The literature in this area is controversial and inconclusive. Keys et al. (1960) observed that diets containing sucrose and lactose to provide 17 percent of total daily calories, produced significantly higher serum cholesterol concentrations in their schizophrenic subjects than diets in which an equivalent amount of carbohydrate was provided by a mixture of fruits, vegetables and legumes. Similarly, Antar and Ohlson (1965), comparing diets containing a 4-to-1 ratio of simple sugars to carbohydrate from potatoes, legumes, vegetables (as well as the reverse) demonstrated an increase in serum total lipids during ingestion of simple sugars. McGandy, et al. (1966), exchanging 23 percent of total daily calories between mono- and disaccharides and carbohydrate from potato and cereal, found higher plasma cholesterol concentrations when the subjects consumed the high simple sugars diet. However, other studies have not demonstrated any differential effect of simple sugars versus starch on plasma lipids (Irwin et al., 1964; Dunnigan et al., 1970). The work with patients with hyperlipidemias (hypertriglyceridemia, hypercholesterolemia, or mixed hyperlipidemia) suggests that isocaloric replacement of sugar for starch adversely affects plasma lipids (Kuo and Bassett, 1965; Kaufmann et al., 1966), but for most persons, this effect occurs only in diets where saturated fat intake is relatively high, i.e., a low P/S ratio (0.1-0.3) and cholesterol intake about 600-800 mg, the usual American intake (Antar et al., 1970; Birchwood et al., 1970; Van Soest and Robertson, 1977).

Complicating the carbohydrate question (simple sugars versus starch) is that there may be significantly more dietary fiber in a high starch versus a high simple sugar diet. Thus, the effect of starch on cholesterol or triglyceride and/or lipoproteins may be a result of the level of dietary fiber in the test diet or the type of dietary fiber. Dietary fiber, i.e., cellulose, hemicellulose, pectins, mucilages, gums, lignin, represents the food components that cannot be digested in the upper gastrointestinal tract of nonruminants. In actuality, many studies report the crude fiber content of the diet and not the dietary fiber. The reason is that only crude fiber values for food are available in present food composition tables, and enough data is not yet available on dietary fiber content of most foods. The term "crude fiber" really represents an analytical technique for determining the dietary fiber of a food; it is the fraction remaining after food materials are treated with fat solvents, strong acid, and alkali. This technique destroys a portion of hemicellulose, cellulose, pectins and gums (Van Soest and Robertson, 1977). Much investigation to date has centered on the effects of crude fiber, or the "true" dietary fiber, *in toto*, or as separate components (pectin, lignin, cellulose) on serum lipid levels in normolipidemic or hyperlipidemic humans. Experimental results have been conflicting probably due to: a) differences in the type of dietary fiber used, b) use of crude fiber versus dietary fiber, c) amount and source of dietary fiber or crude fiber, and d) other components of the diet known to influence plasma lipids, i.e., cholesterol intake, saturated fat, and the initial plasma lipid levels of the subjects. We have found that the source of plant fiber is exceedingly important in determining its hypocholesterolemic effects (Malinow et al., 1976a, 1976; 1978). When cynomologus monkeys are fed wheat bran, rice bran or soya bran, no hypocholesterolemic effect is observed, while addition of alfalfa to the diet is not only hypocholesterolemic but reverses coronary atherosclerosis. Subsequent studies suggest that the alfalfa contains saponins that bind cholesterol in the gut, which prevents its uptake by the gastrointestinal tract - a mechanism different from that of other fiber effects, i.e., binding of bile acids. To date, in humans, it appears that, of the dietary fiber components: (a) bran has no effect on serum cholesterol or triglyceride concentrations (Connell et al., 1975; Eastwood, 1969; Jenkins et al, 1975; Truswell and Kay, 1975; Eastwood et al., 1973; Bremner et al., 1975), (b) large amounts of cellulose (60 to 100 g/d) may lower serum cholesterol in persons ingesting diets with added excessive cholesterol supplements (4 g) (Shurpalekar et al., 1971), (c) pectin (12 to 36 g/d) added to the diet produces a lowering of serum cholesterol (Durrington et al., 1976; Keye et al., 1961) The degree of lipid lowering may also depend on the amount of cholesterol in the diet (Fisher et al., 1965). The mechanism(s) of the hypocholesterolemic effect of fiber have not been completely elucidated. There are at least three possible mechanisms by which fiber may exert its serum cholesterol lowering abilities:

1. saponin-cholesterol binding in the gut.
2. fiber-bile acid binding in the gut.
3. decrease cholesterol absorption due to increased transit time of the intestinal contents.

First described in 1918 (Feigle, 1918), alcoholic hyperlipemia has been subject to extensive investigations over the past two decades (Mistius and Ockner, 1972; Leiber et al., 1962; Schapiro et al., 1965; Lifton and Scheig, 1978). Studies by Losowsky et al. (1963), Schapiro et al. (1965) and Lifton and Scheig (1978) demonstrated that alcohol ingestion in chronic alcoholics can cause hypertriglyceridemia. This lipid abnormality is resolved rapidly with abstention from alcohol and is apparently not provoked by a high-carbohydrate diet. The sera of chronic alcoholics frequently show a Type IV hyperlipoproteinemic pattern with high levels of very-low-density lipoprotein and elevated triglycerides and cholesterol. The alcohol-related Type IV hyperlipoproteinemia appears different from the carbohydrate-induced Type IV and is independent of acute pancreatitis, liver dysfunction, and obesity. Often, normolipidemic individuals who have imbibed heavily on the weekend will show marked hypertriglyceridemia lasting for several days. It is not uncommon to observe concentrations in excess of 1,000 mg/dl in these individuals, particularly if they are hypertriglyceridemic subjects who are sensitive to caloric excess, causing hyper-pre-β-lipoproteinemia (Naito, unpublished report). Bouchier and Dawson (1964) suggested that alcohol might effect serum lipid concentrations by the impairment of plasma triglyceride removal and increased synthesis of triglyceride-rich lipoproteins. The effects of alcohol vary and relate to dose, duration, the individual's sensitivity, and initial blood lipid pattern. Ginsberg, et al. (1974) have shown that the consumption of alcohol (three ounces ethanol daily for seven days) by non-alcoholic, hypertriglyceridemic persons increases their fasting and postprandial plasma triglyceride concentrations whereas in normal persons, plasma triglycerides are increased postprandially only; total plasma cholesterol and HDL-cholesterol levels were unchanged, LDL-cholesterol decreased, and VLDL-cholesterol increased. In their study, alcohol was added on to an already calorically-adequate diet. When a large quantity of alcohol is consumed (up to 8 or 10 ounces of whiskey or the equivalent per day), plasma triglycerides may increase transiently in normals or alcoholics (Connor and Connor, 1977). We have found that a weak positive association exists between daily alcohol intake and HDL-C ($r = 0.15$, $p<0.001$) in the 800 male executives (50 ± 10 years) studied in the northeast Ohio area (Naito et al., 1979). When compared to non- and social drinking (3-4 drinks/wk) groups, moderate drinking (1-2 drinks/day) was associated with a slight elevation (6%) in HDL-C (p 0.025), but total cholesterol ($p<0.005$) and LDL-cholesterol ($p<0.005$) also rose, with no resultant change in the HDL-C/TC ratio. Heavy drinking (>3 drinks/day) was associated with a 15% elevation in HDL-C ($p<0.001$) and an increased HDL-C/TC ratio. However, there were greater incidences of other risk factors, i.e. elevated SGOT, blood pressure, body weight, smoking. Results of our study are not in complete agreement with other published reports. Although a weak association exists between alcohol intake and HDL-C levels, other CHD risk factors appear with alcohol intake, suggesting that serum HDL-C levels alone do not completely explain the reported decreased incidence of CHD.

## Proteins

According to Connor and Connor (1977), the type of protein consumed in mixed human diets has been shown to have no significant effect upon plasma lipid levels in man. However, casein appears to be more cholesterolemic and/or atherogenic than soya protein for chickens (Kritchevsky et al., 1959) and rabbits (Newburgh and Squire, 1920; Meeker and Kesten, 1940). Dietary protein influences blood lipids, particularly cholesterol in growing animals (Leveille and Fisher, 1958; Nishida et al., 1958; Yen and Leveille, 1969). Further studies on dogs and rats that have been fed a protein-free diet is presented in another chapter of this monograph. In individuals with Kwashiorkor, there appears to be decreased serum β-lipoprotein, total cholesterol, triglycerides, and albumin, but α-lipoprotein and lecithin levels are not reduced (Onitiri and Boyo, 1975). It has been suggested that the decreased synthesis of the apoprotein moiety of LDL is the limiting factor for the low lipids (Flores et al., 1970) in protein deficiency. The impaired rate of synthesis of the lipoprotein would cause triglycerides to accumulate in the liver. Human diets with less than eight percent of calories as protein are associated with low serum cholesterol (Brown, 1971) apparently due to the limitation of protein and not caloric intake or not as a result of lowered dietary fat intake.

## NUTRITIONAL MODIFICATION FOR THE PREVENTION AND TREATMENT OF HYPERLIPIDEMIA AND DYSLIPOPROTEINEMIA

With the development of the original concepts of the five recognized types of dyslipoproteinemic conditions by Fredrickson et al. (1967), five types of diets were formulated and prescribed. The different diets have about the same fat composition when available food products are used, except for the Type I diet. All have less than 300 mg of cholesterol and are similar in their ratio of polyunsaturated to saturated fat (P/S). Thus, a single-diet concept for the treatment of hyperlipidemia is emerging that should be simpler and more practical. Connor and Connor (1977) suggested an "alternative diet," in which cholesterol intake is 100 mg/day and protein, carbohydrate and fat intake is 15, 65 and 20% respectively. The P/S ratio is 1.3 and crude fiber intake of 12-15 g/day is recommended. The Inter-Society Commission for Heart Disease Resources recommends the use of a moderate fat diet, suggesting the use of 30-35% of the total calories as fat (<10% saturated fat and >10% as polyunsaturated fat). Cholesterol intake is maintained below 300 mg/day. Originally, the Cleveland Clinic used two different diets (Brown, 1971). One was low in fat, with 15% of the total calories as fat, and the other a vegetable oil diet, with 38% of the total calories as fat. Serum lipids were equally reduced with either diet in hypercholesterolemic subjects (Brown et al., 1973). In hypertriglyceridemic patients(Types IIb and IV) serum lipids were reduced with greater effectiveness with the 38% fat diet than with the 15% diet, i.e., reduction in carbohydrate intake. The Clinic then increased the fat content of the Master Food Plan (semi-low-fat diet) from 33% to 35%, with an elevation of the P/S ratio to 2.0 (Table 6). The Master Food Plan may be used as a guide with which to commence treatment of most patients with the various forms of primary hyperlipoproteinemias. The fat content may be varied as best suits a person's needs. The Master Food Patterns include a range of fat-containing diets, from one with a very low fat content (15%) but high in carbohydrates (65%) to one with a fat content near that of the American diet (40%), but low in carbohydrates (40%).

Patients with hypercholesterolemia will respond to a diet with a fat content of 35% of the total calories and with a P/S ratio of 2:1. The quality of the dietary fat is exceedingly important. The intake of animal fat, which is generally high in saturated fat should be minimized; about 7 ounces of low fat meats (Table 7) a day is the suggested maximum. The polyunsaturated fats (liquid vegetable oils, with the exception of coconut and palm oils) should be substituted for animal fats in the diet. Not all vegetable oils have the same quality of fat. Safflower oil has the highest content of polyunsaturated fats (Table 8), while coconut and palm oils have the lowest. Olive and peanut

Table 6. Nutrient Composition of The Cleveland Clinic Master Food Plan[a]

| Food Patterns | Fat % Calories | Carbohyd. % Calories | Prot. % Calories | P/S Ratio | Cholesterol (mg/day) |
|---|---|---|---|---|---|
| Typical American Diet | 40 | 45 | 15 | 0.4 | 750 |
| Master Food Plan | 40 | 40 | 20 | 1.3 | 300 |
| Semi-Low Fat Diet | 35 | 50 | 15 | 2.0 | 100-300 |
| Low-Fat Diet | 15 | 65 | 20 | 1.3[b] | 300 |

[a] Mean value

[b] The P/S ratio will vary from individual to individual, range = 1.0 - 1.4

Table 7. Classification of Meat, Fish, Poultry by Fat Content

| Low Fat | Medium Fat | High Fat |
|---|---|---|
| Poultry | Roast | Bacon |
| Veal | Steaks | Salt Pork |
| Fish | Hamburger | Sausages |
| | Lamb | Duck |
| | Pork | Goose |
| | Ham | Luncheon Meats |
| | | Frankfurters |
| | | Beef marbled with Fat |
| | | Chicken and Turkey Fat |

Table 8. Fatty Acid Content of Some Food Products[a, b]

| Items | Saturated Fatty Acids | Mono-Unsaturated Fatty Acids | Poly-Unsaturated Fatty Acids |
|---|---|---|---|
| Oils, Salad or Cooking | | | |
| Safflower | 8 | 15 | 72 |
| Corn | 10 | 28 | 53 |
| Soybean | 15 | 20 | 52 |
| Cottonseed | 25 | 21 | 50 |
| Sesame | 14 | 38 | 42 |
| Peanut | 18 | 47 | 29 |
| Olive | 11 | 76 | 7 |
| Coconut | 86 | 7 | -- |
| Beef Products[c] | | | |
| Porterhouse | 36 | 17 | 16 |
| Sirloin | 29 | 14 | 13 |
| Round | 12 | 6 | 5 |
| Hamburger, lean | 10.0 | 5 | 4 |
| Hamburger, regular | 21.2 | 10 | 9 |
| Pork[c] | | | |
| Ham | 10 | 11 | 2 |
| Loin | 25 | 9 | 10 |
| Fish[c] | | | |
| Trout | 3 | 2 | -- |
| Salmon | 5 | 5 | -- |
| Chicken[c] | | | |
| Dark meat with skin | 6 | 2 | 2 |
| Dark meat without skin | 4 | 1 | 1 |
| Cheeses | | | |
| Cheddar | 18 | 11 | 1 |
| Cottage | 2 | 1 | -- |
| Cream | 21 | 12 | 1 |
| Swiss | 15 | 9 | 1 |
| Butter | 46 | 27 | 2 |
| Margarine | 18 | 47 | 14 |
| Milk, whole | 2 | 1 | -- |
| Egg, whole | 4 | 5 | 1 |

a - In percent
b - From Watt and Merrill (1963)
c - Uncooked

Table 9. Cholesterol and Total Fat Content of Some Selected Food Items: Fish, Poultry, Meats[a, b]

| Item | Chol/100 gm | Fat/100 gm |
|---|---|---|
| Haddock | 60 | 0.5 |
| Halibut | 50 | 7.0 |
| Salmon | 47 | 5.9 |
| Tuna | 63 | 0.8 |
| Sardines | 140 | 8.6 (raw) |
| Shrimp | 150 | 0.8 |
| Scallops | 53 | 1.4 |
| Clams | 63 | 1.6 |
| Crab | 101 | 1.9 |
| Lobster | 85 | 1.5 |
| Oysters | 50 | 1.8 (raw) |
| Chicken | | |
| Light | 79 | 3.4 |
| Dark | 91 | 6.3 |
| Turkey | | |
| Light | 77 | 3.9 |
| Dark | 101 | 8.3 |
| Beef | 91 | 5.3 |
| Liver | 438 | 3.8 (raw) |
| Lamb | 100 | 7.5 |
| Pork | 88 | 8.6 |

a - Lean (where possible) and cooked
b - From Feeley et al. (1972) and Watt and Merrill (1963)

oil may be used occasionally. Only special margarines containing large amounts of unhydrogenated liquid vegetable oils are suitable for the diet. Brown et al. (1973) have suggested that knowing the P/S ratio is not sufficient in planning an effective diet because it alone does not indicate the level of fatty acids. When 36-42% of the calories come from fat, the "critical limit" of saturated fat in the diet must be <14% and the polyunsaturated fat must be >15%. In the semi-low fat diet, where 35% of the calories are derived from fat, the maximum allowable amount of saturated fat is 12% and the minimum polyunsaturated fat is 13%. As can be seen, the "critical limits" account for the proportion of fatty acids in relation to the total fat, which will vary from one prescribed diet to another to suit each individual need. Eggs should be limited to 2-3 per week. Careful attention should be given to shellfish, sardines and liver for the high cholesterol content (Table 9). Notice that crab and shrimp are also high in cholesterol content. Liver contains five times as much cholesterol as beef (steaks). Therefore, when adding liver to the diet, careful attention should be given to the cholesterol and fat composition. Type IIa hyperlipoproteinemic subjects, particularly homozygous hyper-β-lipoproteinemic, may be very resistant to this diet and require further reduction in cholesterol intake from 300 to 200 or even 100 mg/day, followed by drug therapy if necessary.

Hypertriglyceridemia and hyper-pre-β-lipoproteinemia (Type IV) are reduced by a diet low in carbohydrate (Ahrens et al., 1961) and require a carbohydrate intake of less than 50%, of which the simple refined sugars should be substituted with complex carbohydrates (starches, dietary fiber). The Cleveland Clinic's basic moderate-fat diet calls for 40% of the total calories as fat and carbohydrates. If hypertriglyceridemia and hypercholesterolemia (Type V) exist, the semi-low-fat diet should be utilized to reduce the exogenous fat intake to 30% of the total calories and the carbohydrate intake to 50%. The other two mixed hyperlipidemias (Types IIb and III) respond to the moderate-fat diet. Table 10 gives suggestions on how the Master Food Plan may be applied to the various dyslipoproteinemic conditions. It is imperative that the proper diagnosis be made and that all acquired disorders are corrected before instituting the prescribed diets. It must be remembered that the caloric intake of the diets is calculated on the basis of obtaining and maintaining ideal body weight. Since the serum lipids and lipoproteins usually respond to the prescribed dietary therapy within three weeks, a follow-up program is necesary. Continuing encouragement and reassurance are needed along with reliable information if the program is to be successful.

When formulating a particular diet, it is important that a conscious effort be made to decrease the simple sugar consumption and replace it with complex carbohydrates. Consumption of fruits and vegetables should be highly encouraged. At present, the average American consumes about 2-4 g of dietary fiber daily, while vegetarians consume about 24 g/day. Perhaps the dietary fiber intake should be increased to about 10-15 g/day for adequate nutrition. During the past century, composition of the average American diet has changed; complex carbohydrates (fruits, vegetables, grain products), which once were the main-stay of the diet, now play a minor role. At the same time, fat and sugar consumption have risen markedly; these two dietary elements alone now comprise more than 60% of our total caloric intake, up 50% from the early 1900's (Dietary Goals for the United States, 1977). Table 11 provides examples of fiber content of selected food items. The data was generated from the Highland View Hospitals-Case Western Reserve University Data Base which has the capability of analyzing more than 2,300 food items and recipes; over 71 specific nutrients of each food item can be determined.

Since plant proteins have a plasma cholesterol lowering effect as compared with animal protein, the diet should be comprised of at least 50% plant protein, as compared to 30% in the standard American diet.

It is important that alcohol intake be restricted, especially by the Type I and V patient, when following the prescribed diets. Patients with the other dyslipoproteinemic forms may tolerate alcohol if taken only occasionally and providing it does not interfere with the weight reduction program or maintaining ideal body weight.

Table 10 summarizes the use of the Master Food Plan for the various hyperlipidemic and hyperlipoproteinemic conditions.

Since elevated blood pressure may accompany hyperlipidemia in 15-20% of the patients, salt intake should also be monitored. Salt (sodium chloride) consumption in the U.S. is estimated to be between 6 and 18 g/day. The Select Committee on Nutrition and Human Needs suggests reducing salt consumption by about 50-85%, to approximately 3 g/day (Dietary Goals for the United States, 1977).

In the last analysis, it is preventive measures through public education that will best enable us to achieve our ultimate goal of minimizing the risk to atherosclerosis. It is particularly important that the parents are educated so that proper behavior patterns and preferences are taught to children by the age of six, an age when acquired traits are probably well-implanted.

Table 10. Summary: Use of The Cleveland Clinic Master Food Plan For Primary Dyslipoproteinemias

| Type of Hyperlipidemia | Elevation of Lipoproteins | Type of Hyperlipoproteinemia | Diet (P/S Ratio) |
|---|---|---|---|
| Exogenous Triglycerides | Chylomicron | I | Low Fat, 15% (P/S=1.3%) |
| Endogenous Triglycerides | Pre-β- | IV | Moderate Fat, 40% (P/S=1.3%) |
| Cholesterol | β- | IIa | Semi-Low Fat, 35% (P/S=2.0%) |
| Mixed Hyperlipidemia | β-<br>Pre-β- | IIb | Moderate Fat, 40% (P/S=2.0%) |
| | | III | Moderate Fat, 40% (P/S=1.3%) |
| | | V | Semi-Low Fat, 30% (P/S=1.3%) |

Table 11. Fiber Content of Selected Foods

| | Amount | Fiber (gm) |
|---|---|---|
| Pear, raw with skin | 1 medium | 2.3 |
| All Bran | 1 oz. | 2.1 |
| Apple, raw with skin | 1 medium | 1.4 |
| Watermelon, raw | 1 wedge | 1.3 |
| Bran Flakes, 40% | 1 oz. | 1.0 |
| Peanuts, roasted | 1 oz. | 0.77 |
| Almonds, shelled | 1 oz. | 0.74 |
| Wheat germ, dry | 1 oz. | 0.71 |
| Orange, raw | 1 medium | 0.65 |
| Shredded Wheat biscuit | 1 oz. | 0.65 |
| Pecans, shelled | 1 oz. | 0.65 |
| Applesauce, canned | 1/2 cup | 0.64 |
| Walnuts, shelled | 1 oz. | 0.60 |
| Banana, raw | 1 medium | 0.60 |
| Corn, cooked (white/yellow) | 1/2 cup | 0.58 |
| Carrot, raw | 1/2 cup | 0.55 |
| Whole wheat bread | 1 slice | 0.45 |
| Pineapple chunks, canned | 1/2 cup | 0.37 |
| Celery, raw | 1/2 cup | 0.31 |
| Pineapple, raw | 1/2 cup | 0.31 |
| Tomato, raw | 1/2 cup | 0.31 |
| Grapes, raw (Concord/Delaware) | 12 | 0.29 |
| Raisins | 2 tbsp. | 0.16 |
| Lettuce, iceberg | 1/2 cup | 0.14 |
| Orange juice, canned | 1/2 cup | 0.12 |
| Apple juice, canned | 1/2 cup | 0.12 |
| Rye bread, light | 1 slice | 0.11 |
| Corn Flakes | 1 oz. | 0.09 |
| White bread, enriched | 1 slice | 0.05 |
| Oatmeal or Rolled Oats, cooked | 1 oz. | 0.05 |

[a] From Highland View Hospital - Case Western Reserve University Nutrient Data Base, Departments of Nutrition and Biometry, Case Western Reserve University, Cleveland, Ohio 44106.

## ROLE OF THE PHYSICIAN, CLINICAL BIOCHEMIST AND DIETITIAN IN DIETARY MANAGEMENT

The ideal approach in attempting to lower a patient's risk for CHD is a team approach, particularly when considering dietary management for treating hyperlipidemia. An intimate working relationship should exist between the physician(s), clinical biochemist, dietitian and patient. It is imperative that the physician and clinical biochemist recognize the importance of nutrition in the overall health care of the patient, have a full understanding and working knowledge of nutritional factors that have been associated with high risk for CHD, and have current knowledge on the most effective way(s) to control the lipid abnormalities via proper dietary management. In the development of this intimate relationship, the first step is close communication with each other. The physician and clinical biochemist must adopt the attitude that they can depend upon the professional knowledge and experiences of the dietitian in helping him achieve the goal of effective patient care. The dietition should be as familiar with the patient as is the physician. Information on the patient's life style, eating habits, likes and dislikes, food practices, and CHD risks should be apparent to both the physician and dietitian in order that they can plan the most effective means of reducing the patient's risk to CHD. With this approach, the responsibility of the dietitian should not only encompass the control of elevated serum lipids through dietary management, but should also include the consideration of diet in the control of elevated blood pressure and body weight, and advice concerning excessive smoking and lack of physical activity as a comprehensive health care therapy for the patient.

For this system of therapy to succeed, it is important that the patient be informed of the seriousness of his physical condition and the logic behind instituting the various approaches to minimize the associated risk factors for premature coronary events. Patient education is sadly lacking in most institutions. Unless the physician, clinical biochemist and dietitian are confident, knowledgeable and enthusiastic about the effectiveness of minimizing the risk of CHD through dietary management, the patient will not develop the attitude necessary to comply with the prescribed diets and possible altered life style. During instruction, it is most important that the patient help in planning his diet; the person buying the food and preparing the meals should also be present. While hyperlipidemia can be genetically determined, in many instances it is also due to environmental conditions, especially improper food habits. If the latter condition applies to the patient, the acquired hyperlipidemic condition of the patient can also be potentially manifested in the other members of the family, particularly the children.

Table 12. The Nutritional Factors Influencing Serum Lipoproteins

| Nutritional Factors | VLDL | LDL | HDL |
|---|---|---|---|
| Total Calories | ↑ | (↑) | (↓) |
| Total Fat | ↑ | ↑ | ↓ |
| Saturated | ↔ | ↑ | (↓) |
| Monounsaturated | ↔ | ↔ | ↔ |
| Polyunsaturated | ↔ | ↓ | (↑) |
| Cholesterol | (↑) | ↑ | (↑) |
| Phytosterols | ↔ | ↓ | ↔ |
| Protein | | | |
| Vegetable | ? | ↓ | (↑) |
| Animal | ? | ↑ | ↔ |
| Carbohydrates | | | |
| Simple | ↑ | ↔ | ↓ |
| Complex | ↔ | ↓ | ↔ |
| Alcohol | ↑ | (↑) | ↑ |
| Vitamin C | ↔ | ↔ | ↔ |

Therefore, nutritional education directed toward all members of the family, particularly those who purchase and prepare the meals is beneficial to all. The prepared meals should, ideally, include the rest of the family members, unless the prescribed diet is drastic and applies only to a person with severe risk factors. It is our experience that group instruction is more effective than individual instruction because of the beneficial effects of interaction among members of the group.

One of the reasons for the poor adherence to diets is the poorly designed or even the lack of design of a follow-up period. This is important for both the physician and patient. The physician needs to monitor the blood lipids and lipoproteins to determine the effectiveness of the prescribed diet, while the patient needs further instructions, information on his progress, and positive reinforcement. In the follow-up period, visits at 2, 4 and 6 weeks after the initial two to three hour instruction should be adequate. After the initial follow-up period, less frequent visits (perhaps once a year) are required to insure adherence to the diet pattern. The continued objective is to follow through with the measurements that test adherence to the diet, including: body weight, blood lipids, lipoproteins and blood pressure. The examination should also include discussion of alcohol consumption and smoking patterns. Finally, the patient must always receive encouragement to stay on his diet and physical exercise program.

## SUMMARY

In summary, the relationship of serum lipids and lipoproteins to CHD has been reviewed. The effect of specific nutritional components in the diet on serum lipid and lipoprotein concentrations has been discussed (and is summarized in Table 12). An approach to dietary management of hyperlipidemic and hyperlipoproteinemic conditions was provided as a basis for future dietary trials. Finally, the role of the physician, clinical biochemist and dietitian in dietary management of the CHD patient has been covered to emphasize the need of a team approach to patient care.

## REFERENCES

Ahrens, E., Jr., Hirsch, J., Insull, W., Tsaltas, T., Blomstrand, R., Peterson, M., 1957, The influence of dietary fats on serum lipid levels in man. Lancet, 1:943.

Ahrens, E., Hirsch, J., Oette, K., Farquhar, J., Stein, Y., 1961, Carbohydrate-induced and fat induced lipemia, Trans. Assoc. Am. Phys., 74:134.

Anderson, J., Grande, F., Keys, A., 1973, Cholesterol-lowering diets, J. Am. Diet. Assoc., 62:133.

Antar, M., Lirrle, J., Lucas, C., Buckley, G., Csima, A., 1970, Interrelationships between the kinds of dietary carbohydrate and fat in hyperlipoproteinemic patients. III. Synergistic effect of sucrose and animal fat on serum lipids, Atherosclerosis, 11:191.

Antar, M., Ohlson, M., 1965, Effect of simple and complex carbohydrates upon total lipids non-phospholipids in young men and women, J. Nutr, 85:329.

Birchwood, B., Little, J., Antar, H., Lucas, C., Buckley, G., Csima, A., Kallos, A., 1970, Interrelationship between the kind of dietary carbohydrate and fat in hyperlipidemic patients. II. Sucrose and starch with mixed saturated and polyunsaturated fats, Atherosclerosis, 11:183.

Blackburn, H., 1974, Progress in the epidemiology and prevention of coronary heart disease, in: "Progress in Cardiology," Yu, P. and Goodwin, J., eds., Lea and Febiger, Philadelphia, pp. 1-36.

Bouchier, I.A. and Dawson, A.M., 1964, The effects of infusions of ethanol on the plasma free fatty acids in man, Clin. Sci., 26:47.

Bremner, W.F., Brooks, P.M., Third, J.L.H.C., Lawrie, T.D.V., 1975, Bran in hypertriglyceridemia: a failure of response, Br. Med. J., 3:574.

Brown, H.B., 1971, Food patterns that lower blood lipids in man, J. Am. Diet. Assoc., 58:303.

Brown, H.B., Lewis, L.A. and Page, I.H., 1973, Mixed hyperlipemia a sixth type of hyperlipoproteinemia, Atherosclerosis, 17:181.

Brunner, D., Altman, S., Loebl, K. and Schwartz, S., 1962, Alpha-cholesterol percentages in coronary patients with and without increased total serum cholesterol levels and in healthy controls, J. Athero. Res., 2:424.

Castelli, W.P., Doyle, J.T., Gordon, T., Hames, C.G., Hjortland, M.C., Hulley, S.B., Kagan, A. and Zukel, W.J., 1977, HDL-cholesterol and other lipids in coronary heart disease, Circulation, 55:767. The Cleveland Clinic Foundation Diet Manual: A Key to Nutrition Care 1978. Cleveland Clinic Foundation.

Connor, W.E., Brown, H.B., Fredrickson, D.S., Steinberg, D., Connor, S.L. and Bickel, J.H., 1973, A maximal approach to the dietary treatment of the hyperlipidemias. Sub-committee on diet and hyperlipidemia, Council on Atherosclerosis, Amer. Heart Assoc., p. 1.

Connor, W.E. and Connor, S.L., 1972, The key role of nutritional factors in the prevention of coronary heart disease, Prev. Med., 1:49.

Connor, W.E. and Connor, S.L., 1977, Dietary treatment of hyperlipidemia in: "Hyperlipidemia: Diagnosis and Therapy," Rifkind, B.M. and Levy, R.I., eds., Gruine and Stratton, New York, NY, p. 281.

Corday, E. and Corday, S.R., 1975, Prevention of heart disease by control of risk factors: The time has come to face the facts, Am. J. Cardiol., 35:330.

Cox, G.E., Taylor, C.B., Cox, L.G. and Counts, M.A., 1958, Atherosclerosis in rhesus monkeys, Arch. Pathol., 66:32.

Davis, C.E. and Havlik, R.J., 1977, Clinical trials of lipid lowering and coronary heart disease prevention, in: "Hyperlipidemia: Diagnosis and Therapy," Rifkind, B.M. and Levy, R.I., eds., Gruine and Stratton, New York, NY, pp. 79-92.

Dayton, S., Pearce, M.L., Hashimoto, S., Dixon, W.J. and Tomiyasu, U., 1969, A controlled clinical trial of a diet high in unsaturated fat is preventing complications of atherosclerosis, Circulation, 40 (Suppl. II): 1.

Dietary Goals for the United States, Select Committee on Nutrition and Human Needs, Second Edition, 1977, U.S. Government Printing Office, Washington, D.C., pp. 1-79.

Dietschy, J.M. and Wilson, J.D., 1970, Regulation of cholesterol metabolism, N. Engl. J. Med., 282:1128, 1179, 1241.

Dunnigan, M., Fyfe, T., McKiddie, M., Crosbie, S., 1970, Effects of isocaloric exchange of dietary starch and sucrose on glucose tolerance, plasma, insulin and serum lipids in man, Clin. Sci., 38:1.

Durrington, P.N., Bolton, C.H., Manning, A.P., Hartog, M., 1976, Effect of pectin on serum lipids and lipoproteins, whole gut transit-time, and stool weight, Lancet, II:394.

Eastwood, M., 1969, Dietary fibre and serum lipids, Lancet, II: 1222.

Eastwood, M.A., Kirkpatrick, J.R., Mitchell, W.D., Bone, A., Hamilton, T., 1973, Effect of dietary supplements of wheat bran and cellulose on feces and bowel function, Br. Med. J., 4:392.

Feeley, R.M., Criner, P.E. and Watt, B.K., 1972, Cholesterol content of foods, J. Am. Diet. Assoc., 61:134.

Fillios, L.C. and Mann, G.V., 1956, The importance of sex in the variability of the cholesterolemic response of rabbits fed cholesterol, Circ. Res., 4:406.

Fisher, H., Griminger, P., Sostmen, E., Brush, M., 1965, Dietary pectin and blood cholesterol, J. Nutr., 86:113.

Flores, H., Sierralta, W. and Monckeberg, F., 1970, Triglyceride transport in protein-depleted rats, J. Nutr., 100:375.

Food and Nutrition Board, National Academy of Sciences--National Research Council and Council on Foods and Nutrition, American Medical Assoc., Diet and Coronary Heart Disease, a joint statement, 1972, Prev. Med., 1:559. Fourth Report of the Director, NHLBI: 1977 National Heart, Blood Vessel, Lung and Blood Program, March 1977, DHEW Publication No. (NIH) 77-1170,

U.S. Department of Health, Education and Welfare, National Institutes of Health, Bethesda, Md.
Fourth Report of the Director: NHLBI, 1977, U.S. Department of Health, Education and Welfare, Superintendent of Documents, U.S. Government Printing Office, Washington, D.C.
Frantz, I.D., Jr., Dawson, E.A., Kuba, K., Brewer, E.R., Gatewood, L.C. and Bartsch, G.E., 1975, The Minnesota Coronary Survey: Effect of Diet on Cardiovascular Events and Deaths, Circulation, 52 (Suppl. II): II-4.
Fredrickson, D.S. and Levy, R.I., 1972, Familial hyperlipoproteinemia, in: "The Metabolic Basis of Inherited Disease," Stanburg, J.B., Wyngaarden, J.B., Fredrickson, D.S., eds., McGraw-Hill, New York, NY, p. 545.
Fredrickson, D.S., Levy, R.I. and Lees, R.S., 1967, Fat transport in lipoproteins - an integrated approach to mechanisms and disorders, N. Engl. J. Med., 276:34, 94, 148, 215, 273.
Galen, R.S., 1977, The normal range. A concept in transition, Arch. Path. Lab. Med., 101:561.
Ginsburg, H., Olefsky, J., Farquhar, J. and Reaven, G., 1974, Moderate ethanol ingestion and plasma triglyceride levels: A study in normal and hypertriglyceridemic persons, Ann. Intern. Med., 80:143.
Glueck, C.J. and Connor, W.E., 1978, Diet-coronary heart disease relationships reconnoitered, Amer. J. Clin. Nutr., 31:727.
Glueck, C.J., Gartside, P., Fallat, R.W., Sielski, J. and Steiner, P.M., 1976, Longevity syndromes: familial hypobeta and familial hyperalpha lipoproteinemia, J. Lab. Clin. Med., 88-941.
Gordon, T., Garcia-Palmieri, M.R., Kagan, A., Kannel, W.B., Schiffman, J., 1974, Differences in coronary heart disease in Framingham, Honolulu and Puerto Rico, J. Chronic Dis., 27:329.
Grundy, S.M., 1977, Treatment of hypercholesterolemia, Amer. J. Clin. Nutr., 30:985.
Grundy, S., 1979, Dietary fats and sterols, in: "Nutrition, Lipids and Coronary Heart Disease," R. Levy, B. Rifkind, B. Dennis,
Grundy, S.M., Ahrens, E.H. and Davignon, J., 1969, The interaction of cholesterol absorption and cholesterol synthesis in man, J. Lipid Res., 10:304.
Hegsted, D.M., McGandy, R.B., Myers, M.L., Stare, F.J., 1965, Quantitative effects of dietary fat on serum cholesterol in man, Am. J. Clin. Nutr., 17:281.
Holzbach, R.T., Corbusier, C., Marsh, M. and Naito, H.K., 1976, The process of cholesterol cholelithiasis induced by diet in the prairie dog: A physicochemical characterization, J. Lab. Clin. Med., 87:987.
Imai, H., Werthessen, N.T., Taylor, C.B. and Lee, K.T., 1976, Angiotoxicity and arteriosclerosis due to contaminants of USP-grade cholesterol, Arch. Pathol. Lab. Med., 100:565.

Imai, H. and Matsumura, H., 1973, Genetic studies on induced and spontaneous hypercholesterolemia in rats, Atherosclerosis, 18:59.

Irwin, M., Taylor, D., Feeley, R., 1964, Serum lipid levels, fat, nitrogen and mineral metabolism of young men associated with a kind of dietary CHD, J. Nutr., 82:338.

Jackson, R.L., Morrisett, J.D. and Gotto, A.M., 1976, Lipoprotein structure and metabolism, Physiol. Rev., 56:259.

Jenkins, D.J.A., Newton, C., Leeds, A.R., Cummings, J.H., 1975, Effect of pectin, guar gum and wheat fibre on serum-cholesterol, Lancet, I:1116.

Johnson, K.G., Yano, K. and Kato, H., 1968, Coronary heart disease in Hiroshima, Japan: A report of a six-year period of surveillance, 1958-1964, Amer. J. Publ. Health, 58:1355.

Kamio, A., Kummerow, F.A., Imai, H., 1977, Degeneration of aortic smooth muscle cells in swine fed excess vitamin $D_3$, Arch. Pathol. Lab. Med., 101:378.

Kannel, W.B., Castelli, W.P., Gordon, T. and McNamara, P.M., 1971, Serum cholesterol, lipoproteins and the risk of coronary heart disease, Ann. Intern. Med., 74:1.

Kaufmann, N., Poznaski, R., Blondheim, S., Stein, Y., 1966, Changes in serum lipid levels of hyperlipemic patients following feeding of starch, sucrose and glucose, Am. J. Clin. Nutr., 18:261.

Keys, A., Anderson, J., Grande, F., 1957a, Serum cholesterol response to dietary fat, Lancet, 1:787.

Keys, A., Anderson, J., Grande, F., 1957b, Prediction of serum cholesterol responses of man to changes in fat in the diet, Lancet, 2:959.

Keys, A., Anderson, J., Grande, F., 1960, Diet-type (fats constant) and blood lipids in man, J. Nutr., 70:257.

Keys, A., Grande, F., Anderson, J.T., 1961, Fiber and pectin in the diet and serum cholesterol concentrations in man, Proc. Soc. Exp. Biol. Med., 106:555.

Kritchevsky, D., Kolman, R.R., Guttmacher, R.M. and Forbes, M., 1959, Serum cholesterol levels in germ-free chicks, Arch. Biochem. Biophys., 85:444.

Kummerow, F.A., 1979, Nutrition imbalance and angiotoxins as dietary risk factors in coronary heart disease, Am. J. Clin. Nutr., 32:58.

Kummerow, F.A., Cho, B.H.S., Huang, W.Y-T., Imai, H., Kamio, A., Deutsch, M.J. and Hooper, W.M., 1976, Additive risk factors in atherosclerosis, Am. J. Clin. Nutr., 29:579.

Kummerow, F.A., Kim, Y., Hull, M., Pollard, J., Ilinov, P., Sorossiev, D.L. and Valek, J., 1977, The influence of egg consumption on the serum cholesterol level in human subjects, Am. J. Clin. Nutr., 30:664.

Kuo, P., Bassett, D., 1965, Dietary sugar in the production of hyperglyceridemia, Ann. Int. Med., 62:1199.

Leren, P., 1966, The effect of plasma cholesterol lowering diet in male survivors of myocardial infarction, Acta Med. Scand. (Suppl.), 466:1.

Lieber, C.S., Leevy, C.M., Stein, S.W., George, W.S., Cherrick, G.R., Abelmann, W.H. and Davidson, C.S., 1962, Effect of ethanol on plasma-free fatty acids in man, J. Lab. Clin. Med., 59:826.

Leveille, G.A. and Fisher, H., 1958, Plasma cholesterol in growing chicken as influenced by dietary protein and fat, Proc. Soc. Exptl. Biol. Med., 98:630.

Lifton, L. and Scheig, R., 1978, Ethanol-induced hypertriglyceridemia. Prevalence and contributing factors, Am. J. Clin. Nutr., 31:614.

Lofland, H.B. Jr., Clarkson, T.B., St. Clair, R.W. and Lehner, N.D.M., 1972, Studies on the regulation of plasma cholesterol levels in squirrel monkeys of two genotypes, J. Lipid Res., 13:39.

Losowsky, M.S., Jones, D.P., Davidson, C.S. and Lieber, C.S., 1963, Studies of alcoholic hyperlipemia and its mechanism, Am. J. Med., 35:794.

McGandy, B., Hegsted, D., Myer, M., Stare, F., 1966, Dietary carbphydrate and serum cholesterol levels in man, Am. J. Clin. Nutr., 18:237.

Mahley, R.W., Weisgraber, K.H. and Innerarity, T., 1974, Canine lipoproteins and atherosclerosis, Circ. Res., 35:722.

Malinow, M.R., McLaughlin, P., Papworth, L., Naito, H.K. and Lewis, L., 1976a, Effect of bran and cholestyramine on plasma lipids in monkeys, Am. J. Clin. Nutr., 29:905.

Malinow, M.R., McLaughlin, P., Papworth, L., Naito, H., Lewis, L. and McNulty, W., 1976b, A model for therapeutic interventions on established coronary atherosclerosis in a non-human primate, in: "Advances in experimental medicine and biology: atherosclerosis drug discovery," C.E. Day, ed., Plenum Press, New York, pp. 3-31.

Malinow, M.R., McLaughlin, P., Naito, H.K., Lewis, L.A. and McNulty, W.P., 1978, Effect of alfalfa meal on shrinkage (regression) of atherosclerotic plaques during cholesterol feeding in monkeys, Atherosclerosis, 30:27.

Mattson, F.H., Erickson, B.A. and Kligman, A.M., 1972, Effect of dietary cholesterol on serum cholesterol in man, Am. J. Clin. Nutr., 25:589.

Mattson, F.H., Volpenhein, K.A. and Erickson, B.A., 1977, Effect of plant sterol esters on the absorption of dietary cholesterol, J. Nutr., 107:1139.

Medalie, J.H., Kahn, H.A., Newfeld, H.N., Riss, E. and Goldbourt, U., 1973, Five year myocardial infarction incidence - II. Association of single variables to age and birthplace, J. Chronic Dis., 26:329.

Meeker, D.R. and Kesten, H.D., 1940, Experimental atherosclerosis and high protein diets, Proc. Soc. Exptl. Biol Med., 45:543.

Miettinen, M., Turpeinen, O., Karvonen, M.J., Elosuo, R. and Paavilainen, E., 1972, Effect of cholesterol lowering diet on mortality from coronary heart disease and other causes, Lancet, 2:835.

Miller, N.E., Førde, O.H., Thelle, D.S. and Mjøs, O.D., 1977, The Tromsø heart study. High density lipoprotein and coronary heart disease: A prospective case control study, Lancet, 1:8019.

Minamisono, T., Wada, M., Akamatsu, A., Okabe, M., Handa, J., Morita, T., Asagami, C., Naito, H.K., Nakamoto, S., Lewis, L.A. and Mise, J., 1978, Dyslipoproteinemia (a remnant lipoprotein disease) in uremic patients on hemodialysis, J. Clin. Chim. Acta, 84:163.

Mistius, S.P. and Ockner, R., 1972, Effects of ethanol on endogenous lipid and lipoprotein metabolism in small intestine, J. Lab. Clin. Med., 80:34.

Mjøs, O.D., Thelle, D.S., Førde, O.H. and Vik-Mo, H., 1977, Family study of high density lipoprotein cholesterol and the relation to age and sex. The Tromsø heart study, Acta Med. Scand., 201:323.

Naito, H.K. and Gerrity, R.G., 1979, Unusual resistance of the ground squirrel to the development of dietary-induced hypercholesterolemia and atherosclerosis, Exptl. Mol. Path., 31: 452-467.

Naito, H.K., Holzbach, R.T. and Corbusier, C., 1977, Characterization of serum lipids and lipoproteins of prairie dogs fed a chow diet or cholesterol supplemented diet, Exptl. Mol. Pathol., 27:81.

Naito, H.K., Taylor, A.M., David, J.A. and Gerrity, R.G., 1979, Association between alcohol intake and blood lipid and lipoprotein concentrations in men, Circulation, 60 (Suppl. II):14.

The National Diet Heart Study Final Report, National Diet Heart Study Research Group, 1968, Circulation, 37 (Suppl. I):1.

Newburgh, L.H. and Squire, T.L., 1920, High protein diets and arteriosclerosis in rabbits: A preliminary report, Arch. Intern. Med., 26:38.

Nichols, A.B., Ravenscroft, C., Lamphier, D.E. and Ostrander, L.D., 1977, Daily nutritional intake and serum lipid levels. The Tecumseh Study, J. Clin. Nutr., 29:1384.

Nikkila, E., 1953, Studies on lipid-protein relationships in normal and pathological sera and effect of heparin on serum lipoproteins, Scand. J. Clin. Lab. Invest., 5 (Suppl. 8):1.

Nishida, T.F., Takenaka, F. and Kummerow, F.A., 1958, Effect of dietary protein and heated fat on serum cholesterol and betalipoprotein levels and on the incidence of experimental atherosclerosis in chicks, Circ. Res., 6:194.

Onitiri, A.C. and Boyo, A.E., 1975, Serum lipids and lipoproteins in children with kwashiorkor, Brit. Med. J., 3:630.

Osborne, J.C. and Brewer, B., Jr., 1977, The plasma lipoproteins, Adv. Prot. Chem., 31:253.

Peng, S.K., Taylor, C.B., Tham, P., Werthessen, N.T. and Mikkelson, B., 1978, Effect of auto-oxidation products from cholesterol on aortic smooth muscle cells, Arch. Pathol. Lab. Med., 102:57.

Rhoads, G.G., Gulbrandsen, C.L. and Kagan, A., 1976, Serum lipoproteins and coronary heart disease in a population study of Hawaii Japanese men, N. Engl. J. Med., 294:293.

Schapiro, R.H., Scheig, R.L., Drummey, G.D., Mendelson, J.H. and Isselbacher, K.J., 1965, Effect of prolonged ethanol ingestion on the transport and metabolism of lipids in man, N. Engl. J. Med., 272:610.

Shurpalekar, K.S., Doraiswamy, T.R., Sundaravalli, O.E., Narayana Rao, M., 1971, Effect of inclusion of cellulose in an 'atherogenic diet' on blood lipids of children, Nature, 232:554.

Stone, N.J. and Levy, R.I., 1972, Hyperlipoproteinemia and coronary heart disease, Prog. Cardiovasc. Dis., 14:341.

Truswell, A. and Kay, R., 1975, Absence of effect of bran on blood lipids, Lancet, I:922.

Van die Redaskie, 1973, Coronary heart disease in China, S. African Med. J., 47:1485.

Van Soest, P., Robertson, J., 1977, What is fibre and fibre in food? Nutr. Rev., 35:12.

Watt, B.K. and Merrill, A.L., 1963, Composition of foods. Agriculture handbook No. 8, U.S. Government Printing Office, Washington, D.C.

Wilhelmsen, L., Wedel, H. and Tibblin, G., 1973, Multivariate analysis of risk factors for coronary heart disease, Circulation, 48:950.

Yeh, Y.Y. and Leveille, G.A., 1969, Effect of dietary protein on hepatic lipogenesis in the growing chick, J. Nutr., 98:356.

ACKNOWLEDGEMENTS

This study was supported by Grant No. HL-6835 from the National Heart, Lung and Blood Institute, Grant No. CRP-433 from The Cleveland Clinic Foundation, Grant No. 8557 from the Bleeksma Fund of The Cleveland Clinic Foundation and Grant No. 3085R from The American Heart Association, Northeast Ohio Affiliate, Inc. I am grateful to Mrs. Jeannette Goodman for typing this manuscript.

DIETARY MANAGEMENT OF HYPERLIPIDEMIA: THE PRUDENT POLYUNSATURATED FAT DIET

Francine M. Hoerrmann, R.D., Donna B. Rosenstock, R.D.
Department of Nutritional Services, The Cleveland Clinic Foundation and Herbert K. Naito, Ph.D., Head, Lipid-Lipoprotein Laboratories, Division of Laboratory Medicine and Division of Research, The Cleveland Clinic Foundation and Clinical Professor, Cleveland State University

## INTRODUCTION

One of the most dramatic and exciting findings in recent years in the scientific literature is the observed decline in coronary heart disease (CHD) mortality rates in the United States. This has been a steady phenomenon seen in white and non-white, men and women since 1968, which included the entire age span from 35-74 years (Stammler, 1979).

Correspondingly, several trends have been proceeding in parallel over these years:

1. Positive developments in emergency care techniques of CHD patients (both acute and chronic).

2. Improvements in early detection methods for the potential CHD candidate.

3. Better equipped and staffed coronary care units.

4. Improvements in cardiac assist instrumentation.

5. More effective antiarrhythmia, antihypertensive and cardiopulmonary drugs.

6. Increased interest in physical exercise, better nutrition, decreased smoking and other behavior modification in controlling major risk factors.

7. More wide-spread education of the general population on medical information.

There is reason to believe that diet modification for the control of hypertension and hyperlipidemia has played an important role in reducing the incidence of CHD (Stammler, 1979). Numerous diet plans have been suggested by physicians, associations, and expert committees over the past 20 years to decrease CHD risk. No less than 20 expert scientific committees have recommended dietary modification as a means for controlling blood lipid levels and reducing the risk of CHD. Table 1 summarizes their recommendations. Some countries, such as Canada, have adopted these recommendations as a national policy.

The Select Committee on Nutrition and Human Needs has recommended dietary goals for the American general population (Select Committee on Nutrition and Human Needs, 1977). Although the diet recommendations differ in some details, the basic fat-modified diet as suggested by the Inter-Society Commission for Heart Disease Resources is the common base; it is essentially a calorie controlled, moderately low cholesterol and moderately low saturated fat diet. It is intended to provide all essential nutrients in ample amounts and to fit well into the life styles of industrialized, relatively sedentary populations.

Turning our attention to the high-risk group for heart attacks, how does one effectively use dietary means to control CHD risk factors? The objective of nutritional modification is to lower serum lipids and low-density lipoproteins sufficiently so that hyperlipidemia and hyperlipoproteinemia as risk factors in the development of CHD are eliminated or greatly reduced. Clearly, the most important criteria of the therapeutic success in the use of these diets are the long term serum lipid and lipoprotein responses obtained in the individual patient.

What is needed is a detailed description of how therapeutic diets for hyperlipidemia are formulated and written. How is a diet developed for patients with hyperlipidemia? What are the necessary tools when constructing a therapeutic diet? The following is a detailed discussion of the developmental procedure of the Cleveland Clinic Foundation's Prudent Polyunsaturated Fat Diet (PPFD). This chapter will include the parameters of the diet, why they were chosen, tools needed to develop a diet, the dietary considerations needed to enhance patient adherence and patient education.

Table 1. Recommendations of Expert Committees on Moderate Dietary Change to Reduce Risk of Coronary Heart Disease

| Country and organization of origin | For gen. populations (GP) or high risk (HR) Groups | As % of Total Calories | | | Chol. (mg/day) | Decr. Sugar |
|---|---|---|---|---|---|---|
| | | Fat | PUFA | SFA | | |
| Scand. countries, 1968 Official Collective Recommend. | GP | 25-35 | Increase | Decrease | - | Yes |
| Un. States, 1972 Inter-Soc. Commission for Heart Dis. Resources | GP HR | <35 | up to 10 | <10 | <300 | - |
| Un. States, 1972 Am. Health Fndn. | GP | <35 | 10-12 | 10-12 | 300 | Yes |
| Un. States, 1972 Am. Med. Assoc. & Nat. Acad. of Sciences | HR | Substan. decrease in sat. fat | Substit. PUFA for SFA | Decrease | Reduce | - |
| Internat. Soc. of Cardiol. 1973 | HR | <30 | - | Decrease | <300 | Yes |
| The Netherlands, 1973 Nutrition Council | GP | 30-35 Ideally 33 | 10-13 | Decrease Substan. | 250-300 | Yes |
| Australia, 1974 Nat. Heart Fndn. | GP HR | 30-35 | Increase | Decrease | <300 | - |
| United Kingdom, 1974 Dept. of Health & Social Security | GP | Reduce | - | Reduce | - | - |
| Germany, 1975 | GP | 20-25 | 10-15 | Reduce | 300 | - |
| New Zealand, 1976 | GP | 35 | Subst. PUFA for SFA | Decrease | 300-600 | Yes |

Table 1. Recommendations of Expert Committees on Moderate Dietary Change to Reduce Risk of Coronary Heart Disease (Continued)

| Country and organization of origin | For gen. populations (GP) or high risk (HR) Groups | As % of Total Calories | | | Chol. (mg/day) | Decr. Sugar |
|---|---|---|---|---|---|---|
| | | Fat | PUFA* | SFA* | | |
| U.K., 1976 Royal Coll. of Physicians & Br. Cardiac Soc. | GP | Toward 35 | Subst. PUFA for SFA | Reduce | Reduce | Yes |
| Norway, 1976 Ministry of Agriculture | GP | <35 | Increase | Decrease | - | Yes |
| Canada, 1977 Dept. of Nat. Health & Welfare | GP | 30-35 | Subst. PUFA for SFA | Decrease Substan. | <400 | Yes |
| Ireland, 1977 An Foras Taluntais | GP | Reduce | - | - | - | - |
| Un. States, 1977 Dietary Goals | GP | 30 | up to 20 | <10 | <300 | Yes |
| Un. States, 1978 Am. Heart Assoc. | GP | 30-35 | up to 10 | <10 | <300 | - |
| Europ. Soc. of Cardiol., 1978 | HR | Toward 30 | Subst. PUFA for SFA | 10 | <300 | Yes |
| Food and Agriculture Organ./ World Health Organ., 1978 | GP | 30-35 | 10-12 | Decrease | <300 | Yes |

PUFA = polyunsaturated fatty acid; SFA = saturated fatty acid; Chol. = Cholesterol; decr. = decrease

RATIONALE OF THE DIET PARAMETERS

The initial dietary treatment for hyperlipidemia should meet the following criteria for maximum effectiveness:

1. Reduce cholesterol intake to ∿300 mg/day.

2. Increase polyunsaturated fat to saturated fat (P/S) ratio to 1.0-1.3.

3. Reduce total fat intake to 80-100 gms/day (35-40% of total kilocalories).

4. Maintain caloric intake between 1800-2500 kilocalories/day (applies to the average adult, doing normal routine work).

5. Limit sodium intake to 4-6 gms/day.

6. Decrease consumption of simple sugars.

It is generally recognized that CHD is caused by a combination of factors. Some of the known risk factors for CHD are high concentrations of blood lipids, hypertension, smoking, overweight, imbalanced and excessive food intake, diabetes mellitus, gout and renal disease, as discussed in the previous paper by Naito. Some of these risk factors can be modified or controlled by altering certain dietary habits. Some of these modifications include an increase in fiber intake, a decrease in the amount of dietary sodium, and a reduction of calorie intake until ideal body weight is achieved. Almost all lipids and lipoprotein disorders can be altered through appropriate diets (Keys, et al. 1965).

According to the American Heart Association (1973), the average American diet contains approximately 600 mg of cholesterol/day and 145 gms of fat with a P/S ratio of about 0.3. Consumption of this quantity and quality of fat is considered to be unhealthy for the general public, let alone for the high risk CHD individual with hyperlipidemia and hyperlipoproteinemia.

The Cleveland Clinic Foundation's Prudent Polyunsaturated Fat Diet was designed as a therapeutic regimen for hypercholesterolemia and concentrates on Brown's (1976) Master Food Plan which utilizes lean meat, fish and poultry, low fat or skim milk dairy products and polyunsaturated oils and margarines (see Table 2).

The PPFD can be modified to be used for the dietary management of those individuals with elevated triglycerides and/or who

Table 2. Master Food Patterns[a,b,c]

| Food Pattern | Fat %Cal. | CHO[d] %Cal. | Sat. F.A.[d] %Cal. | Poly F.A.[d] %Cal. |
|---|---|---|---|---|
| Basic Low Fat* | 12 | 73 | 5 | 1 |
| Moderate Low Fat* | 20 | 65 | 5 | 8 |
| Moderate Fat* | 33 | 52 | 7 | 12 |
| Customary Fat* | 38 | 47 | 10 | 15 |

a - Mean daily protein intake: 15% of total calories (range = 15-20%)
b - Mean daily cholesterol intake: 300 mg (range = 75-300 mg)
c - Adapted from Brown (1976)
d - Carbohydrate, saturated fatty acids, polyunsaturated fatty acids

are above their ideal body weight. Since obesity is correlated to a high consumption of refined, simple sugars, alcohol and an excessive caloric intake (Kannel, et al. 1979), it seems prudent to design a diet that controls these factors. Often a reduction in dietary cholesterol and saturated fat result in a deficit of kilocalories. For those individuals at their ideal body weight, this deficit is supplemented with an increased intake of complex carbohydrates such as starches, fruits, vegetables and vegetable proteins.

The sodium intake as salt and high sodium foods can affect the control of hypertension. The average daily diet of Americans contains approximately 7-15 gms of sodium (Select Committee on Nutrition and Human Needs, 1977). However, these figures will vary depending on the quantity of convenience food items used in the diet. The sodium content of these foods is available in a handbook published by the West Suburban Dietetic Association (Rezabek, 1979). When food selection is derived according to the Basic Four Food Groups, the Recommended Daily Allowances (National Research Council, 1974) are met for protein, vitamins and minerals. The PPFD was designed with the above considerations in mind.

## DEVELOPMENT OF THE THERAPEUTIC DIET

Certain basic considerations must be met when planning a therapeutic diet. They include a profile of the patient providing age, socioeconomic conditions, number of meals eaten away from home per day, physical activity and limitations and the presence of any chronic illness which may necessitate additional dietary modification, i.e. diabetes mellitus, gout, hypertension. Diet histories and a 24-hour recall (Figures 1-3) are tools that are used in obtaining food preferences and details of an individual's eating pattern. Setting up a summary containing this information will aid the dietitian in planning a therapeutic diet that is adaptable to an individual's life-style, needs and preferences. This information will ensure optimal nutritional care, enhance patient dietary adherence and provide a base for individual instructions.

What tools are available to the dietitian when planning a therapeutic diet? Reliable information on nutrients can be obtained from food composition handbooks such as Church and Church (1975) and the United States Department of Agriculture (USDA) Handbook #456 (Adams, 1975), or from Handbook #8 (Watt and Merrill, 1963), or a nutrient data bank. Food values are listed in either grams or in common household weights and measures. Access to a nutrient data bank greatly facilitates the dietitian's work load when developing individually tailored diets.

The data bank at Case Western Reserve University is a computerized food table data storage facility which contains more than 2,300 food items and recipes. The food items, including brand name items, in the data base have been selected as a result of analysis of several thousand 24-hour intake records from several areas of the United States. For each food item, there is a storage space for 71 nutrients. (Table 3) The nutrient composition of foods is derived from a variety of sources (Adams, 1975; American Home Economics Association, 1975; Anderson, 1976; Anderson et al., 1977; Anderson et al., 1975; Brignòli et al., 1976; Dicks, 1965; Diem and Lentner, 1973; Dunbar and Stunkard, 1979; Ehrheart and Mason, 1967; Exler et al., 1977; Exler and Weihrauch, 1976; Exler and Weihrauch, 1977; Feeley et al., 1972; Feeley and Watt, 1970; Fetcher, 1967; Fristrom et al., 1975; Fristrom and Weihrauch, 1976; Gormican, 1970; Hardinge, 1961; Hardinge, 1965; Leung, 1961; Leverton and Odell, 1959; Marsh et al, 1977; Matthews and Garrison, 1975; Mattice, 1950; McCarthy, 1968; Murphy et al., 1975; National Research Council, 1974; Nutrient Composition of Common Sizes and Measures of Foods, 1970; Organization of the United Nations, 1970; Orr, 1969; Orr and Watt, 1957; Osis, 1972; Perloff and Butrum, 1977; Posati et al., 1975; Positi et al., 1975; Positi and Orr, 1976; Streiff, 1971;

THE CLEVELAND CLINIC FOUNDATION
NUTRITION PROFILE FORM

Name:______________________________ Age:_____ Height:_____ Weight:_______
Address:___________________________ Occupation:_______________________
___________________________ Scheduled Work Hours:_______________
Phone:(home)________________________
(work)__________ __________ Date:_____________________________

1. List other household members:

| | NAME | RELATION | AGE | SPECIAL DIET FOLLOWED |
|---|---|---|---|---|
| a. | | | | |
| b. | | | | |
| c. | | | | |
| d. | | | | |
| e. | | | | |

2. Do you now or have you ever followed a "special diet"? Yes_____ No_____
If yes, what type of diet?_________________________________
How long did you follow the diet?___________________________
For what reason did you follow the diet?______________________
From what source did you recieve the information about your diet?
Physician_____ Dietitian_____ Book or Magazine_____ Title___________
Special diet of health food store_____ Other_____________________
3. Do you take vitamin/mineral/diet supplements? Yes_____ No_____
If yes, list:_________________________________________
_________________________________________________
4. Who is responsible for home food preparation?___________________
5. Who is responsible for food purchasing?_______________________
6. Which meals are most often eaten away from home:_________________
Breakfast_____ Lunch_____ Dinner_____ Snacks_____
7. Do you have trouble chewing or swallowing food Yes_____ No_____
If yes, explain:_______________________________________
_________________________________________________
8. Do you have any food allergies? Yes_____ No_____
If yes, explain:_______________________________________
_________________________________________________

Fig. 1. The Cleveland Clinic Foundation Nutrition History Form

NUTRITION PROFILE FORM - Page 2

9. Do you have frequent occurrences of the following?

| | Yes | No | | Yes | No |
|---|---|---|---|---|---|
| a. indigestion | ____ | ____ | d. diarrhea | ____ | ____ |
| b. nausea | ____ | ____ | e. constipation | ____ | ____ |
| c. vomiting | ____ | ____ | | | |

10. Have you had any change in weight within the: a) past 3 months? b) past year? c) past 2 years? d) past 5 years? (please circle).
If yes, explain: ______________________________

11. Do you exercise or participate in athletic activities? Yes____ No_____
If yes, what type? ______________________________
How often? ______________________________

12. What do you usually eat? Please list all foods that you usually eat between the time that you get up in the morning and the time you to to bed at night. Remember, everything you eat or drink is food. Cookies, nuts, sweets, salad dressings, gravy, cocktails, highballs, beer and soft drinks are food items just as are meat and potatoes. Indicate how the foods are prepared, such as broiled, baked, fried, boiled, etc.

| At Breakfast | At Lunch | At Dinner |
|---|---|---|
| | | |
| Mid-Morning Snack | Mid-Afternoon Snack | Before Bed Snack |
| | | |

During the night: ______________________________

Fig. 2. The Cleveland Clinic Foundation Nutrition History Form

NUTRITION PROFILE FORM - Page 3

Please indicate below how often you eat or drink the following items.
Include amount and where indicated, circle the type of food eaten.
Example Milk: whole skim low fat.

FOOD ITEM

Milk: whole skim low fat chocolate:________
Cheese buttermilk yogurt:________
Pudding cottage cheese custard ice cream:________
Beef lamb pork: baked broiled fried:________
Chicken Turkey veal: baked broiled fried ________
Fish shellfish: baked broiled fried________
Bacon sausage luncheon meat hot dogs:________
Liver organ meats (kidney, tongue, etc.):________
Eggs egg substitutes:________
Dried beans peas legumes:________
Peanut butter nuts seeds:________
Cereals: whole grain sugar coated other:________
Breads: whole grain white other ________
Potatoes: baked mashed fried________
Rice macaroni noodles spaghetti:________
Dark green or yellow vegetables:________
Other vegetables (including lettuce):________
Citrus fruit or juice (not fruit drink):________
Other fruit or juice: sweetened unsweetened ________
Cake cookies doughnuts pastries pie "Jello":________
Sugar jelly syrup honey hard candy chocolate candy:________
Pretzels potato chips peanuts salty snack foods:________
Soups: cream other________
Butter lard shortening bacon fat:________
Margarine vegetable oil (list brand names):________
Salad dressings mayonnaise:________
Gravy cream sauces:________
Alcoholic beverages: Type beer/wine other alcohol:________
Kool-Aid fruit drinks lemonade punch:________
Soft drinks ("pop") Diet soft drinks:________
Coffee tea decaf. coffee Postum:________

Which of the following do you add to your coffee or tea? Indicate amount per cup:

Nothing____ Cream____ Nondairy Cream____ Sugar____

Saccharin____ Honey____ Other____

Fig. 3. The Cleveland Clinic Foundation Nutrition History Form

Table 3. Nutrients - In HVH-CWRU Nutrient Data Base

| | |
|---|---|
| Calories | Minerals and Trace Elements |
| Total Protein | Iron |
| Animal Protein | Calcium |
| Plant Protein | Phosphorous |
| Total Fat | Sodium |
| Animal Fat | Potassium |
| Plant Fat | Magnesium |
| Total Carbohydrate | Chlorine |
| Refined Carbohydrate | Chromium |
| Natural Carbohydrate | Cobalt |
| Alcohol | Copper |
| Ash | Iodine |
| Fiber | Manganese |
| Water | Molybdenum |
| Caffeine | Selenium |
| | Sulphur |
| Cholesterol | Zinc |
| Fatty Acids | |
| Total Saturated | Amino Acids |
| Total Unsaturated | Cysteine |
| Oleic | Cystine |
| Linoleic | Histidine |
| | Isoleucine |
| Vitamins | Leucine |
| Vitamin D | Methionine |
| Total Vitamin A | Phenylalanine |
| Performed Vitamin A | Threonine |
| Beta-Carotene | Tryptophan |
| Total Tocopherol | Tyrosine |
| Alpha Tocopherol | Valine |
| Other Tocopherol | |
| Ascorbic Acid | Sugars |
| Thiamin | Glucose |
| Riboflavin | Fructose |
| Niacin | Lactose |
| Pyridoxal B6 | Maltose |
| Vitamin B12 | Sucrose |
| Folic Acid | Reducing Sugars |
| Pantothenic Acid | |
| Biotin | |
| Choline | |

HVH-CWRU = Highland View Hospital-Case Western Reserve University

Sturderant et al., 1973; Toepfer et al., 1951; Umbarger, 1965; Watt and Merrill, 1963; Weihrauch et al., 1976; Zook and Lehmann, 1968). All analytical data were substantiated and verified for appropriate analytical methodology and evaluated for accuracy by research dietitians and an experienced technical assistant before incorporation into the data base. Manufacturers' data were accepted only when they were received directly from the manufacturers and based on their chemical analyses. Data were subjected to several validation processes including programmed verification before insertion in the table. The data base is reviewed and updated on a continuing basis with the addition of new items or nutrient composition values as information becomes available. For many nutrients, the only available data are from the various scientific journals. This information is carefully evaluated for methodology and reliability before incorporation into the data base. It must be recognized that for some nutrients, i.e. Vitamin D, Vitamin $B_6$, iodine, copper, biotin and pantothenic acid, only a limited amount of reliable food composition data are available. All nutrient data for food items are calculated on the basis of 100 gram portions. However, analysis for all food items may be retrieved in a variety of weights (i.e. from grams to ounces or pounds and, for many items, analysis may be retrieved for volume equivalents and for descriptive terms). An example of a computer printout is illustrated in Figures 4-6.

NUTRIENT ANALYSIS SUMMARY
NAME:
HVH-CWRU NUTRIENT DATA BASE DATE: Friday, November 30, 1979 ID: CC 002 05

| | | | Kcal | T-Pro Gm | T-Fat Gm | T-CHO Gm | Chol mg | PUFA Gm | SFA Gm | Fiber Gm |
|---|---|---|---|---|---|---|---|---|---|---|
| Breakfast | | | | | | | | | | |
| 6 | Vol Oz | ORANGE JUICE UNSW, CANNED | 90 | 1.5 | 0.4 | 20.9 | | | | .1867 |
| 1 | Large | EGG WHOLE, POACHED | 79 | 6.0 | 5.6 | 0.6 | 273 | .72 | 1.665 | 0 |
| 2 | Wt Oz | CANADIAN BACON, BROILED OR FRIED, DRAINED, SERVING WT | 157 | 15.7 | 9.9 | 0.2 | 50 | 1.527 | 3.365 | 0 |
| 1 | Slice | WHOLE WHEAT TOAST | 55 | 2.4 | 0.7 | 10.8 | | | .1121 | .361 |
| 2 | Teasp | MARGARINE REGULAR, FLEISCHMANN | 68 | 0.1 | 7.6 | 0.0 | 0 | 2.026 | 1.231 | |
| Lunch | | | | | | | | | | |
| 2 | Wt Oz | BEEF ROUND RUMP ROAST, BONELESS LM ROASTED, SERVING WT | 118 | 16.5 | 5.3 | 0.0 | 52 | .681 | 2.213 | 0 |
| 2 | Slice | WHOLE WHEAT BREAD | 112 | 4.8 | 1.4 | 21.9 | | | .2714 | .736 |
| .5 | Cup | PINEAPPLE CHUNKS, CANNED, SOLIDS & LIQUIDS, JUICE PACK | 71 | 0.5 | 0.1 | 18.6 | | | | .369 |
| 8 | Vol Oz | SKIM MILK | 86 | 8.4 | 0.4 | 11.9 | 5 | .0171 | .2867 | 0 |
| .33 | Cup | SUGAR GRANULATED, WHITE | 244 | 0.0 | 0.0 | 63.0 | 0 | 0 | 0 | 0 |
| .167 | Teasp. | CINNAMON, GROUND | 1 | 0.0 | 0.0 | 0.3 | 0 | .0020 | .0024 | .0921 |
| .5 | Cup | WALNUTS ENGLISH (HALVES) | 326 | 7.5 | 32.0 | 7.9 | 0 | 21 | 3.47 | 1.05 |

Fig. 4. Example of the HVH-CWRU Nutrient Data Base Computer Printout.

| | | | Kcal | T-Pro Gm | T-Fat Gm | T-CHO Gm | Chol mg | PUFA Gm | SFA Gm | Fiber Gm |
|---|---|---|---|---|---|---|---|---|---|---|
| Dinner | | | | | | | | | | |
| 4 | Vol Oz | APRICOT NECTAR, VITAMIN C ADDED | 72 | 0.4 | 0.1 | 18.3 | | | | .251 |
| 3 | Wt Oz | FLOUNDER BAKED, SERVING WT | 119 | 25.5 | 1.1 | 0.0 | 43 | .5022 | .2384 | 0 |
| .5 | Cup | POTATO PARED, BOILED & DICED OR SLICED | 50 | 1.5 | 0.1 | 11.2 | | | | .3875 |
| .5 | Cup | WHITE SAUCE (MEDIUM) | 203 | 4.9 | 15.6 | 11.0 | 16 | | 8.75 | |
| .75 | Cup | BEANS GREEN, CUT STYLE CANNED, DRAINED SOLIDS | 24 | 1.4 | 0.2 | 5.3 | | | | 1.013 |
| 1 | Slice | WHOLE WHEAT BREAD | 56 | 2.4 | 0.7 | 11.0 | | | .1357 | .368 |
| .5 | Cup | CABBAGE RAW, COARSELY SHREDDED OR SLICED | 8 | 0.5 | 0.1 | 1.9 | | | | .28 |
| 2 | Teasp | MAYONNAISE | 67 | 0.1 | 7.5 | 0.2 | 7 | | 1.307 | |
| 2 | Teasp | MARGARINE REGULAR, FLEISCHMANN | 68 | 0.1 | 7.6 | 0.0 | 0 | 2.026 | 1.231 | |
| Evening Snack | | | | | | | | | | |
| 1 | Cup | Skim Milk | 86 | 8.4 | 0.4 | 11.9 | 5 | .0171 | .2867 | 0 |
| .5 | Medium | BANANA YELLOW RAW, EP | 51 | 0.7 | 0.1 | 13.2 | | | | .2975 |
| 6 | Number | STRAWBERRIES, WHOLE FRESH UNSW | 41 | 0.8 | 0.6 | 9.4 | | | | 1.453 |

Fig. 5. Example of the HVH-CWRU Nutrient Data Base Computer Printout

| NUTRIENT SUMMARY | | | | % RDA | % CAL |
|---|---|---|---|---|---|
| Kilocalories | 2250 | | | 83 % | |
| Total Protein | 109.8 | Gm | | 196 % | 20 % |
| Total Fat | 97.4 | Gm | | | 39 % |
| Total Carbohydrate | 249.6 | Gm | | | 44 % |
| Cholesterol | 449 | mg | * | | |
| Total Polyunsat FA | 28.52 | Gm | * | | |
| Total Saturated FA | 24.56 | Gm | * | | |
| Fiber | 6.844 | Gm | * | | |
| Ascorbic Acid | 252.6 | mg | * | 561 % | |
| Thiamin | 1.724 | mg | * | 153 % | |
| Niacin | 15.26 | mg | * | 264 % | |
| Riboflavin | 1.788 | mg | * | 132 % | |
| Pyridoxal B6 | 1389 | ug | * | 69 % | |
| Vitamin B12 | 2.469 | ug | * | 82 % | |
| Folic Acid | .3154 | mg | * | 79 % | |
| Total Vitamin A | 4853 | IU | * | 97 % | |
| Vitamin D | 199.9 | IU | * | | |
| Iron | 15.72 | mg | | 157 % | |
| Calcium | 1105 | mg | | 138 % | |
| Phosphorus | 1820 | mg | | 228 % | |
| Magnesium | 424.1 | mg | * | 121 % | |
| Copper | 1.347 | mg | * | | |
| Zinc | 13.64 | mg | * | 91 % | |

HVH-CWRU Nutrient Data Base

Fig. 6. Example of the HVH-CWRU Nutrient Data Base Computer Printout

The computer system of the Nutrient Data Base is a Digital Equipment Corporation PDP-11/45 minicomputer running under the UNIX operating system. The system is a 16-bit computer with 124K words of core and semiconductor memory and 100 MB of on-line secondary storage.

Information on food composition and nutrient adequacy through the use of computers is a growing, time saving and valuable tool for the dietitian. If a computer is not available the menus developed from the designed diet will have to be hand calculated to determine nutrient adequacy.

After the parameters of the diet have been established, educational materials should be developed. The materials should emphasize a positive approach to dietary change and why the change is recommended. Written diet instruction materials can be constructed using various formats. Having peers or patients review the proposed material will help to ensure that the information was covered in an understandable, comprehensive and accurate way. The following format outlines the diet instruction material used at The Cleveland Clinic Foundation:

1. Introduction.

2. Definition of key terms, i.e. cholesterol, saturated fat and polyunsaturated fat.

3. Summary of the key principles of the diet.

4. Discussion of food groups.

5. Special instructions on label reading, cooking tips, shopping tips, dining out in restaurants and cookbook references.

Audiovisual aids and slide presentations can help to supplement the written materials. It has been our experience that the most effective way to teach the PPFD is in a classroom type of environment. This method of teaching promotes participation from patients as well as from their family members. Often patients who are already familiar with the diet offer helpful suggestions on meal preparation and shopping suggestions to those who are receiving their initial dietary instructions. A short slide presentation emphasizing the key principles of the PPFD is also provided in the instruction. This presentation leads into a group discussion monitored by the dietitian. Feedback regarding the effectiveness of our diet class and patient compliance will be discussed later in this paper.

After the parameters of the therapeutic diet were established, The Cleveland Clinic Foundation's Prudent Polyunsaturated Fat Diet was developed by manipulating the use of saturated fat, polyunsaturated fat and cholesterol content of meat products, dairy products, margarines and oils (Table 4). The use of 2 egg yolks per week was calculated on a weekly basis (71 mg cholesterol/ day). To achieve the indicated P/S ratio, special consideration was placed on margarines. The PPFD recommends using margarines that list either liquid safflower, liquid sunflower, or liquid corn oil as the first ingredient and that the ratio of polyunsaturated fat to saturated fat is 2.0 (Figures 7 and 8).

NUTRITION INFORMATION PER SERVING

| | |
|---|---|
| SERVING SIZE | 1 TABLESPOON (14 GRAMS) |
| SERVINGS PER CONTAINER (ONE POUND) | 32 |
| CALORIES | 100 |
| PROTEIN | 0 GRAMS |
| CARBOHYDRATE | 0 GRAMS |
| FAT | 11 GRAMS |
| PERCENT OF CALORIES FROM FAT† | 99% |
| **POLYUNSATURATED†** | **4 GRAMS** |
| **SATURATED†** | **2 GRAMS** |
| CHOLESTEROL† | (0 MG./100 GM.) 0 MILLOGRAMS |
| SODIUM | (815 MG./100 GM.) 115 MILLOGRAMS |

INGREDIENTS: **MAZOLA LIQUID CORN OIL**, Partially Hydrogenated Soybean Oil, Skim Milk, Salt, Lecithin, Monoglyceride, Natural Flavor, Artificial Flavor and Color (Carotene). Idopropyl Citrate and Calcium Disodium EDTA Added to Protect Flavor. Vitamins A and D added.

Fig. 7. Example of an unacceptable margarine. Note that the primary ingredient is liquid corn oil. The P/S ratio of the margarine is about 2.

This number is determined by dividing the grams of polyunsaturated fat by the grams of saturated fat listed on the Nutrition Information label. A margarine with a P/S ratio of less than 2.0 will not produce a total diet P/S ratio of 1.0 to 1.3.

The meats allowed in the diet have an average cholesterol content of 30 mg or less per ounce. This includes the use of Brown's (1976) recommendations of low and medium fat meats such as lean beef, pork, ham, Canadian bacon, lamb, veal, fish and poultry without the skin. The meat group is limited to a total intake of 7 ounces/day or 210 mg cholesterol.

| NUTRITION INFORMATION PER SERVING | |
|---|---|
| SERVING SIZE | 14g |
| SERVING PER CONTAINER | 32 |
| CALORIES | 100 |
| CARBOHYDRATES | 0g |
| FAT (100% OF CALORIES) | 11g |
| **POLYUNSATURATES*** | **4g** |
| **SATURATES*** | **2g** |
| CHOLESTEROL* (0 mg/100g) | 0mg |

**PREPARED FROM PARTIALLY HARDENED SOYBEAN OIL AND COTTONSEED OIL, PASTEURIZED SKIMMILK, SALT, MONO & DIGLYCERIDES; POTASSIUM SORBATE ADDED AS A PRESERVATIVE, CITRIC ACID ADDED TO PROTECT FLAVOR, ARTIFICIALLY COLORED WITH BETA CAROTENE. VITAMIN A PALMITATE ADDED.**

* Information of fat and cholesterol content is provided for individuals who, on the advice of a physician, are modifying their total dietary intake of fat & cholesterol.

Fig. 8. Example of an unacceptable margarine. Note that the primary ingredient is a hydrogenated vegetable oil. This makes the product unacceptable, even though the P/S ratio is 2.

Skim milk and foods made with skim milk may be used as desired. If 2% milk is preferred, 8 ounces may be incorporated into the daily meal plan or 16 ounces of 1% milk may be substituted.

The desired P/S ratio is achieved by using 2 to 3 tablespoons of margarine or acceptable oils each day. The acceptable oils are liquid safflower, liquid soybean, liquid corn, liquid cottonseed, liquid sesame seed and liquid sunflower seed oil. The use of salad dressings, mayonnaise and stir fried vegetables is suggested to reach the recommended amount (Figures 9 and 10). Most brands of commercial mayonnaise have a high P/S ratio. However, these products do contain egg yolk which will affect the cholesterol content of the diet if used on a regular basis. Therefore, our patients are advised to use mayonnaise sparingly. An excessive intake of margarines or oils, i.e. >3 tablespoons/day has been associated with the potential risk of cholelithiasis (Sturderant, 1973). Therefore, we strongly suggest that to achieve the desired daily P/S ratio, the saturated fat intake be drastically reduced and the polyunsaturated intake increased up

INGREDIENTS: **Soybean Oil**, Glucose Syrup, Sugar, Water, Vinegars, Tomato Paste, Salt, Paprika, Dehydrated Onion and Garlic, Natural Flavorings and Spices, Algin Derivative Added.

Figure 9. Ingredients of an acceptable oil useful for achieving the desired P/S ratio.

INGREDIENTS: **PARTIALLY HYDROGENATED SOYBEAN OIL,** WATER, SUGAR, CORN SYRUP, RED WINE VINEGAR, TOMATO PASTE, SALT, LEMON JUICE CONCENTRATE, DRIED ONION AND GARLIC, XANTHAN GUM (IMPROVES POURABILITY), SORBIC ACID (A PRESERVATIVE), SPICES, NATURAL FLAVORING, CALCIUM DISODIUM EDTA (A PRESERVATIVE), ARTIFICIAL COLORING.

Figure 10. Ingredients of an acceptable oil useful for achieving the desired P/S ratio.

to 3 tablespoons of acceptable vegetable oils.

Table 4. Prudent Polyunsaturated Fat Diet Calculations[a]

| | Fat gms | Sat. FA gms | Poly FA gms | Chol mg |
|---|---|---|---|---|
| 7 oz. lean beef (7 x 3 gms) | 21 | 9[b] | 1.2[b] | 196[c] |
| 2 c. 1% milk (2 x 2.5 gms) | 5 | 3.1[d] | 0.2[d] | 14[e] |
| 4 tsp. tub marg. (4 x 4 gms) | 16 | 2.4[f] | 4.8[f] | - |
| 4 tsp. corn oil (4 x 5 gms) | 20 | 2.6[g] | 11.6[g] | - |
| Total: | 62 | 17.1 | 17.8 | 210 |
| Calories: | 558 | 153.9 | 160.2 | |
| % Total Calories Based on 1800 Calories | 31 | 8.5 | 8.8 | |
| Total P/S Ratio | | | 1.0 | |

a - Adapted from Brown, 1978.
b - Beef fat Sat. = 43% Poly. = 6%
c - Beef chol. = 28 mg/oz.
d - Butter fat Sat. = 62% Poly. = 4%
e - Milk 1% chol. = 7 mg.
f - Tub marg. Sat.= 15% Poly. = 30%
g - Corn oil Sat. = 13% Poly. = 58%

The PPFD excludes the use of various meat items such as convenience luncheon meats, frankfurter, sausage, bacon, poultry skin, commercially prepared meats and fish, TV dinners, commercially canned soups, stews and similar products. The reasons that they are excluded are twofold. The first is due to the high saturated fat content and/or cholesterol content greater than 30 mg per ounce. The second reason is due to the lack of nutrient information regarding the total fat, saturated fat and cholesterol content of commercially prepared convenience food items. (Table 5)

Table 5 illustrates examples of the cholesterol content of selected food items. Degree of hydrogenation is also important. Hydrogenation changes the chemical structure of an oil from an

Table 5. Cholesterol Content of Selected Foods

| Cholesterol/mg | Food | Amount |
|---|---|---|
| over 1500 | Brains | 3 oz. raw |
| 500-1500 | Kidneys | 3 oz. cooked |
| 200-500 | Liver (beef, calf, hog, lamb) | 3 oz. cooked |
| | Heart, beef | 3 oz. cooked |
| | Egg yolk | 1 |
| 100-200 | Clams, halibut, tuna | 3 oz. cooked |
| | Chicken, turkey (light meat) | 3 oz. cooked |
| | Beef, pork, lobster, chicken, turkey dark meat) | 3 oz. cooked |
| | Lamb, veal, crab | 3 oz. cooked |
| 20-50 | Whole milk | 8 oz. |
| | Cream, light | 1 oz. |
| | Creamed cottage cheese | 1/2 cup |
| | Processed American cheese | 1 oz. |
| | Processed Swiss cheese | 1 oz. |
| | Ice cream | 1/2 cup |
| | Cheddar cheese | 1 oz. |
| | Frankfurter | 1 |
| | Butter | 1 Tbsp. |
| | Cake (from commercial mix) | 2-1/2 oz. |
| | Oysters, salmon | 3 oz. cooked |
| under 20 | Half and half | 1 oz. |
| | Low fat yoghurt | 8 oz. |
| | Lard | 1 Tbsp. |
| | Mayonnaise | 1 Tbsp. |
| | Uncreamed cottage cheese | 1/2 cup |
| | Skim milk | 8 oz. |

unsaturated fat to a saturated fat. If a vegetable oil is solid at room temperature or lists a partially hardened or hydrogenated oil as the first word on the list of ingredients, it should be avoided because it is now a saturated fat. All vegetable oils are not high in polyunsaturated fats. (Table 6) Most nondairy creamers and whipped toppings are made from coconut oil or palm oil which are saturated fats. Labels of products that state either "vegetable fat" or "vegetable oil" are usually one of these two oils and should be avoided.

Table 6. Fatty Acid Composition of Various Oils (percent distribution)[a]

| Fatty Acid | Coconut Oil | Palm Oil | Olive Oil | Corn Oil | Safflower Oil |
|---|---|---|---|---|---|
| $C_{8:0}$ | 5.40 | - | - | - | - |
| $C_{10:0}$ | 8.40 | - | - | - | - |
| $C_{12:0}$ | 45.42 | 0.13 | - | - | - |
| $C_{14:0}$ | 18.10 | 0.91 | 0.08 | - | - |
| $C_{16:0}$ | 10.53 | 38.42 | 11.91 | 13.1 | - |
| $C_{16:1}$ | 0.40 | - | 1.05 | - | 0.90 |
| $C_{18:0}$ | 2.30 | 4.57 | 2.48 | 4.07 | - |
| $C_{18:1}$ | 7.51 | 42.44 | 73.24 | 29.01 | 17.30 |
| $C_{18:2}$ | - | 12.42 | 9.07 | 54.00 | 79.00 |
| $C_{18:3}$ | - | 0.71 | 1.01 | - | 0.13 |
| Others | 1.94 | 0.40 | 1.16 | - | 2.67 |

a - Naito, H.K. (Unpublished report, determined by gas-liquid chromatography.)

## PRACTICAL CONSIDERATIONS OF THE DIET

Information on brand names, label reading and food preparation may permit their use and help the palatability and variety of the diet. It is important that the diet be a positive change, encouraged by physicians, dietitians and family members. By providing information on label reading, food preparation and dining out, patients are offered a liveable diet. The food industry continues to provide new food products which dietitians must evaluate for the cholesterol and saturated fat content. The dietitian's professional judgement, in many instances, is the only guideline for evaluating the product.

By law, ingredients are listed on the label in descending order of the content by weight. Many items such as breads and cereals contain small amounts of saturated fats. Such products are allowed if the saturated fat is not listed as one of the first five ingredients (Figures 11 and 12).

INGREDIENTS: UNBLEACHED ENRICHED WHEAT FLOUR (FLOUR, NIACIN, REDUCED IRON, THIAMINE MONONITRATE (VITAMIN $B_1$), RIBOFLAVIN (VITAMIN $B_2$). CRACKED WHEAT, WATER, 100% STONE GROUND WHOLE WHEAT FLOUR, CORN SYRUP, UNSULPHURED MOLASSES, **PARTIALLY HYDROGENATED VEGETABLE SHORTENING** (SOYBEAN, PALM AND/OR COTTONSEED OILS), YEAST, WHEAT GLUTEN, SALT, HONEY, VINEGAR AND MONO- AND DIGLYCERIDES (FROM HYDROGENATED VEGETABLE OIL).

Fig. 11. Example of an unacceptable grain product. Although the type of fat used is saturated, note that it is not the primary ingredient.

INGREDIENTS Enriched Flour (Flour, Niacin, Reduced Iron, Thiamine Mononitrate, Riboflavin), Sugar, **Vegetable Shortening (contains one or more of the following partially hydrogenated oils: Soybean, Cottonseed, Palm, Peanut)**, Graham Flour, High Fructose Corn Syrup, Honey, Salt, Baking Soda, Artificial Flavor

Fig. 12. Example of an unacceptable grain product. Note that the saturated fat used is listed 3rd and is a primary ingredient.

Several low cholesterol cookbooks are available that our patients have found to be helpful. Not all cookbooks contain accurate information but we recommend three to our patients (Brown, 1968; American Heart Association, Inc., 1973; Jones, (1975).

Correct food preparation encourages creativity and increases variety in the diet. Table 7 provides a substitution list for high-cholesterol containing food products. With the elimination of many convenience food items, meal preparation will become more time consuming. Soups, casseroles and baked goods must be home-made with acceptable ingredients. The amount of counseling and cooking suggestions given by the dietitian to the patient will vary depending on a patient's previous eating pattern. Suggestions include trimming the fat from all meat before cooking and removing the skin from all poultry. The meat should be weighed after cooking with the bone and the fat removed. Meats can be baked, broiled or roasted on a rack. This will keep the meat above the fat drippings. Meats can be made moist by adding tomato juice, lemon juice, wine, broth, water or an acceptable oil combined with herbs and spices. Low fat gravies are made by refrigerating the meat drippings until the fat has hardened. The hardened fat can be skimmed off and removed and the gravy may be made from the remaining natural juice which contains a negligible amount of fat. Any acceptable vegetable oil or margarine may be added to the natural juices.

Combination dishes such as soups, stews and casseroles are prepared by cooking the meat separately and removing the fat either by draining the meat or chilling it overnight.

The vegetables should be prepared in the usual manner using acceptable margarines or oils in place of other unacceptable fats. Stir frying can be tried for variety. Recipes using egg whites, cholesterol free egg substitutes and skim milk products should be encouraged.

Dining out in restaurants is an important part of the American lifestyle. This is an area that is frequently overlooked and the dietitian needs to give the patients practical suggestions to help them maintain their diet. Most restaurants and cafeterias offer enough variety of foods so that a prescribed meal pattern can be selected. Restaurants that specialize in one product, i.e. omelettes or ice cream are not the wisest choice. Waiters and waitresses are usually helpful in explaining how an item is prepared. Many times they can have the food item prepared without the butter or cream sauce. Counseling patients regarding food choices in a restaurant is a difficult aspect of dietary management. In many instances the frequency of meals eaten in restaurants will need to be reduced but not omitted totally from the dietary protocol. Although dining out tends to violate the dietary recommendations of the PPFD, the suggestions offered in Table 8 list food choices which can be substituted on limited occasions. Cooperation of the restaurant industry regarding food preparation is not only needed but necessary in maximizing the therapeutic effectiveness of diets.

Table 7. Recipe Substitutions for Cholesterol Containing Foods[a,b]

| Foods High in Cholesterol and/or Saturated Fat | Recipe Substitutions |
|---|---|
| Marbled red meats, bacon, sausage, hot dog, luncheon | Lean red meats, pork, ham, Canadian Bacon, poultry, fish and peanut butter* Limit use to 7 oz./day. *Old fashioned or homemade may be used as desired. |
| Egg yolk Limit to 2/week, including those used in cooking | Cholesterol free egg substitutes and egg whites. |
| Whole milk | Skim milk, evaporated skim milk, reconstituted non-fat dry milk. Limited amounts of 1% or 2% milk. |
| Cream, half and half, non-dairy creamers | Cream substitutes made from acceptable oils, evaporated skim milk or non-fat dry milk mixed to double or triple strength. |
| Cheese | Low fat cheese containing 5% milk fat or less. May use 1 oz. of regular cheese in place of 2 oz. of meat, fish or poultry. |
| Sour cream | 8 oz. low fat plain yoghurt + 1 teaspoon cornstarch. |
| Cream cheese | Low fat cottage cheese blended with low fat yoghurt or skim milk to desired consistency. |
| If a recipe calls for: | Use instead: |
| 1 egg yolk | 1 egg white or 1 egg white + 1 teaspoon acceptable oil |
| 1 whole egg | 2 egg whites or 1 egg white + 1 teaspoon acceptable oil, or 1/4 cup cholesterol free egg substitutes. |

Table 7. (Cont.)

| | |
|---|---|
| 2 whole eggs | 3 egg whites or 2 egg whites + 2 teaspoons acceptable oil, or 1/2 cup cholesterol free egg substitutes |
| Evaporated milk | Evaporated skim milk |
| 1 oz. baking chocolate | 3 tablespoons unsweetened cocoa powder + 1 tablespoon acceptable fat |
| If a recipe calls for: | Use instead: |
| Butter or shortening | Safflower, sunflower, corn cotton-seed, soybean and sesame seed oils or acceptable margarines |
| Whipped toppings | Evaporated skim milk or non-fat dry skim milk powder |
| 1 oz. bacon (2 strips) | 1 oz. lean Canadian bacon or boiled ham; or 2 tablespoons of imitation bacon bits; or 2 strips of imitation bacon strips. |

a, b - Adapted from: Cleveland Clinic Foundation Diet Manual, 1979. American Heart Association Cookbook, 1973.

Table 8. Restaurant Suggestions

Appetizers: Consomme or bouillon, fruit cup, vegetable or fruit juice, seafood cocktail with lemon juice or acceptable sauce.

Entrees: Lean meat, poultry or fish baked or broiled without fat or gravy. Remove poultry skin and trim fat. Limit the amount and kind of meat according to the diet plan.

Vegetables: Any vegetable prepared without fat. Avoid all creamed vegetables and those containing butter, cheese or sour cream.

Salads: Vegetable, fruit and gelatin salads with dressings or mayonnaise allowed in the diet. Avoid those containing coconut, cream or cheese.

Breads: Sliced bread, hard rolls, English Muffins, bread sticks, plain melba toast or crackers. Avoid rich dinner rolls, biscuits, cornbread, popovers, muffins, pancakes, waffles and French toast.

Sandwiches: Sliced chicken, turkey, veal, lean beef and ham. Chicken, turkey or tuna salad without chopped egg. (Ask that the butter or margarine be omitted).

Desserts: Fresh or canned fruit, gelatin, angel food cake, fruit ice.

Beverage: Coffee, tea, decaffeinated coffee, fruit or vegetable juices, carbonated beverages, skim milk.

## MODIFICATIONS OF THE PRUDENT POLYUNSATURATED FAT DIET

If reduction to ideal body weight is necessary, the diet can be divided into exchange groups. At The Cleveland Clinic Foundation, the current PPFD has not been developed in this manner. However, for those individuals who are above their ideal body weight and/or have elevated triglycerides, a modified version of the American Diabetes Association's 'Exchange Lists For Meal Planning' (1976) which incorporates the parameters of the PPFD is used. If the patient is at his or her ideal body weight, a caloric restriction is not used, but rather the patient is counseled on guidelines for maintaining his/her weight. For those patients with elevated triglycerides and at their ideal body weight, a diet was designed restricting the use of concentrated carbohydrates, i.e. sugar, honey, jams, jellies, soda pop and alcohol (Department of Nutritional Services, 1977a). This is used in conjunction with the PPFD.

If a patient has hypertension, a 3 to 4 gm sodium restriction is generally used. This level of restriction eliminates the use of smoked and cured meats and fish, sauerkraut, pickles, olives and condiments such as ketchup, mustard and steak sauce, to list a few examples. (Department of Nutritional Services, 1977b). Commercially prepared dinners contain a high fat content as well as excessive levels of sodium.

In counseling pregnant or lactating women on the prudent diet, attention must be given to their increased nutrient and caloric needs. Vegetarians, depending on the degree of dietary restriction, will need information on foods available and maintaining an adequate nutrient intake. If indicated, the dietitian must educate the vegetarian on how to combine foods to achieve a protein intake of high biological value.

Children's diets need special consideration. The Pediatric Lipid Clinic at the Cleveland Clinic Foundation follows children from family members who are at high risk for premature CHD. This clinic handles the children's diets separately from the adult population. Because of space limitations, this topic will not be covered here.

To ensure any guarantee of patient adherence to the dietary protocol, dietary follow up with a dietitian is necessary and critical. This phase is usually the weakest link of most dietary programs. The initial diet instruction cannot cover all aspects of the diet. Questions will arise when the individual attempts to incorporate, at home, the diet into his or her lifestyle. Also, additional dietary counseling can be used as a supportive mechanism in optimizing patient adherence.

It is generally recognized that ongoing dietary follow up protocols increase patient adherence and success with changes in lifestyle. At the present time, diet follow up is patient initiated at the Cleveland Clinic Foundation. Being a tertiary care center, a large percentage of our patients are from out of town. At the time of the dietary instruction, the patients are encouraged to write or call if questions arise. We realize that this is not an adequate system. Although the Cleveland Clinic Foundation lacks rigid structure and consistency in the follow-up program as carried out by the "high risk" cardiology patients, we feel that the dietary program has been effective. Preliminary study (see discussion below) indicates that patient adherence to the PPFD appears to be high.

## PATIENT ADHERENCE

A retrospective study was done on 49 males and 2 females ranging in age from 34-66 years, returning to The Cleveland Clinic Foundation for their six week post-operative check up during January and February, 1979. All had attended the Prudent Polyunsaturated Fat Diet class prior to hospital discharge. During the post-operative check up, a diet survey sheet was distributed to the patients. Questions included a 24-hour recall designed to assess dietary adherence to the PPFD. Eighty-five percent admitted that the diet was a change from their previous eating habits. Sixty-nine percent stated they do not have trouble following the diet and 65% did not find the diet too restrictive. It was reported by 81% that their meat consumption was seven ounces or less per day. Sixty-two percent said they could not reduce their meat consumption below the seven ounce allowance. Twenty percent avoided egg yolks in their diets completely and 69% ate one to two egg yolks per week. Fifty-four percent reported using skim milk as a beverage or in cooking and 35% indicated using 2% milk. Fifty-eight percent used the recommended amount of polyunsaturated oils and margarines per day.

The results of the study indicate that the Cleveland Clinic's Prudent Diet has practical applications to the hypercholesterolemic individual in the American population. The demands of the diet are not so restrictive that they cannot be incorporated into a given lifestyle. It appears that the transition to the PPFD is not particularly difficult. This is advantageous to ensure high adherence to the dietary protocol. A therapeutic diet should be developed in a format that meets the demands of the individual's lifestyle or better yet, that of a given population. It is up to the professional staff to provide the educational tools and support mechanisms needed to alter these lifestyle changes. Ultimately, the success of the PPFD is judged by the lowering of the individual's blood lipid (i.e. cholesterol) levels, and our follow up studies indicate that with good dietary adherence the blood lipid levels of most individuals do decrease and even "normalize."

## SUMMARY

The vast number of deaths resulting from atherosclerosis clearly identify CHD as a leading public health problem. The etiology of CHD is complex and multifaceted. The process of atherosclerosis can be manipulated by controlling certain nutritional risk factors. Three decades after the Framingham Study, The Dietary Goals of the United States (Select Committee on Nutrition and Human Needs, 1977), identify the need for increasing the American population's awareness regarding the usefulness of modifying the nutritional risk factors for CHD. Altering the current American food habits to incorporate a prudent intake of cholesterol and saturated fats, to limit total fat consumption and to achieve and maintain ideal body weight by controlling the quantity and quality of total calories is an important concept to promote. A prudent diet regimen can be used for screened individuals with atherosclerotic heart disease. The decline in the number of deaths from cardiovascular disease is due to a combination of many factors, one being more effective education of the American population in controlling the primary and secondary risk factors. It is the responsibility of physicians, dietitians, nurses, public health officials and allied health scientists to reinforce lifestyle changes by providing the educational tools to those individuals already at high risk as well as to the general population. We have discussed the rationale of the Cleveland Clinic Foundation's PPFD, a step-by-step instruction on means of developing a hypocholesterolemic therapeutic diet and considerations needed to enhance patient adherence for a successful dietary program.

The following are samples of menus designed to meet the parameters of the Prudent Polyunsaturated Fat Diet discussed in the above section. The menus were written for the reference 50 year old male, 175 cm in height, at his ideal body weight of 70 kg with a caloric requirement of 2000 kilocalories per day. The Case Western Reserve University Nutrient Data Bank analyzed the menus for nutrient adequacy, cholesterol content, P/S ratio, fiber content and the percent of total calories derived from carbohydrate, protein and fat.

## DAY 1

| | Serving Size |
|---|---|
| **Breakfast** | |
| Apricot Nectar | 6 oz. |
| Farina | 1 cup |
| Low cholesterol egg substitute | 1/2 cup |
| Whole wheat toast | 2 slices |
| Margarine | 3 tsp. |
| Skim milk | 8 oz. |
| **Lunch** | |
| Corned beef sandwich (on rye bread) | 1 (with 3 oz. meat) |
| Three bean salad | 1 cup |
| Apple | 1 large |
| **Dinner** | |
| Salmon steak | 4 oz. |
| Herb rice | 1 cup |
| Beets | 1/2 cup |
| Tossed salad | 1 cup |
| Italian dressing | 3 tsp. |
| Whole wheat bread | 2 slices |
| Margarine | 2 tsp. |
| Angel food cake | 1-1/2" slice |
| Skim milk | 8 oz. |
| **Evening Snack** | |
| Old fashioned peanut butter | 2 tbsp. |
| Crackers | 8 |

## DAY 2

| | |
|---|---|
| **Breakfast** | |
| Grapefruit | 1/2 |
| Raisin bran | 1 cup |
| *Pancakes (*homemade with acceptable ingredients) | 3 |
| Margarine | 3 tsp. |
| Skim milk | 8 oz. |

Lunch

| | |
|---|---|
| Syrian sandwich | 1 (with 3 oz. meat) |
| Celery sticks | |
| Orange | 1 medium |
| Skim milk | 8 oz. |

Dinner

| | |
|---|---|
| Pork chop | 4 oz. |
| *Bread dressing (*homemade with acceptable ingredients) | 1 cup |
| Wax beans | 1 cup |
| Pan roll | 2 |
| Margarine | 3 tsp. |
| **Bananas Flambe | |

DAY 3

Breakfast

| | |
|---|---|
| Cranapple juice | 6 oz. |
| *Fried egg (*made with margarine) | 1 |
| English muffin | 1 |
| Margarine | 3 tsp. |

Lunch

| | |
|---|---|
| Chicken salad sandwich (on whole wheat bread) | 1 (with 2 oz. meat) |
| Mayonnaise | 2 tsp. |
| Carrot sticks | |
| Apple | 1 large |
| Skim milk | 8 oz. |

Dinner

| | |
|---|---|
| Roast beef | 5 oz. |
| Mashed potatoes | 1 cup |
| Corn | 3/4 cup |
| Whole wheat bread | 2 slices |
| Margarine | 3 tsp. |
| **Pink Lemonade pie | 1/8 pie |
| Skim milk | 8 oz. |

## DAY 4

| Breakfast | Serving Size |
|---|---|
| Orange juice | 6 oz. |
| Farina | 1 cup |
| Low cholesterol egg substitute | 1/2 cup |
| Whole wheat toast | 2 slices |
| Margarine | 3 tsp. |
| Skim milk | 8 oz. |

| Lunch | |
|---|---|
| Onion soup | 1 cup |
| Ham sandwich on roll | 1 (with 2 oz. meat) |
| Sliced tomato salad | |
| Mayonnaise | 2 tsp. |
| Pear | 1 medium |
| Skim milk | 8 oz. |

| Dinner | |
|---|---|
| Lamb chop | 5 oz. |
| Rice Pilaf | 1 cup |
| Peas | 3/4 cup |
| Mixed greens | 1 cup |
| Italian dressing | 2 tsp. |
| Whole wheat bread | 2 slices |
| Margarine | 2 tsp. |

## DAY 5

| Breakfast | |
|---|---|
| Pineapple juice | 6 oz. |
| *Omelet (*made with low cholesterol egg substitute) | 1/2 cup |
| Rye toast | 2 slices |
| Margarine | 3 tsp. |
| Skim milk | 8 oz. |

| Lunch | |
|---|---|
| Cottage cheese fresh fruit plate | 1/2 cup (low fat cheese) |
| Melba toast | 8 |
| **Whole wheat cupcake | 2 |

** printed with permission © American Heart Association

| | Serving Size |
|---|---|
| **Dinner** | |
| Veal Italian style | 6 oz. |
| Noodles | 1 cup |
| Zucchini | 1 cup |
| Fresh spinach salad | 1 cup |
| French dressing | 3 tsp. |
| Kaiser roll | 2 |
| Margarine | 2 tsp. |
| Fresh fruit cup | 1 cup |

## DAY 6

| | |
|---|---|
| **Breakfast** | |
| Orange juice | 6 oz. |
| Poached egg | 1 |
| Canadian bacon | 2 oz. |
| Whole wheat toast | 2 slices |
| Margarine | 3 tsp. |
| **Lunch** | |
| Roast beef sandwich (on whole wheat bread) | 1 (with 2 oz. meat) |
| Pineapple chunks | 3/4 cup |
| **Cinnamon nuts | 1 cup |
| Skim milk | 8 oz. |
| **Dinner** | |
| Apricot nectar | 6 oz. |
| Baked fillet of flounder | 3 oz. |
| *Delmonico potatoes (*made with acceptable ingredients) | 1 cup |
| Green beans | 3/4 cup |
| Coleslaw | 3/4 cup |
| Whole wheat bread | 2 tsp. |
| Margarine | 3 tsp. |
| **Evening Snack** | |
| *Banana/Strawberry milkshake (*made with skim milk) | |

## DAY 7

| | Serving Size |
|---|---|
| **Breakfast** | |
| Banana | 1 medium |
| Special K | 1 cup |
| *French toast (*made with low cholesterol egg substitute and wheat bread | 2 |
| Margarine | 3 tsp. |
| Skim milk | 8 oz. |
| **Lunch** | |
| Tomato juice | 6 oz. |
| Baked perch | 4 oz. |
| Herb rice | 1 cup |
| Parslied carrots | 1/2 cup |
| Rye roll | 2 |
| Margarine | 2 tsp. |
| Skim milk | 8 oz. |
| **Dinner** | |
| Spaghetti with meat sauce | 2 cups (with 3 oz. meat) |
| Tossed salad | 1 cup |
| Italian dressing | 3 tsp. |
| Italian bread | 2 (1" slices) |
| Margarine | 1 tsp. |
| Melon | 1 cup |

## DAY 8

| | |
|---|---|
| **Breakfast** | |
| Orange juice | 6 oz. |
| Life cereal | 1 cup |
| Low cholesterol egg substitute | 1/4 cup |
| English cup | 1 |
| Margarine | 3 tsp. |
| Skim milk | 8 oz. |
| **Lunch** | |
| Hamburger on bun | 1 (3 oz. pattie) |
| Lettuce and tomato slices | |
| Grapes | 20 |

| | Serving Size |
|---|---|
| Iced tea | |
| **Dinner** | |
| Ham steak | 4 oz. |
| Sweet potatoes | 1 cup |
| Cauliflower | 3/4 cup |
| *Biscuit (*homemade with acceptable ingredients) | 3 |
| Lettuce wedge | |
| French dressing | 2 tsp. |
| Margarine | 3 tsp. |
| Skim milk | 8 oz. |

DAY 9

| | |
|---|---|
| **Breakfast** | |
| Grapefruit sections | 1/2 cup |
| Oatmeal with raisins | 1 cup (with 1/4 cup raisins) |
| Bagel | 1 |
| Margarine | 2 tsp. |
| Skim milk | 8 oz. |
| **Lunch** | |
| Tuna salad sandwich (on whole wheat bread) | 1 (2 oz. tuna) |
| Mayonnaise | 2 tsp. |
| Fresh fruit cup | 1/2 cup |
| Angel food cake | 1-1/2" slice |
| Skim milk | 8 oz. |
| **Dinner** | |
| Roast pork | 5 oz. |
| Mashed potatoes | 1 cup |
| Broccoli | 1/2 cup |
| Marinated beet and onion salad | 3/4 cup |
| Whole wheat bread | 1 slice |
| Margarine | 3 tsp. |
| *Pudding (*made with skim milk) | 1/2 cup |
| **Evening Snack** | |
| Popcorn, popped | 2 cups |
| Iced tea | 2 cups |

DAY 10

| | Serving Size |
|---|---|
| **Breakfast** | |
| Orange juice | 6 oz. |
| 40% Bran Flakes | 1 cup |
| Low cholesterol egg substitute | 1/4 cup |
| Whole wheat toast | 2 slices |
| Margarine | 3 tsp. |
| Skim milk | 1 cup |
| **Lunch** | |
| Roast turkey sandwich (on whole wheat bread) | 1 (3 oz. meat) |
| Mayonnaise | 2 tsp. |
| Carrot and celery sticks | |
| Apple | 1 |
| **Dinner** | |
| Broiled steak | 4 oz. |
| Baked potato | 1 large (7 oz.) |
| Asparagus | 1/2 cup |
| Tossed salad | 1 cup |
| Oil and vinegar dressing | 2 tsp. |
| French bread | 1" slice |
| Margarine | 3 tsp. |
| Strawberries | 1 cup |

## REFERENCES


Adams, C., 1975. Nutritive Value of American Foods in Common Units, United States Department of Agriculture Handbook No. 456, Washington, D.C.

American Diabetes Association, Inc. and The American Dietetic Association, 1976. Exchange Lists for Meal Planning.

American Health Foundation Committee on Food and Nutrition, 1972. Position Statement on Diet and Coronary Heart Disease, Preventive Med., 1:255.

American Heart Association, Inc. 1973. Diet and Coronary Heart Disease, New York, N.Y.

American Heart Association, Inc., 1973b. American Heart Association Cookbook. David McKay Co., Inc., New York.

American Heart Association Committee on Nutrition, 1978. Diet and Coronary Heart Disease, Circulation, 58:762A.

American Home Economics Association, 1975. Handbook of Food Preparation, 7th edition, Washington, D.C.

American Medical Association Council on Foods and Nutrition, 1972. Diet and Coronary Heart Disease, J. Amer. Med. Assoc., 222: 1697.

Anderson, B., 1976. VII. Pork Products: Comprehensive Evaluation of Fatty Acids in Foods, J. Am. Diet Assoc., 69:44.

Anderson, B., Fristrom, G. and Weihrauch, J., 1977. X. Lamb and Veal: Comprehensive Evaluation of Fatty Acids in Foods, J. Am. Diet Assoc., 70:53.

Anderson, B., Kinsella, J. and Watt, B., 1975. II. Beef Products: Comprehensive Evaluation of Fatty Acids in Foods, J. Am. Diet Assoc., 67:35.

Brignoli, C., Kinsella, J. and Weihrauch, J., 1976. V. Unhydrogenated Fats and Oils: Comprehensive Evaluation of Fatty Acids in Foods, J. Am. Diet Assoc., 68:224.

Brown, H.B., 1968. Low Fat Vegetable Recipes, Cleveland Clinic Research Division, The Cleveland Clinic Foundation, Cleveland, Ohio.

Brown, H.B., 1976. Diet Management of Hyperlipidemia, Practical Cardiology, 2(4): 19.

Brown, H.B., 1978. Fat-Modified Animal Products. Proceedings of an International Symposium, Primary Prevention in Childhood of Atherosclerosis and Hypertensive Diseases (In print), Chicago, Illinois.

Church, C. and Church, H., 1975. Food Values of Portions Commonly Used, (12th edition), J.B. Lippincott Company, Philadelphia.

Committee on Diet and Cardiovascular Disease, 1977. Recommendations for Prevention Programs in Relation to Nutrition and Cardiovascular Disease. Report of the Committee on Diet and Audiovascular Disease as Amended and Adopted by the Department of National Health and Welfare, Canada, June.

Connolly, J.F., 1977. The Health Advisory Committee of An Foras Taluntais. The Prevention of Coronary Heart Disease, Dublin, Ireland, An Foras Taluntais, December, 1977.

Council on Rehabilitation, International Society of Cardiology, 1973. Myocardial Infarction: How to Prevent, How to Rehabilitate, The Boehringer Manheim Co.

Department of Nutritional Services, 1977a. Prudent Polyunsaturated Fat, No Added Salt Diet Instruction, The Cleveland Clinic Foundation.

Department of Nutritional Services, 1977b. Prudent Polyunsaturated Fat, Carbohydrate Controlled Diabetic Diet Instruction, The Cleveland Clinic Foundation.

Dicks, M., 1965. Vitamin E Content of Foods and Foods for Human and Animal Consumption, Bulletin 435, Agricultural Experimental Station, University of Wyoming, Laramie, Wyoming.

Diem, K. and Lentner, C., (eds.), 1973. Documenta Geigy, Scientific Tables, 7th edition, Ciba-Geigy Limited, Basel, Switzerland.

Dunbar, J. and Stunkard, A., 1979. Adherence to Medical Regimen, In: Nutrition, Lipids and Coronary Heart Disease (R.I. Levy,

B.M. Rifkind, B.H. Dennis, N.D. Ernst, eds.), Raven Press, New York, p. 391.

Editorial Board, European Society of Cardiology, 1978. Preventing Coronary Heart Disease: A Guide for the Practising Physician. Assm, The Netherlands, Van Gorcim & Comp., B.V.

Ehrheart, J. and Mason, B., 1967. Sugar and Acid in the Edible Portion of Fruits, J. Am. Diet Assoc., 50:130.

Exler, J., Avens, R. and Weihrauch, J., 1977. XI. Leguminous Seeds: Comprehensive Evaluation of Fatty Acids in Foods, J. Am. Diet Assoc., 71:412.

Exler, J. and Weihrauch, J., 1976. VIII. Finfish: Comprehensive Evaluation Fatty Acids in Food, J. Am. Diet Assoc., 69:243.

Exler, J. and Weihrauch, J., 1977. XII. Shellfish: Comprehensive Evaluation of Fatty Acids in Foods, J. Am. Diet Assoc., 71: 518.

Feeley, R., Criner, P. and Watt, B., 1972. Cholesterol Content of Foods, J. Am. Diet Assoc., 61:134.

Feeley, R. and Watt, B., 1970. Nutritive Value of Foods Distributed Under U.S.D.A. Food Assistance Programs, J. Am. Diet Assoc., 57:528.

Fetcher, E., 1967. Quantitative Estimation of Diets to Control Serum Cholesterol, Am. J. Clin. Nutr., 20:475.

Food and Agriculture Organization/World Health Organization, 1978. Dietary Fats and Oils in Human Nutrition: Report of an Expert Consultation, Rome: FAO Food and Nutrition Paper No. 3.

Fristrom, G., Stewart, B., Weihrauch, J. and Posati, L., 1975. IV. Nuts, Peanuts and Soups: Comprehensive Evaluation of Fatty Acids in Foods, J. Am. Diet Assoc., 67:351.

Fristrom, G. and Weihrauch, J., 1976. IX. Fowl: Comprehensive Evaluation of Fatty Acids in Foods, J. Am. Diet Assoc., 69:517.

Gormican, A., 1970. Inorganic Elements in Foods Used in Hospital Menus, J. Am. Diet Assoc., 56:397.

Hardinge, M., 1961. Lesser Known Vitamins in Foods, J. Am. Diet Assoc., 38:240.

Hardinge, M., 1965. Carbohydrates in Foods, J. Am. Diet Assoc., 46:197.

H.V.H.-C.W.R.U., 1979. Nutrient Data Base, Department of Nutrition and Biometry, Case Western Reserve University, Cleveland, Ohio.

Inter-Society Commission for Heart Disease Resources, Atherosclerosis Study Group and Epidemiology Study Group, revised 1972. Primary Prevention of the Atherosclerotic Disease, Circulation, 42:53A.

Jones, J., 1975. Diet for a Happy Heart, 101 Productions, San Francisco, CA.

Kannel, W.B., Gordon, T. and Castelli, W.P., 1979. Obesity, Lipids and Glucose Intolerance. The Framingham Study, Am. J. Clin. Nutr., 32:1238.

Keys, A., 1968. Official Collective Recommendation on Diet in the Scandinavian Countries, Nutr. Rev., 26:259.

Keys, A., Anderson, J., Grande, F., 1965. Serum cholesterol response to changes in the diet, Metabolism, 14:747.
Leung, W., 1961. Food Composition Table for Use in Latin America, The Institute of Nutrition of Central America and Panama and National Institutes of Health, Bethesda.
Leverton, R. and Odell, G., 1959. The Nutritive Value of Cooked Meat, Agricultural Experimental Station, Oklahoma State University, Publication MP-49.
Marsh, A., Moss, M. and Murphy, E., 1977. Composition of Foods: Spices and Herbs - Raw, Processed, Prepared, Consumer and Food Economics Institute, Agriculture Handbook No. 8-2, U.S.D.A., Washington, D.C.
Matthews, R. and Garrison, Y., 1975. Food Yields Summarized by Different Stages of Preparation, Consumer and Food Economics Institute, Northeastern Region, Agricultural Research Service, U.S.D.A. Handbook No. 102.
Mattice, M., 1950. Bridges' Food and Beverage Analyses (3rd ed.), Lea and Febiger, Philadelphia.
McCarthy, M., 1968. Phenylalanine and Tyrosine in Vegetables and Fruits, J. Am. Diet Assoc., 52:130.
Murphy, E., Willis, B. and Watt, B., 1975. Provisional Tables on the Zinc Content of Foods, J. Am. Diet Assoc., 66:345.
National Heart Foundation of Australia, Committee on Diet and Heart Disease, 1974-i. Dietary Fat and Coronary Heart Disease. A Review. Med. J. Australia, pp. 575, 616, 663.
National Research Council, 1974. Recommended Dietary Allowances, Food and Nutrition Board, National Academy of Science, Washington, D.C.
Netherlands Nutrition Council, 1973. Recommendations on Amount and Nature of Dietary Fats, Voeding, 34:552.
Nutrient Composition of Common Sizes and Measures of Foods, 1970. Internal Publication, Highland View Hospital and Case Western Reserve University, Cleveland, Ohio (March, 1970).
Organization of the United Nations, 1970. Amino Acid Content of Foods and Biological Data on Proteins, Food and Agriculture, Rome, Italy.
Orr, M., 1969. Pantothenic Acid, Vitamin B6 and Vitamin B12 in Foods, Home Economics Research Report No. 36, Consumer and Food Economics Research Division, Agricultural Research Service, U.S.D.A., Washington, D.C.
Orr, M. and Watt, B., 1957. Amino Acid Content of Foods, Home Economics Research Report No. 4, U.S.D.A., Washington, D.C.
Osis, D., 1972. Dietary Zinc Intake in Man, Am. J. Clin. Nutr., 25:582.
Pahlke, G., 1975. Dietary Fat and Degenerative Diseases, Nutr. Metab., 18:113.
Perloff, B. and Butrum, R., 1977. Folacin in Selected Foods, J. Am. Diet Assoc., 70:161.

Posati, K., Kinsella, J. and Watt, B., 1975a. I. Dairy Products: Comprehensive Evaluation of Fatty Acids in Foods, J. Am. Diet Assoc., 66:482.

Positi, L., Kinsella, J. and Watt, B., 1975b. III. Eggs and Egg Products: Comprehensive Evaluation of Fatty Acids in Foods, J. Am. Diet Assoc., 67:111.

Positi, L. and Orr, M., 1976. Composition of Foods: Dairy and Egg Products - Raw, Processed, Prepared, Consumer and Food Economics Institute, Agriculture Handbook, No. 8-1, U.S.D.A. Washington, D.C.

Rezabek, K., 1979. Nutritive Value of Convenience Foods (2nd ed.), West Suburban Dietetic Association.

Royal College of Physicians of London and the British Cardiac Society Joint Working Party, 1976. Prevention of Coronary Heart Disease, J. Royal Coll. Physicians, 10:214.

Royal Norwegian Ministry of Agriculture, 1975-1976. On Norwegian Nutrition and Food Policy, Report No. 32 to the starting.

Select Committee on Nutrition and Human Needs, 1977. Dietary Goals for the United States, United States Senate, Washington, D.C.

Stamler, J., 1979. Population Studies, In: Nutrition, Lipids and Coronary Heart Disease (R.I. Levy, B.M. Rifkind, B.H. Dennis, N.D. Ernst (eds.) Raven Press, New York, p. 25.

Streiff, R., 1971. Folate Levels in Citrus and Other Juices, J. Am. Clin. Nutr., 24:1390.

Sturderant, R.A., Pearce, M.L. and Dayton, S., 1973. Increased Prevalence of Cholelithiasis in Men Ingesting a Serum-Cholesterol-Lowering Diet, N. Eng. J. Med., 288(1):24.

The National Heart Foundation of New Zealand, 1976. Coronary Heart Disease: A Progress Report, The National Heart Foundation of New Zealand.

Toepfer, E., Zook, E., Orr, M. and Richardson, L., 1951. Folic Acid Content of Foods, Agriculture Handbook No. 29, U.S.D.A. Washington, D.C.

Umbarger, B., 1965. Phenylalanine Content of Foods, The Children's Hospital Research Foundation and the University of Cincinnati College of Medicine, Cincinnati, Ohio.

United Kingdom Department of Health and Social Security Advisory Panel of Committee on Medical Aspects of Food Policy (Nutrition), 1974. Diet and Coronary Heart Disease, Report on Health and Social Subjects, No. 7.

Watt, B. and Merrill, A., 1963. Composition of Foods - Raw, Processed, Prepared, Agriculture Handbook No. 8, U.S.D.A. Washington, D.C.

Weihrauch, J., Kinsella, J. and Watt, B., 1976. VI. Cereal Products: Comprehensive Evaluation of Fatty Acids in Foods, J. Am. Diet Assoc., 68:335.

Zook, E. and Lehmann, J., 1968. Mineral Composition of Fruits, J. Am. Diet Assoc., 52:225.

ACKNOWLEDGEMENTS

We gratefully acknowledge the expert advice of Helen B. Brown, Ph.D., Elizabeth A. Linn, R.D. and Charlene B. Krejci, R.D. during the preparation of this manuscript. We also would like to express our gratitude to Jeannette Goodman for typing this manuscript. This study was supported by Grant No. CRP 400 from the Cleveland Clinic Foundation, Grant No. HL-6835 from NHLBI and Grant No. 8557 from the Bleeksma Fund of The Cleveland Clinic Foundation.

# EFFECTS OF HYPOPROTEINEMIA ON SERUM LIPOPROTEIN COMPOSITION OF DOGS AND RATS

Lena A. Lewis, Herbert K. Naito, Irvine H. Page and Bruce A. Sebek

Biochemistry Department
The Cleveland Clinic Foundation
9500 Euclid Avenue
Cleveland, Ohio 44106

## INTRODUCTION

The significance of protein composition of serum in relation to the pattern of serum lipids and lipoproteins has not been clearly elucidated previously. Clinical investigations of protein depleted human beings have included extensive studies of patients with nephrosis and with malnutrition. In hypoproteinemic patients with nephrosis the increased concentration of serum cholesterol and triglyceride is carried in the -S 20-40 and -S40-70 low density (LDL) and -S 70-400 very low density (VLDL) and chylomicron fractions. The concentration of the -S 0-10 high density (HDL) fraction is usually not increased and may be decreased (Lewis, L.A., 1980). Similarly in rats with the experimental nephrotic syndrome, the increased concentration of serum cholesterol is chiefly in the LDL fraction (Lewis and Heyman, 1954). In both species the hypoproteinemia accompanies renal damage with proteinuria. These lipid and lipoprotein changes are in contrast with those reported in individuals with low serum protein levels due to malnutrition. Coward and Whitehead (1972) found in children with protein-calorie malnutrition a decrease in serum total cholesterol, triglyceride and β-lipoprotein concentrations. To determine what effect hypoproteinemia not due to renal disease exerts on the serum lipoprotein pattern dogs with normal renal function were depleted of protein reserves by a combination of plasmapheresis and feeding of a protein-free diet, and rats depleted by feeding of a protein-free diet. The serum lipoproteins and cholesterol were studied during the period of depletion and of recovery, when a nutritionally adequate diet was reinstituted.

METHODS

Dog Study

Normal adult mongrel dogs weighing between 10 and 15 kg were studied. The animals were fed a diet of dog chow* and cooked beef. After control blood studies were obtained after an 18-hour fast, the dogs were placed on a protein-free diet, described by Allison and Anderson (1945). One hundred fifty to 200 ml of blood, depending on the size of the dog, was removed, the cells were repeatedly washed and then returned to the animals. This procedure was repeated every other day for three bleedings. Blood samples were collected weekly for biochemical analyses. After four weeks on the protein-free diet the dogs were returned to the complete ration on which they were before protein-depletion and studies continued for nine weeks.

Rat Study

Pathological changes, described over many years, that occur in young children during protein deprivation vary greatly from those that occur in the adult with protein restriction. To learn more about the changes induced in the lipoprotein pattern during a period of very slow growth and of much more rapid development, two groups of rats were studied. The first was a group of old males weighing between 500 and 600 g at the beginning of the study, the second group comprised young, sexually mature, but still rapidly growing, male rats. The experimental procedures used for each group follow:

Large old rats. Seven adult male Sprague-Dawley rats weighing between 500 and 575 g, which had been maintained on chow diet, were bled from the tail after an 18 hour fast, for serum cholesterol and protein determination. They were then placed on the above protein-free diet (Allison and Anderson, 1945) with supplementary vitamins being given twice weekly. The rats had free access to food at all times except the 18 hours before collection of blood samples. Small samples of blood were taken after the rats had received the special diet for four weeks and eight weeks, and after ten weeks they were sacrificed and blood collected for analyses. (See below, Analytical Procedures C).

Young adult rats. (a) Effects of protein free diet on serum lipoprotein and lipids. A group of 12 young, male rats, average weight 285 g, was studied. After initial blood analyses when they were on the chow diet**, eight of the rats were placed on the protein-

---

* Ken-L-Ration, Quaker Oats Co., Chicago, Illinois 60654

** Wayne rat chow, Allied Mills, Special Feed Department, N. Wacker Drive, Chicago, Illinois 60606

Table 1. Experimental Diet for Rats (Diet I).

| High Fat Diet | gm | % of Calories |
|---|---|---|
| egg yolk | 10 | P = 21.3<br>F = 50.4 |
| soya oil | 21 | C = 28.3 |
| nonfat milk solid | 65 | |
| salt mixture (Osborn Mendel) | 3 | |
| Vitamins NBC - Vitamin diet fortification mix contains 100 mg choline/100 gm | 1 | |

Table 2. Solutions for Parenteral Nutrition for Rats

| Base Solution | Solution A Fat-free | Solution B Fat-Supplemented |
|---|---|---|
| 8.5% Freamine | 500 ml | 500 ml |
| 50% Dextrose | 500 ml | 500 ml |
| Sterile $H_2O$ | 109 | - |
| 10 Intralipid | - | 309 |
| Additives | | |
| 10% Ca glucomate | 10 (4.6 mEq) | 10 |
| 50% $MgSO_4$ | 0.2 (0.8 mEq) | 0.2 |
| 14.9% KCl | 35 ( 70 mEq) | 35 |
| 23.4% NaCl | 8 ( 32 mEq) | 8 |
| 16.4% Na acetate | 8 ( 16 mEq) | 8 |
| MVI | 5 | 5 |

free diet and four were continued on the chow. After six and ten weeks, blood studies were made.

(b) Effects of protein-adequate diet after protein deprivation on serum lipoprotein and lipids. A second group of 20 young rats, average weight 298 g, were studied. The plan of study was the same as described above except that the rats were returned to one of two nutritionally complete diets after protein depletion. Diet I was a high energy diet, high in fat containing 50.4% fat, 21.3% protein and 28.3% carbohydrate (Table I). Diet II was the standard rat chow*. Six weeks after the rats were returned to nutritionally adequate diet they were sacrificed and blood lipids, lipoproteins and proteins determined.

(c) Effects of adequate amino acid nutrient given parenterally after protein deprivation on serum lipoprotein and lipids of rats. A third group of 24 young mature rats, average weight 260 g, was similarly depleted of protein by dietary means for six weeks. At that time eight were sacrificed and the remaining received intravenous infusions of fluid to provide complete nutrition parenterally. Eight received solution A, basic solution which was fat free, eight received solution B, which contained additional intralipid (Table 2). After seven days of parenteral protein repletion the rats were sacrificed. In addition to serum analyses, liver samples were studied by electron microscopy.

## Analytical Procedures

Lipoproteins were determined ultracentrifugally at a density of 1.063 (Gofman, 1945) and/or of 1.21 (Lewis, 1952), or by electrophoresis on paper (Lees, 1960), acrylamide gel electrophoresis (Naito, et al., 1973) or on agarose gel (Pfizer System, 1974). Cholesterol was measured by the method of Abell et al. (1952) or by the autoanalyzer-II technique, by the protocol of the Lipid Research Clinics (1974); total protein by the biuret method (Sols, 1947). Starch gel electrophoresis was performed by the thin-layer modification (Lewis, 1966) of the method of Smithies (1962), using a discontinuous system of buffers (Poulek, 1957). Gels were stained for protein or lipid using amidoschwartz 10B or oil red-O. Electrophoretic studies of serum proteins were made on paper by the method of Durrum (1958) using barbiturate buffer, pH 8.6, or acetate by the microzone technique (1973). The livers of the rats were studied by light microscopy and some also by electron microscopy (EM). Details of the EM procedure have been summarized by Blanchard et al. (1979).

*Wayne rat chow, Allied Mills, Special Feed Department, N. Wacker Drive, Chicago, Illinois 60606

## RESULTS

The lipids in the serum or plasma are transported bound with protein in the form of lipoproteins. The lipoproteins have been separated and characterized chiefly by two methods, ultracentrifugation and electrophoresis. By ultracentrifugation of serum after increasing the density by addition of salts to an appropriate level, the lipoproteins are floated to the top of the tube in a way similar to the separation of cream in a separator. The lipoprotein concentrate is removed and studied in the analytical ultracentrifuge, where lipoproteins of different densities determined by the proportion of lipid to protein in the molecule are demonstrated to have different flotation rates. Most frequently used densities for lipoprotein studies are d 1.063 (Gofman, 1949) at which lipoproteins of lower densities are floated (VLDL and LDL) and d 1.21 (Lewis, 1952) at which both lower density and high-density lipoproteins are concentrated. Table 3 represents the comparison between the terminology of fractions resolved at d 1.063 and d 1.21, and also includes the electrophoretic characteristics of the fractions.

### Dog Study

Normal Dog Lipoprotein Pattern. Normal mongrel dog's serum has a cholesterol concentration which varies greatly from animal to animal, but irrespective of the cholesterol level the lipoproteins normally are concentrated predominantly in the high density, α-lipoprotein fraction with flotation rate between -S 0 and 10. The -S 0-10 fraction which is present in the amount of 400 to 1000 mg per dl is usually resolved in the ultracentrifuge as a double peak with maximum deflection at -S 4 and -S 7. The -S 4 is the more concentrated. There is often a small amount of material with flotation rate between -S 10 and 20; this fraction may be referred to as intermediate density lipoprotein (IDL). The lower density lipoproteins, -S 20-40, may be resolved as a single or double peak with maximum deflection at -S 23, and -S 28. This fraction when studied by paper electrophoresis migrates in the β-globulin area. The concentration of this fraction is normally less than 100 mg/dl. The fractions with flotation rate between -S 20 and 40 at d 1.21, are resolved in the $S_f$ 0-12 class at a density of 1.063.

When studied by starch-gel electrophoresis, dog lipoproteins may show four clearly resolved fractions with relatively fast mobility in the range of the α-globulins, and one or two fractions with very slow mobility. The most concentrated and deeply stained fraction with amido schwartz protein stain migrates most rapidly and has a mobility approximately 0.90 X that of dog albumin in starch gel when tris-borate buffer is used.

Table 3. Nomenclature of Lipoproteins: Electrophoretic vs. Ultracentrifugation Techniques

| Method of Study | Name of Fraction |
|---|---|
| Electrophoresis | α-lipoprotein - lipoprotein with electrophoretic mobility of $\alpha_1$-globulin. They may be resolved as single or multiple bands depending on species from which serum was derived and on type of support media. |
| Ultracentrifugation | HDL, high density lipoproteins of d 1.063-1.21.<br>$HDL_1$ subclass of HDL, d 1.063 - 1.080<br>$HDL_2$ subclass of HDL, d 1.080 - 1.125<br>$HDL_3$ subclass of HDL, d 1.125 - 1.21<br>HDL fraction has flotation rate at d 1.21 of -S 0-10.<br>HDL does not float at d 1.063 |
| Electrophoresis | Pre β-lipoprotein - lipoproteins with electrophoretic mobility of $\alpha_2$-globulins, when separated on agarose, paper or starch powder. Pre βs may be resolved as a single or as multiple bands depending on nature of serum being studied. |
| Ultracentrifugation | VLDL, very low density lipoproteins, d<1.006 g/ml). At d 1.21, fraction with flotation -S70-400 includes part, but not all of VLDL. At d 1.063, $S_f$20-400 is VLDL. The more rapidly floating fractions are less dense, have larger molecules. |
| Electrophoresis | β-lipoprotein, lipoprotein with electrophoretic mobility of β-globulin. It may be resolved as single or double band, depending on support medium and buffer system used. |
| Ultracentrifugation | LDL, low density lipoprotein d 1.006 - 1.063 g/ml<br>$LDL_1$ subclass of LDL, d 1.006 - 1.019 g/ml<br>$LDL_2$ subclass of LDL d 1.019 - 1.063 g/ml<br>LDL has flotation rate at d 1.21 of -S 25-70 or at d 1.063 $S_f$0-20.<br>May be divided into subfractions, for example, -S 40-70 or $S_f$12-20 and -S 25-40 or $S_f$0-12. |

Table 3. Nomenclature of Lipoproteins: Electrophoretic vs. Ultracentrifugation Techniques (Continued)

General Comments

In summary, low-density lipoproteins when studied by electrophoresis have electrophoretic mobility of β-globulin and immunologic properties of β-lipoprotein. High-density lipoproteins have electrophoretic mobility of $\alpha_1$-globulin and immunologic characteristics of α-lipoproteins. Very-low-density lipoproteins have electrophoretic mobility of $\alpha_2$-globulin.

Table 4. Lipoprotein Changes During Protein Depletion of Dog

DOG G

| Day of Study | T.P.[a] g/dl | Alb.[b] | Chol.[c] mg/dl | Lipoprotein -S 20-40 | 10-20 | 5-10 | 0-5 |
|---|---|---|---|---|---|---|---|
| 1 | 6.4 | - | 210 | 75 | 10 | 210 | 600 |
| 2 | 6.38 | 3.7 | 222 | | | | |
| 150 ml blood plasmapheresis protein-free diet started. | | | | | | | |
| | 5.2 | | 297 | 182 | 157 | 225 | 400 |
| 9 | 200 ml | | | | | | |
| 11 | 150 ml | | | | | | |
| 16 | 4.5 | - | 361 | 300 | 177 | 258 | 600 |
| 23 | 4.3 | 1.60 | 397 | 250 | 208 | 300 | 575 |
| 30 | 4.2 | 1.39 | 484 | 325 | 242 | 218 | 678 |
| Complete diet resumed. | | | | | | | |
| 37 | 5.1 | 1.58 | 326 | 236 | 232 | -552- | |
| 44 | 5.5 | 2.63 | | 135 | 164 | 295 | 450 |
| 69 | 6.0 | 3.91 | 318 | 189 | 166 | 400 | 782 |
| 75 | 5.3 | 3.84 | 313 | 141 | 54 | 345 | 465 |
| 83 | 5.8 | 3.70 | 287 | 101 | 90 | 402 | 402 |
| 90 | 5.8 | 3.87 | | | | | |
| 97 | 6.1 | 3.39 | 270 | 124 | 71 | 276 | 500 |

a - T.P. = Total Protein; b - Alb. = Albumin; c - Chol = Total Cholesterol.

Effect of Protein Depletion on Serum Lipoproteins of Dogs. During the period of protein depletion, the dogs lost between 4 and 9 percent of their initial body weight. Their serum total cholesterol concentration usually increased, in some cases by as much as 200 percent (Fig. 1). There was no close correlation between the degree of hypoalbuminemia and the cholesterol increment. The serum triglyceride level was slightly but not significantly increased from a pretreatment level of 62 $\pm$ 18 mg/dl to a level of 85 $\pm$ 23 mg/dl ($P > 0.05$) at the time of most marked hypoproteinemia. The lipoprotein pattern was greatly modified. The increased cholesterol was carried in all normally occurring lipoprotein fractions (-S 0-40), but a relatively much greater proportion occurred in -S 10-20 (IDL) and -S 20-40 (LDL) components than in the normal (Table 4). The -S 5-10 (HDL) fraction also increased relatively more than the -S 0-5 (HDL) fraction. There was no increase in -S 40-70 (LDL) fraction or less dense, -S>70, lipoproteins. In the starch-gel electrophoretic studies, increased concentration of slower migrating fractions in the slow $\alpha$ and $\beta$ range was observed. The slow $\alpha$ component in some cases was greatly augmented.

In one dog the serum cholesterol concentration decreased to the normal level, while the animal was still receiving the protein-free ration. The lipoprotein pattern showed a lower -S 0-5 concentration, and higher -S 5-10 and -S 10-20 (IDL) fraction than when the animal was eating a complete ration, but there was no increment in -S 20-40 lipoproteins. The dog had a long, lean body build, and lost only 4 percent body weight while on the protein-free diet.

Effect of Protein Repletion on Serum Lipoproteins of Dogs after Protein Deprivation. When the animals were returned to a complete ration the lipoprotein pattern gradually returned to the original normal pattern. The speed of return to normal varied, but was usually complete in six weeks. Serum total protein and protein electrophoretic pattern became normal at about the same time as the lipoprotein (Table 4).

## Rat Study

Serum lipoprotein pattern of healthy, mature rats. The serum lipids of rats as in most mammalian species are transported in lipoprotein fractions with the electrophoretic mobility of $\alpha_1$ and $\beta$-globulins and flotation rate at d 1.21 between -S 0 and -S 40. Unlike the dog ultracentrifugal lipoprotein pattern the HDL ($\alpha$-lipoproteins) have a more diffuse contour. Both species unlike most adult human beings have a greater proportion of lipids transported in HDL than LDL. In the healthy rat usually less than 15 percent of the total serum lipoprotein is present as LDL.

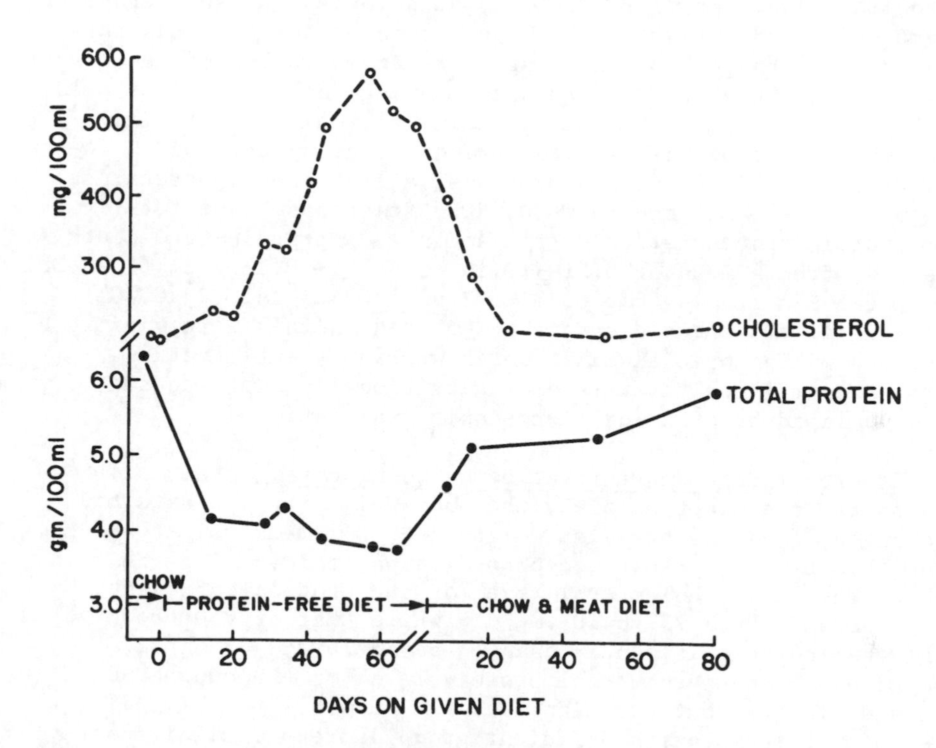


Fig. 1. Serum total cholesterol and total protein changes during canine protein depletion and repletion.

Effect of Protein Deprivation on Lipoprotein Distribution of Large, Old Rats. After four weeks of protein deprivation, the serum cholesterol levels of four of the rats increased slightly (15 percent) while three did not change from the pretreatment level. During this period, six of the seven rats each lost between 15 and 20 percent body weight, while that of the seventh was stable. The serum total protein concentration did not change. After ten weeks on the diet, the serum cholesterol levels decreased 11 percent from the initial level of 85.4 $\pm$ 7.6 to 76.1 $\pm$ 9.8 P >0.05, and serum triglyceride increased 93 percent from 72.0 $\pm$ 7.0 to 138.9 $\pm$ 4.5 mg/dl P <0.001. The animals lost an average of 34 percent of body weight. The lipoprotein patterns of six rats were markedly changed, 6/6 showing appreciably increased concentration -S 70-400 VLDL, and 4/6 also showing increase in -S 40-70 LDL; 0/6 showed altered HDL-lipoprotein. Normal rats of similar weight when fed chow diet showed insignificant amounts of -S 70-400 VLDL and -S 40-70 LDL lipoproteins, all of the lipoproteins being in -S 20-40 LDL and -S 0-20, HDL, fractions. The total serum protein concentration of all animals on protein-free diets decreased, from a mean of 6.21 $\pm$ 0.19 to 4.13 $\pm$ 0.33 g/dl (P <0.001) and the albumin from 4.60 $\pm$ 0.33 to 2.93 $\pm$ 0.45 g/dl (P <0.001). The livers of all the old, protein depleted animals were visibly fatty, and microscopic examination showed fatty infiltration. Livers of normal healthy rats on a nutritionally adequate diet showed no lipid by gross or microscopic examination.

Effects of Protein Deprivation on Young Rats. After six weeks on the protein-free diet, the body weight of the rats had not changed, (average pre-diet weight 286, six week weight 287 g). During this period the rats on chow diet had gained 34 percent in weight. The serum cholesterol level of the protein deficient group increased from 73 to 101 mg/dl, while that of the chow fed animals was not significantly changed being 76 and 82 mg/dl, respectively. The serum total protein level of the chow group increased slightly but not significantly from 6.4 to 6.7 g/dl while that of the protein deficient group decreased significantly from 6.2 $\pm$ 0.10 to 5.3 $\pm$ 0.12 g/dl (P <0.001) during the ten weeks. The albumin decreased from 3.4 $\pm$ 0.2 to 2.3 $\pm$ 0.5 g/dl and there was an increase in the concentration of proteins of $\alpha_1$-globulin mobility from 0.9 $\pm$ 0.15 to 1.4 $\pm$ 0.18 g/dl (P <0.02). The concentration of serum albumin and $\alpha_1$-globulin of the chow fed animals had not changed significantly from the initial levels.

The lipoprotein pattern of normal rat serum when studied by acrylamide gel electrophoresis showed multiple lipid stained bands of $\alpha$-globulin mobility and smaller amounts of lipid stained material with mobility between that of $\alpha$ and $\beta$-lipoprotein, and a small well resolved fraction with mobility of $\beta$-lipoprotein. The most rapidly migrating part of the $\alpha$-lipoprotein was present in greatest concentration. After six weeks on the protein deficient diet the relative concentration of the fast-migrating α-lipoprotein fraction decreased while that of slower migrating α-lipoprotein, and the fraction of intermediate mobility increased. The concentration of the β-lipoprotein was not changed. After ten weeks of protein depletion the percent of α-lipoprotein of fast mobility ($\alpha_1$-globulin) by the acrylamide gel technique had decreased being 34 percent (range 28-44%) of the total α-lipoprotein in the depleted animal and 56 percent (range 50-68%) in the chow fed. The concentration of fractions of slow electrophoretic mobility (i.e., β and pre-β lipoprotein) was greater in the depleted than in the chow-fed animals.

## Protein Depleted Young Rats When Returned to a Nutritionally and Protein Adequate Diet

The rats in the second study of protein deprivation of younger animals showed the same type and degree of change as group 2 described in the previous section which included a significant increase in the concentration of serum cholesterol after ten weeks of protein deprivation (Table 5). During the first week on the diet the animals continued to gain weight, but after ten weeks on the PFD had lost, so that they were near their weight when protein depletion was started. There was no change in triglyceride concentration. The rats that were returned to a chow diet had regained after six weeks the weight lost during the last weeks of protein deprivation while those on the high fat diet did not gain quite as rapidly. The serum cholesterol level of both groups during protein repletion decreased, but neither fell to the predepletion level of 74 mg/dl. Normal, continuously chow fed rats of similar age showed serum cholesterol concentrations of 77 $\pm$ 0.8 and triglyceride of 77 $\pm$ 3 mg/dl. The serum lipoprotein patterns of both groups of rats after six weeks of protein repletion were less abnormal than at the period of maximum protein depletion. There was considerable variation between patterns of animals in each group. The changes as in the maximum depletion period included an increased proportion of slower migrating α-lipoprotein on acrylamide gel than was found in the continuously chow fed animal.

A distinct relationship between the proportion of slow migrating α-lipoprotein and serum cholesterol concentration (Fig. 2) was found, which could be expressed as

$$\% \text{ slow } \alpha\text{-lipoprotein} = 7/8 \text{ (choles. concen.} - 47.5 \pm 8.75).$$

During the six weeks of protein repletion the total serum protein concentration of the rats returned to the normal range for rats on chow diet continuously, being 6.8, 7.3 and 6.7 g/dl for chow-

Table 5. Change in Body Weight, Serum Cholesterol and Triglyceride of Rats During Protein Depletion and Repletion

| | Body Weight, Grams | | | | Serum Cholesterol, mg/dl | | | Serum Triglyceride, mg/dl | | |
|---|---|---|---|---|---|---|---|---|---|---|
| | Initial | 4 wks PFD[1] | 10 wks PFD | 6 wks CD[2] | Initial Control | 10 wks PFD | 6 wks CHD | Initial Control | 10 wks PFD | 6 wks CD |
| Chow Group N=7 | 257 | 344 | 239 | 385 | 74 | 125 | 99 | 80 | 89 | 69 |
| | | | | | | | | range | 56-120 | 32-95 |
| % change from prev. level | | | -30% | +61% | | +65% | -27% | | +11% | -19% |
| % change from start PFD | | | -30% | +12% | | +65% | +23% | | +11% | -19% |
| | | | | 6 weeks HED[3] | | | | | | 6 weeks HED |
| High energy diet group N=10 | 262 | 394 | 244 | 360 | 74 | 115 | 85 | 78 | 73 | 72 |
| | | | | | | | | range | 37-127 | 32-95 |
| % change from prev. level | | | -30% | +51% | | +55% | -26% | | -6% | 0% |
| % change from start PFD | | | -30% | -9% | | +55% | +15% | | -6% | -8% |

1 - PFD = protein deficient diet
2 - CD = chow, nutritionally complete diet
3 - HED = high energy, high fat diet

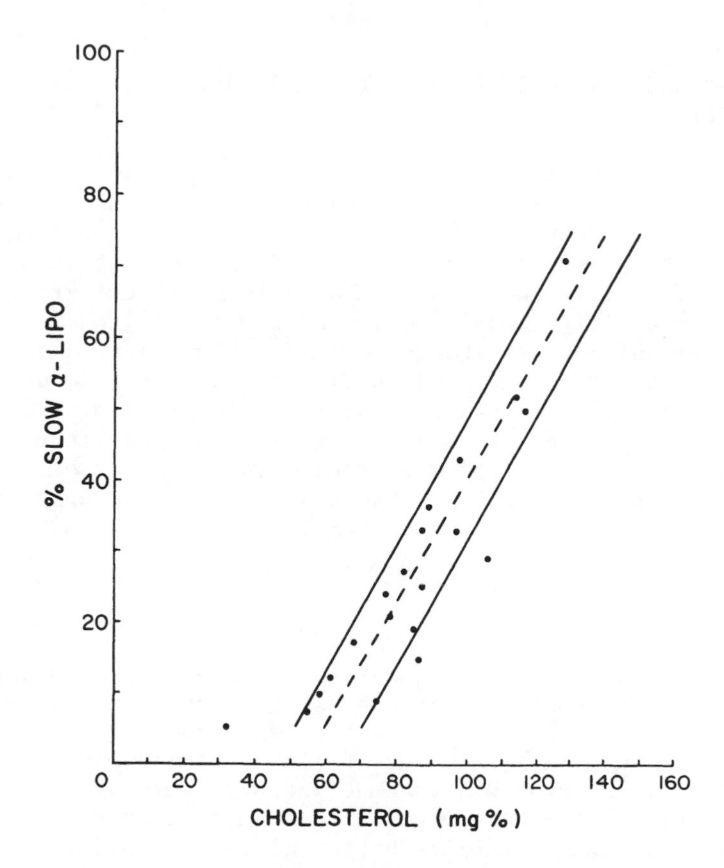


% slow α-lipo = 7/8 (Cholesterol Concentration) - 47.5±8.7

Fig. 2. Relation of slow α-lipoprotein concentration to total cholesterol concentration in rats during protein depletion.

repletion, high energy diet repletion, and continuous chow diet, respectively.

Fatty infiltration of the liver was found in the rats studied after ten weeks of protein deprivation. After six weeks on nutritionally complete diet, rats on the chow diet showed normal hepatic architecture, while those on the high energy, fat diet continued to show fatty infiltration.

RESULTS OF YOUNG RATS PROTEIN DEPLETED WHEN GIVEN ADEQUATE NUTRITION PARENTERALLY

As in the preceding section when rats after protein depletion were returned to a nutritionally adequate program, given in this group of rats (see Methods, Effect of adequate amino acid in Section c parenterally, they showed rapid increase in serum protein and albumin levels. The lipid and lipoprotein patterns of the group receiving the fat free formula became nearly normal, while those on the fat-supplement formula showed significantly elevated serum cholesterol levels. (These studies will be published in detail in a separate publication). Electron microscopic studies on livers of the protein depleted rats showed a marked increase in cytoplasmic fat and smooth endoplasmic reticulum and a marked decrease in rough endoplasmic reticulum. Protein repletion reversed these ultrastructural and biochemical abnormalities (Figs. 3-5).

## Discussion

In the protein depleted dogs of this study the serum cholesterol concentration usually increased as the total protein and albumin decreased. The correlation between serum protein and cholesterol concentration, however, failed in some animals in which a low serum protein level was reached where it remained during continuation of the protein-free diet; serum cholesterol reached the highest level when the protein had fallen to the lowest level, then decreased despite continued low protein levels. Variations like these depending partly on duration of protein deprivation may help explain some of the varied effects of protein deprivation on serum lipid levels that have been reported.

The serum lipid and lipoprotein changes associated with deficient protein nutrition vary greatly in the animal or the human being. Thus in Kwashiorkor in which calories are usually adequate but protein deficient β-lipoprotein is reduced and according to some authors cited in the review by Roberts (1980) α-lipoprotein levels are maintained. The changes in lipoprotein pattern are thought to be due to a deficiency of apoproteins for synthesis of VLDL. The lipoprotein changes are associated with fatty infil-

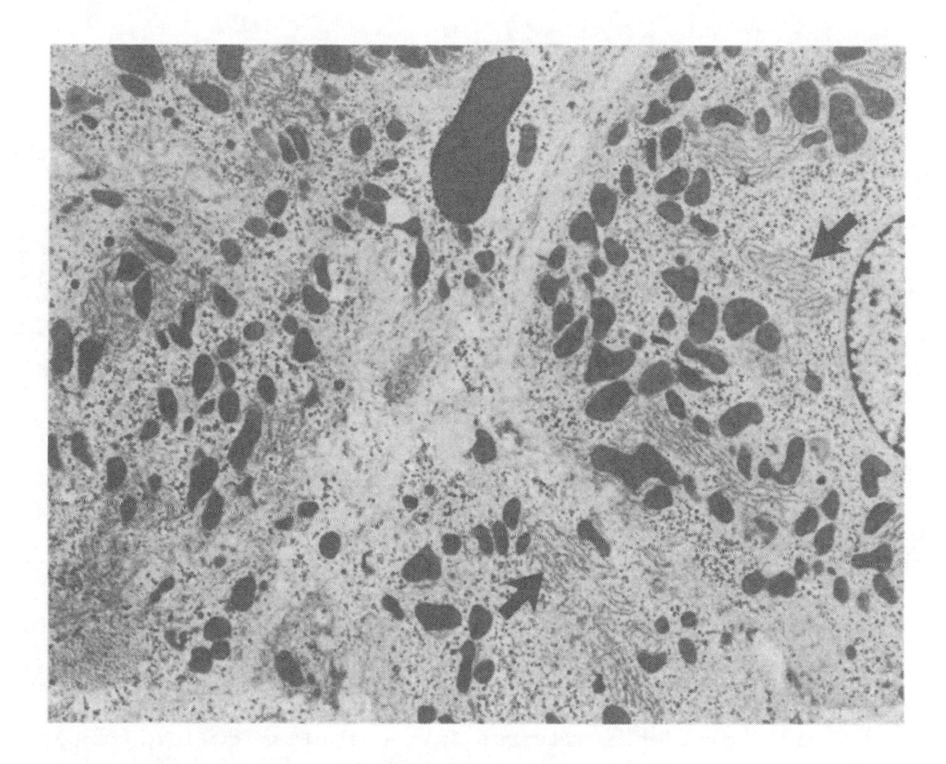

Fig. 3. Control. Note the abundant, roughly parallel profiles of rough endoplasmic reticulum (arrows) in close association with the mitochondria (X7020).

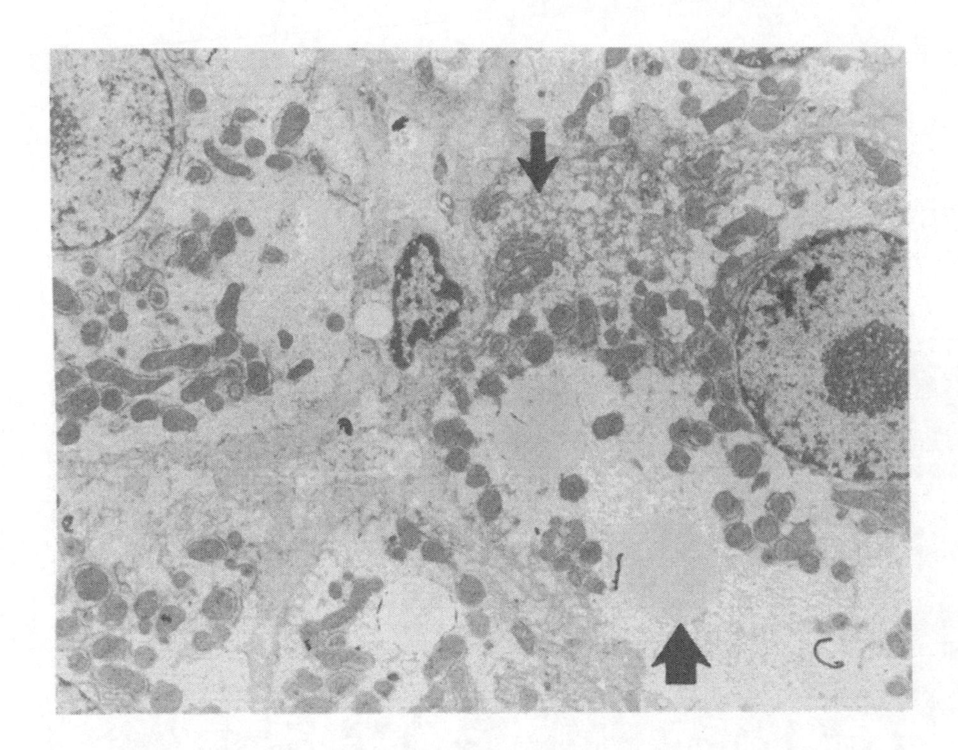

Fig. 4. Protein depletion. Numerous intracytoplasmic lipid droplets (large arrow) are present. There is a reduction in the amount of mitochondrial associated rough endoplasmic reticulum and a proliferation of branching, smooth endoplasmic reticulum (small arrow) compared to the control (X7020).

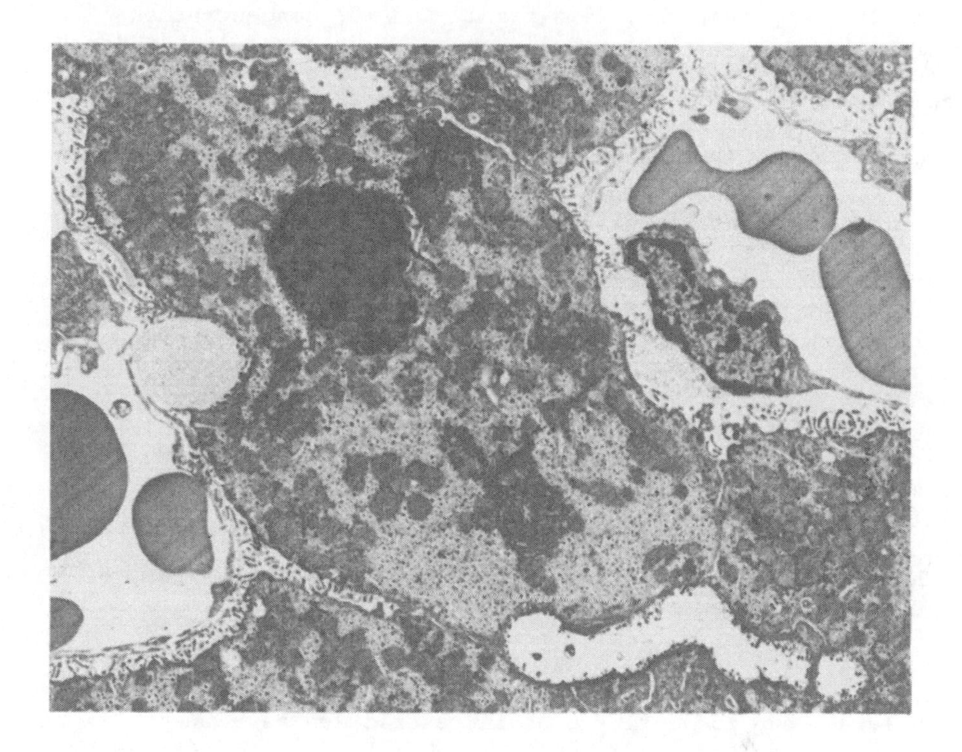

Fig. 5. TPN repletion. Return of ultrastructure to a state comparable to that of the control (X7020).

tration of the liver and both abnormalities are corrected in the human being after restoration of adequate protein intake. Lipoprotein changes are less marked in human subjects with Marasmus which is the picture of total starvation. Many patients represent a combination of Kwashiorkor and Marasmus and the changes are probably determined by which is playing a dominant role. The relative permanence of HDL levels in our protein depleted dogs, shown by the fact that HDL, -S 0-5 and -S 5-10, concentrations were nearly as high during protein depletion as normally, indicates the preferential use of available protein building materials to maintain these fractions rather than albumin and some other proteins. It is also possible, however, that the catabolism of HDL is slower in the protein depleted than in the healthy dog.

The mechanism utilized for the transport of additional amounts of cholesterol in the protein depleted dog seems to be similar to that which operates in normal dogs fed large amounts of added cholesterol (2 g per day). In both, the lipoprotein fractions with flotation rates between -S 0 and -S 20 (HDL, IDL) are greatly increased and the -S 20-40, LDL, fraction is also augmented. The -S 70-400, VLDL and -S >400, chylomicron, are not increased in the cholesterol fed dog even when the serum cholesterol is as high as 400 mg/dl (Mahley, 1974, 1977). Neither were they increased when protein supplies for synthesis of plasma proteins (present study) were limited.

The present studies emphasize the difference in the form of lipid transport in various species and type of alteration caused by different pathologic mechanisms. In man (Lewis, 1980) and rats (Lewis, 1954) with nephrosis, when plasma proteins are low there is a large increment in VLDL. Also in man and rats low serum protein levels due to dietary protein restriction have been reported to cause decrease in TG, VLDL, TC, LDL and HDL (Roberts, 1980).

In nephrectomized dogs maintained by peritoneal dialysis (Lewis, 1958) serum cholesterol and LDL and IDL fractions usually increased. Little or no increase in VLDL (-S 70-400) or in -S 40-70 part of LDL, lipoprotein fractions occurred. The HDL, -S 0-10, fraction was slightly decreased.

The dog maintains a relatively normal lipoprotein pattern under various adverse conditions such as protein deprivation and following nephrectomy when maintained by peritoneal dialysis. This suggests that in this species the building of lipoproteins is a "high priority" function. This preferential use of protein building materials for synthesis of plasma proteins was also evident in studies of Garrow (1959) using tracer techniques. In protein depleted dogs he found the turnover rate of proteins to be less than that of the normals, but found a great increase in the plasma protein specific activity as depletion progressed. This indicated a preferential use of protein for plasma protein synthesis. The preferential use of limited protein supplies for specific purposes has recently been demonstrated in the brain growth of undernourished neonatal rats by Freedman et al. (1980). The undernutrition was produced by feeding the lactating dams a low protein (12% casein) diet. The body weight and brain weights of the pups were 50% and 88% of normal. All areas of the brain were not equally supplied with amino acids as indicated by alterations in serotonin binding and alterations of enzyme activity in norepinephrine and dopamine areas.

Studies by Alusi (1975) on children with protein-caloric malnutrition, showed that of the serum proteins tested all except $C_4$ and immunoglobulins were lower than in controls. During refeeding the first serum protein to show significant rise was $C_3$, followed by transferrin, pre-albumin and lastly albumin. Thus in children as in the dogs, protein building materials are selectively used, and albumin is not high on the priority list.

Two conditions in dogs in which transport of cholesterol primarily in $\alpha$-lipoproteins and $\beta$-lipoproteins (-S 20-40) lipoproteins fail are diabetic acidosis and extreme hypothyroidism accompanied by a high-fat, cholesterol rich diet. In these circumstances appreciable amounts of lipids, (i.e., cholesterol, triglycerides, phospholipids) are transported in VLDL lipoprotein (Lewis, unpublished data).

The protein moiety of the canine -S 10-20, (ID) fraction differs from that of the -S 0-10 ($\alpha$-lipoprotein) as has been demonstrated by the studies of Mahley (1974 and 1979). The apoprotein demonstrated in the -S 10-20 fraction contains apo E, the arginine-rich polypeptide (Shore et al., 1980) which confers on it not only unique physical-chemical properties but also unique biological characteristics. The apo-E containing HDL fraction is capable of binding to the high affinity cell surface receptor of fibroblasts. This receptor also reacts with LDL as was demonstrated by Brown et al. (1976). HDL fractions that do not contain apo E do not have this unique property. Unique chemical properties of the HDL fractions containing apo E include their ability to be precipitated by heparin-manganese, and the relatively high cholesterol content of these fractions.

Clinical and experimental protein-calorie malnutrition are known to be associated with morphological and functional alterations that occur in the liver. A fatty liver in children suffering from Kwashiorkor is a well known feature of the disease and many attempts have been made to define the cause.

Triglyceride is the main lipid that accumulates in the liver of these patients, while their serum β-lipoprotein concentration is decreased. Pre-β-lipoprotein may also be low. Alpha-lipoprotein levels are frequently, but not always, decreased. Reduced hepatic synthesis of proteins, particularly of some of the apolipoproteins, may be a major mechanism causing the accumulation of fat in the liver (Flores et al., 1970). Our electron microscopic studies on rat livers support this concept. Our results demonstrated that protein depleted rats showed marked increase in cytoplasmic fat and smooth endoplasmic reticulum and a marked decrease in rough endoplasmic reticulum suggesting that protein synthesis is diminished. Seven days of protein repletion reversed these ultrastructural and biochemical abnormalities indicating that the process is readily reversible. In studies on human beings the importance of nutritional deficiency as a cause of liver injury was stressed by Tandon et al. (1974). They found that in protein-caloric malnutrition hepatomegaly occurred in six of fourteen patients studied although liver function studies including bilirubin, glutamic pyruvic transaminase and alkaline phosphatase were normal. The mean serum protein concentration was decreased to 4.9 and albumin to 2.6 g/dl. In all fourteen patients ultrastructural studies revealed organelle pathology of hepatocytes including conspicuous changes of rough endoplasmic reticulum and mitochondria. Repeat studies on seven of the patients after recovery from protein-caloric malnutrition showed that the ultrastructural abnormalities of the liver were nearly completely reversed to normal. These observations indicate that the liver changes due to protein deprivation are reversible. Our EM results agree completely with this concept. Our studies on the dogs which showed a restoration of normal serum lipoprotein and protein pattern after restoration of protein adequate nutrition also suggest an improved hepatic function. In contrast with protein deficient dogs which showed no increase in concentration of serum lower density (i.e. $-S > 40$) lipoproteins some protein depleted rats showed large increments in these fractions. Nephrotic rats also exhibited increased concentrations of serum VLDL and of chylomicron (Lewis and Heymann, 1954).

In protein malnutrition studies on rats Blanchard et al. (1979) observed as we have in these experiments that decreases in serum total protein and albumin are accompanied by an increased $\alpha_1$-globulin level. In rats treated by parenteral nutrition after protein depletion Steiger et al. (1977) noted that fatty livers persisted if the nutrient solution contained 40% of the non-amino acid calories as Intralipid $^R$ *, whereas the hepatic lipids decreased promptly when the nutrient solution contained only carbo-

*Cutter Laboratories, Berkeley, CA

hydrate for supplemental calories.

In total parenteral nutrition of the human being who is nutritionally depleted at the beginning of treatment, one of the complications which may occur is an essential fatty acid deficiency. It is manifest by classical clinical signs and the presence of eicosatrienoic acid in plasma (Richardson and Sgoutas, 1975). In the protein depleted animal no essential fatty acid deficiency could occur as the diets given were adequate in essential fatty acid.

The effect of amino acid diets upon serum lipids in man was studied by Olson et al. (1970). In the original study (Olson, 1958) in which the diet consisted of 25 g rice and mixed vegetable protein a fall of 20 percent in the serum cholesterol level was achieved. Addition of methionine, the limiting amino acid, caused an increase in serum cholesterol in some but not all of the subjects. In later studies (Olson et al., 1961; Bazzano et al., 1968; Olson et al., 1964; Olson, 1966) the essential amino acids were supplied in needed amounts and glutamate was given as the sole source of nonessential nitrogen. With this combination a very significant decrease in serum cholesterol was achieved, but was not observed if glycine replaced glutamic acid in the mixture. In the study negative nitrogen balance occurred in one dietary period in which the amino acids and glutamate were omitted. On this diet the serum cholesterol increased. It was suggested that in this period necessary nitrogen was obtained by catabolism of body protein. This was the situation in our animals during protein depletion, where nitrogen was only available from body protein.

The importance of the quantity and quality of the protein in the diet in determining serum lipid and lipoprotein levels and the hepatic lipid levels is not a unique property of protein among components of foods. The effect of a given level of protein may be greatly modified by the type of carbohydrate in the diet as was clearly demonstrated in dietary studies on rats by Cohen and Teitelbaum (1966). They found a higher incorporation of acetate into liver fatty acids and cholesterol _in vivo_ and _in vitro_ in rats fed a sucrose rich diet than in those receiving a high starch diet. Reduction of the dietary protein from 18 to 11% and replacing the calories with sucrose in the sucrose diet group caused increased hepatic fatty acid and cholesterol synthesis. In contrast, increase in starch content of the diet of the starch fed group caused only an increased hepatic fatty acid synthesis but a decreased acetate incorporation into cholesterol.

In a study on young cebus monkeys subjected to diets low in protein, energy or a combination of protein and energy deficiency, it was concluded that an energy restriction superimposed on a limited protein intake did not accelerate evidence of protein deficiency (Samonds and Hegsted, 1978). High energy diets have frequently been administered to people following extreme caloric and protein deprivation. Our finding that restoration of serum lipid and lipoprotein levels to near normal in the protein depleted rats was equally effective whether the well balanced chow diet or the high energy high fat diet was used indicates the validity for use of such diets in human nutrition. In both the importance of adequate protein is indicated. A further study is needed, however, to determine how long a balanced chow diet would be needed to reduce hepatic lipid levels of the high energy diet animals to normal.

Studies of two men who were on a raft for fifty-six days without food, and ingesting only suboptimal amounts of water during the last twenty-six days of the trip showed some abnormality of liver function for a short period during refeeding (Lee et al., 1977). In our study rats receiving a high energy diet for refeeding showed fatty infiltration of the liver while those receiving a conventional rat chow diet low in fat had histologically normal hepatic architecture. These results raise the possibility that careful consideration of the type of diet during refeeding after protein deprivation may be of importance. The fact that normal liver function ultimately was observed in the Lee study (1977), however, is indicative of the great recuperative power of the liver.

The changes in serum lipids and lipoproteins caused by protein deprivation observed in the dogs and rats of this study were obviously affected by many factors including age of animal, duration of deprivation of protein and composition of the diets ingested. These results emphasize the complex nature of studies involving dietary protein deficiencies and why experts still after years of investigation find large gaps in knowledge about malnutrition and why "the role of neither protein nor calories is clearly defined" (Hegsted, 1978).

## SUMMARY

The ultracentrifugal and starch-gel electrophoretic patterns of the serum lipoproteins of the normal dog have been described and contrasted with those of the protein-depleted animal. The serum cholesterol and lipoprotein concentrations of dogs were usually increased during protein depletion produced by plasmapheresis and feeding of a protein-free diet. The increased cholesterol was transported primarily in -S 10-20 and -S 20-40 lipoprotein fractions, with some increase in the more dense -S 0-5 and -S 5-10 lipoprotein fractions. There was no increase in concentration of VLDL (-S > 40).

The ability of the protein depleted dog to maintain high concentrations of protein-rich lipoproteins was noteworthy.

In protein depleted, old rats, there was some increase in the concentration of very low density -S 70-400 and -S 40-70 lipoprotein fractions.

Young mature rats on a protein deficient diet showed significant alteration in their serum lipoprotein patterns. The concentration of α-lipoprotein of fast electrophoretic mobility decreased, while that of slower mobility increased. The concentration of β-lipoprotein also increased. Livers of both young and old, protein depleted rats showed fatty infiltration. Serum lipoprotein and hepatic structure returned toward normal after the young rats were returned to a nutritionally complete diet.

It is concluded that the serum lipoprotein patterns of both dogs and rats are altered during prolonged periods of protein deprivation. The changes are reversible, the patterns returning toward normal with a nutritionally complete ration. Changes in hepatic structure were noted in the protein depleted rat; these changes also were reversible.

ACKNOWLEDGEMENTS

The excellent technical assistance of Mr. Joseph Paksi and Ms. Jeannette Tanos during the experimental part of this study is greatly appreciated. Mr. Manuel Glynias, the recipient of a Northeast Ohio summer student grant, also helped very efficiently during the study. This study was partially supported by Grant No. 5835 from NHLBI, Bethesda, Md. We gratefully acknowledge Mrs. Helen Brewster, Mrs. Phyllis Pittman and Mrs. Jeannette Goodman for their excellent assistance in typing this manuscript.

REFERENCES

Abell, L.L., Levy, B.B., Brodie, B.B. and Kendall, F.E., 1952, A simplified method for the estimation of total cholesterol in serum and demonstration of its specificity, J. Biol. Chem., 195:357.

Allison, J.B. and Anderson, J.A., 1945, The relation between absorbed nitrogen, nitrogen balance and biological value of proteins in adult dogs, J. Nutrition, 29:413.

Bazzano, G. and Olson, R.E., 1968, Effect of milk and milk components upon serum lipids and lipoprotein levels in man, in: "Proc. Symp. Dairy Lipids, Lipid Metabolism," Avi, Westport, Connecticut.

Blanchard, J., Steiger, E., O'Neal, M., Naito, H., Sebek, B., 1979, Effects of protein depletion and repletion on liver, structures, nitrogen content and serum proteins, Annals Surgery,

Brown, M.S., Ho, Y.K. and Goldstein, J.L., 1976, The low density lipoprotein pathway in human fibroblasts: Relation between cell surface receptor binding and endocytosis of low density lipoproteins, Ann. N.Y. Acad. Sci., 275:244.

Cohen, A.M. and Teitelbaum, A., 1966, Effect of different levels of protein in sucrose and starch diets on lipid synthesis in the rat, Israel. J. Med. Sci., 2:727.

Coward, W.A. and Whitehead, R.G., 1972, Changes in serum β-lipoprotein concentration during the development of Kwashiorkor and in recovery, Brit. J. Nutrition, 27:383.

De Lalla, O.F. and Gofman, J.W., 1954, Ultra-centrifugal analysis of serum lipoproteins, Methods Biochem. Anal., 1:459.

Durrum, E.L., 1958, in: "A Manual of Paper Chromatography and Paper Electrophoresis," Second Edition, R.J. Block, E.L. Durrum and G. Zweig, eds., Academic Press, New York, pp. 523-528.

Flores, H., Sierralta, W. and Monckcherg, F., 1970, Triglyceride transport in protein-depleted rats, J. Nutrition, 100:375.

Freedman, L.S., Samulls, S., Fish, I., Schwartz, S.A., Lange, B., Katz, M., Morgano, L., 1980, Sparing of the brain in neonatal undernutrition: Amino acid transport and incorporation into brain and muscle, Science, 207:902.

Garrow, J.S., 1959, Effect of protein depletion on distribution of protein synthesis in the dog, J. Clin. Invest., 38:1241.

Gofman, J.W., Lindgren, F.T. and Elliot, H., 1949, Ultracentrifugal studies of serum lipoproteins of human serum, J. Biol. Chem., 179:973.

Hegsted, M., 1978, Protein-calorie malnutrition, American Scientist, 66:61.

Lee, P.A., Wallin, J.D., Kaplowitz, N., Burkhartsmeier, L., Kane, J.P. and Lewis, S.B., 1977, Endocrine and metabolic alterations with food and water deprivation, Am. J. Clin. Nutr., 30:1953.

Lees, R.S. and Hatch, F.T., 1963, Sharper separation of lipoprotein species by paper electrophoresis in albumin-containing buffer, J. Lab. Clin. Med., 61:518.

Lewis, L.A., Green, A.A. and Page, I.H., 1952, Ultracentrifuge lipoprotein pattern of serum of normal, hypertensive and hypothyroid animals, Am. J. Physiol., 171:391.

Lewis, L.A. and Heymann, W., 1954, Ultracentrifugal analysis of serum lipoproteins in nephrotic syndrome of rats, Proc. Soc. Exper. Biol. and Med., 86:766.

Lewis, L.A., Page, I.H. and Kolff, W.J., Serum lipoprotein and cholesterol changes in nephrectomized dogs maintained by peritoneal dialysis.

Lewis, L.A., 1966, Thin-layer starch-gel electrophoresis, a simple accurate method for characterization and quantitation of protein components, Clin. Chem., 12:596.

Lewis, L.A., 1980, The kidney, in: "Handbook of Electrophoresis and Chromatography," Vol. II, The Lipoproteins, L.A. Lewis and J.J. Opplt, eds., CRC Press, Boca Raton, Florida.

Lewis, L.A., unpublished data, 1980, Am. J. Physiol., 195:161, 1958.

Lipid Research Clinics Program 1, Manual of Laboratory Operations, DHEW Publications No. (NIH) 75-688, 1974, CDC, Atlanta, Georgia.

Mahley, R.W. and Kuisgrabe, K.H., 1974, Isolation and characterization of plasma lipoproteins from control dogs, Circ. Res., 35:713.

Mahley, R.W., Innerarity, T.L., Bersot, T.P., Lipson, A. and Margolis, S., 1978, Alterations in human high-density lipoproteins with or without increased plasma-cholesterol, induced by diets high in cholesterol, Lancet, II:807.

Mahley, R.W. and Weisgraber, K.H., 1979, Subfractionation of high density lipoproteins into two metabolically distinct subclasses by heparin affinity chromatography and Geon-Pevikon electrophoresis, in: "Report of the High Density Lipoprotein Methodology Workshop," U.S. Department of Health, Education and Welfare, Public Health Service, National Institutes of Health, NIH Publication No. 79-1661.

Chapter 2, A microzone method for serum proteins, 1973, in: Microzone R Electrophoresis Manual," M.D. Gebalt, ed., Beckman Instruments, Inc., Fullerton, California.

Naito, H.K., Wada, M., Ehrhart, L.A. and Lewis, L.A., 1973, Polyacrylamide-gel disc electrophoresis as a screening procedure for serum lipoprotein abnormalities, Clin. Chem., 19:223.

Olson, R.E., Vester, J.W., Gursey, D., Davis, N. and Longman, D., 1958, The effect of low-protein diet upon serum cholesterol in man, Am. J. Clin. Nutr., 6:310.

Olson, R.E., Ito, S., Tripathy, K. and Eagles, J., 1961, Studies of the mechanism and specificity of the hypocholesterolemic action of low protein diets in man, Am. J. Clin. Nutr., 9:24.

Olson, R.E., Nichaman, M.Z., Nittka, J. and Dorman, L., 1964, The effect of amino acid intake upon serum cholesterol in man, J. Clin. Invest., 43:1233.

Olson, R.E., 1966, Effect of glutamate upon fat transport in man, Proc. Intern. Congr. Nutr. VIIth, Hamburg, Verlag, Friedr. Vol. 2.

Olson, R.E., Nichaman, M.Z., Nittka, J. and Eagles, J.A., 1970, Effect of amino acid diets upon serum lipids in man, Am. J. Clin. Nutr., 23:1614.

Olusi, S.O., McFarlane, H., Osunkoya, B.O. and Adesina, H., 1975, Specific protein assays in protein-calorie malnutrition, Clin. Chim. Acta, 62:107.

Pfizer Pol-E-Film System, 1974, Manual: Serum lipoprotein phenotyping.

Poulek, M.D., 1957, Starch-gel electrophoresis in a discontinuous system of buffers, Nature, 180:1477.

Richardson, T.J. and Sgoutas, D., 1975, Essential fatty acid deficiency in four adult patients during total parenteral nutrition, Am. J. Clin. Nutr., 28:258.

Roberts, D.C.K. and Carroll, 1980, Lipoprotein changes in undernutrition and overnutrition, in: "Handbook of Electrophoresis and Chromatography," Vol. I, The Lipoproteins, CRC Press, L.A. Lewis and J.J. Opplt, eds.

Sals, A., 1947, Improved biuret reaction of proteins and two standard colorimetry, Nature (London), 60:89.

Samonds, K.W. and Hegsted, D.M., 1978, Protein deficiency and energy restriction in young cebus monkeys, Proc. Natl. Acad. Sci., U.S.A., 75:1600.

Shore, V.G., Garcia, R.E., Penn, A.L. and Shore, B., 1980, Polyacrylamide gel electrophoresis and isoelectric focusing of plasma lipoproteins, in: "Handbook of Electrophoresis and Chromatography," Vol. I, The Lipoproteins, CRC Press, L.A. Lewis and J.J. Opplt, eds.

Smithies, O., 1962, Molecular size and starch-gel electrophoresis, Arch. Biochem. Biophys., Suppl. 1, 125.

Steiger, E., Naito, H.K., O'Neil, M. and Cooperman, A., 1977, Serum lipids in total parenteral nutrition (TPN) - effect of fat, Surg. Forum, 28:83.

Tandon, B.N., Ramanujan, R.A., Tandon, H.D., Pur, B.K. and Gandhi, P.C., 1974, Liver injury in protein-calorie malnutrition. An electron microscopic study, Am. J. Clin. Nutr., 27:550.

Weisgraber, K.H., Mahley, R.W. and Assmann, G., 1977, The rat arginine rich apoprotein and its redistribution following injection of iodinated lipoproteins into normal and hypercholesterolemic rats, Atherosclerosis, 28:121.

# DIETARY FIBER AND DISEASE

Jon A. Story
Associate Professor
Department of Foods and Nutrition
Purdue University
West Lafayette, IN 47907

## INTRODUCTION

Dietary fiber is defined as plant cell wall materials which are not subject to hydrolysis by human digestive enzymes (Trowell, 1972b). This group of materials includes a large number of polysaccharides, including cellulose, hemicellulose, pectins, gums, as well as lignin. In plants these polysaccharides serve as either a storage form of carbohydrate or as structural components of the plant. The physical characteristics of these polysaccharides and lignin in relation to dietary fiber have been expertly reviewed by others (Cummings, 1976; Southgate, 1976).

In spite of their almost total unavailability in humans, the components of dietary fiber play an important role in human nutrition, both normal and as possible preventative factors in several disease states of major proportion in the Western word (Burkitt, et al., 1974). Interest in the role of dietary fiber in human nutrition has been rekindled in recent years as a result of these epidemiologic and experimental observations relating this component of the diet to the incidence of a large number of diseases. This review will examine the evidence for the relationship of dietary fiber to three specific disease states (atherosclerosis, colon cancer and diabetes) in an attempt to clarify the current status of our understanding concerning the role of dietary fiber in causation of these diseases.

## ATHEROSCLEROSIS

The relationship between intake of a high, complex carbohydrate diet and lower incidence of coronary heart disease in humans was observed by Walker et al. (1961). They found that, in addition to the Bantu, whites who consumed a diet similar to that of the Bantu had an extremely low incidence of coronary heart disease as evidenced by electrocardiographic examination. This diet was described as very high in carbohydrate (1-1.5 pounds of dry cereal per day) and low in animal fat and protein. They also noted that white prisoners who consumed large amounts of cereal products had an extremely low incidence of death from coronary disease.

Trowell (1972c) brought attention to these studies and noted other reports (Antonis and Bersohn, 1962) which dealt with bile acid excretion when Western diets were compared with high-fiber African diets. High-fiber diets resulted in increased bile acid excretion which was suggested by Trowell as a possible mechanism for decreased cholesterol levels by a mechanism similar to that observed with bile acid sequestrants. Trowell (1972a) also reported concomitant changes in the incidence of heart disease and diverticular disease in Great Britain during the previous 40 years. Periods during the second World War when dietary fiber intake was increased were accompanied by a slowing of the increase in incidence of both diseases. In addition other populations which eat diets high in dietary fiber have lower serum cholesterol levels and a decreased incidence of coronary heart disease (Trowell, 1972a).

Most of this epidemiological work was brought together in a report by Burkitt et al. (1974). The incidence of a large number of diseases prevalant in Western civilizations was compared with the incidence of these diseases in lesser developed populations. Dietary fiber intake was discussed as a major determinant in these differences. Changes in nutrient availability, transit time, intracolonic pressure, intestinal flora, and bile salt metabolism were suggested as mechanisms responsible for the changes in incidence of disease. Ischemic heart disease was described as "virtually unknown" in native African populations while being responsible for one-third of all deaths in the United States.

Studies in experimental animals had also indicated an involvement of dietary fiber in cholesterol metabolism and experimental atherosclerosis. Portman (1960) reported increased cholic acid excretion and reduced half-life of cholic acid in rats fed a commercial ration compared to those fed a semi-purified diet with cellulose added for fiber. His experiments also found that some fraction of the grain component of the commercial ration was responsible for these changes.

In rabbits, Kritchevsky and Tepper (1968) found that a component of commercial rations other than lipid resulted in a resistance to high saturated fat levels which, in a semi-purified diet, resulted in cholesteremia and atherosclerosis. Cookson et al. (1967) had also reported a regression of dietary cholesterol induced cholesteremia in rabbits using a diet supplemented with large amounts of alfalfa. Atheroma was also shown to be prevented by some sources of dietary fiber when added to an atherogenic semi-purified, cholesterol-free diet in rabbits. Wheat straw and peat appeared to reduce the cholesteremic and atherogenic effects of the diet with cellophane spangles added for bulk (Moore, 1967).

Bile acid metabolism and cholesterol absorption had also been shown to be altered by some sources of dietary fiber. Alfalfa decreased absorption of an oral dose of radioactive cholesterol and increased excretion of bile acids synthesized from this radioactive precursor (Kritchevsky et al., 1974). Pectin had also been reported to increase fecal bile acid excretion (Lin et al., 1957; Leveille and Sauberlich, 1966). Eastwood and Boyd (1967) began the investigation of a mechanism for these changes in excretion when examining the distribution of bile salts along the intestine. They observed a significant quantity of bile salt associated with the solid material in the small intestine and not in free solution. Later experiments (Eastwood and Hamilton, 1968) indicated an ability of dietary fiber sources to adsorb bile salts *in vitro*. Experiments indicated that lignin was an important component of dietary fiber in bile salt adsorption and that the adsorption was hydrophobic in nature. Others (Kritchevsky and Story, 1974; Story and Kritchevsky, 1976) have also observed adsorption of bile salts by a large number of dietary fiber sources and components.

Subsequent experiments in man and animals have investigated the cholesterol lowering capacity of dietary fiber and many of its components and have examined several possible mechanisms for the observed changes. In experimental animals, lignin has been shown to reduce cholesterol levels or prevent their increase in rats due, in part at least, to increases in bile acid excretion (Judd et al., 1976; Story et al., 1977). This correlated well with the *in vitro* finding of bile acid adsorption *in vitro*. In man lignin has not been extensively examined but has been shown to reduce human serum cholesterol levels by some workers (Thiffault et al., 1970) and has resulted in significant increases in the hands of others (Londner and Moller, 1973).

Cellulose has been used extensively in both animal and human studies and, as evidenced by its lack of interaction with bile salts, has not been found to result in many changes in cholesterol metabolism. In rats cellulose results in little change in cholesterol levels of liver and/or serum (Looney and Lei, 1978). When fed with cholesterol, cellulose does not prevent the liver accumulation of cholesterol in rats (Tsai et al., 1976) but reduces the accumulation in comparison to a fiber-free diet (Story et al., 1977). In rabbits, cellulose is an important part of a cholesterol-free, semipurified atherogenic diet as shown by the lack of effectiveness of the diet when other sources of dietary fiber are used (Hamilton and Carroll, 1976; Kritchevsky et al., 1977). In man a similar lack of effect has been observed at reasonable levels of administration (Eastwood et al., 1973).

Hemicellulose has been reported as an effective adsorbant of bile salts (Birkner and Kerm, 1974) but studies in man and animals have been hindered by a lack of availability of pure hemicellulose. Psylium seed has been reported to prevent cholesterol accumulation in cholesterol fed rats (Beher and Casazza, 1971) and to lower serum cholesterol levels in man (Garvin et al., 1965; Forman et al., 1968). Fecal bile acid excretion was reported to increase in both cases.

Recent attention has focused on several soluble polysaccharides and their cholesterol lowering effects. Pectin has long been reported to lower serum cholesterol in humans (Keys et al., 1961). In experimental animals many workers have reported reductions or lack of accumulation of cholesterol in rats (Ershoff, 1963; Tsai et al., 1976; Chang and Johnson, 1976; Judd et al., 1976; Story et al., 1977), rabbits (Berensen et al., 1975), and chickens (Fisher et al., 1965; Fahrenback et al., 1966). Guar gum also reduces cholesterol levels in experimental animals (Tsai et al., 1976).

A recent review (Truswell, 1979) examined 12 experiments using a variety of levels of pectin supplementation (2-35 g/day) in man. Significant reductions in plasma cholesterol were reported in all but two experiments and in those lower, but not significant, levels were also reported. Examination of the mechanism involved (Kay and Truswell, 1977) indicated an increase in both neutral and acidic steriods in the feces of the pectin-fed subjects. Guar gum also reduces plasma cholesterol levels in man (Jenkins et al., 1975). A recent experiment in which guar was incorporated into crispbread and fed to hyperlipidemic subjects (Types II and IV) at a level of 13 g of guar/day for 8 weeks (Jenkins et al., 1980). The guar treatment reduced plasma cholesterol levels 13%, with high density lipoprotein cholesterol (HDL) levels remaining constant.

Wheat bran, another source of dietary fiber which is used extensively by humans, has been extensively examined for cholesterol-lowering capability. A summary of these reports (Truswell and Kay, 1976) indicated a lack of effect of widely varied dose (14-100 g/day) on plasma cholesterol or triglyceride levels in man. More recently hard red spring wheat bran (39 g/day) has been shown to lower plasma cholesterol in humans (Munoz et al., 1979). The earlier studies used the more common white wheat bran. There are no apparent differences in composition between the two sources of bran, but some difference is apparently responsible for the observed difference in function.

Another source of dietary fiber for humans which has recently been examined more thoroughly is oats, and more specifically, oat bran. The cholesterol-lowering effects of whole oats (rolled) has been observed in experimental animals (DeGroot et al., 1963; Fisher and Griminger, 1967) and man (DeGroot et al., 1963; Lukyen et al., 1965). Recent evidence (Chen and Anderson, 1979) indicates that oat bran reduces low density lipoprotein (LDL) cholesterol while increasing HDL cholesterol in the rat. In man they have also reported a selective lowering of LDL (Anderson and Chen, 1979).

The effectiveness of dietary fiber in lowering risk of atherosclerotic heart disease has been examined primarily in relation to its ability to lower blood cholesterol levels. The first mechanism to attempt to explain the ability of some sources of dietary fiber to lower cholesterol levels dealt with the ability of these dietary fibers to increase fecal steroid excretion. Adsorption of bile salts by dietary fiber increases their excretion by inhibiting reabsorption. In addition adsorption of bile salts in the small intestine also interferes with cholesterol absorption which results in increased fecal neutral steroid excretion from both dietary and biliary sources. The net result of these changes is to increase fecal steroid excretion sufficiently so as to decrease body pools of cholesterol. Reduction of serum cholesterol is considered a decrease in risk of developing coronary heart disease.

Recently a second possible mechanism of action for reduction of risk has emerged. This involves the observation that reductions in serum cholesterol resulting from changes in dietary fiber appear selectively in the LDL fraction with no changes or possibly increases in HDL cholesterol levels. Since LDL cholesterol is positively correlated with atherosclerosis and HDL cholesterol is negatively correlated, these selective changes seem to offer a second positive effect of dietary fiber on cholesterol levels. Mechanisms for the specific changes in LDL and HDL have not been elucidated at this time.

## COLON CARCINOGENESIS

An epidemiological association between dietary fiber intake and cancer of the colon was enumerated and a hypothesis concerning the mechanism for this relationship offered by Burkitt (1971). The relationship began as a comparison of the high fiber, low cancer incidence populations of Africa with the low fiber high incidence populations of the Western world. The mechanism derived from several observations concerning the effects of high dietary fiber diets. The amount of dietary fiber consumed is related inversely with the transit time of material through the intestine (Walker, 1961). In addition, differences in fecal bacteria in relation to the level of dietary fiber in the diet had been observed (Hill et al., 1971; Aries et al., 1971). Diets low in dietary fiber resulted in bacteria with a greater ability to 7α-dehydroxylate bile acids and likewise a larger amount of secondary bile acids (deoxycholic and lithocholic), the products of this 7α-dehydroxylation. The observation that deoxycholic acid could be converted to a potent carcinogen in the laboratory (Haddow, 1958) resulted in speculation concerning the possibility of these reactions taking place in the colon. Using these facts, Burkitt hypothesized that low fiber diets resulted in increased concentrations of the potentially carcinogenic secondary bile acids in the colon and, as a result of increased time of exposure as a result of slowed transit, these bile acids would have more of a chance to interact with the intestinal epithelium.

The relationship of transit time with the incidence of colon cancer in humans has not been resolved. Burkitt et al. (1974) found a strong inverse correlation between transit time and fecal weight and suggested a relationship between increased stool weight and decreased transit time and a lower incidence of colon cancer. Other studies have indicated that transit time is not as well related. Eastwood et al. (1976) noted that, in spite of a higher incidence of constipation in females, the occurrence of colon cancer is slightly lower in females than in males. Glober et al. (1974) observed an almost two-fold greater transit time in Caucasian Hawaiians than in those of Japanese descent while colon cancer rates were very similar. The role of transit time in susceptibility to colon cancer remains to be fully understood.

Bile acids have become a major focus of study in understanding differences in susceptibility to colon cancer caused by changes in diet. Subsequent to the studies implicating bile acids mentioned above, several workers have more thoroughly examined the role of bile acids in colon carcinogenesis. Hill and Aries (1971) first suggest this relationship by correlating fecal bile acid concentrations with the incidence of colon cancer in several countries. Colon cancer patients were also found to have a higher concentration of bile acids in their feces than non-cancer patients (Hill et al., 1975).

Animal studies have also indicated a role of bile acids as co-carcinogens. Narisawa et al. (1974) used intrarectally instilled N-methyl-N'-nitro-N-nitrosoguanidine as a carcinogen and found that lithocholate and taurodeoxycholate applied repeatedly after the carcinogen resulted in more tumors than when they were absent. These two bile acids did not cause tumors when administered alone. Lithocholic has also been found to be co-mutagenic with two carcinogens in the Ames assay (Kelsey and Pienta, 1979). Lithocholic was also mutagenic in the hamster embryo cell transformation assay (Silverman and Andrews, 1977). Increasing concentration of colonic bile acids also increases susceptibility to chemical carcinogens in rats when these changes are caused by drugs such as cholestyramine (Nigro et al., 1973), by transplantation of the bile duct (Chomchai et al., 1974), or by feeding a high fat diet (Nigro et al., 1975).

Experiments examining directly the effects of various sources of dietary fiber on the susceptibility of rats to chemical carcinogens have had mixed results. Wheat bran generally has been shown to decrease the incidence of colon tumors in this model. Wilson et al. (1977) fed 20% wheat bran and either beef fat or corn oil and administered dimethylhydrazine (DMH) subcutaneously. Fewer animals had tumors with either dietary fat when fed wheat bran, although the number of tumors per tumor bearing animal was slightly higher. Fleiszer et al. (1978) added 28 g wheat bran per 100 g diet for rats and also administered DMH and also found a much smaller number of animals with tumors in the bran fed group. Nigro et al. (1979) and Watanabe et al. (1979) also observed reduced tumor incidence when wheat bran was fed to animals given azoxymethane (AOM) injections to induce tumors. Nigro et al. (1979) also observed that cellulose and alfalfa resulted in fewer tumors than a fiber-free diet when a low-fat diet (10%) was fed but observed little difference with a high-fat (35%) diet. Using an intrarectally administered carcinogen (methylnitrosourea) Watanabe et al. (1979) found that alfalfa (15%) increased tumor incidence while wheat bran (15%) reduced incidence slightly and pectin (15%) dramatically.

These data seem to be consistent with the above discussion of bile acid concentration and carcinogen susceptibility. Those sources of dietary fiber which adsorb bile salts, e.g. alfalfa, result in smaller reductions or increases in tumor incidence while those which do not bind bile acids but dilute the feces by holding water result in lower tumor incidence. Nigro et al. (1979) measured bile acid concentration in the feces of animals used in the above experiments and found a direct relationship between bile acid concentration and susceptibility to colonic tumors induced by AOM. Thus experimental evidence to date would indicate an involvement of dietary fiber in colon carcinogenesis mediated, at least in part, through its effects on bile acid excretion.

## DIABETES MELLITUS

Trowell (1973, 1974) reported an epidemiologic association between consumption of complex carbohydrate diets, especially high in dietary fiber, and the incidence of diabetes mellitus. He correlated incidence of diabetes mortality in women in England and Wales with the crude fiber content of flour during the years prior to and during the second World War. While the crude fiber content of flour increased during the war and sugar intake decreased the number of deaths decreases. Cleave (1966) had also associated the concomitant increase in simple carbohydrates and decrease in complex carbohydrate intake with increase in a number of diseases including diabetes.

Several experiments in man have indicated the validity of these arguments. Kiehm et al. (1976) utilized a high carbohydrate, high plant fiber diet in patients with diabetes. Reductions in fasting glucose and in insulin or oral hypoglycemic doses was observed in all patients. Anderson et al. (1979) have recently extended these experiments to a larger number of patients all of whom were receiving exogenous insulin. The diet contains 70% of kilocalories from carbohydrate, 71% of which were considered "complex" carbohydrates, 9% of kilocalories from lipid and 21% of kilocalories from protein. Weight did not change significantly while dependance on insulin was reduced in all cases and was eliminated in all cases where 20 units/day or less was required prior to diet change. Plasma glucose concentrations, both fasting and post-prandial, were reduced and urinary glucose was also reduced. Serum cholesterol was also reduced significantly.

Jenkins et al. (1977) have reported lower post-prandial glucose and serum insulin levels in normal subjects to which guar gum and pectin were added to the test meal. Subsequent studies (Jenkins et al., 1978) have indicated that viscosity of the added dietary fiber is a critical characteristic with the more viscous materials being more effective in altering glucose metabolism.

Similar findings have been reported using test meals (60 g carbohydrate) given to normal subjects as apples, apple puree or apple juice (Haber et al., 1977). The more complex the associated carbohydrate the lower the peak plasma glucose level and the smaller the hypoglycemic rebound after this peak level. Plasma mortality levels were also reduced when apples or puree were consumed, compared to apple juice. These changes in insulin response appear to be related to an alteration in site of absorption of glucose to a more distal location which would alter secretion of gastric inhibitory polypeptide (Thomas et al., 1977) and, correspondingly, reduce insulin secretion. The long-term effects of these changes in insulin secretion appear to reduce the insulin requirement as is evidence by the work of Anderson discussed above.

## CONCLUSIONS

The study of the involvement of dietary fiber in both normal and disease oriented nutrition has evolved as an important part of nutrition. Use of various sources or components of dietary fiber for prevention or alteration of risk of atherosclerotic heart disease or colon cancer and in treatment of diabetes have been examined. In all cases a great deal of variation is found in the effects observed with various dietary fiber sources. However theoretical mechanisms of the involvement of dietary fiber have been proposed and experimental data seem to indicate that many of these mechanisms are valid with selected dietary fibers.

Atherosclerosis in experimental animals and selected risk factors for atherosclerosis in humans are modified by some components of dietary fiber, i.e. pectins, lignin and some gums. These components appear to alter cholesterol metabolism by increasing excretion of bile salts and/or neutral steroids in the feces, a mechanism suggested very early in the study of dietary fiber (Eastwood, 1969). Data presented appear to support the above mechanism but do not appear to explain more recent observations concerning changes in lipoprotein cholesterol levels.

The mechanism of involvement of dietary fiber in colon cancer is also tied to its interaction with bile acids. Increased concentration of bile acids in the colon apparently results in increased susceptibility to chemical carcinogens. The effects of various dietary fibers seems to be related to their ability to increase or decrease bile acid concentrations by adsorption of bile acids or absorption of water, respectively. The role of altered transit time or direct effects on carcinogen absorption or activation have not been fully examined.

Diabetus mellitus can be treated using a high complex carbohydrate and low fat diet. This diet has the added benefit of lowering serum cholesterol levels in a group of people already at a higher risk for heart disease. Alteration of the absorption site of glucose appears to alter the hormonal response to glucose and improve the metabolism of the absorbed nutrients.

In all cases, changes in absorption of nutrients or other materials is an integral part of the effects of dietary fiber on the parameters in question. Dietary fiber components, either through direct adsorption, dilution, changes in transit speed, or alteration of the absorptive surface change the amount of material absorbed or the location in the intestine of absorption. One or both of these changes can obviously have marked effects on the eventual metabolic fate or effect of the absorbed constituent.

It would seem essential to gain a better understanding of the effects of these changes in absorption. Knowledge of physical properties of components of dietary fiber and resultant physiological changes will allow selection of dietary fiber for specific desired changes without other unwanted side effects.

Dietary fiber has progressed from one of the least examined parts of our diet to one of faddish popularity. Our ability to ascertain answers to many of the above questions and thus gain an understanding of the role of dietary fiber in nutrition will eventually place dietary fiber in proper perspective in relation to other diet components. Dietary fiber is an important part of our overall nutrition, but until we do understand its place in nutrition, its use as a therapeutic agent must be limited to a few well examined situations.

ACKNOWLEDGEMENT

Supported, in part, by the Indiana Agricultural Experiment Station (paper #8093) and the Northeast Indiana Chapter of the American Heart Association-Indiana Affiliate.

REFERENCES

Anderson, J.W., and Chen, W-J.L., 1979, Plant fiber. Carbohydrate and lipid metabolism, Am. J. Clin. Nutr. 32:346.

Anderson, J.W., and Ward, K., 1979, High-carbohydrate, high-fiber diets for insulin-treated men with diabetes mellitus, Am. J. Clin. Nutr. 32:2312.

Antonis, A., and Bersohn I., 1962, The influence of diet on fecal lipids in South African white and Bantu prisoners, Am. J. Clin. Nutr. 11:142.

Aries, V.C., Crowther, J.S., Drasar, B.S., Hill, M.J., and Ellis, F.R., 1971, The effect of a strict vegetarian diet on the faecal steroid concentration, J. Pathol. 10:54.

Beher, W.T., and Casazza, K.K., 1971, Effects of psyllium hydrocolloid on bile acid metabolism in normal and hypophysectomized rats, Proc. Soc. Exp. Biol. Med. 136:253.

Berenson, L.M., Bhandaru, R.R., Radhakrishnamurthy, B., Srinivasan, S.B., and Berenson, G.S., 1975, The effect of dietary pectin on serum lipoprotein cholesterol in rabbits, Life Sci. 16:1533.

Birkner, H.J., and Kern, F. Jr., 1974, In vitro adsorption of bile salts to food residues, salicylayo-sulfapyridine and hemicellulose, Gastroenterol. 67:237.

Burkitt, D.P., 1971, Epidemiology of cancer of the colon and rectum, Cancer 28:3.

Burkitt, D.P., Walker, A.R.P., and Painter, N.S., 1972, Effect of dietary fibre on stools and transit times and its role in the causation of disease, Lancet 2:1408.

Burkitt, D.P., Walker, A.R.P., and Painter, N.S., 1974, Dietary fibre and disease, J. Am. Med. Assoc. 229:1068.

Chang, M.W., and Johnson, M.A., 1976, Influence of fat level and type of carbohydrate on the capacity of pectin in lowering serum liper lipids of young rats, J. Nutr. 106:1562.

Chen, W-J.L., and Anderson, J.W., 1979, Effects of plant fiber in decreasing plasma total cholesterol and increasing high-density lipoprotein cholesterol, Proc. Soc. Exp. Biol. Med. 162:310.

Chomchai, C., Bhadrachari, N., and Nigro, N.D., 1974, The effect of bile on the induction of experimental intestinal tumors in rats, Dis. Colon Rectum 17:310.

Cleave, T.L., 1974, "The Saccharine Disease," John Wright and Sons, Bristol, England.

Cookson, F.B., Altschul, R., and Fedoroff, S., 1967, The effects of alfalfa on serum cholesterol and in modifying or preventing cholesterol-induced atherosclerosis in rabbits, J. Atheroscler. Res. 7:69.

Cummings, J.H., 1966, What is fiber?, in: "Fiber in Human Nutrition," G.A. Spiller and R. J. Amen, eds., Plenum Press, New York, p. 1.

Eastwood, M.A., 1969, Dietary fiber and serum lipids, Lancet 2:1222.

Eastwood, M.A., and Boyd, G.S., 1967, The distribution of bile salts along the small intestine of rats, Biochim. Biophys. Acta 137:393.

Eastwood, M.A., and Hamilton, D., 1968, Studies on the adsorption of bile salts to non-absorbed components of diet, Biochim. Biophys. Acta 152:165.

Eastwood, M.A., Kirkpatrick, J.R., Mitchell, W.D., Bone, A., and Hamilton, T., 1973, Effects of dietary supplements of wheat bran on faeces and bowel function, Brit. Med. J. 4:392.

Eastwood, M.A., Eastwood, J., and Ward, M., 1976, Epidemiology of bowel disease, in: "Fiber in Human Nutrition," G.A. Spiller and R.J. Amen, eds., Plenum Press, New York, p. 207.

Ershoff, B.H., 1963, Comparative effects of a purified and stock diet on DBH (2,5-ditertbutylhydroquinone) toxicity in the rat, Proc. Soc. Expt. Biol. Med. 112:362.

Fahrenbach, M.J., Riccardi, B.A., and Grant, W.C., 1966, Hypocholesteremic activity of mucilaginous polysaccharides in white leghorn cockrels, Proc. Soc. Exp. Biol. Med. 123:321.

Fisher, H., Griminger, P., Sostman, E.R., and Brush, M.K., 1965, Dietary pectin and blood cholesterol, J. Nutr. 86:113.

Fleiszer, D., MacFarlane, J., Murray, D., and Brown, R.A., 1978, Protective effect of dietary fibre against chemically induced bowel tumors in rats, Lancet 2:552.

Forman, D.T., Garvin, J.E., Forestner, J.E., and Taylor, C.B., 1968, Increased excretion of fecal bile acids by an oral hydrophillic colloid, Proc. Soc. Exp. Biol. Med. 127:1060.

Garvin, J.E., Forman, D.T., Eiseman, W.R., and Phillips, C.R., 1965, Lowering of human serum cholesterol by an oral hydrophillic colloid, Proc. Soc. Exp. Biol. Med. 120:744.

Glober, G.A., Klein, K.L., Morre, J.O., and Abba, B.C., 1974, Bowel transit times in two populations experiencing similar colon cancer risks, Lancet 2:80.

Haber, G.B., Heaton, K.W., Murphy, D., and Burroughs, L., 1977, Depletion and disruption of dietary fibre. Effects on satiety, plasma-glucose, and serum-insulin, Lancet 2:679.

Haddow, A., 1958, Chemical carcinogens and their modes of action, Brit. Med. Bull. 14:79.

Hamilton, R.M.G., and Carroll, K.K., 1976, Plasma cholesterol levels in rabbits fed low fat, low cholesterol diets: Effect of dietary proteins, carbohydrates and fibre from different sources, Atheroscler. 24:47.

Hill, M.J., and Aries, V.C., 1971, Faecal steroid composition and its relationship to cancer of the large bowel, J. Path. 104:129.

Hill, M.J., Crowther, J.S., Drasar, B.S., Hawksworth, G., Aries, V., and Williams, R.E.O., 1971, Bacteria and the aetiology of cancer of large bowel, Lancet 1:95.

Hill, M.J., Drasar, B.S., Williams, R.E.O., Meade, T.W., Cox, A.G., Simpson, J.E.P., and Morson, B.C., 1975, Faecal bile acids and clostridia in patients with cancer of the large bowel, Lancet 2:535.

Jenkins, D.J.A., Leeds, A.R., Newton, C., and Cummings, J.H., 1975, Effect of pectin, guar gum and wheat fibre on serum cholesterol, Lancet 1:1116.

Jenkins, D.J.A., Leeds, A.R., Gassull, M.A., Cochet, B., and Alberti, G.M., 1977, Decrease in postprandial insulin and glucose concentrations by guar and pectin, Ann. Intern. Med. 86:20.

Jenkins, D.J.A., Wolever, T.M.S., Leeds, A.R., Gassull, M.A., Haisman, P., Dilawari, J., Goff, D.V., Metz, G.L., and Alberti, K.G.M.M., 1978, Dietary fibres, fiber analogues, and glucose tolerance. The importance of viscosity, Brit. Med. J. 1:1392.

Jenkins, D.J.A., Reynolds, D., Salvin, B., Leeds, A.R., Jenkins, A.L., and Jepson, E.M., 1980, Dietary fibre and blood lipids: Treatment of hypercholesterolemia with guar crispbread, Am. J. Clin. Nutr. 33:575.

Judd, P.A., Kay, R.M., and Truswell, A.S., 1976, The cholesterol-lowering effect of pectin, Nutr. Metab. 20:181.

Kay, R.M., and Truswell, A.S., 1977, Effect of citrus pectin on blood lipids and fecal steroid excretion in man, Am. J. Clin. Nutr. 30:171.

Kelsey, M.I., and Pienta, R.J., 1979, Transformation of hamster embryo cells by cholesterol, α-epoxide and lithocholic acid, Cancer Letters 6:143.

Keys, A., Grande, F., and Anderson, J.T., 1961, Fiber and pectin in the diet and serum cholesterol concentration in man, Proc. Soc. Exp. Biol. Med. 106:555.

Kritchevsky, D., and Story, J.A., 1974, Binding of bile salts in vitro by non-nutritive fiber, J. Nutr. 104:458.

Kritchevsky, D., and Tepper, S.A., 1968, Experimental atherosclerosis in rabbits fed cholesterol-free diets: Influence of chow components, J. Atheroscler. Res. 8:357.

Kritchevsky, D., Tepper, S.A., and Story, J.A., 1974, Isocaloric, isogravic diets in rats. III. Effects of non-nutritive fiber (alfalfa or cellulose) on cholesterol metabolism, Nutr. Rep. Int. 9:301.

Kritchevsky, D., Tepper, S.A., Williams, D.E., and Story, J.A., 1977, Experimental atherosclerosis in rabbits fed cholesterol-free diets. Part 7. Interaction of animal of vegetable protein with fiber, Atheroscler. 26:397.

Leveille, G.A., and Sauberlich, H.E., 1966, Mechanism of the cholesterol depressing effect of pectin in the cholesterol-fed rat, J. Nutr. 88:208.

Lin, T.M., Kim, K.S., Karvinen, E., and Ivy, A.C., 1957, Effect of dietary pectin, "protopectin," and gum arabic on cholesterol excretion in rats, Am. J. Physiol. 188:66.

Looney, M.A., and Lei, K.Y., 1978, Dietary fiber, zinc and copper effects on serum and liver cholesterol levels in the rat, Nutr. Rep. Int. 17:329.

Moore, J.H., 1967, The effect of the type of roughage in the diet on plasma cholesterol levels and aortic atherosis in rabbits, Brit. J. Nutr. 21:207.

Munoz, J.M., Sandstead, H.H., Jacob, R.A., Logan, G.M., Reck, S.J., Klevay, L.M., Dintzis, F.R., and Inglett, G.E., 1979, Effects of some cereal brans and textured vegetable protein on plasma lipids, Am. J. Clin. Nutr. 32:58.

Narisawa, T., Magadia, N.E., Weisburger, J.H., and Wynder, E.L., 1974, Promoting effect of bile acid on colon carcinogenesis after intrarectal instillation of N-methyl-N'-nitro-N-nitrosoguanidine in rats, J. Nat. Cancer Inst. 55:1093.

Nigro, N.D., Bhadrachari, N., and Chomchai, C., 1973, A rat model for studying colonic cancer: Effect of cholestyramine on induced tumors. Dis. Colon Rectum 16:438.

Nigro, N.D., Singh, D.V., Campbell, R.L., and Pak, M.S., 1975, Effect of dietary beef fat on intestinal tumor formation by azoxymethane in rats, J. Nat. Cancer Inst. 54:439.

Nigro, N.D., Bull, A.W., Klopfer, B.A., Pak, M.S., and Campbell, R.L., 1979, The effect of dietary fiber on azoxymethane induced intestinal carcinogenesis in rats, J. Nat. Cancer Inst. 62:1097.

Portman, O.W., 1960, Nutritional influences on the metabolism of bile acids, Am. J. Clin. Nutr. 8:462.

Silverman, S.J., and Andrews, A.W., 1977, Bile acids: Comutagenic activity using the Salmonella/mammalian-microsome mutagen test, J. Nat. Cancer Inst. 59:1557.

Southgate, D.A.T., 1976, The chemistry of dietary fiber, in: "Fiber in Human Nutrition," G.A. Spiller and R.J. Amen, eds., Plenum Press, New York, p. 31.

Story, J.A., and Kritchevsky, D., 1976, Comparison of the binding of various bile acids and bile salts in vitro by several types of fiber, J. Nutr. 106:1292.

Story, J.A., Czarnecki, S.K., Baldino, A., and Kritchevsky, D., 1977, Effect of components of fiber on dietary cholesterol in the rat, Fed. Proc. 36:1134.

Thomas, F.B., Shook, D.F., O'Dorisio, T.M., Cataland, S., Mekhjian, H.S., Caldwell, J.H., and Massaferri, E.L., 1977, Localization of gastric inhibitory polypeptide release by intestinal glucose perfusion in man, Gastroenterol. 72:49.

Trowell, H., 1972a, Ischemic heart disease and dietary fibre, Am. J. Clin. Nutr. 25:926.

Trowell, H., 1972b, Crude fibre, dietary fiber and atherosclerosis, Atheroscler. 16:138.

Trowell, H., 1972c, Dietary fibre and coronary heart disease, Eur. J. Clin. Biol. Res. 17:345.

Trowell, H., 1973, Dietary fibre, ischaemic heart disease and diabetes mellitus, Proc. Nutr. Soc. 32:151.

Trowell, H., 1974, Diabetes mellitus death-rates in England and Wales 1920-1970 and food supplies, Lancet 2:998.

Truswell, A.S., 1979, Effects of different types of dietary fibre on plasma lipids, in: "Dietary Fibre: Current Developments of Importance to Health," K.W. Heaton, ed., Food and Nutrition Press, Westport, Conn., p. 105.

Truswell, A.S., and Kay, R.M., 1976, Bran and blood lipids, Lancet 1:367.

Tsai, A.C., Elias, J., Kelly, J.J., Lin, R.S.C., and Robson, J.R.K., 1976, Influence of certain dietary fibers on serum and tissue cholesterol levels in rats, J. Nutr. 106:118.

Walker, A.R.P., 1961, Crude fibre, bowel motility and pattern of diet, S. Afr. Med. J. 35:114.

Walker, A.R.P., Mortimer, K.L., Kloppers, P.J., Botha, D., Grusin, H., and Seftel, H.C., 1961, Coronary heart disease in South African poor whites and white prisoners habituated to a Bantu type of diet, Am. J. Clin. Nutr. 9:643.

Watanabe, K., Reddy, B.S., Weisburger, J.H., and Kritchevsky, D., 1979, Effect of dietary alfalfa, pectin and wheat bran on azoxymethane or methylnitrosourea-induced colon carcinogenesis in F344 rats, J. Nat. Cancer Inst. 63:141.

Wilson, R.B., Hutcheson, D.P., and Wideman, L., 1977, Dimethylhydrazine-induced colon tumors in rats fed diets containing beef fat or corn oil with and without wheat bran, Am. J. Clin. Nutr. 30:176.

# SOME MANIFESTATIONS OF MALABSORPTION IN DISEASE

James D. Jones, Ph.D.

Section of Clinical Chemistry, Department of Laboratory Medicine, Mayo Clinic

Mary A. Jones, Ph.D.

Department of Dietetics, Saint Marys Hospital
Rochester, Minnesota

## INTRODUCTION

The term malabsorption has come to include both defective digestion and absorption. All ingested nutrients, i.e., fat, carbohydrate, proteins, minerals, $H_2O$, and vitamins, normally enter the body via the gastrointestinal tract. It must be remembered that the absorbing surface of the digestive tract is not a one-way street. Therefore, in addition to oral food intake, a significant load of 'endogenous nutrients' is added to the intestinal contents of the alimentary canal. Complex nutrient molecules of both exogenous and endogenous origin undergo stepwise degradation to absorbable form and are transported into the intestinal cell also in a stepwise manner. Consideration of these steps and appreciation that absorbed nutrients exit from the intestinal cell, via either the portal vein or the lymphatic system, are essential to understanding malabsorption.

Equally important, though not as well appreciated, are the nutrient interactions and discriminations which occur not only in the G.I. tract but in other tissues. It would be highly desirable from the laboratory point of view to be able to relate specific tests to specific problems. However, malabsorption problems frequently present as multiple nutrient deficiencies of chronic syndromes, seldom as deficiencies of single nutrients, which have been characterized so well by the nutritional biochemists of the 40's and 50's.

## REVIEW OF ASSIMILATION OF NUTRIENTS

Ingested food has been subjected to maceration, acidification, denaturation, and dilution by the time it leaves the stomach. Little digestion occurs until it is exposed in the small intestine to pancreatic enzymes and, in the case of dietary fat, solubilized with conjugated bile acids. The final steps in digestion of carbohydrates and protein occur on and within intestinal epithelial cells yielding simple compounds to be removed via the portal system. The micelles formed within the intestinal lumen from bile salts, fatty acids and β-monoglycerides disaggregate at the epithelial surface. The bile salts remain within the lumen, available for reuse, while lipid components readily pass the lipophilic brush border membrane. Products of fat digestion, long chain fatty acids ($>C_{10}$) and their mono and diglycerides, must be modified to triglycerides, packed with other lipids within a lipoprotein envelope to form chylomicra before they can exit from the cell to be carried away via the lymph. Short chain fatty acids (called medium chain in medical literature) are absorbed directly without modification and removed via the portal vein.

Interference with any of these steps essential to digestion, absorption and transport usually results in malabsorption with accompanying steatorrhea and diarrhea. Dehydration due to electrolyte and water loss may become rapidly critical, particularly in infants and young children because of their high extracellular water. In addition there is evidence that the bowel of the infant is more permeable and susceptible to osmotic loads than the bowel of adults (Younoszai et al., 1978).

## ABNORMALITIES ASSOCIATED WITH MALABSORPTION

Most abnormalities associated with malabsorption are readily classified into those of intraluminal phase, mucosal cell transport and intestinal lymphatic transport of fat. Each of these plus an unclassified category will be considered separately.

### Intraluminal Phase

Hydrolysis of dietary triglyceride by pancreatic lipase must precede normal absorption. Lipase at pH 6 to 7 in the presence of conjugated bile salts solubilizes the lipid in the form of monoglyceride, fatty acids and fat soluble vitamins into micelles. A defect in any one of these steps leads to defective fat absorption which is evident by an increased excretion of fecal fat (steatorrhea).

Along with malabsorption of fat, one expects malabsorption of fat soluble vitamins which may be seen in increased prothrombin time where vitamin K is not absorbed, and in hypocalcemia and tetany where vitamin D deficiency occurs resulting in decreased calcium absorption. Formation of insoluble calcium soaps of unabsorbed fatty acids may also interfere with calcium absorption.

Bacteria in the colon may act upon the increased amount of fat entering the colon to produce hydroxy fatty acids which are irritants and cathartic and further aggravate the diarrhea and decrease transit time (Soong et al., 1972).

Estimation of serum carotene is a simple, commonly used laboratory test for fat malabsorption.

Apparent pancreatic insufficiency may occur from a number of causes such as a low pH in the duodenum which irreversibly inactivates lipase. An example appears in the Zollinger-Ellison syndrome where there is grossly excessive secretion of gastric HCl (Shimoda et al., 1968). Surgical alterations may alter the relationships between gastric emptying and pancreatic and biliary secretion and cause malabsorption. Subtotal gastric resection and gastrojejunostomy divert food stuffs from the normal pathway in the upper small intestine and thus blunt or even eliminate the humoral stimulus for pancreatic secretion which occurs in response to the presence of chyme in the upper small intestine. Further, mixing of bile salts, pancreatic juice and chyme may be inadequate, and, with a probable decrease in transit time, contribute to malabsorption.

Diagnostic criteria for cystic fibrosis are relevant to this classification. In 1970, we (Jones et al., 1970) determined the frequency distribution of sweat sodium values in a large group of patients of varying ages including some with vague G.I. problems. The analysis yielded three groups: those with clearly elevated sweat sodium, diagnosed as having cystic fibrosis; those with normal sweat sodium, diagnosed as free of cystic fibrosis and those with sweat sodiums in between in a gray area. Some of the patients in the gray area presented with G.I. symptoms suggesting that the intermediate sweat sodium concentration could have been a positive for cystic fibrosis. di Sant'Agnese and Davis (1979) reviewed symptoms in 307 patients with cystic fibrosis 18-47 years of age. They found pulmonary disease in 97% of the patients and 95% were identified as suffering from pancreatic insufficiency, but were "seldom symptomatic although steatorrhea and azotorrea were massive". Although many physicians were reluctant to classify such adult patients as having cystic fibrosis, it is now apparent that many patients survive cystic fibrosis into adulthood and have malabsorption as a result of pancreatic fibrosis. The message, here, is that sweat sodium or chloride is an appropriate test for adults and may be essential in making the correct diagnosis of malabsorption.

Synthesis of bile acids from cholesterol and their conjugation occurs in the liver. The bile salts are reabsorbed in the distal ileum, and patients with diseased or surgically removed small bowel will suffer from a reduced bile acid pool and subcritical concentration of conjugated bile acids (1-2 mM) in the upper jejunum. Thus fat malabsorption is the normal consequence of a diseased liver and of a diseased or absent ileum with reduced bile acid pool.

Malabsorption caused by the contaminated small bowel syndrome, CSBS, is classified as a intraluminal problem. In that it presents complications in a number of the situations in other classifications, it will be addressed later.

## Mucosal Cell

Diseases in which specific abnormalities in mucosal cell transport occur include cystinuria, Hartnups, disaccharidase deficiencies, and monosaccharide transport defects. Those genetic diseases which relate to amino acid assimilation support the currently accepted concept that peptides are hydrolyzed after uptake from the intestinal lumen. Glycine is absorbed more readily, as evidenced by the more rapid rise and greater concentration of plasma glycine, when consumed as a di- or tripeptide than as the free amino acid (Craft et al., 1968). Oral loads of free tryptophan in the Hartnups' patients are poorly absorbed (Asatoor et al., 1970). However, in both Hartnups and normals, tryptophan in peptides and proteins is readily absorbed and confirms that the absorption process involves two steps.

Similar experiments on the absorption of cystine and dibasic amino acids by the cystinuric yielded similar data (Milne et al., 1961). In fact, ingestion of L-cysteine caused plasma levels of cystine to increase more in cystinurics than in normals. However, cystine absorption *per se* appears to be more complex and to involve reduction to cysteine during transport into the cell. These genetic disease of amino acid transport create no gastrointestinal problem for the patient. As the transport defect is also in the kidney, the patient loses the amino acids in the urine. Cystinurics get kidney stones; some Hartnup's patients appear to lose enough tryptophan to have symptoms of niacin deficiency (Jepson, 1972).

Patients with disaccharidase deficiencies do have intestinal symptoms. The dietary carbohydrate is not hydrolyzed, not absorbed and remains in the G.I. tract. Water is retained by the osmotic effect of the unabsorbed disaccharide. The bacteria of the lower tract metabolize the disaccharide and produce organic acids and gas, resulting in bloating, distention discomfort and diarrhea. Lactase deficiency is the most common and most thoroughly studied, and it does occur in children. Barr and coworkers (1979) brought attention to the role played by lactose intolerance in chidren with recurrent abdominal pain.

The lactose-breath $H_2$ test has been used in a number of these studies. Newcomer et al. (1975) concluded from a prospective study of 25 lactose deficient subjects that breath $H_2$ is the method of choice for screening large populations for lactase deficiency. It is sensitive, non-invasive and simple to perform but requires a gas-liquid chromatograph for the determination. The lactose breath $H_2$ test was reported by these investigators to compare very favorably with three other indirect tests, $^{14}CO_2$ from I-$^{14}C$ lactose and increase in plasma glucose after an oral lactose load and the rise in plasma galactose after lactose and ethanol were given together.

Two conditions occur in children, primary alactasia and severe lactose intolerance (Lifshitz, 1966 and Berg, 1969). Lactose is found in the urine in both cases and can be identified by high voltage electrophoresis or thin-layer chromatography. Glucose-galactose malabsorption which presents during the first week of life is an example of a specific defect in transport of glucose unrelated to Na transport. Na transport is normal in this condition. The disease presents as diarrhea and has been diagnosed by identification of the two monosaccharides in feces (Menzies and Seakins, 1974).

Vitamin $B_{12}$ deficiency has no G.I. symptoms per se; it may be due to a lack of intrinsic factor which is usually accompanied by achlorhydria. The vitamin is normally actively absorbed in the ilieum, and resection or disease can eliminate active absorption.

Non-specific abnormalities of mucosal cell transport include gluten sensitive enteropathy, tropical sprue, and effects of radiation and drugs. The gut is very sensitive to radiation and drugs because of the high rate of turnover of the mucosal epithelium. A characteristic flattening of the villi is seen in nontropical sprue. The mucosal structure obtained by peroral biopsy shows flattening of villi in other diseases too (Katz and Grand, 1979) with similar effects to the patient. The lack of normal structure results in loss of mucosal hormones, enzymes, and specific absorptive capacity and, thus, may interfere with absorption of fat, carbohydrate, and protein and, in fact, all nutrients.

Very complex nutritional problems are seen in Crohn's disease of the small bowel which almost always involves the ileum in segmented areas but also involves the jejunum and may be diffuse in its most severe form. Not only does granulomatous infiltration of the ileum and jejunum depress the ability of the columnar cell to absorb multiple nutrients, but intestinal lymphatics are extensively involved, severely impairing fat transport and causing fat to remain in the lumen (Donaldson, 1978). Fistula formation is common. Where enterocolic fistulas occur, there is recirculation of bowel contents with stasis and bacterial overgrowth into the small bowel, leading to deconjugation of bile salts and less effective solubilization of lipids for absorption. Whatever abnormalities lead to fat malabsorption, the resulting diarrhea and steatorrhea may cause demineralization of bone and hypocalcemic tetany due to vitamin D malabsorption and deficiency followed by decreased calcium absorption and to hypoprothrombinemia due to vitamin K malabsorption. Abnormal xylose tolerance tests also demonstrate the reduced capacity for absorption of water soluble nutrients such as iron, calcium, vitamins and others. Clinically, megaloblastic anemia from depressed folate absorption occurs as does iron deficiency anemia from depressed iron absorption. Persistent loss of small amounts of blood via the intestinal wall may contribute to iron deficiency anemia. Surprisingly, with a diseased ileum, the vitamin $B_{12}$ absorption is usually good unless bacterial overgrowth interferes. Weight loss due to failure to assimilate adequate calories is characteristic. Protein losing enteropathy may be severe leading to hypoalbuminemia. The diarrhea and steatorrhea and presence of bile salts in the colon contribute to the development of hyponatremia, hypokalemia and dehydration. Also contributing to final inanition from lack of adequate absorption are anorexia and fear of eating because of expected aggravation of crampy pain and distention. Absorptive capacity may be further impaired by surgical treatment to remove varying amounts of diseased ileum and jejunum.

A relatively rare disease often affecting the terminal ileum and producing diarrhea is the carcinoid syndrome. The tumor may rob the host of 50% of the available tryptophan. Its coexistence with Crohn's disease has been reported (Wood et al., 1970).

Architecture of the small intestine is also disturbed in about 70% of the cases of generalized amyloidosis. Malabsorption syndrome with steatorrhea appears to be relatively rare having been found in only 5 to 6% of cases reported (Kyle and Bayrd, 1975).

Scleroderma may involve the small bowel and even the colon resulting in dilatation and stasis leading to bacterial overgrowth. About 50% of patients with small bowel involvement exhibit malabsorption with increased fecal fat and impaired D-xylose and $B_{12}$ absorption (Kahn et al., 1966).

## Lymphatic Transport

The diseases that have been associated with abnormal lymphatic transport of fat are primary lymphatic disease, Whipple's disease, constrictive pericarditis and congestive heart disease. At first lipids are removed from the G.I. lumen but do not pass from the lymphatics into the circulation preventing continued uptake of lipids. Steatorrhea appears with all its sequela.

## ACTIVITY OF INTESTINAL MICROFLORA

### Contaminated Small Bowel Syndrome (CSBS)

The significance of the contaminated small bowel syndrome in many diseases is becoming appreciated. The normal gut flora consists of high numbers of anaerobes in a highly reducing medium in the colon, and very low numbers of aerobic flora in the upper tract (Gracey, 1979). This changes in many clinical situations which are complicated by the overgrowth of bacteria. Anatomical abnormalities of the gut and situations with abnormal gut motility are frequently complicated by bacterial overgrowth as are many postoperative states. These conditions may cause stasis resulting in higher counts and shifts in kind of bacteria found in the upper tract. A case report on CSBS as a complication of disordered intestinal motility due to amyloidosis was recently described (Battle et al., 1979).

A number of the diseases that can be classified in previous categories are complicated by CSBS. In these diseases, the bacterial overgrowth is probably due to the stasis as in Crohn's disease, or to the increased concentration of nutrients that occur in the tract. Solutes left in the gut lumen stimulate bacterial growth. The microflora then compete with the host for essential nutrients while producing products that alter the function of the more distal portions of the tract. Unconjugated bile acids formed from bacterial deconjugation are believed to alter water transport (Teem and Phillips, 1972). The osmotic load can be increased by the action of colon bacteria on carbohydrates yielding monosaccharides and organic anions. Water absorption is inhibited by deoxycholate and hydroxy fatty acids formed by colon bacteria from conjugated cholate and dietary fat (Soong et a., 1972). In CSBS, $B_{12}$ absorption may be compromised in that the microflora produce inactive cyanocobalamin analogues from vitamin $B_{12}$ which are absorbed, displace $B_{12}$ from liver stores (Brandt et al., 1975) and interfere with some binding assays (Kohlhouse et al., 1978); inactive vitamin $B_{12}$ analogues that are not absorbed or are poorly absorbed are also formed in the gut where there is bacterial overgrowth and might interfere with absorption of vitamin $B_{12}$ (Brandt et al., 1977). Phillips (1972) has summarized the sequence of events now believed to occur in diarrhea.

### Other Activities of Microflora

Not all activities of the microflora are detrimental to the host. Vitamin K and folate are synthesized in the gut and are available to the host. Bacterial metabolism of compounds secreted into the gut allows excretion of unwanted compounds via the feces.

The metabolic activity of the flora and enteric cycling of urea, uric acid and creatinine has been investigated in detail. Efforts have been made to determine the quantity cycled, the factors which alter the quantities cycled and the products formed. An estimated 9 liters of water, 2 from the diet and 7 from intestinal secretions, are cycled through the G.I. tract per day (Phillips, 1972). Urea and creatinine are distributed in body water and thus, readily secreted into the G.I. tract. Interest in enteric cycling of these compounds is from two directions - reutilization of nitrogen, which would be expected to improve the state of protein nutrition, and elimination of metabolic products (metabolites can be trapped in the gut by use of sorbents). The clinical situations where this applies are in liver and kidney failure.

In hepatic coma, the object is to reduce the ammonia load to the liver, as 25% of the urea produced is enterically cycled and subject to conversion to $NH_3$ by gut bacterial urease. The removal of such a "toxin" could be of clinical significance. In uremia, the patient also cycles his body fluids through the G.I. tract. In this case, not only are the serum concentrations of those compounds normally excreted via the urine increased, but many of the products of bacterial metabolism are absorbed and are also retained.

Our studies on creatinine were prompted by the observation of Enger and Blegen (1964) that, as the serum concentration of creatinine increased, urinary excretion decreased. This decrease was not related to decreased muscle mass and suggested that creatinine was subject to enteric cycling similar to that of urea (Walser and Bodenlos, 1959) and uric acid (Sorenson, 1959).

In a series of experiments, we demonstrated:

1. Induction of "creatininase" in the intestinal flora of the rat fed creatinine but not creatine (Jones and Burnett, 1972).

2. Enteric cycling of up to 60% or 700 mg of creatinine formed per day in uremics (Jones and Burnett, 1974).

3. That the proposed pathways for creatinine degradation by gut bacteria shown in Fig. 1 are followed in uremic patients (Jones and Burnett, 1975).

The uremic patient has bacterial overgrowth in his small bowel (Simenhoff et al., 1976) and thus forms these and other compounds in the G.I. tract in quantities sufficient to permit absorption from the gut.

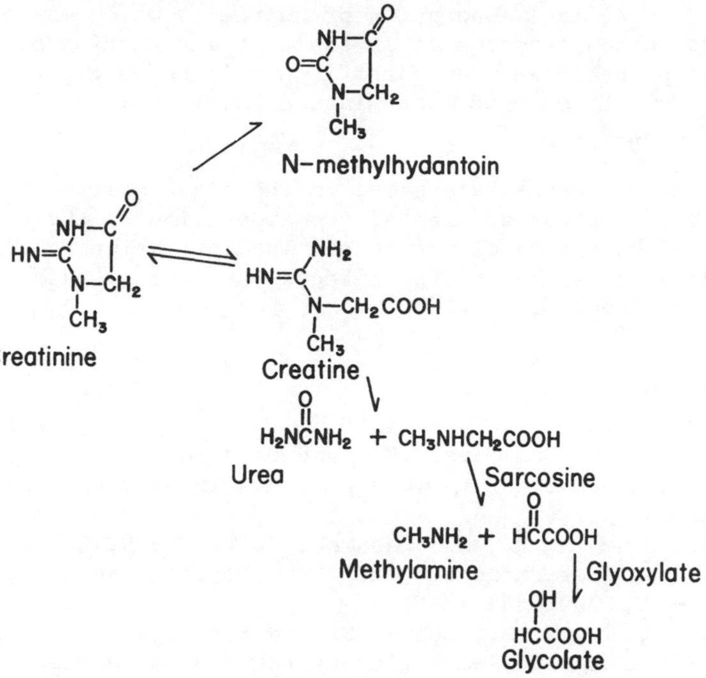

N-methylhydantoin

Creatinine

Creatine

$H_2NCNH_2$ + $CH_3NHCH_2COOH$

Urea

Sarcosine

$CH_3NH_2$ + HCCOOH

Methylamine

Glyoxylate

HCCOOH

Glycolate

Fig. 1. Metabolism of Creatinine by the Microflora of the Intestine in Uremia. (Reprinted from Kidney International, Jones and Burnett (1975), with permission.)

The quantitative significance of enteric cycling has been determined by direct measurement of urea, creatinine and uric acid. The biochemical significance waits on the elucidation of the fate of the metabolites and of any affects the metabolites have on the host.

SUMMARY

Efficient digestion and absorption of ingested foodstuffs requires an intact, functional gastrointestional tract. Interferences with digestive processes may cause loss of specific dietary nutrients and also loss of nutrients secreted into the tract. The diseases associated with malabsorption may be classified according to:

Abnormalities of the intraluminal phase
Abnormalities of mucosal cell transport
Abnormalities of intestinal lymphatic transport.

Impaired intestinal absorption of nutrients which may present as the malabsorption syndrome as a result of interaction of dietary components or by deficiency of specific nutrients, e.g., vitamin D, are reviewed. Effects of contaminated bowel in many disease states is emphasized.

Diagnostic approaches are based on the biochemistry of digestion and transport of nutrients. Consideration of the processes involved in malabsorption allows better interpretation of diagnostic test and understanding of the effects of these disease processes on other laboratory values.

REFERENCES


Asatoor, A. M., Cheng, B., Edwards, D. G., Lant, A. F., Matthews, D. M., Milne, M. D., Navab, F., and Richards, A. J., 1970, Intestinal absorption of two dipeptides in Hartnup disease, Gut, 11:380.

Barr, R. G., Levine, M. D., and Watkins, J. B., 1979, Recurrent abdominal pain of childhood due to lactose intolerance, N. Eng. J. Med., 300:1449.

Battle, W. M., Rubin, M. R., Cohen, S. and Snape, W. J., Jr., 1979, Gastrointestinal-motility dysfunction in amyloidosis, N. Eng. J. Med., 301:24.

Berg, N. O., Dahlqvist, A., Lindberg, T., and Studnitz, W. V., 1969, Severe familial lactose intolerance - a gastrogen disorder?, Acta Paediat. Scand., 58:525.

Brandt, L. J., Bernstein, L. H., Efron, F., and Wagle, A., 1975, Vitamin $B_{12}$ analogue production in the experimental blind loop, (abstract), Gastroenterology, 68:863.

Brant, L. J., Bernstein, L. H., and Wagle, A., 1977, Production of vitamin $B_{12}$ analogues in patients with small-bowel bacterial overgrowth, Ann. Intern. Med., 87:546.

Craft, I. L., Geddes, D., Hyde, C. W., Wise, I. J., and Matthews, D. M., 1968, Absorption and malabsorption of glycine and glycine peptides in man, Gut, 9:425.

di Sant'Agnese, P. A. and Davis, P. B., 1979, Cystic fibrosis in adults - 75 cases and a review of 232 cases in the literature, Amer. J. Med., 66:121.

Donaldson, R. M., Jr., Crohn's disease of the small bowel in "Sleisenger and Fordtran Gastrointestinal Disease", 1978, pp. 1052-1076, W. B. Saunders, Philadelphia.

Enger, E. and Blegan, E. M., 1964, The relationship between endogenous creatinine clearance and serum creatinine in renal failure, Scand. J. Clin. Lab. Invest., 16:273.

Gracey, M., 1979, The contaminated small bowel syndrome: pathogenesis, diagnosis and treatment, Amer. J. Clin. Nutr., 32: 234.

Jepson, J. B., in "The Metabolic Basis of Inherited Disease", Eds: J. B. Stanbury, J. B. Wyngaarden and D. S. Fredrickson, pp. 1486-1503, third edition, McGraw-Hill, New York, 1972.

Jones, J. D. and Burnett, P. C., 1972, Implication of creatinine and gut flora in the uremic syndrome: induction of "creatininase" in colon contents of the rat by dietary creatinine, Clin. Chem., 18:280.

Jones, J. D. and Burnett, P. C., 1974, Creatinine metabolism in humans with decreased renal function: creatinine deficit, Clin. Chem., 20:1204.

Jones, J. D. and Burnett, P. C., 1975, Creatinine metabolism and toxicity, Kidney Internatl., 7:S294.

Jones, J. D., Steige, H., and Logan, G. B., 1979, Variations of sweat sodium values in children and adults with cystic fibrosis and other diseases, Mayo Clin. Proc., 45:768.

Kahn, I. J., Jeffries, G. H., and Sleisenger, M. H., 1966, Malabsorption in intestinal sclerodema. Correction by antibiotics, N. Eng. J. Med., 274:1339.

Katz, A. J. and Grand, R. J., 1979, All that flattens are not "sprue", Gastroenterology, 76:375.

Kolhouse, J. F., Kondo, H., Allne, N. C., Podell, E., and Allen, R. H., 1978, Cobalamin analogues are present in human plasma and can mask cobalamin deficiency because current radioisotope dilution assays are not specific for true cobalamin, N. Eng. J. Med., 299:785.

Kyle, R. A. and Bayrd, E. D., 1975, Amyloidosis: a review of 236 cases, Medicine, 54:271.

Lifshitz, F., 1966, Congenital lactase deficiency, J. Peds., 69:229.

Menzies, I. S. and Seakins, J. W. T., 1969, Sugars, in "Chromatographic and Electrophoretic Techniques", Vol. 1, 3rd edition, (I. Smith, ed.), pp. 310-329, Interscience Publications, New York.

Milne, M. D., Asatoor, A. M., Edwards, K. D. G., and Loughridge, L. W., 1961, The intestinal absorption defect in cystinuria, J. Br. Soc. Gastroenter., 2:323.

Newcomer, A. D., McGill, D. G., Thomas, P. J., and Hormann, A. F., 1975, Prospective comparison of indirect methods for detecting lactase deficiency, N. Eng. J. Med., 293:1232.

Phillips, S. F., 1972, Diarrhea: a current view of the pathophysiology, Gastroenterology, 63:495.

Shimoda, S. S., Saunders, D. R., and Rubin, C. E., 1968, The Zollinger-Ellison syndrome with steatorrhea. Mechanisms of fat and vitamin $B_{12}$ malabsorption, Gastroenterology, 55:705.

Simenhoff, M. L., Saukkonen, J. J., Burke, J. F., Wesson, L. G., and Schaedler, R. W., 1976, Amine metabolism and the small bowel in uraemia, Lancet, 2:818.

Soong, C. S., Thompson, T. B., Poley, J. R., and Hess, D. R., 1972, Hydroxy fatty acids in human diarrhea, Gastroenterology, 63:748.

Sorenson, L. B., 1959, Degradation of uric acid in man, Metabol. Clin. and Exper., 8:687.

Teem, M. V. and Phillips, S. F., 1972, Perfusion of the hamster jejunum with conjugated and unconjugated bile acids: inhibition of water absorption and effects on morphology, Gastroenterology, 62:261.

Walser, M. and Bodenlos, L. J., 1959, Urea metabolism in man, J. Clin. Invest., 38:1617.

Wood, W. J., Archer, R., Schaeffer, J. W., Stevens, C., and Griffen, Jr., W. O., 1970, Coexistence of regional enteritis and carcinoid tumor, Gastroenterology, 59:265.

Younoszai, M. K., Sapario, R. S., and Laughlin, M., 1978, Maturation of jejunum and ileum in rats, J. Clin. Invest., 62:271.

# AMINO ACIDS IN HEALTH AND DISEASE

Paul M. Tocci, Ph.D.

Director, Biochemical Genetics Laboratory, Mailman Child Development Center, Department of Pediatrics University of Miami, Miami, Florida

## INTRODUCTION

The proteins of the human body are in a constant state of flux with some proteins being synthesized and catabolized continually. Dietary proteins and tissue protein breakdown products provide a pool of amino acids which provide the building blocks for formation of new protein for growth and for rebuilding tissue protein. Essential amino acids are those not synthesized by man or not produced sufficiently rapidly enough to support this activity. Therefore, man requires a source of nitrogen and essential amino acids to support growth, protein synthesis and biosynthesis of important nitrogenous metabolites. The differences in value of food proteins are due to variations in amino acid pattern of each protein. Since the body can form only some of the amino acids, nine "essential" ones must be supplied in the diet (Table I) (McLaughlan et al., 1963, Schelling, 1975). Protein normally is not absorbed by the human intestine, but its hydrolysis products peptides and free amino acids are readily absorbed by the small intestine and transported via the portal blood. A diet low in protein or food protein which is deficient in one or more essential amino acids will restrict protein synthesis. Nonessential amino acids are generally metabolized to pyruvate which leads to gluconeogensis. Therefore, in times of great need (starvation) body proteins can provide energy via glucose. The problem here is that the conversion of protein to glucose is only about 50% efficient (Bessman, 1975) so that 100 g of glucose will require 200 g of protein or the equivalent of 1 kg of lean body weight. Most of the essential amino acids are converted to acetyl CoA and their nitrogen as well as that of the amino acids converted to glucose is lost to the body primarily as ammonia.

Table I. Essential Amino Acid Requirements in Healthy Human Infants

| Amino Acid | Mean (mg/kg/day) |
|---|---|
| Histidine | 34 |
| Isoleucine | 126 |
| Leucine | 150 |
| Lysine | 103 |
| Methionine | 45 |
| Phenylalanine | 90 |
| Threonine | 87 |
| Tryptophan | 22 |
| Valine | 105 |

(From Holt et al., 1960)

PROTEIN REQUIREMENTS

A well balanced diet containing 8-10% protein from various sources will satisfy the needs of the adult for nitrogen and essential amino acids as well as provide sufficient caloric content to prevent utilization of amino acids for energy (National Academy of Sciences, 1968).

Infants up to a year of age require 1.8-2.2 g of protein/kg/day. After a year of age, the gain in body weight is about 3% nitrogen or 18% protein and the requirements to maintain this increment plus the obligatory loss ranges from 2.1 g to 2.3 g of protein/kg/day. The gain in body protein per day then is equal to the daily weight gain multiplied by 0.18. Growth increases the protein requirements enormously, from 2 g at birth to 60 g for adolescent girls, and 80 g for adolescent boys (Wehr and Lewis, 1966, National Academy of Sciences, 1968). In adulthood, about 1 g protein/kg/day is sufficient to maintain body weight (Holt et al., 1960). In healthy persons heavy muscular activity does not change the need for protein unless hypertrophy of muscle is in progress or heavy sweat losses are sustained. Pregnancy requires increases of 0.5 g of protein/day in the first trimester and up to 5.4 g/day in the last. Lactation increases the protein requirement 16 g/day. Menstruation causes very little loss of nitrogen (0.3 mg protein/day/kg). The normal stress encountered in daily life does not appreciably change the need for protein but the stress of serious infection, burns, or anemia temporarily raises the protein needs of the individual. Severe food restrictions, acquired and genetic disease effects will be discussed later in this work.

## FACTORS AFFECTING PLASMA AMINO ACID CONCENTRATIONS

Three things must be considered when examining amino acid concentrations in the human; methodology, intermediary metabolism and genetics.

### Methodology

Analytical methods for amino acids go back to the beginning of the 20th century, when Garrod (1908) introduced the concept of inborn errors of metabolism. At that time only two disorders involving amino acids were known (alcaptonuria and cystinuria) and were diagnosed by simple tests applied to the patients urine. Two more disorders were discovered by similar methods, phenylketonuria and tyrosinosis (Medes type) in the next 30 years. The advent of partition chromatography coupled with the ninhydrin strain in the 40's spawned the discovery of some 50 "ninhydrin-positive" traits (Dent, 1946). The application of gas chromatography and mass spectrometry in the 60's helped reveal "ninhydrin-negative" traits of amino acid metabolism (Horning and Horning, 1971). The choice of the several methods for examining amino acid metabolism in humans depends upon the goal of the analyst; whether qualitative or quantitative, specific or comprehensive, manual or automated. Qualitative assay of amino acid mixtures may be achieved by chemical (Hill et al., 1965), Hochella, 1967), microbiological (Guthrie and Susi, 1963), chromatographic (Efron et al., 1964, Levy, 1968, Raine et al., 1972, Tocci, 1967, Jackson et al., 1968) and electrophoretic (Efron, 1959, 1968) techniques. Individual amino acids are generally estimated by fluorometric methods (Waalkes and Udenfriend, 1957, McCamen and Robins, 1962). Liquid elution chromatography utilizing the ninhydrin reaction has been the standard for quantitative analysis of complex amino acid mixtures (Moore and Stein, 1954) though recently gas chromatography (Gehrke et al., 1968) and liquid-elution chromatography utilizing fluorescence are becoming popular (Benson and Hare, 1975).

All methods of analyses of the amino acids have pitfalls of which one must be aware in order to recognize an abnormal finding when it is encountered: venipuncture will lower the taurine and glutamic acid concentration for up to one hour (Rouser et al., 1962); plasma is preferred to serum since it can be processed faster (serum may change its amino acid composition while standing at room temperature during clot retraction), the use of excess heparin as an anti-coagulant may cause hemolysis which will alter the amino acid pattern, and impurities in some lots of EDTA can produce spurious peaks in the liquid elution chromatogram (Perry and Hansen, 1969); plasma specimens can be contaminated by platelets and leukocytes which contain large amounts of taurine and dicarboxylic acid (Rouser et al., 1962; Soupart, 1962) hemolysis causes the appearance of glutathione, increases ornithine, and reduces arginine and possibly cystine (Winter and Christensen, 1964; Levy and Barkin, 1971).

Immediate deproteinization is important to avoid losses of disulfide amino acids which bind to plasma proteins at room temperature or in the refrigerator (Moore and Stein, 1954). Deproteinization may be accomplished with 3% sulfosalycylic acid (Gerritson et al., 1965) or picric acid (Moore and Stein, 1954). After deproteinization, storage for long periods of time at -20°C raises tryptophan, glutamine and aspartic acid concentrations (Scriver and Rosenberg, 1973). Other factors which may affect amino acid concentrations are continued fasting and food intake (Weller et al., 1969; Adibi, 1970), while physical activity and choice of arm for venipuncture have no influence (Armstrong and Stave, 1973).

## Intermediary Metabolism

After birth, the amino acid concentrations of plasma fall in the first few hours of life (Dickenson et al., 1965; Linblad and Baldeston, 1969). Placental insufficiency can also lower amino acid concentrations in the newborn. The maternal plasma amino acid concentrations are reduced significantly in the first half of pregnancy, especially citrulline, ornithine, and arginine. The concentrations rise toward normal at term (Reid et al., 1971). Maternal protein deprivation can be devastating to the developing fetus (Naeye et al., 1973; Mestyan et al., 1969; Bessman, 1972). There is circadian periodicity to the concentrations of plasma amino acids (Wurtman et al., 1968; Feigin et al., 1968) which is not great, but can effect the interpretation of plasma amino acid concentrations in research protocols where repeated sampling of blood for plasma amino acids may be necessary (Feigin and Haymond, 1970), or in the identification of heterozygotes for phenylketonuria (Rosenblatt and Scriver, 1968). Sex-related differences are minor (Dirren et al., 1975). Amino acid deficiency in man is usually a function of protein malnutrition. The use of synthetic or modified diets has added to this classification because of the attendant hazard of deficiencies involving the essential amino acids. These problems are particularly severe in the neonate, since at birth skeletal muscle comprises only 20-25% of the birth weight and although amino acids can be metabolized to provide calories, neither glucose nor fat can act as precursors for essential amino acids. Special care must be taken in the interpretation of amino acid results from infants on amino acid supplementation, given either orally or intravenously (Logan et al., 1974; Brans et al., 1974; Ghadimi et al., 1971; Ghadimi, 1973; Abitol et al., 1975; Heird and Winters, 1975) and adults given intravenous amino acid solutions after surgical operations (Tweedle et al., 1973; Furst et al., 1978). Starvation results in cyclic changes in the amino acid concentrations of plasma that reflect the changing amino acid pools and the splanchnic handling of amino acids (Felig et al., 1969; Adibi et al., 1970; Cahill, 1970). Branched chain and other amino acids are the first to leave the muscle and enter the plasma as a result of the loss of insulin control in muscle pools. Then, the concentrations of alanine and other glycogenic amino acids begin to fall in the plasma because the liver utilizes them for gluconeogenesis. Later as starvation progresses alanine remains low and the drain of protein is diminished as the brain adapts to operating on ketone bodies instead of glucose (Owen et al., 1967). When this occurs the kidney assumes the larger role of gluconeogenesis and ammoniagenesis. It utilizes glycine, alanine and proline to manufacture glucose and glutamine for ammoniagenesis. Plasma glycine concentration steadily rises unexplainably as alanine and the branched chain amino acids steadily decline (Cahill, 1970).

In protein-calorie malnutrition, which differs from starvation, the concentrations of proline, serine, glycine, an phenylaline rise in the plasma while leucine, iso-leucine, valine, tryptophan, methionine, threonine, arginine and citrulline are in relative deficiency (Holt et al., 1963). There are large oscillations of the plasma amino acid concentrations during the early stages (Heard et al., 1969; Whitehead, 1967) which make treatment difficult, probably due to the loss of tissue pools for storage and enzymes for metabolism of the amino acids, as well as intestinal changes which compromise the availability of dietary nitrogen. Maternal malnutrition causes changes in the amino acid patterns of the fetus which persist after birth (Mestyan et al., 1969).

Several attempts to determine the extent of protein and calorie deprivation have been based on the aminogram (quantitative amino acid analysis results); in marasmus (Heard et al., 1969; Arroyove et al., 1962; Whitehead, 1967; Arroyove and Bowering, 1968) malnutrition (Baertle et al., 1964; Westall et al., 1958), and kwasiorkor (Whitehead and Dean, 1964; Rutishauser and Whitehead, 1969; Holt et al., 1963). However, the aminogram is of little use in diagnosing the state of the protein economy in the normal human, since the free amino acids constitute only 0.1% of the total body protein. Normal variations of the amino acids are often as much as 200-300% as compared with approximately 10% variation in blood electrolytes. Obesity also alters plasma amino acid concentrations in non-diabetic people (Felig et al., 1969). Glycine is depressed while tyrosine, phenylalanine and the branched-chain amino acids are significantly elevated. This is probably a reflection of the ineffectiveness of insulin to control amino acid transportation into and out of skeletal muscle. Diabetes in its various manifestations also alters the aminogram probably by shifting of amino acids from muscle to liver (Felig et al., 1970 and Aoki et al., 1975). Disorders affecting the liver and kidney (McGale et al., 1972; Nyhan and Tocci, 1966;

Hilty et al., 1974; Furst et al., 1978) as well as endocrine and muscular disorders (Felts, 1971; Kisser, 1974) very often disturb the homeostasis of amino acid metabolism.

Genetics

For clinical purposes the most important application of amino acid analyses is the determination of free amino acids in physiological fluid and tissue as a key to the diagnosis of the more than 50 inborn errors of amino acid metabolism. These diseases are relatively rare and are seen mainly in the newborn. Primary aminoacidopathies are caused by the lack of an enzyme or enzyme activity for intermediary metabolism and are usually of autosomal recessive inheritance. They affect only one or a few amino acids. The clinical signs and symptoms of genetic amino acid disorders are varied and diffuse so that the differential diagnosis of amino acid disorders is extensive. It is important to recognize these disorders since many of them are treatable if diagnosed early. Most of the inborn errors of amino acid metabolism cause severe mental retardation, sickness and early death in the untreated state. For this reason, there are large screening programs for detecting them in the neonate (Jackson et al., 1968; Levy et al., 1968). Even when the opportunity is missed for early diagnosis the correct diagnosis of a genetically determined aminoacidopathy is of great importance for genetic counseling of the family, the early diagnosis of abnormalities of newborns in the family, and the determination of the carrier state in siblings. There are genetic disorders involving amino acid metabolism for which treatment may be vitamin therapy, rather than restriction or addition of protein or specific amino acids (Sherwood and Paris, 1964). Final confirmation of many of those disorders is often difficult since a liver biopsy may be needed (Scriver and Rosenberg, 1973) or may be impossible since all the enzymes are not yet known.

High urinary excretion of amino acids can occur in the absence of aminoacidemia due to a defect in the renal tubular reabsorption mechanism. Some of these disorders are also characterized by a disturbance of intestinal reabsorption of certain amino acids (Segal, 1976; Scriver, 1969).

SUMMARY

In summary, the plasma amino acid concentration in human beings are amazingly constant in healthy adults. On the other hand, there are methodological problems, variations between individuals, small changes due to acquired disease or due to fad diets or very large changes due to genetic disorders or amino acid supplementation of which one must be aware to interpret correctly the results of amino acid analyses and intelligently apply this information in the treatment of patients.

REFERENCES

Abitol, C. L., Feldman, D. B., Ahmann, P. and Rudman, D., 1975, Plasma amino acid patterns during supplemental intravenous nutrition of low-birth weight infants, Pediat., 86:766.

Adibi, S. A., Drash, A. L., and Livi, E. D., 1970, Hormone and amino acid levels in altered nutritional states, J. Lab. Clin. Med., 76:722.

Aoki, T. T., Manzano, M., Kozak, G. P., Cahill, G. F., Jr., 1975, Plasma and cerebrospinal fluid amino acid levels in diabetic ketoacidosis before and after corrective therapy, J. Amer. Diabetes Ass., 24:463.

Armstrong, M. D. and Stave, U., 1973, A study of free amino acid levels I. Study of factors affecting the validity of amino acid analysis, Metab., 22:549.

Arroyove, G., Wilson, D., DeFunes, C. and Behar, M., 1962, The free amino acids in blood plasma of children with kwashiorkor and marasmus, Am. J. Clin. Nutr., 11:517.

Arroyove, G. and Bowering, J., 1968, Plasma free amino acids as an index of protein nutrition, Arch. Latin. deNutr., 23:341.

Baertle, J. M., Placko, R. P. and Graham, G. G., 1974, Serum proteins and free amino acids in severe malnutrition, Am. J. Clin. Nutr., 27:733.

Bessman, S. P., 1975, Interrelationships of various food materials in total parenteral nutrition, (H. Ghadimi, ed.) John Wiley and Sons, Inc.

Bessman, S. P., 1972, Genetic failure of fetal amino acid justification: A common basis for many forms of metabolic, nutritional and "nonspecific" mental retardation, J. Pediat., 81:834.

Brans, Y. W., Sumners, J. E., Dweck, H. S. and Cassady, G., 1974, Feeding the low birth weight infant: Orally or parenterally. Preliminary results of a comparative study, Pediat., 54:15.

Cahill, G. F., Jr., 1970, Starvation in man, N. Eng. J. Med., 282:668.

Dent, C. E., 1946, Detection of amino acid in urine and other fluids, Lancet, (ii):637.

Dickinson, J. S., Rosenblum, H., Hamilton, P. B., 1965, Ion exhange chromatography of the free amino acids in the plasma of the newborn infant, Pediat., 36:2.

Dirren, H., Robinson, A. B., Pauling, L., 1975, Sex related patterns in the profiles of human urinary amino acids, Clin. Chem., 21, 1970.

Efron, M. L., 1959, Two-way separation of amino acids and other ninhydrin reacting substances by high voltage electrophoresis followed by paper chromatography, Biochem. J., 72:691.

Efron, M. L., Young, D., Moser, H. W. and McCready, R. A., 1964, A simple chromatographic screening test for the detection of disorders of amino acid metabolism, N. Eng. J. Med., 270:1378.

Efron, M. L., 1968, High voltage paper electrophoresis in (I. Smith, ed.) Chromatographic and Electrophoretic Techniques, Vol. II, Second Edition (pp. 166-193) Interscience, New York.

Feigin, R. D., Klainer, A. S., Beisel, W. R., 1968, Factors affecting ciradian periodicity of blood amino acids in man, Metab. J.

Feigin, R. D. and Haymond, M. W., 1970, Circadian periodicity of blood amino acids in the neonate, Pediat., 45:782.

Felig, P., Owen, P. E., Warrem, J. and Cahill, G. F., Jr., 1969, Amino acid metabolism during prolonged starvation, J. Clin. Invest., 48:584.

Felig, P., Marliss, E. and Cahill, G. F., Jr., 1969, Plasma amino acid levels and insulin secretion in obesity, N. Eng. J. Med., 281:811.

Felig, P., Marliss, E., Ohman, L. and Cahill, G. F., Jr., 1970, Plasma acid levels in diabetic ketoacidosis, Diabetes, 19:412.

Felts, J. H., King, J. J., Jr., 1971, Enhanced secretion of free amino acids by hyperthyroidism patients, Clin. Chem., 17:388.

Furst, P., Ahlberg, M., Alvestrand, A., Bergstrom, J., 1978, Principles of essential amino acid therapy in uremia. Am. J. Clin. Nutr., 31:1744.

Garrod, A. E., 1908, The croonian lectures, Lancet, (ii).

Gehrke, C. W., Roach, D., Zumwalt, R. W. and Stalling, D. L., 1968, Quantitative gas liquid chromatography of amino acids in proteins and biological substances, analytical biochemistry, Inc., Columbia.

Gerristen, T., Rekberg, M. L., Waisman, H., 1965, On the determination of free amino acid levels in serum, Anal. Biochem., 11:460.

Ghadimi, H., Abaci, F., Kuman, S. and Rathi, M., 1971, Biochemical aspects of intravenous alimentation, Pediatr., 48:955.

Ghadimi, H., 1973, A review: Current status of parental amino acid therapy, Pediat. Res., 7:169.

Guthrie, R. and Susi, A., 1963, A Simple method for determining phenylalanine in large populations of newborn infants, Pediat., 32:338.

Heard, C. R. C., Kriegsman, S. M. and Pratt, B. S., 1969, The interpretation of plasma amino acid ratios in protein calorie deficiency, Br. J. Nutr., 23:203.

Heird, W. C. and Winters, R. W., 1975, Total Parenteral Nutrition, J. Pediat., 86:2.

Hill, J. B., Summer, G. K., Pender, M. W. and Roszel, N. O., 1965, An automated procedure for blood phenylalanine, Clin. Chem., 11:541.

Hilty, M. D., Romshe, C. A. and Delameter, P. V., 1974, Reyes snydrome and hyperaminoacidemia, J. Pediat., 84:362.

Hochella, N. J., 1967, Automated fluorometric determination in blood, Anal. Biochem., 21:227.

Holt, L. E., Jr., Gregory, P., Pratt, E. L., Snyderman, S. E., Wallace, W. M., 1960, Protein and amino acid requirements in early life, New York University Press, New York.

Holt, L. E., Snyderman, Norton, P. M., Roitman, E. and Finch, J., 1963, The plasma aminogram in kwashiorkor, Lancet, (ii) 1344.

Horning, E. C. and Horning, M. G., 1971, Human metabolic profiles obtained by GC and GC/MS, J. Chrom. Sci., 9:129.

Jackson, H., Sardharwalla, I. B. and Ebers, G. C., 1968, Two systems of amino acid chromatography suitable for mass screening, Clin. Biochem., 2:163.

Kisser, W., 1974, Amino acid determinations in the serum of patients with multiple sclerosis, Anal. Chem., 86:3.

Levy, H. L., Shih, V. E., Madigan, P., Karolkewics and MacCready, R. A., 1968, Results of a screening method for amino acids. I. Whole Blood, Clin. Biochem., 1:200.

Levy, H. L. and Barkin, E., 1971, Comparison of amino acid concentration between plasma and erythrocytes, Lab. Clin. Med., 78:517.

Linblad, B. S., 1970, The venous plasma free amino acid levels during the first several hours of life. I. After normal and short gestation complicated by hypertension with special reference to the small dates syndrome, Acta. Paediat. Scan., 59:13.

Linblad, B. S. and Baldesten, A., 1969, Time studies on free amino acid levels of venous plasma during the neonate period, Acta. Paediat. Scan., 58:252.

Logan, R. W., Young, D. G., Ross, D. A., Stewart, B. R., Kubo, M. and Tryfonas, G., 1974, Comparison of an oral and intravenous regimen in the newborn, Arch. Dis. or Childhood, 49:200.

McCaman, M. W., and Robins, E., 1962, Fluorometric method for the determinations of phenylalanine in serum, J. Lab. Clin. Med., 59:885.

McGale, E. H. F., Pickard, J. C., Aber, G., 1972, Quantitative changes of amino acids in patients with renal disease, Clin. Chem., 38:395.

McLaughlan, J. M., Noel, F. J., Morrison, A. B. and Campbell, J. A., 1963, "Methodology of protein evaluation" in (Munroe, A. N. and Allison, J. B., eds.), Mammalian protein in metabolism, II. 391-423, Academic Press, New York, New York.

Mestyan, J., Feketa, M., Jarari, I., Sulyak, Imkoff, S. and Soltzberg, G. Y., 1969, The post natal changes in the circulating free amino acid pool in the newborn infant. II. The plasma amino acid ratio in intra-uterine malnutrition, Biol. Neonat., 14:164.

Moghissi, K. S., Churchill, J. A., Kurrie, D., 1975, Relationships of maternal amino acids and proteins to fetal growth and mental development, Amer. J. Obstet. Gynecol., 123:398.

Moore, S. and Stein, W. H., 1954, Chromatography of amino acids on sulfonated polystyrene resins, J. Biol. Chem., 211:893.

Monaco, F., Mutani, R., Durrelli, L., Delsedime, M., 1975, Free amino acids of patients with epilepsy: Significant increases of taurine, Epilepsia, 16:245.

Naeye, R. L., Blan, W. L., Paul C., 1973, Effects of maternal nutrition on the human fetus, Pediat., 54:494.

National Academy of Sciences, 1968, (National Research Council): Recommended Dietary Allowances. The Academy-The Council: Wash.

Nyhan, W. L., Tocci, P. M., 1966, Aminoaciduria, Ann. Rev. Med., 17:133.

Owen, O. E., Morgan, A. P., Kemp, H. G., Sullivan, J. M., Herrera, M. G., Cahill, C. G., Snyderman, S. E. and Norton, P. M., 1967, Brain metabolism during fasting, J. Clin. Invest., 46:15:89.

Perry, T. L., and Hansen, S., 1969, Technical pitfalls leading to errors in the quantitation of plasma amino acids, Clin. Chem. Acta, 25:43.

Raine, D. N., Cooke, J. R., Andrews, W. A. and Mahon, D. F., 1972, Screening for inherited metabolic diseases by plasma chromatography in a large city, Brit. Med. J., 3:7.

Reid, D. W. J., Campbell, D. J., and Yakymyshyn, L. Y., 1971, Quantitative amino acids in amniotic fluids and maternal plasma in early and late pregnancy, Am. J. Obstet. Gynec., 111:251.

Rosenblatt, D. and Scriver, C. P., 1968, Heterogeneity in genetic control of phenylalanine metabolism in man, Nature (London), 218:677.

Rouser, G., Jeliner, B., Samuels, A. J., Kinugasa, K., Holden, J. T., 1962, "Free amino acids in the blood of man and animals" in Amino Acid Pools; (J. T. Holden, ed.) Elsevier, New York.

Rutishauser, I. H. E. and Whitehead, R. G., 1969, Field evaluation of two biochemical tests which may reflect nutritional status in three areas of Uganda, Br. J. Nutr., 23:1.

Schelling, G. T., 1975, "An efficient procedure for complete evaluation of dietary proteins" in Nutrition and Clinical Nutrition, Vol. I, Protein Nutritional Quality of Foods and Feeds, Part I, (M. Friedman, ed.) Marcel Deker, Inc., New York.

Scriver, C. R., 1969, The human biochemical genetics of amino acid transport, Pediatr., 44:348.

Segal, S., 1976, Disorders of renal amino acid transport, N. Eng. J. Med., 294:1044.

Sherwood, L. M., Parris, E. E., 1969, Inherited aminoacidopathies demonstrating vitamin dependency, N. Eng. J. Med., 281:145.

Soupart, P., 1962, "Free amino acids of blood and urine in the human" in Amino Acid Pools, (J. T. Holden, ed.), Elsevier, New York.

Turner, B., and Brown, D. A., 1970, Amino acid excretion during infancy and early childhood, Med. J. Aus., 1:11.

Tweedle, D. E. F., Spivey, J., and Johnson, I. D. A., 1973, Choice of intravenous amino acid for use after surgical operations, Metab., 22:173.

Wehr, R. F., and Lewis, G. T., 1966, Amino acids in blood plasma of young and aged adults, Soc. for Exper. Bio. Med., 121:349.

Weller, L. A., Margen, S., Calloway, D. H., 1969, Variation in fasting and postprandial amino acids of men fed adequate or protein free diets, Am. J. Clin. Nutr., 22:1557.

Westall, R. G., Roitman, E., de la Pena, C., Rasmussen, H. R., Cravioto, J., Gomez, F., and Holt, L. E., 1958, The plasma amino acids in malnutrition, Arch. Dis. Child, 23:499.

Whitehead, R. G. and Dean, R. F. A., 1964, Serum amino acids in kwashiorkor. II. An abbreviated method of estimation and its application, Am. J. Nutr., 14:320.

Whitehead, R. G., 1967, Biochemical tests in differential diagnosis of protein and calorie deficiencies, Arch. Dis. Child., 42:479.

Wurtman, R. J., Rose, C. M., Chou, Chaun, C., Larin, F. F., 1968, Daily rhythms in the concentration of various amino acids in human plasma, N. Eng. J. Med., 279:171.

---

This review was supported in part by MCH grant MCT #00028-16.

# AMINOGRAMS IN VARIOUS TYPES OF LIVER DISEASE

Alvin Dubin, M.S.

Director of Biochemistry, Hektoen Institute for
Medical Research and Associate Professor, Biochemistry,
Rush Medical School, Chicago, Illinois

Shibban Ganju, M.D.

Fellow in Gastroenterology, Cook County Hospital,
Chicago, Illinois

Paul B. Szanto, M.D.

Professor of Pathology, Chicago Medical School,
Director of Pathology, Cook County Hospital, and
Scientific Director of Hektoen Institute for Medical
Research, Chicago, Illinois

Ann Poulos, B.S.

Research Assistant, Hektoen Institute, Chicago, Illinois

Frederick Steigmann, M.D.

Director of Gastroenterology, Cook County Hospital and
Hektoen Institute for Medical Research and Professor
Emeritus, University of Illinois School of Medicine,
Chicago, Illinois

## INTRODUCTION

The concentration of amino acids in the serum is significantly altered in various liver diseases (Fisher, 1975; Iber, 1957; Mellinkoff, 1955; Ning, 1967; Rosen, 1977). While there is considerable information about the amino acid concentrations in cirrhosis and hepatic encephalopathy, there is only limited information about the serum amino acid concentration in liver diseases other than cirrhosis and hepatic encephalopathy (Mellinkoff, 1955; Ning, 1967). Accordingly, we studied the amino acid concentrations in the serum of patients with different liver diseases for the following purposes:

(a) To recognize consistent derangements, if any, of the amino acid concentrations in different types of liver disease.

(b) To make a comparative study of the amino acid concentrations in various liver diseases.

(c) To define the diagnostic and prognostic significance of alterations in amino acid concentrations by correlating these changes with the severity of the histological findings.

## MATERIALS

Sera from 68 fasting patients with different forms of liver diseases and 9 healthy hospital workers were studied. The various types of liver disease and the number of patients in each group are described in Table I. The clinical diagnosis of the type of liver disease in all patients was confirmed by the histological examination of liver tissue obtained by needle biopsy, except in the patients with coma due to hepatic or non-hepatic causes. All patients in coma were in stage IV when the blood was drawn for amino acid determination. The five patients with coma of non-hepatic origin had no liver disease by history, physical examination or routine biochemical tests of liver function. The patients with hepatic coma had clinical evidence of hepatic disease-ascites, jaundice, asterixis, neurological findings and gross abnormalities of liver function.

## METHODS

Venous blood samples were collected in dry tubes. The serum was immediately separated by centrifugation, frozen and stored at -18°C until used. One ml of this serum was deproteinized by the addition of 0.1 ml of 50% (w/v) sulfosalicylic acid. The supernatant was recentrifuged (Beckman Microcentrifuge) and 25 μl of 4% (w/v) lithium hydroxide was added to 400 μl of the supernatant to adjust to a proper pH. One hundred μl of this mixture was injected onto a 460 x 6 mm column with Beckman's spherical resin, Type W-3P (Beckman Instrument Model 119 CL Amino Acid Analyzer) together with 100 μl of α-amino, β-guanidino, propionic acid hydrochloride. The final concentration of various amino acids was calculated and integrated by the automated Beckman Model 126 and reported in μmoles/liter.

The mean value of each amino acid of each disease group was compared to the mean normal value of that amino acid by discriminatory analysis. The following chemical analyses were used as part of liver function: Total protein, albumin, globulin, gammaglobulin, alkaline phosphatase, cholesterol, total and direct bilirubin, aspartate aminotransferase, alanine aminotransferase, isocitric dehydrogenase and lactate dehydrogenase.

Table I. Numbers of Patients in the Various Groups Studied

| GROUP | NUMBER |
|---|---|
| Normal | 9 |
| Hepatic Coma | 10 |
| Non-Hepatic Coma | 5 |
| Acute Viral Hepatitis | 11 |
| Chronic Active Hepatitis | 6 |
| Fatty Liver | 7 |
| Fatty Cirrhosis | 9 |
| Advanced Cirrhosis | 20 |
| TOTAL | 77 |

## RESULTS

### Normal

The results of the amino acid determinations in normal individuals are presented in Table II.

### Fatty Liver

Methionine, phenylalanine and serine were the only amino acids whose concentration increased significantly ($p < 0.05$). All other amino acids showed no significant change (Table III).

### Fatty Cirrhosis

Patients with moderate cirrhosis associated with mild to moderate fatty changes showed a significant increase ($p < 0.05$),

Table II. Normal Fasting Concentrations of Amino Acids in Serum

| AMINO ACID | MEAN (μMOLES/LITER) | STD. DEV. | RANGE |
|---|---|---|---|
| Phosphoserine | 5.2 | 0.8 | 3.6- 6.3 |
| Taurine | 106.3 | 25.3 | 69.1-147.4 |
| Aspartic Acid | 24.9 | 3.1 | 19.1- 28.1 |
| Hydroxyproline | 13.9 | 8.1 | 8.7- 31.4 |
| Threonine | 107.0 | 21.2 | 77.7-135.0 |
| Serine | 111.2 | 21.0 | 68.2-136.0 |
| Asparagine | 54.4 | 13.1 | 36.4- 72.5 |
| Glutamic Acid | 55.6 | 13.8 | 28.0- 76.8 |
| Glutamine | 494.3 | 113.6 | 317.0-645.0 |
| Proline | 196.5 | 69.9 | 92.1-301.0 |
| Glycine | 200.7 | 30.3 | 146.0-228.0 |
| Alanine | 390.4 | 92.3 | 300.7-518.0 |
| Citrulline | 17.0 | 14.4 | 5.4- 51.3 |
| α-Amino-n-Butyric Acid | 12.7 | 6.9 | 7.1- 26.3 |
| Valine | 166.9 | 33.0 | 122.4-212.0 |
| 1/2 Cystine | 65.6 | 15.4 | 47.1- 94.0 |
| Methionine | 13.0 | 6.5 | 5.2- 17.0 |
| Cystathionine | 0.83 | 0.24 | 0.50-1.10 |
| Isoleucine | 45.8 | 9.9 | 30.2- 58.0 |
| Leucine | 100.9 | 16.7 | 77.0-125.0 |
| Tyrosine | 48.1 | 6.5 | 35.2- 54.1 |
| Phenylalanine | 49.4 | 6.5 | 39.7- 57.6 |
| Tryptophan | 36.9 | 7.2 | 27.5- 50.5 |
| Ammonia | 372.1 | 41.2 | 319.8-426.0 |
| Ornithine | 80.7 | 16.7 | 57.6-117.0 |
| Lysine | 150.5 | 21.1 | 118.4-181.1 |
| Histidine | 65.0 | 8.9 | 53.2- 76.2 |
| Arginine | 66.6 | 12.8 | 47.7- 70.9 |

Table III. Differences in Serum Amino Acids in Fatty Liver and Liver Cirrhosis

| | μMOLES/LITER | | | |
|---|---|---|---|---|
| | NORMAL | FATTY LIVER | FATTY CIRRHOSIS | ADVANCED CIRRHOSIS |
| Aspartic Acid | 24.9 | 45.6 | 45.6(+) | 42.3(+) |
| Serine | 111.2 | 145.1(+) | 155.1(+) | 152.7(+) |
| Asparagine | 54.4 | 72.1 | 70.7 | 99.3(+) |
| Glutamic Acid | 55.6 | 230.5 | 357.0 | 270.3(+) |
| Glutamine | 494.3 | 374.6 | 154.8(-) | 197.7(-) |
| 1/2 Cystine | 65.6 | 31.0 | 7.6(-) | 25.4(-) |
| Methionine | 13.0 | 35.6(+) | 21.9(+) | 43.5(+) |
| Isoleucine | 45.8 | 65.6 | 59.2(+) | 57.3 |
| Tyrosine | 48.1 | 62.9 | 65.9(+) | 102.3(+) |
| Phenylalanine | 49.4 | 69.0(+) | 73.0(+) | 82.4(+) |
| Ammonia | 372.1 | 385.2 | 422.7 | 453.6(+) |
| Arginine | 66.6 | 93.1 | 91.4 | 108.4(+) |

(+) and (-) indicate significant increase and decrease, respectively, from normal ($p < 0.05$).

in aspartic acid, serine, methionine, and in the aromatic amino acids, tyrosine and phenylalanine concentrations. The only branched chain amino acid that showed an increased concentration was isoleucine.

The concentrations of glutamine and half-cystine were markedly decreased. All other amino acids studied showed no significant changes from the normal (Table III).

Advanced Cirrhosis

Phenylalanine and tyrosine concentrations were significantly increased ($p < 0.05$).

The concentrations of methionine, serine, asparagine, glutamic acid, aspartic acid, arginine and ammonia were increased to a lesser degree. Glutamine and half-cystine levels were significantly decreased. The other amino acids showed no significant changes (Table III).

Hepatitic Coma

The concentrations of phenylalanine, methionine, glutamic acid and tyrosine showed the most marked elevation. The concentrations of serine, arginine and ammonia were significantly increased ($p < 0.05$). The concentrations of leucine and valine were markedly decreased while those of isoleucine and tryptophan remained unchanged.

The other amino acids studied showed no statistically significant changes (Table IV).

Non-Hepatic Coma

In non-hepatic coma the concentrations of phenylalanine, glycine, serine, asparagine and glutamic acid were significantly increased, while those of tyrosine, tryptophan, valine, leucine, isoleucine and others showed no significant change (Table IV).

Acute Viral Hepatitis

Phenylalanine, tyrosine, methionine, taurine, threonine, serine, lysine and arginine were significantly elevated. The concentration of tryptophan did not increase significantly.

The concentrations of the branched chain amino acids leucine and isoleucine were increased while valine was not significantly altered.

There was marked decrease in the concentration of half-cystine (Table V).

Chronic Active Hepatitis

Patients with chronic active hepatitis showed a mild increase in the concentration of many amino acids, though in only a few of them was the increase significant. The concentrations of tyrosine, phenylalanine, methionine and glutamic acid were markedly increased

Table IV. Differences in Serum Amino Acids Between Hepatic Coma and Non-Hepatic Coma

| | HEPATIC COMA | | NON-HEPATIC COMA | |
|---|---|---|---|---|
| | μMOLES/LITER | /NORMAL | μMOLES/LITER | /NORMAL |
| Phosphoserine | 9.9 | +1.895 | 5.2 | +0.989 |
| Hydroxyproline | 6.6 | -0.476 | 14.5 | -1.046 |
| Serine | 157.6 | +1.417 | 141.3 | +1.270 |
| Asparagine | 73.6 | -1.351 | 81.5 | -1.497 |
| Glutamic Acid | 196.0 | +3.523 | 145.1 | +2.608 |
| Proline | 210.4 | +1.071 | 149.8 | +0.762 |
| Glycine | 317.7 | +1.583 | 270.6 | +1.348 |
| Citrulline | 21.5 | +0.722 | 15.4 | +0.858 |
| Valine | 101.5 | -0.608 | 172.1 | -1.031 |
| Methionine | 61.3 | +4.708 | 16.8 | +1.292 |
| Leucine | 72.2 | -0.716 | 103.5 | -1.026 |
| Tyrosine | 70.3 | +1.460 | 46.1 | +0.973 |
| Phenylalanine | 102.2 | +2.068 | 73.2 | +1.482 |
| Ammonia | 494.2 | +1.328 | 326.2 | +0.877 |
| Ornithine | 82.3 | -1.020 | 100.6 | -1.247 |
| Arginine | 98.3 | +1.476 | 62.5 | +0.938 |
| T = 3.291, p <0.01 | | +16.580 ± 6.091 | | +6.495 ± 3.671 |

(p <0.05). Leucine, valine, threonine, asparagine and lysine were only slightly (but not significantly) increased. The only amino acid that was markedly decreased was that of half-cystine (Table V).

Table V. Differences in Serum Amino Acids in Acute and Chronic Active Hepatitis

| | μMOLES/LITER | | |
|---|---|---|---|
| | NORMAL | ACUTE HEPATITIS | CHRONIC ACTIVE HEPATITIS |
| Threonine | 106.7 | 217.3(+) | 138.8(+) |
| Serine | 111.2 | 206.2(+) | 156.8 |
| Asparagine | 54.4 | 104.5 | 83.0 |
| Glutamic Acid | 55.6 | 265.8 | 286.6(+) |
| Valine | 166.9 | 204.1 | 218.6(+) |
| 1/2 Cystine | 65.6 | 18.4(-) | 1.3(-) |
| Methionine | 13.0 | 51.5(+) | 32.6(+) |
| Isoleucine | 45.8 | 68.8(+) | 74.8 |
| Leucine | 100.9 | 138.1(+) | 147.0(+) |
| Tyrosine | 48.1 | 92.5 | 76.8(+) |
| Phenylalanine | 49.4 | 93.4(+) | 95.2(+) |
| Ammonia | 372.1 | 448.9(+) | 430.4 |
| Lysine | 150.5 | 217.0(+) | 208.6(+) |
| Arginine | 66.6 | 102.5(+) | 82.7 |

(+) and (-) indicate significant increase and decrease, respectively, from normal ($p < 0.05$).

Fatty Liver and Fatty Cirrhosis

Glutamine, half-cystine and methionine were considerably more decreased in fatty cirrhosis than in fatty liver.

When comparing phosphoserine, taurine, methionine, cystathionine and lysine between these two groups, statistically significant differences were seen (Table VI).

Table VI. Elevations in Serum Amino Acids Above Normal in Fatty Liver and Fatty Cirrhosis

| | FATTY LIVER | | FATTY CIRRHOSIS | |
|---|---|---|---|---|
| | μMOLES/LITER | /NORMAL | μMOLES/LITER | /NORMAL |
| Phosphoserine | 7.3 | +1.393 | 4.7 | +0.895 |
| Taurine | 128.2 | +1.206 | 111.1 | +1.045 |
| Methionine | 35.6 | +2.739 | 21.9 | +1.683 |
| Cystathionine | 2.0 | +2.424 | 1.7 | +2.024 |
| Lysine | 176.4 | +1.172 | 163.1 | +1.084 |
| Normal* = 5.000 ± 0.778 | | +8.934 ± 2.614 (T = 3.949, p <0.01) | | +6.731 ± 1.626 (T = 2.296, p <0.05) |

*The algebraic sum of this discriminant in the normal population in which the mean value of each amino acid is assigned the value 1.00.

Fatty Cirrhosis and Advanced Cirrhosis

The concentrations of tyrosine, methionine, cystathionine, arginine, half-cystine, proline and asparagine were higher in advanced cirrhosis than in fatty cirrhosis. In fatty cirrhosis, taurine, glutamic acid and lysine were more elevated than in advanced cirrhosis (Table VII).

Hepatic and Non-Hepatic Coma

In patients with hepatic coma, methionine was increased to more than four times normal, while it was only slightly elevated in non-hepatic coma.

Tyrosine, phenylalanine and glutamic acid were also elevated much more in hepatic than in non-hepatic coma.

Leucine, valine and hydroxyproline were significantly lower in hepatic than in non-hepatic coma (Table IV).

Table VII. Differences in Serum Amino Acids Between Fatty Cirrhosis and Advanced Cirrhosis

| | FATTY CIRRHOSIS | | ADVANCED CIRRHOSIS | |
|---|---|---|---|---|
| | μMOLES/LITER | /NORMAL | μMOLES/LITER | /NORMAL |
| Taurine | 111.1 | +1.045 | 90.1 | +0.848 |
| Threonine | 109.8 | -1.026 | 130.5 | -1.220 |
| Asparagine | 70.7 | -1.299 | 99.3 | -1.824 |
| Glutamic Acid | 357.0 | +6.417 | 270.3 | +4.859 |
| Proline | 148.2 | -0.754 | 203.1 | -1.034 |
| 1/2 Cystine | 7.6 | -0.116 | 25.4 | -0.387 |
| Methionine | 21.9 | -1.683 | 43.5 | -3.342 |
| Cystathionine | 1.7 | -2.024 | 4.6 | -5.576 |
| Tyrosine | 65.9 | -1.370 | 102.3 | -2.126 |
| Lysine | 163.1 | +1.084 | 137.7 | +0.915 |
| Histidine | 60.3 | -0.928 | 68.0 | -1.047 |
| Arginine | 91.4 | -1.372 | 108.4 | -1.628 |
| T* = 2.842, p <.01 | | -2.026 ± 3.632 | | -11.562 ± 9.714 |

*The T statistic refers to the difference between the mean discriminant values of the two disease groups.

## Acute Viral Hepatitis and Chronic Active Hepatitis

Most of the amino acid levels tended to be significantly higher than normal in both of these conditions. No statistically significant differences were found between these two groups (Table VIII).

Table VIII. Differences in Serum Amino Acids Between Acute Viral Hepatitis and Chronic Active Hepatitis

| | ACUTE VIRAL HEPATITIS | | CHRONIC ACTIVE HEPATITIS | |
|---|---|---|---|---|
| | μMOLES/LITER | /NORMAL | μMOLES/LITER | /NORMAL |
| Taurine | 160.5 | +1.510 | 111.0 | +1.044 |
| Hydroxyproline | 7.7 | -0.551 | 13.1 | -0.942 |
| Threonine | 217.3 | +2.031 | 138.8 | +1.297 |
| Serine | 206.2 | +1.854 | 156.8 | +1.410 |
| Asparagine | 104.5 | +1.920 | 83.0 | +1.526 |
| Glutamine | 384.6 | +0.778 | 305.0 | +0.617 |
| Glycine | 304.8 | +1.519 | 270.4 | +1.347 |
| 1/2 Cystine | 18.4 | +0.280 | 1.3 | +0.020 |
| Methionine | 51.5 | +3.958 | 32.6 | +2.504 |
| Tyrosine | 92.5 | +1.922 | 76.8 | +1.596 |
| Tryptophan | 57.4 | +1.554 | 40.8 | +1.105 |
| Arginine | 102.5 | +1.539 | 82.7 | +1.241 |
| T* = 1.677, N.S. | | +18.314 ±6.344 | | +12.765 ± 3.159 |

*The T statistic refers to the difference between the mean discriminant values of the two disease groups.

## DISCUSSION

The concentrations of serum amino acids are not only altered in various liver diseases but the pattern of change also seems to vary from one pathological group to another.

The patients with hepatic coma showed consistent increase in methionine, phenylalanine and tyrosine. Tryptophan was unaltered. Branched chain amino acids (BCAA), leucine and valine were decreased while isoleucine was unaffected. Similar changes accompanied by decreased isoleucine concentrations have been shown in other studies (Fisher, 1976; Iber, 1957; Rosen, 1977; Silverman, 1968; Steigmann, 1970). Taurine concentrations have been reported to be increased in hepatic encephalopathy (HE) (Rossi-Fanelli, 1976).

It has been proposed that patients with hepatic coma have decreased BCAA because of hyperinsulinemia which enhances the uptake of BCAA by the muscle and adipose tissue (Munro, 1975; Soeters, 1976). The aromatic amino acids (AAA) are elevated because of the failure of the liver to metabolize them as shown by Munro (1975) and Soeters (1976). BCAA compete with AAA for transport across the blood brain barrier so that a decrease in the BCAA would facilitate the passage of AAA into the cerebrospinal fluid (James, 1978). Excesses of AAA will inhibit the synthesis of catecholamines (Wurtman, 1971). Octopamine is produced as a result from phenylalanine and tyrosine as shown by Meredith (1979). Excess of tryptophan leads to increased production of serotonin (Fernstrom, 1972). Increased methionine levels may lead to increased methanethiol which has a synergistic effect with ammonia (Condon, 1970; Merino, 1975). These end products may be related to the pathogenesis of hepatic coma.

In patients with advanced cirrhosis but without coma, the concentrations of BCAA were near normal, though phenylalanine, tyrosine and methionine were increased. It would therefore appear that patients with advanced cirrhosis would be triggered into HE by the decrease in BCAA and conversely, the patients with hepatic coma would improve by increase in their BCAA. This has indeed been shown by Fischer et al. (1976).

The patients with coma of non-hepatic etiology showed an increase in phenylalanine while other AAA and BCAA were near normal. The pathogenesis of non-hepatic coma may be related to factors other than the ratio of AAA to BCAA.

The amino acid pattern seems to go from bad to worse as histological changes progress from fatty liver to advanced cirrhosis and coma. The change is both qualitative and quantitative. In fatty liver only three amino acids were altered from the normal while in fatty cirrhosis and advanced cirrhosis 8 and 12 amino acids were affected. Among the altered amino acids the concentration was most markedly different from the normal in the most severe liver diseases, i.e., hepatic coma and advanced cirrhosis. Similarly, the quantitative changes were the least in those with less severe disease, i.e., fatty liver and fatty cirrhosis.

The elevation of amino acid concentrations such as methionine, tyrosine, asparagine, threonine and tryptophan is marked in acute viral hepatitis while it is less so in chronic active hepatitis. This elevation may be mainly due to release of amino acids from the necrosed hepatocytes, though muscle catabolism may contribute to a small extent (Rosen, 1977). Tyrosine concentrations have been shown to correlate well with serum aspartate aminotransferase activities and it has been suggested that AAA concentrations, especially that of tyrosine, can be used to predict the extent of hepatic necrosis (Rosen, 1977).

The AAA are markedly elevated in patients with coma due to "fulminant hepatitis" while BCAA are near normal, thereby altering the ratio of AAA to BCAA (Rosen, 1977). This alteration of ratio may be related to causation of coma in fulminant hepatitis. In our patients with acute viral hepatitis, both BCAA and AAA were elevated, keeping the ratio near normal. Thus, it may be that the concentrations of BCAA may bear some prognostic significance and will determine the outcome of a case of viral hepatitis (Dubin, 1979).

SUMMARY

Patients with advance cirrhosis showed significantly increased phenylalanine and tyrosine, while methionine, serine, asparagine, glutamic acid, aspartic acid, arginine and ammonia were increased to a lesser degree, and glutamine and half-cystine levels were significantly decreased. In hepatic coma phenylalanine, methionine, glutamic acid and tyrosine showed the most marked elevation, arginine and ammonia were significantly increased, and leucine and valine were markedly decreased. In contrast, tyrosine, tryptophan, valine, leucine, isoleucine showed no significant change in non-hepatic coma.

The alteration in the amino acid pattern, both qualitatively and quantitatively correlates with the etiology and pathology of liver disease. Different patterns of change can be seen in different liver diseases. It may be possible to predict the severity and prognosis of liver disease by the changes seen in amino acids. Therapy may accordingly be modified to correct the imbalance of amino acids.

REFERENCES

Condon, R. E., Bombeck, C. T., Steigmann, F., 1970, Heterologous bovine liver perfusion therapy of acute hepatic failure, Am. J. Surg., 119:145.

Conn, H. O., and Lieberthal, M. M., 1979, Chapter 2, Subtitle - False neurotransmitters hepatic coma syndromes and lactulose, Williams & Wilkins Co., Baltimore.

Dubin, A., Ganju, S., Szanto, P. B., and Steigmann, F., 1979, Plasma aminogram in acute viral hepatitis (Abstract), Gastroen., 76:1978.

Fernstrom, J. D., Wurtmart, R. S., 1972, Brain seratonin content physiological regulation by plasma neutral amino acids, Science, 178:414.

Fisher, J. E., Funovice, J. M., Aguirre, A., James, J. H., Keane, J. M., Wesdorp, R. I. C., Yoshimura, N., Westman, T., 1975, The role of plasma amino acids in hepatic encephalopathy, Surg., 78:276.

Fisher, J. E., Rosen, H. M., Ebeid, A. M., James, H., Keane, J. M., and Soeters, P. B., 1976, The effect of normalization of plasma amino acids on hepatic encephalopathy in man, Surg., 80:77.

Iber, F. L., Rosen, H., Levenson, S. M., and Chalmers, T. C., 1957, The plasma amino acids in patients with liver failure, J. Lab. Clin. Med., 50:417.

This work supported in part by the Gwen Dubin Fund.

James, J. H., Escourrou, J., and Fischer, J. E., 1978, Blood-brain amino acid transport activity is increased after porta caval anastomosis, Science, 200:1395.

Mellinkoff, S. M., Boyle, D., and Frankland, M., 1955, The effect of amino acid glucose infusion upon the serum amino acid and blood sugar concentrations in viral hepatitis, J. Lab. Clin. Med., 46:560.

Merino, G. E., Jetzer, T., Doizaki, W. M. D., and Najarian, J. S., 1975, Methionine-induced hepatic coma in dogs, Am. J. Surg., 130:41.

Munro, H. N., Fernstrom, J. D., and Wurtman, R. J., 1975, Insulin, plasma amino acid imbalance and hepatic coma, Lancet, 1:722.

Ning, M., Lowenstein, L. M., and Davidson, C. S., 1967, Serum amino acid concentrations in alcoholic hepatitis, J. Lab. Clin. Med., 70:554.

Rosen, H. M., Yoshimura, N., Hodgman, J. M., and Fischer, J. E., 1977, Plasma amino acid patterns in hepatic encephalopathy of differing etiology, Gastro., 72:483.

Rossi-Fanelli, F., Capocacci, L., and Fischer, J. E., 1976, Correlation of plasma taurine levels with grades of hepatic encephalopathy in man (Abstract), Gastro., 72:1181.

Silverman, D. A., Dubin, A., Alavi, I., and Steigmann, F., 1968, "Possible new pathogenic factors in hepatic coma: Plasma and cerebrospinal fluid amino acids alterations". Presented at the meeting of the American Association for the Study of Liver Diseases, Chicago, IL (October, 1968).

Soeters, P. B., and Fischer, J. E., L976, Insulin, glucagon, amino acid imbalance and hepatic encephalopathy, Lancet, II:880.

Steigmann, F., Condon, R. E., Silverman, D. A., Bombeck, C. T., Alavi, I., and Dubin, A., 1970, Hepatic coma: newer pathogenetic and therapeutic factors, Am. J. Gastro., 54:355.

Wurtman, R. J., Larin, F., Mostafapour, S., and Fernstrom, J. D., 1971, Brain catechol synthesis: control of brain tyrosine concentration, Science, 185:183.

# THE ROLE OF ERYTHROCYTE AMINO ACIDS IN ENERGY METABOLISM

John E. Sherwin, Ph.D.

Department of Pathology, Valley Children's Hospital
and Guidance Clinic, Fresno, California 93703

## INTRODUCTION

Although there is much information regarding free amino acids in plasma, the amino acids of erythrocytes have been little studied (Felig, 1973; Felig et al., 1970; Levy and Barkin, 1971; McMenamy et al., 1960). However, accurate comparisons of the amino acid concentrations of plasma and erythrocytes can result in a better understanding of the relative role of these amino acid pools in energy metabolism.

### Amino Acid Content of Erythrocytes and Plasma

The concentrations of amino acids in erythrocytes are quite different from their concentrations in plasma (Table 1). While many amino acids do not exhibit plasma: erythrocyte ratios different from 1.0, aspartate, glutamate, serine and glycine are present in erythrocytes at higher concentrations than in serum (McMenamy et al., 1960). On the other hand, at least in humans, arginine is more concentrated in plasma than in erythrocytes (Levy and Barkin, 1971). The presence of essentially equal concentrations of many amino acids in plasma and erythrocytes, even in metabolic disorders resulting in elevated concentrations, indicates that amino acid transport is either passive or that the transport systems have very high saturation points (Levy and Barkin, 1971). It has become increasingly clear that the erythrocytes do not passively reflect all amino acid concentrations. However, the functional role of erythrocyte amino acids remains unclear, particularly since mature red blood cells do not synthesize protein. This is reflected in the fact that amino acid concentrations in reticulocytes, which synthesize protein, are quite different from

Table 1. Amino Acid Concentrations in Plasma and Erythrocytes

| | Concentration (μmole/L) | |
|---|---|---|
| | Plasma | Erythrocytes |
| Essential Amino Acids | | |
| Isoleucine | 49 ± 9 | 43 ± 8 |
| Leucine | 114 ± 9 | 100 ± 5 |
| Lysine | 164 ± 11 | 143 ± 18 |
| Methionine | 24 ± 1 | 11 ± 3 |
| Phenylalanine | 60 ± 7 | 61 ± 16 |
| Threonine | 134 ± 34 | 154 ± 55 |
| Tryptophan | --- | --- |
| Nonessential Amino Acids | | |
| Alanine | 258 ± 34 | 250 ± 31 |
| Arginine | 55 ± 12 | 25 ± 15 |
| Aspartate | 0 | 264 ± 59 |
| Citrulline | 26 ± 14 | 25 ± 8 |
| Cystine | 73 ± 16 | 7 ± 8 |
| Glutamate | 56 ± 20 | 220 ± 49 |
| Glutamine | 283 ± 79 | 201 ± 38 |
| Glycine | 192 ± 21 | 264 ± 25 |
| Histidine | 117 ± 17 | 115 ± 8 |
| Ornithine | 129 ± 23 | 149 ± 2.8 |
| Serine | 171 ± 44 | 265 ± 39 |
| Tyrosine | 75 ± 19 | 78 ± 28 |
| Valine | 164 ± 15 | 157 ± 12 |

those in erythrocytes (Allen, 1960; McMenamy et al., 1960; Yunis and Arimura, 1965).

### Effects of Exercise on Amino Acid Concentrations

Plasma amino acids are extracted by the liver and utilized for energy metabolism via the gluconeogenic pathway (Mallette et al., 1969; Sladek and Snarr, 1971; Snell, 1974). While there is little reason to expect that the amino acids of the erythrocyte are unavailable for gluconeogenic energy production, there is little evidence for an active role.

Maintenance of glucose homeostasis and energy output during prolonged exercise requires an increase in gluconeogenesis since depletion of hepatic glycogen stores occurs during the initial 30-40 minutes of exercise. Ahlborg et al., (1974) demonstrated significant increases in splanchnic uptake of glycine, alanine, threonine, serine and proline following a 240 minute exercise period. The relative contribution of plasma amino acids to total glucose production rises from 5% at rest to approximately 12% after prolonged exercise. This data also suggests that alanine accounts for 55%, and glycine 21%, of this splanchnic amino acid exchange.

During exercise, the liver also releases amino acids (Felig and Wahren, 1971). Only the branch chain amino acids valine, leucine and isoleucine are released in significant amounts. The release of these amino acids from the liver suggests that they cannot be metabolized directly to glucose in the liver and that they can be more readily metabolized to glucose by other organs. The conversion of branch chain amino acids to alanine in muscle is consistent with this hypothesis (Odessy et al., 1974).

Glutamate and glutamine are extracted by the liver to about 25% of the extent of alanine (Felig et al., 1973; Garber et al., 1976a, 1976b; Karl et al., 1976). These amino acids are converted to alpha-ketoglutarate and subsequently to oxaloacetate for gluconeogenesis. The consistent uptake of glutamate and glutamine by the gastrointestinal tract suggests that the splanchnic uptake of these amino acids may be merely a direct reflection of extraction of glutamate and glutamine by the intestinal tract. This is a significant difference from alanine and glycine which are absorbed directly by the liver and actually released by the gastrointestinal tract.

## UTILIZATION OF AMINO ACIDS BY THE LIVER AND KIDNEY FOR GLUCONEOGENESIS

### The Glucose-Alanine Cycle and Energy Production

It has become clear over the past ten years that gluconeogenesis has a significant role in energy metabolism (Daniel et al., 1977; Felig, 1973; Felig and Wahren, 1974; Felig et al., 1970).

The liver and to a lesser extent, the kidneys are the principle sites of gluconeogenesis. Three amino acids account for more than 80% of splanchnic gluconeogenesis. Alanine has been identified as the primary amino acid extracted by the splanchnic bed in both the postabsorptive state and after prolonged fasting or exercise. Alanine accounts for more than 50% of gluconeogenesis. Glycine and glutamate/glutamine account for approximately 20% and 15% of gluconeogenesis respectively. These amino acids are particularly important as gluconeogenic substances during prolonged exercise. This gluconeogenic pathway is depicted in Figure 1.

The branch chain amino acids are converted to alanine in the muscle (Felig and Wahren, 1971; Paul and Adibi, 1978). Net synthesis of alanine from amino acids occurs both by transamination and by oxidative deamination. The resultant alpha ketoacid is converted to malate and then to pyruvate. Transamination with glutamate results in net formation of alanine. These gluconeogenic entry reactions are summarized in Figure 2.

From these reactions, it is apparent that alanine and glutamine are related through the common substrate glutamate. Thus, in muscle metabolism, conditions resulting in increased pyruvate availability yield increased alanine synthesis and decreased levels of glutamine. Alternatively, factors increasing ammonia production divert carbon and nitrogen from alanine formation toward glutamine synthesis. Since glutamine is the principle substrate for renal gluconeogenesis and alanine is the major substrate for hepatic gluconeogenesis, muscle amino acid metabolism may have significant regulatory impact upon gluconeogenesis in these organs (Felig et al., 1977) (Figure 3).

Since the pyruvate required for net alanine synthesis does not result from glucose metabolism in skeletal muscle, the decarboxylation of malate by malic enzyme represents a major source of pyruvate formation (Goldstein and Newsholme, 1976). Thus, factors affecting the flux of intermediates in and out of the tricarboxylic acid cycle probably are significant in the regulation of alanine and glutamine synthesis in muscle.

Studies of plasma amino acids during starvation suggest that obese and normal subjects respond to fasting initially with a decrease in alanine and an increase in branched chain amino acids (Felig et al., 1969, 1972; Haymond et al., 1974; Wahren et al., 1973). The marked decrease in alanine is a result of hepatic depletion of this substance for gluconeogenesis. The increase in branched chain amino acids is due to hepatic release in an attempt to provide the muscle with amino acid substrates for alanine and glutamine synthesis so that renal and hepatic gluconeogenesis can be maintained. Hypoalaninemia persists in prolonged starvation. The regulatory importance of this hypoalaninemia is suggested by

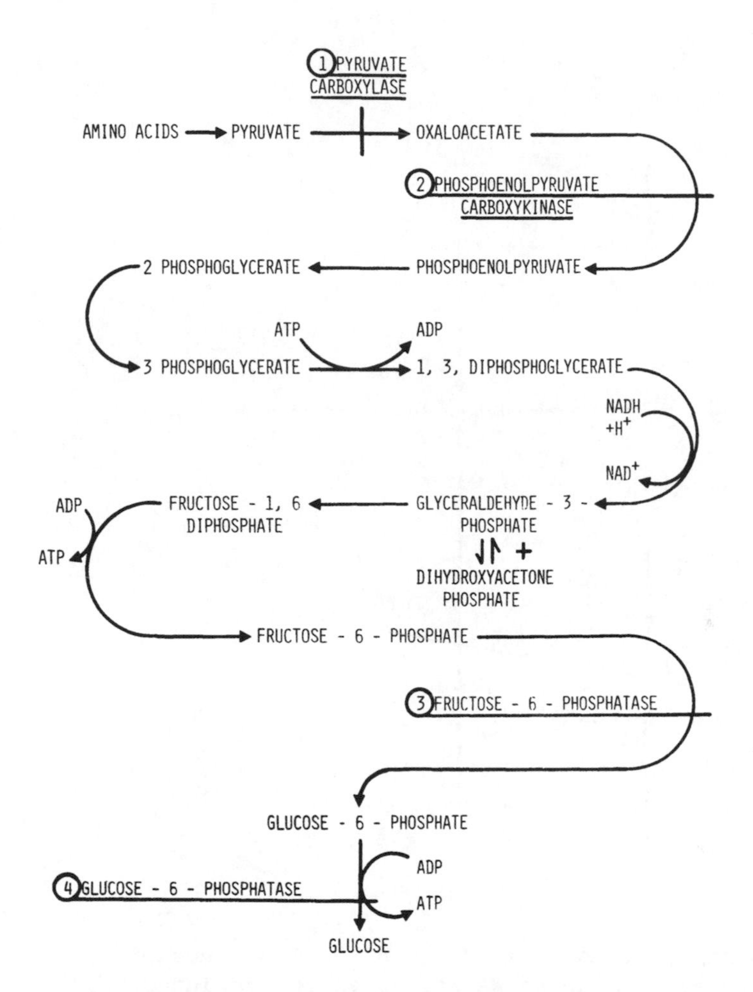


Fig. 1. The Gluconeogenic Pathway for Production of Glucose From Amino Acids. Gluconeogenesis requires 4 enzymes not present in the glycolytic pathway because of thermodynamic irreversibility. They are numbered 1-4. All other enzymatic reactions are reversals of glycolytic enzyme reactions due to substrate availability.

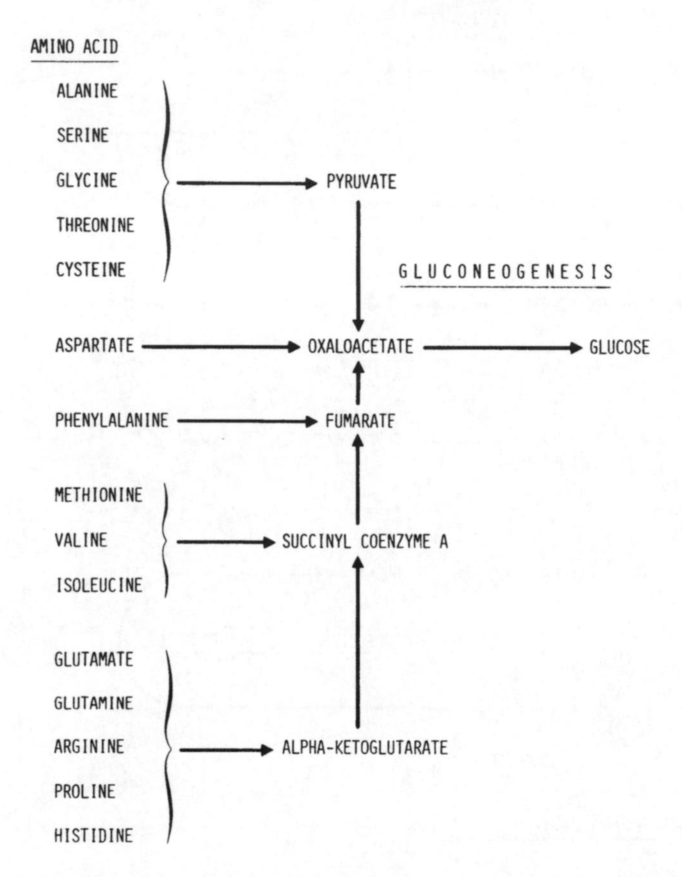


Fig. 2. Entrance of Amino Acids Into the Gluconeogenic Pathway. Amino acids are converted to intermediates of the tricarboxylic acid cycle and thence to oxaloacetate. Oxaloacetate is converted to glucose via the gluconeogenic pathway of Figure 1.

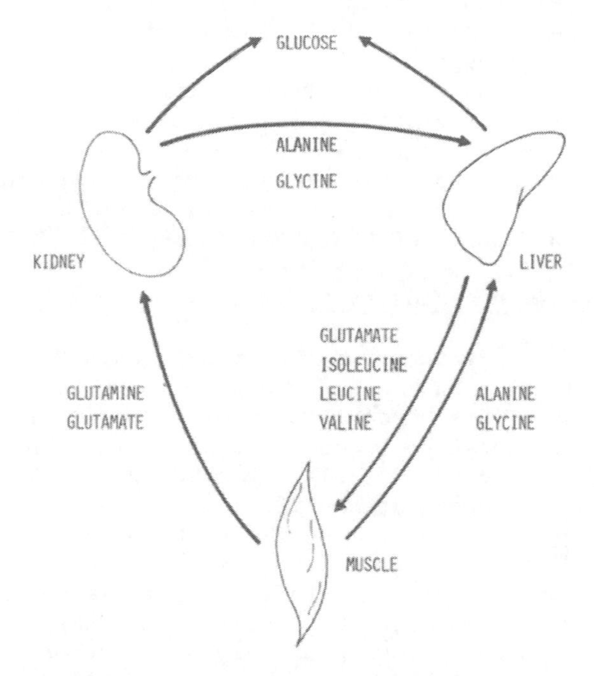


Fig. 3. Interorgan Transport of Amino Acids for Gluconeogenesis. Glucose is produced by the kidney and liver during interorgan exchange of amino acids. Branched chain amino acids are transported to the muscle for conversion to alanine before returning to the liver for gluconeogenesis. Glutamate is converted by muscle to glutamine and transported by the blood to the kidney for gluconeogenesis.

the fact that intravenous infusion of alanine into starved subjects results in a prompt hyperglycemic response.

Energy Production From Amino Acids Other Than Alanine

In contrast to the behavior of other amino acids, the plasma concentration of glycine, threonine and to a lesser extent, serine, exhibit a delayed increase in starvation (Felig et al., 1969). This is consistent with the hypothesis that as alanine is depleted, secondary gluconeogenic amino acids become important metabolic sources of energy. Although hepatic utilization of these gluconeogenic substrates is minimal, renal gluconeogenesis is augmented by extraction of these amino acids (Mallette et al., 1969). The low level of plasma serine may be due to its intermediate role in the conversion of glycine and threonine to gluconeogenic substrates.

## TRANSPORT OF AMINO ACIDS FOR GLUCONEOGENESIS

Active Transport of Amino Acids by the Blood

It has been generally accepted that plasma rather than erythrocytes is the vehicle of amino acid exchange between muscle and gluconeogenic organs. This hypothesis stems primarily from *in vitro* studies of equilibration times for amino acid transport across the erythrocyte membrane (Felig et al., 1973). The results of these studies have demonstrated uniformly slow equilibration times. Initial suggestions of the possible importance of erythrocytes in amino acid transport stem from comparison of results from studies using whole blood and studies using plasma. Van Slyke and Meyer (1912) demonstrated a 2-3 fold increase in total blood amino acid nitrogen following a meal, while others using only plasma could not demonstrate such an increase. Further, the fact that the cell/plasma ratio for amino acids is relatively constant has been suggested as evidence of a diffusion controlled process of amino acid accumulation by erythrocytes. However, if the process were diffusion controlled, the cell/plasma ratio should be approximately 1.0 for all amino acids. At least for arginine and glutamine this is not the case. Arginine concentration in human erythrocytes is quite low relative to plasma, less than 20%, while glutamate concentrations in human erythrocytes are thirteen fold higher than plasma concentrations (Levy and Barkin, 1971).

Aoki et al. (1972) demonstrated an apparent dynamic role for the transport of glutamate. They infused insulin into the forearm and examined the subsequent distribution of glutamate. Plasma arterio-venous differences across the forearm muscles demonstrated little glutamate uptake. However, whole blood arterio-venous differences revealed that while resting muscle glutamate uptake was small, insulin markedly increased the uptake of glutamate by muscle from whole blood This is consistent with the hypothesis that in the resting state, much of the plasma glutamate is taken up by the erythrocytes. However, during insulin infusion this uptake by the erythrocytes is blocked and shunted into the muscle. In addition to this redirection of glutamate, the erythrocyte releases a portion of its glutamate to the muscle (Figure 4). This uptake and release of glutamate is rapid (a single passage through the muscle).

The evidence for generalized amino acid transport by erythrocytes stems from research in dogs (Elwyn et al., 1972). All evidence suggests a rapid exchange of amino acids between plasma and liver. For glycine this has been calculated to be 350 μmoles/min/kg of liver. Henriques et al., (1955) concluded that no membrane barrier exists to diffusion of glycine between plasma and liver. Thus, the principle determinant of this type of exchange

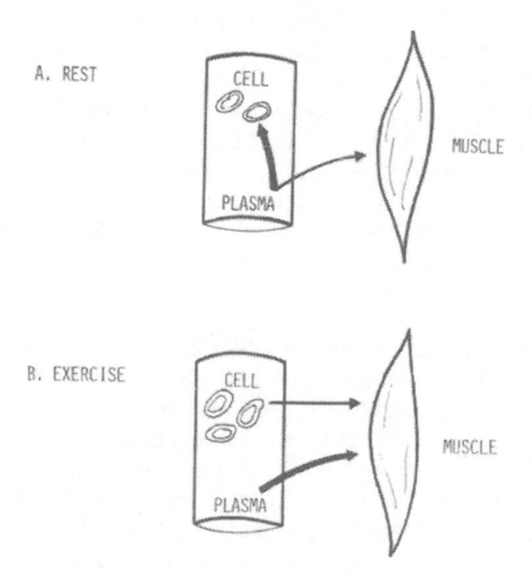


Fig. 4. The Role of Plasma and Erythrocytes in Maintenance of Muscle Glutamate. Muscle glutamate concentrations are maintained during exercise by a shift of accumulated glutamate out of both the plasma and the erythrocytes. The width of the arrow indicates the relative contribution (Aoki et al., 1972).

is blood flow. For glutamate flux, the liver output is negative into the plasma but positive into erythrocytes. While glycine and serine exhibit smaller fluxes than glutamate, the output of these amino acids from liver are in opposite directions for cells and plasma as is the case for glutamate. The magnitude and rapidity of these fluxes, 40-100 μmoles/min/kg liver, makes the explanation of these movements on the basis of a tissue → plasma → erythrocyte sequence unsatisfactory. In order to explain the observed changes in concentration solely in terms of red cell-plasma amino acid transfer, very rapid changes in transport velocities and enormous rates of production of glycine and serine must be postulated. At the present, no evidence for these production rates and velocities has been reported.

In vitro studies of amino acid transport have elucidated several amino acid specific transport systems which are energy and sodium dependent (Christensen, 1962). Additionally, passive transport in erythrocytes has been proposed for some amino acids. Passive transport is probably most important quantitatively in the movement of amino acids between the luminal content of the small bowel and the plasma postprandially. Although passive transport of amino acids may account for a significant percent of amino acid absorption from the gut, it seems likely that active transport is the dominant factor in erythrocyte amino acid translocation. The fact that erythrocyte amino acid transport is frequently spared in genetic disorders suggests that renal, intestinal and erythrocyte active transport systems are genetically dissimilar (Levy and Barkin, 1971; Marliss et al., 1972; Wahren et al., 1973). This relative genetic stability may be a reflection of the quantitative importance of active transport of amino acids in each tissue.

Active transport of amino acids requires sodium (Christensen, 1962). The membrane bound amino acid carrier forms a ternary complex with the amino acid and sodium. This complex traverses the membrane and releases both the amino acid and the sodium. This transport requires energy. Available evidence suggests actual complex formation rather than an interaction of the sodium pump, ATPase and the amino acid carrier complex. Arginine produces a sodium independent inhibition of glycine transport apparently because of the guanidinium portion of arginine occupies the sodium site while the remainder occupies the amino and carboxyl sites of the amino acid binding site. The observed stereospecificity of the amino acid transport system suggests that amino acid side chain recognition may also be a part of the system. While glutathione has been suggested as a participant in erythrocyte amino acid transport, in vitro evidence for this is lacking (Hagenfeldt et al., 1978). Patients with glutathione synthetase deficiency have presented divergent patterns of amino acid accumulation. Thus, at present, there is no convincing evidence for a significant role of glutathione in the active transport of amino acids by the erythrocyte.

## Developmental Changes in Blood Cell Amino Acid Transport

Mature human red cells accumulate both alanine and glycine, but are apparently unable to transport methionine, valine, isoleucine, or leucine. Reticulocytes transport all of these amino acids (Antonioli and Christensen, 1969). These differences are probably the result of continuing protein synthesis in the reticulocyte. The erythrocyte continues to accumulate glycine for glutathione synthesis and glycine is probably transported for splanchnic or renal gluconeogenesis while alanine is transported principally for splanchnic gluconeogenesis (Figure 3). Two separate systems have been postulated for glycine transport (Christensen, 1962). A transport agent serving principally for the transport of alanine is entirely separate from that handling cationic amino acids. While there is some evidence for sodium independent amino acid active transport in the erythrocyte, the evidence for this is equivocal (Antonioli and Christensen, 1969; Christensen, 1962). Christensen (1962) has concluded that interaction of neutral amino acids in erythrocytes is restricted to a single sodium dependent system.

The greatest difficulty in understanding the role of the erythrocyte in inter-organ amino acid transport is the observed dichotomy between *in vitro* amino acid transport rates and the rate of transport observed *in vivo*. The transport rates *in vivo* are magnitudes larger than the rates *in vitro*. Both direct transfer of amino acids between erythrocytes and tissue cells, and rapid increases in erythrocyte transport velocities upon entry of the cells into the capillary beds have been proposed to account for this discrepancy. Any rapid *in vitro* transfer of amino acids has been obscured by experimental design in studies of erythrocyte accumulation of amino acids. Therefore, the evidence for these postulates is circumstantial and principally results from a failure to provide more reasonable explanations. These hypotheses will be difficult if not impossible to test.

## SUMMARY

Gluconeogenesis from amino acids occurs primarily in the liver and kidney. Alanine, glycine and glutamate are the principal gluconeogenic amino acids but all common amino acids can potentially be utilized for gluconeogenesis. Amino acid transport to the liver or kidney occurs from the gut where amino acids are absorbed following a meal or from the muscle where amino acids are produced as a result of transamination, amination or protein degradation. Transport of amino acids in the blood is mediated by both the plasma and the cells. Quantitatively, approximately 75% of amino acid transport appears to be mediated by the plasma. Red blood cells are responsible for the remaining 25%. Specific membrane bound carrier proteins are responsible for the accumulation of amino acids by the red blood cells. These carrier proteins require sodium and energy for proper function. While many specific amino acid carrier proteins are lost during reticulocyte maturation, alanine, glycine and glutamate carrier proteins are retained in the erythrocyte. Translocation of amino acids from the erythrocyte to the tissue is very rapid, 300 µmoles/min. This observed velocity precludes a sequence of transfer from the erythrocyte to the tissue via the plasma by known transfer mechanisms. Direct transfer between the erythrocyte and tissue has been proposed to account for the observed velocities. However, the mechanism by which this transfer occurs remains to be elucidated.

REFERENCES

Ahlborg, G., Felig, P., Hagenfeldt, L., Hendler, R., and Wahren, J., 1974, Substrate turnover during prolonged exercise in man, J. Clin. Invest., 53:1080.
Allen, D. W., 1960, Amino acid accumulation by human reticulocytes, Blood, 16:1564.
Antonioli, J. A., and Christensen, H. N., 1969, Differences in schedules of regression of transport systems during reticulocyte maturation, J. Biol. Chem., 244:1505.
Aoki, T. T., Brennan, M. F., Muller, W. A., Moore, F. D., and Cahiu, Jr., G. F., 1972, Effect of insulin on muscle glutamate uptake, J. Clin. Invest., 51:2889.
Christensen, H. N., 1962, Biological Transport, W. A. Benjamin, Inc., New York.
Daniel, P. M., Pratt, O. E., and Spargo, E., 1977, The metabolic homeostatic role of muscle and its functions as a store of protein, Lancet, 2(8035):446.
Elwyn, D. H., Launder, W. J., Parish, H. C., and Wise, Jr., E. M., 1972, Roles of plasma and erythrocytes in interorgan transport of amino acids in dogs, Amer. J. Physiol., 222:1333.
Felig, P., 1973, The glucose-alanine cycle, Metabolism, 22:179.
Felig, P., 1976, Amino acid metabolism in exercise, Ann. N. Y. Acad. Sci., 301:56.
Felig, P., and Wahren, J., 1971, Influence of endogenous insulin secretion of splanchnic glucose and amino acid metabolism in man, J. Clin. Invest., 50:1702.
Felig, P., and Wahren, J., 1974, Protein turnover and amino acid metabolism in the regulation of gluconeogenesis, Fed. Proc., 33:1092.
Felig, P., Owen, O. E., Wahren, J., and Cahill, Jr., G. F., 1969, Amino acid metabolism during prolonged starvation, J. Clin. Invest., 48:584.

Felig, P., Pozefsky, T., Marliss, E., Cahill, Jr., G. F., 1970, Alanine: key role in gluconeogenesis, Science, 167:1003.

Felig, P., Marliss, E., Pozefsky, T., and Cahill, Jr., G. F., 1970, Amino acid metabolism in the regulation of gluconeogenesis in man, Am. J. Clin. Nutr., 23:986.

Felig, P., Kim, Y. J., Lynch, V., and Hendler, R., 1972, Amino acid metabolism during starvation in human pregnancy, J. Clin. Invest., 51:1195.

Felig, P., Wahren, J., and Raf, L., 1973, Evidence of interorgan amino acid transport by blood cells in humans, Proc. Nat. Acad. Sci., 70:1775.

Felig, P., Wahren, J., Karl, I., Cerasi, E., Luft, R., and Kipnis, D. M., 1973, Glutamine and glutamate metabolism in normal and diabetic subjects, Diabetes, 22:573.

Felig, P., Wahren, J., Sherwin, R., and Palarologos, G., 1977, Amino acid and protein metabolism in diabetes mellitus, Arch. Intern. Med., 137:507.

Garber, A. J., Karl, I. E., and Kipnis, D. M., 1976, Alanine and glutamine synthesis and release from skeletal muscle. II. The precursor role of amino acids in alanine and glutamine synthesis, J. Biol. Chem., 251:836.

Garber, A. J., Karl, I. E., and Kipnis, D. M., 1976, Alanine and glutamine synthesis and release from skeletal muscle. IV. Beta-adrenergic inhibition of amino acid release, J. Biol. Chem., 251:851.

Goldstein, L., and Newsholme, E. A., 1976, The formation of alanine from amino acids in diaphragm muscle of the rat, Biochem. J., 154:555.

Hagenfeldt, L., Larsson, A., Andersson, R., 1978, The gamma-glutamyl cycle and amino acid transport, N. Engl. J. Med., 299:587.

Haymond, M. W., Karl, S. E., and Pagliara, A. S., 1974, Increased gluconeogenesis in the small-for-gestational-age infant, N. Engl. J. Med., 291:322.

Henriques, O. B., Henriques, S. B., and Neuberger, A., 1955, Quantitative aspects of glycine metabolism in the rabbit, Biochem. J., 60:409.

Karl, I. E., Garber, A. J., and Kipnis, D. M., 1976, Alanine and glutamine synthesis and release from skeletal muscle. III. Dietary and hormonal regulation, J. Biol. Chem., 251:844.

Levy, H. L., and Barkin, E., 1971, Comparison of amino acid concentrations between plasma and erythrocytes. Studies in normal human subjects and those with genetic disorders, J. Lab. Clin. Med., 78:517.

Mallette, L. E., Exton, and Park, C. R., 1969, Control of gluconeogenesis from amino acids in the perfused rat liver, J. Biol. Chem., 244:5713.

Marliss, E. B., Aoki, T. T., Toews, C. J., Felig, P., Connor, J. J., Kyner, J., Huckabee, W. E., and Cahill, Jr., G. F., 1972, Amino acid metabolism in lactic acidosis, Am. J. Med., 52:474.

McMenamy, R. H., Lund, C. C., Neville, G. J., Wallach, D. F. H., 1960, Studies of unbound amino acid distribution in plasma, erythrocytes, leucocytes and urine of normal human subjects, J. Clin. Invest., 39:1675.

Odessey, R., Khairallah, E. A., and Goldberg, A. L., 1974, Origin and possible significance of alanine production by skeletal muscle, J. Biol. Chem., 249:7623.

Paul, H. S., and Adibi, S. A., 1978, Leucine oxidation in diabetes and starvation: Effects of ketone bodies on branched-chain amino acid oxidation in vitro, Metabolism, 27:185.

Sladek, C. D., and Snarr, J. F., 1971, Effect of the exogenous amino acid concentration on the rate of gluconeogenesis in liver slices, Proc. Soc. Exp. Biol. Med., 138:181.

Snell, K., 1974, Pathways of gluconeogenesis from L-serine in the neonatal rat, Biochem. J., 142:433.

Van Slyke, D. D., and Meyer, G. M., 1912, The amino acid nitrogen of the blood. Preliminary experiments on protein assimilation, J. Biol. Chem., 12:399.

Wahren, J., Felig, P., Havel, R. J., Jorfeldt, L., Pernow, B., and Saltin, B., 1973, Amino acid metabolism in McArdle's syndrome, N. Engl. J. Med., 288:774.

Wahren, J., Felig, P., Hendler, R., and Ahlborg, G., 1973, Glucose and amino acid metabolism during recovery after exercise, J. Appl. Physiol., 34:838.

Yunis, A. A., and Arimura, G. K., 1965, Amino acid transport in blood cells. II. Patterns of transport of some amino acids in mammalian reticulocytes and mature red blood cells, J. Lab. Clin. Med., 66:177.

## CONTRIBUTORS

Jeffrey Bland
Department of Chemistry, University of Puget Sound and Bellevue-Redmond Medical Laboratories,
Tacoma, Washington

Alvin Dubin*
Hektoen Institute for Medical Research and Department of Biochemistry, Rush Medical School,
Chicago, Illinois

Shibban Ganju
Department of Gastroenterology, Cook County Hospital,
Chicago, Illinois

Donald E. Hill*
Departments of Pediatrics and Physiology, University of Arkansas for Medical Sciences,
Little Rock, Arkansas

Francine M. Hoerrmann
Department of Nutritional Services, The Cleveland Clinic Foundation,
Cleveland, Ohio

James D. Jones*
Department of Laboratory Medicine, Mayo Clinic,
Rochester, Minnesota

Mary A. Jones
Department of Dietetics, Saint Marys Hospital,
Rochester, New York

Surat Komindr
Department of Medicine, Ramathodi Hospital, Mahidol University,
Bangkok, Thailand

Lena A. Lewis*
Department of Biochemistry, The Cleveland Clinic Foundation and Department of Chemistry, Cleveland State University, Cleveland, Ohio

Herbert K. Naito*
Department of Biochemistry, The Cleveland Clinic Foundation and Department of Chemistry, Cleveland State University, Cleveland, Ohio

George A. Nichoalds
Department of Obstetrics and Gynecology, University of Tennessee Center for the Health Sciences, Memphis, Tennessee

Irvine H. Page
Department of Biochemistry, The Cleveland Clinic Foundation, Cleveland, Ohio

Shi-Kaung Peng
Veterans Administration Medical Center and Albany Medical College of Union University, Albany, New York

Ann Paulos
Hektoen Institute for Medical Research, Chicago, Illinois

Donna B. Rosenstock
Department of Nutritional Services, The Cleveland Clinic Foundation, Cleveland, Ohio

Bruce A. Sebek
Department of Biochemistry, The Cleveland Clinic Foundation, Cleveland, Ohio

Raymond J. Shamberger*
Department of Biochemistry, The Cleveland Clinic Foundation, Cleveland, Ohio

John E. Sherwin
Department of Pathology, Valley Childrens Hospital and Guidance Clinic, Fresno, California

Frederick Steigmann
Department of Gastroenterology, Cook County Hospital, Hektoen Institute for Medical Research, and University of Illinois School of Medicine, Chicago, Illinois

Ruth C. Steinkamp
Bureau of Cancer and Special Services, Arkansas State Department of Health (retired),
Little Rock, Arkansas

Jon A. Story
Department of Foods and Nutrition, Purdue University,
West Lafayette, Indiana

Paul B. Szanto
Hektoen Institute for Medical Research, and Departments of Pathology, Chicago Medical School and Cook County Hospital,
Chicago, Illinois

C. Bruce Taylor
Veterans Administration Medical Center and Albany Medical College of Union University,
Albany, New York

Paul M. Tocci*
Department of Pediatrics, University of Miami, and Biochemical Genetics Laboratory, Mailman Child Development Center,
Miami, Florida

---

*Summary papers presented at the National Academy of Clinical Biochemistry symposium "Nutritional Elements in Clinical Biochemistry", July, 1979.

INDEX